CELL CULTURE ENGINEERING IV

Current Applications of Cell Culture Engineering

VOLUME 1

Cell Culture Engineering IV

Improvements of Human Health

Edited by

BARRY C. BUCKLAND

Merck Research Laboratories, Biochemical Process R&D,
Rahway, New Jersey, U.S.A.

Co-editors

JOHN G. AUNINS
THEODORA A. BIBILA
WEI-SHOU HU
DAVID K. ROBINSON
WEICHANG ZHOU

Reprinted from Cytotechnology, Volume 15, 1994

Kluwer Academic Publishers
Dordrecht / Boston / London

A C.I.P. Catalogue record of this book is available from the Library of Congress.

ISBN 978-94-010-4114-0 ISBN 978-94-011-0257-5 (eBook)
DOI 10.1007/978-94-011-0257-5

Printed on acid-free paper

Contents

Cytotechnology **15:** 1–2, 1994.

Introduction

Applications of Cell Culture for the improvement of human health

Abstract

Current applications of Cell Culture Engineering which have a major beneficial impact for the improvement of human health range from a great variety of vaccines (examples include measles, mumps, rubella, polio, and Hepatitis A) made using this technology to a whole new range of therapeutic proteins dependent for their expression on animal cell culture hosts (examples include EPO, TPA and γ-interferon). Novel applications "in the pipeline" include cell therapy products and gene therapy products.

In recognition of the importance of this technology to current practice and in anticipation of its future potential, Professors Wei-Shou Hu (University of Minnesota) and Sinskey (M.I.T.) had the vision to initiate the first Conference in the Cell Culture Engineering series in 1988. They went on to organize a second Conference two years later and a third Conference in this series was chaired in 1992 by Prof. Flickinger (University of Minnesota). These Conferences were all sponsored by the Engineering Foundation.

In retrospect it is fascinating to consider the important contribution that this series of Conferences has had on the practice of Cell Culture Engineering. They have provided an ideal forum for the presentation of the latest developments in this emerging technology as well as for the formation of numerous personal contacts nurtured by the collegial environment generated by the Engineering Foundation style of Conference. The challenge in putting together the fourth Conference was to continue this atmosphere of excitement and innovation and to combine discussions of what have now become well established technologies with discussions of novel areas such as Somatic Cell Therapy.

This special volume of Cytotechnology includes a number of the papers presented at Cell Culture Engineering IV held March 7–12, 1994 in San Diego, California, USA. This Conference continued the tradition established in previous Conferences of a strong interdisciplinary representation with focus on recent developments of animal cell culture. Approximately 240 people attended this conference with approximately 160 from industry and with representation from 22 different countries. Lively discussions, which are not captured here, resulted from many of these presentations and especially from the Poster Sessions.

Papers presented in this special edition include discussions on the latest approaches toward improving the cell host through improved understanding of the molecular biology, for example through improved understanding of some of the physiological responses of cell culture such as adaptation to suspension culture and adaptation to serum free medium.

There is renewed interest in the development of novel vaccines and this topic was featured prominently in Cell Culture Engineering IV. Although a "traditional" area, there are opportunities for both applying current technologies to "old" products as well as developing new technologies for future areas of interest. A novel area which attracted considerable discussion was that of the application of Cell Engineering for Somatic Cell Therapy.

Important breakthroughs have recently been made in the understanding of the effects of mixing and aeration on cell cultures. These range from understanding made through visual observations to those which come from mathematical modelling. This gain in knowledge allows for a more rational approach to bioreactor design and operation.

As in microbial systems, the key to more sophisticated approaches to process control lies in improved monitoring techniques. Approaches at the moment are for the most part at an early stage of development, but rapid improvements are expected over the next several years. Certainly, the computer tools are available. It is more a question of improving "input" into the computer.

Continued interest is focussed on quality related topics. Therefore, a large number of papers and posters were presented on glycosylation of therapeutic proteins. Whereas these papers presented a "snapshot" of current technology, a keynote talk by Dr. Middaugh provided us a glimpse into the future of techniques which may become available over the next 10 years. If we can get to the same level of sophistication as for "small" molecules, then we can imagine a situation in which it will become possible to make process changes for biologics without the need for confirmation in the clinic. Instead, we would be able to provide analytical data to reassure ourselves and regulatory authorities that a process change can be made without harm to product quality.

Plans are already being made for the next Conference in this series (Cell Culture Engineering V) to be held in San Diego in 1996. This Conference will be chaired by Dr. Bibila and myself with assistance of two co-chairs (Prof. J. Chalmers and Dr. R. Arathoon). Already, we are trying to anticipate topics for the future. Ironically, a key to future advances may well lie in the study of dying cells (apoptosis) in contrast to our current focus on the living. We would predict a continued high level of interest on quality related issues which encompass the following: process monitoring and control, bioreactor design, analytical characterization, and regulatory issues. Finally, we try to anticipate the technology needed to make the next wave of products. For example, what method is best to make products to be used for Gene Therapy?

Barry C. Buckland

Cytotechnology **15:** 3–9, 1994.

Cultural and physiological factors affecting expression of recombinant proteins

J.B. Griffiths and A.J. Racher
CAMR, Porton Down, Salisbury, Wiltshire SP4 OJG, UK

Key words: Chimeric antibody, culture systems, glutamine, growth rate, porous carrier

Abstract

The variability in expression of recombinant proteins has been analyzed with regard to (a) comparison of clones from the same transfection experiment; (b) comparison of the same genetic construct in different cell lines; (c) the effect of the culture system used (free suspension, aggregate suspension, and microcarrier); and (d) physicochemical parameters in long-term (100d) culture in a macroporous fixed bed bioreactor (FBR).

Differences in product expression between clones were accompanied by differences in growth rates, metabolic kinetics, and ability to grow in suspension as opposed to attached culture. The single most important factor affecting product expression when comparing constructs (for SEAP and IgG), cell lines (BHK 21 and myeloma), and culture systems was whether cells were grown in an attached or suspension mode. Thus key factors could be related to cell morphology (suspension versus monolayer), the presence of microenvironments and physiological stress to control growth rate.

The relationship of key process parameters to volumetric and specific rAb productivity of the FBR was investigated in a partial factorial experiment with a rBHK cell line. The highest productivity levels are associated with a combination of the highest values tested for re-cycle (195 ml min^{-1}) and dilution rates (1 d^{-1}) and glutamine concentration (2.5 mmol l^{-1}), plus the lowest values for bead size (2 mm) and inoculum density (10^7 ml^{-1}). Together with data from fluidised bed cultures, these results suggest that higher productivity is not primarily the result of greater cell numbers within the system but more the physicochemical definition of the system.

Abbreviations: FlBR – fluidised bed bioreactor; FBR – fixed bed reactor; STR – stirred tank reactor; SEAP – secreted alkaline phosphatase; rAb – recombinant antibody

Introduction

The use of animal cells for the industrial production of recombinant proteins has increased dramatically over the last ten years. This emphasises the necessity of defining and controlling culture conditions to ensure maximum yields of high quality products. Considerable attention has been paid to understanding the biochemical and physiological factors affecting both growth and production kinetics of cultured animal cells. However, this has been largely restricted to mAb production by hybridomas (e.g. Al-Rubeai *et al.*, 1992). To obtain the full benefit from the use of recombinant DNA technology in animal cells, similar studies need to be made for recombinant cell lines. Reports now appearing in the literature which include studies on the kinetics of recombinant protein secretion are, generally, for a particular product by a given cell line (Cockett *et al.*, 1990; Hayter *et al.*, 1991; Robinson and Memmert 1991; Pendse *et al.*, 1992). There are fewer reports describing the expression of either several recombinant proteins in one cell line or the same recombinant protein in several cell lines, or the use of different culture systems (Wagner *et al.*, 1988; Conradt *et al.*, 1989; Ryll *et al.*, 1990). However these studies do not usually describe the kinetics of product formation, which has been addressed in this work, and has led to some clear indications of the factors most important for maximising recombinant product expression in cell culture.

Materials and methods

Cell lines and media

The rAb expressed by the cell lines BHK.IgG and F3b10 (derived from BHK21 and Sp2/0 cell lines respectively) consists of the variable portions of a murine antibody, immunoglobulin class IgG1, linked to the constant parts of the human IgGl molecule (Kaluza *et al.*, 1991). Expression is driven by the murine IgG promoter. The cell line BHK.SEAP expresses the secreted form of alkaline phosphatase. Construction of these cell lines is detailed elsewhere (Racher *et al.*, 1994).

The microcarrier and aggregate culture of the BHK cell lines were done in high glucose DMEM: the FBR and FIBR cultures were with 1:1 DMEM/Hams F12. Serum was supplied at 5% v/v. The F3b10 cells were cultured in RPMI 1640 containing the serum-replacement Nutridoma-SR (Boehringer-Mannheim) at 1% v/v.

Culture conditions

Two different STR systems were used in this study. Firstly, spinner flasks (250 ml working volume) (Corning) fitted with a paddle impeller (diameter 5.4 cm; height 1.8 cm) operated at 50 rpm. The spinner flasks were kept in a 7% CO_2–93% air atmosphere. The second STR was a 2 l (1 l working volume) vessel (Applikon BV, Schiedam, The Netherlands) fitted with a 4.0 cm diameter, 3.3 cm high impeller. The pH was controlled at 7.1 $\pm$ 0.1 and dissolved oxygen at 60% air saturation. The impeller speed was 100 rpm for the F3b10 clone and 60 rpm for BHK.IgG.

The BHK cell lines were grown on Cytodex 3 microcarriers (Pharmacia AB, Uppsala Sweden) at 3 g l^{-1}. The BHK lines were also grown in suspension culture as natural aggregates (Moreira *et al.*, 1992). The F3b10 cell line was only grown as a suspension culture. For both suspension and aggregate cultures, the final viable cell density after inoculation was 1.5-2.0 $\times$ 10^5 cells ml^{-1}.

Design and operation of the FBR with porous Siran (Schott Glaswerke) beads has been described in detail previously (Racher *et al.*, 1990; Racher and Griffiths, 1993). Briefly the FBR was comprised of a 150 ml vessel containing 100 ml of beads. For the FIBR, the reactor vessel volume was 200 ml and contained 80 ml, settled volume, of beads. Both reactor vessels were connected to a 1.5 l (1.0 l working volume) reactor which was used for medium conditioning. The FBR and FIBR were both operated as batch cultures for the first 72–96h before continuous medium feed was started by means of a peristaltic pump. A constant volume was maintained in the reservoir by a weir in the vessel wall. The system was operated in continuous perfusion mode for 240h, or for 2400h.

The characteristics of the borosilicate beads used in the FBR were: diameter, 1–2 or 5–6 mm; porosity, 60%; pore size, < 200 μm. The beads used in the FIBR were smaller, with a diameter of 0.7 mm.

Analytical methods

Cell numbers were determined using a haemocytometer. Viable cells were distinguished by the trypan blue dye exclusion method. IgG1 was measured by ELISA as described previously (Racher *et al.*, 1990), except anti-human Ig $F(ab')_2$ (Amersham International plc, Amersham, UK) was used with human IgG1 as the standard. SEAP activity was determined spectrophotometrically (Berger *et al.*, 1988).

Results and discussion

1. *Clonal variation of Sp2/0 cells expressing SEAP*

The variability in both expression of recombinant proteins and a range of cultural characteristics has been analyzed using cell lines from a single transfection experiment. High producing transfectants of Sp2/0 expressing SEAP (including HP62D, HPGC and HP35B) at different maximum cell specific rates were isolated (Hauser *et al.*, unpublished). Additional differences were: the minimum inoculum size needed to initiate a new culture, the serum-dependence of growth and SEAP production, lactate production kinetics (this difference was only found in shake-flask and not in controlled bioreactor cultures) and growth kinetics. This variability, also seen in hybridomas, indicates the potential of screening for useful scale-up culture characteristics as well as productivity levels. One consistent feature (Fig. 1) was the correlation of high SEAP production with high glutamine catabolism and

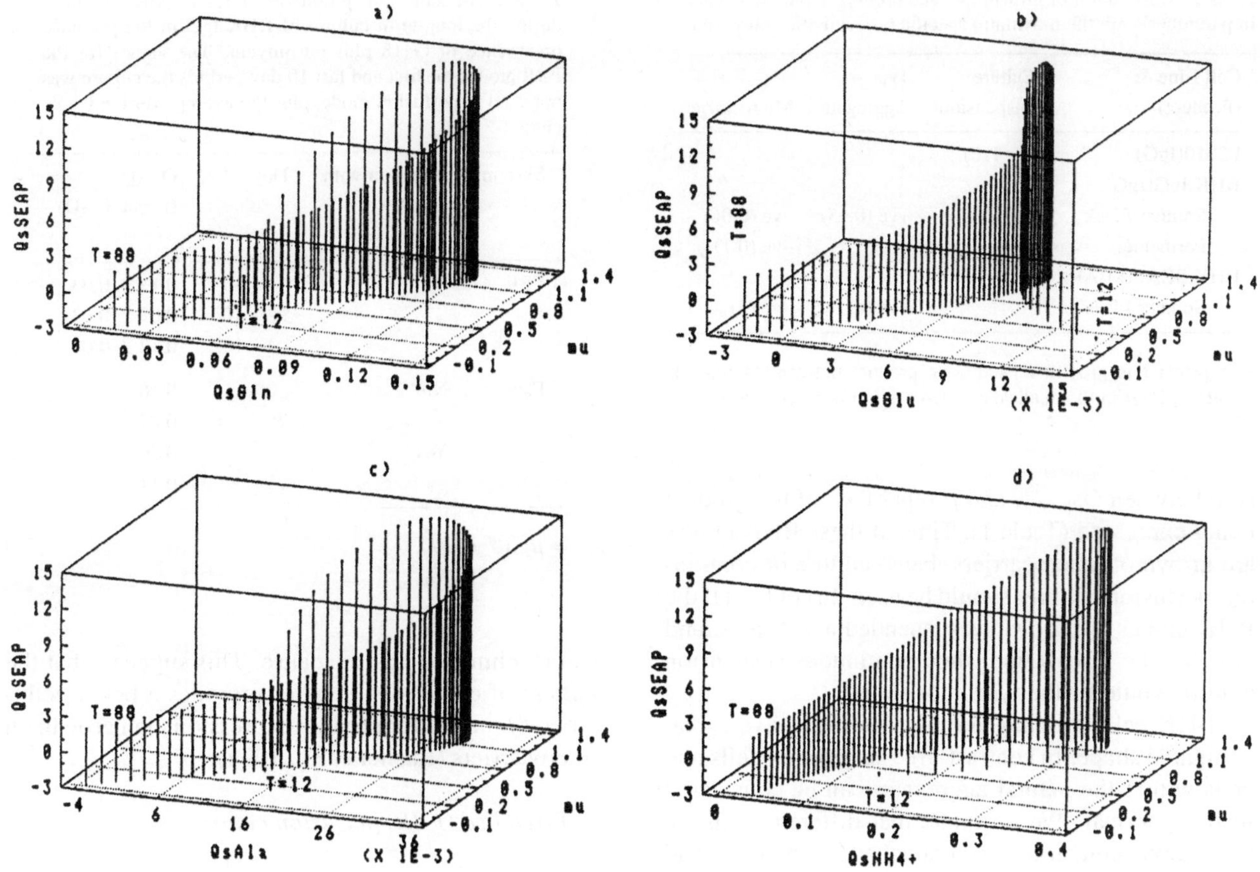

Fig. 1. Plot of SEAP production rate against metabolite rates and specific growth rate. Clone HP35B was grown in DMEM + 5% FCS using a 2 l (1 l working volume) STR operated in batch mode. Data points were calculated using the differential derivatives of the polynomial equations used to smooth the experimental data: glutamate and ammonium rates have been corrected for glutamine decay. The period shown in the figure is from lag through to early stationary growth phase. Cartoon a: glutamine uptake; b glutamate production; c alanine production; d ammonium production. Symbols: T = time (h) at start and end of plot. Units: Qs_{SEAP}, ng 10^{-6} cells d^{-1}; Qs_{Gln} etc μmol 10^{-5} cells d^{-1}; mu d^{-1}.

the production of glutamate, alanine and ammonium and high specific growth rate.

2. *Correlation of growth rates (μ) and product expression*

Analysis of the relationship between $Qs_{product}$ and μ for recombinant Sp2/0 and BHK 21 cells grown in two different STR systems is shown in Table 1. Markedly higher maximum values for Qs_{rAb}, were seen for F3b10 compared to BHK.IgG. As these cell lines were generated using the same plasmids, the observed differences are not caused by the plasmids *per se*. Therefore the intracellular environment must be having a marked effect upon rAb production. A possible explanation is that the intracellular environment is affecting promoter activity. The murine IgG promoter used is known to be active in B-cells but not in fibroblast or epithelial cells such as BHK. However, other explanations, such as differences in the secretion rate of mature rAb cannot be excluded.

When F3b10 cells were grown in single cell suspension and the BHK cell lines as natural aggregates, $Qs_{product}$ increased as μ increased (Racher *et al*, 1994). Summary of this data is shown in Table 1. This was observed for both rAb and SEAP. In contrast, when cells were grown on microcarriers, $Qs_{product}$ increased when μ decreased, again for both products. Therefore it appears that the choice of growth mode can have a profound effect upon the relation-

Table 1. Correlation of growth rate and product expression. Values in parenthesis are the maximum specific productivities observed

Cell Line & (Product)	Culture Suspension	Type Aggregate	Microcarrier
F3b10(IgG)	+ve (10)		
BHK.IgG(IgG)			
Spinner Flask		+ve (0.15)	−ve (0.06)
Fermenter			−ve (0.1)
BHK.SEAP(SEAP)			
Spinner Flask		+ve (8)	−ve (3)

+ve product is growth related; −ve product not growth related. (Units: IgG -μ/million cells/day; SEAP ng/million cells/day).

Table 2. Specific rAb production rates at different times during the long-term culture of BHK.IgG in the presence or absence of G418 plus puromycin. The values for the FBR are for the first and last 10 day periods the culture was operated in perfusion mode, plus the average for the whole culture

System	Selection with G418 + Puromycin	Day	Qs_{rAb} (mean ± SD)
FBR	No	4 – 10	0.12 ± 0.03
		92 – 102	0.09 ± 0.03
		4 – 102	0.12 ± 0.06
Flask	No	0	0.16
		75	0.07
	Yes	0	0.16
		75	0.11

* μg/10^6 cells/d.

ship between $Qs_{product}$ and μ regardless of the product being expressed (Table 1). Thus, if these BHK clones are grown on microcarriers, batch culture or continuous perfusion cultures should be used. In contrast if the BHK clones are grown as suspended aggregates, and also for the F3b10 line, then continuous suspension culture would be the system of choice.

BHK cells growing on microcarriers have a flat, elongated shape. In contrast, free suspension cells and cells within the natural aggregates can be considered to be spherical. We postulate that differences in the microenvironment between attached, suspension and aggregate cells are associated with, if not the cause of, alteration in the relationship between $Qs_{product}$ and μ. Other workers have reported association between change in cell shape and alteration of metabolism (Benecke *et al.*, 1978; Folkman and Moscona 1978; Cacan *et al.*, 1993) and the ordered structure of the cytoskeleton (Pollack *et al.*, 1975; Tucker *et al.*, 1978). Modulation of integrin activity by extracellular matrix components and other cells leads to phosphorylation of proteins involved in cell adhesion and spreading, and modulation of gene expression (Hynes, 1992; Zachary and Rozengurt, 1992). Therefore, evidence exists to support the hypothesis that the change in cell shape is associated with, if not the cause, of alteration in the relationship between Qs_{rAb} and μ.

Growth of both BHK cell lines in spinner-flask cultures, either as aggregates or on microcarriers, produced similar maximum viable cell densities. However, values for $Qs_{product}$ were 2-3-fold higher using aggregates compared to microcarriers. As this result was observed in the same culture system with particles of similar sizes (mean diameter of aggregates, 130 μm; Cytodex 3 175 μm), hydrodynamic differences can be eliminated as the cause. This suggests that the culture of cells as aggregates provides a better cellular environment for product expression than growth on microcarriers (Racher *et al.*, 1994).

3. *Fixed bed (FBR) perfusion cuture*

The FBR system has been operated continuously for 2450 h using the BHK.IgG cell line (Racher and Griffiths, 1993). The initial growth phase of the culture lasted until about 500 h, achieving a bed cell density calculated at 2.3 × 10^7 cells/ml bed, before declining to a steady state of about 2.0 × 10^7 cells/ml bed.

Comparison of data from the FBR and flask cultures (Table 2) shows conditions brought about by immobilisation improve the stability of recombinant protein production over extended culture periods. The percentage loss over the 100 day FBR culture was less than that seen upon repeated sub-culturing in T-flasks, for a shorter period. The FBR culture in Table 2, unlike the one in Table 4, was not grown using the best combination of parameter levels (described in Table 3).

An aim was to identify, not necessarily optimise as this would vary between different cell lines, parameters important in modulating productivity, and assign qualitative values to them using a model cell line (BHK.IgG).

Using a partial factorial experimental design (Racher and Griffiths, 1993), limited combinations of the parameters at their different levels (either high or

Table 3. BHK.IgG cell line in FBR

Variable	Range	Optimum condition
Bead Size	X5	Small (1–2mm)*
Circulation Rate	X4	High (195 ml/min)
Dilution Rate	X3	High (1 l/l/day)*
Glutamine	X3	High (2.5 mmol/l)
Inoculum Size	X5	Low (2 million/ml)

(rAb Concn. 4.13 [Range 0.99–4.13] μg/l Media Feed)
* Most Significant [t-test]

Table 4. Productivity data from different culture systems. Values for the FBR and FlBR systems are the means from the period the system was operated in continuous mode. Data for the batch culture are the maximum rates observed before release of rAb due to cell lysis became a problem. The values for Qs_{rAb} were all calculated using cell numbers estimated from the glucose uptake rate. (Data from Racher and Griffiths, 1993)

System	Qv_{rAb} μg/ml/day	Qs_{rAb} $\mu g/10^6$ cells/day)	C_{rAb} (μg/ml)
FBR	0.46	0.32	0.46
FlBR	0.23	0.15	0.24
Batch Microcarrier	0.09	0.07	0.27

low) were tested. Individually none of the parameters tested had a significant affect upon productivity. The work indicated that to achieve high productivity in FBR systems, the system should be operated using a combination of high circulation and dilution rates, small bead and inoculum sizes and high feed glutamine concentration (Table 3).

Park and Stephanopoulous (1993) showed that nutrient transfer into porous particles by intra-particle convection could be modulated by varying either the total flow through the bed or the bead size. The findings that productivity was associated with smaller beads and higher circulation rates is in agreement with the predictions made for nutrient transfer by Park and Stephanopoulous (1993). The characteristics of FlBR systems are smaller beads and higher circulation rates, so that the beads move through the working volume of the reactor vessel. Therefore FlBR systems are associated with greater intraparticle convection, which suggests that higher productivities could be achieved with a FlBR compared to a FBR system. Data in Table 4 allow such a comparison to be made.

The data in Table 4 show that Qv_{rAb} was 2-fold greater in the FBR compared to the FlBR, although cell densities were comparable in both systems (1.7–1.8×10^7 cells/ml bed). However, the value of Qs_{rAb} in the FBR was about 2-fold greater than that for the FlBR. Therefore the difference in Qv_{rAb}, results from the difference in Qs_{rAb}. The difference in Qs_{rAb} could arise from differences in cell turnover, affecting the value of the true μ rather than the apparent μ. It has been shown in batch microcarrier cultures of BHK.IgG that Qs_{rAb}, increases as μ decreases (Racher *et al.*, 1994). In the FlBR, beads are continually colliding with one another leading to cell loss and replacement, whereas in the FBR the beads are stationary and so the replacement rate should be lower. This argues that the true μ should be higher in the FlBR. Therefore the FlBR should have a lower value for Qs_{rAb} because it has a higher value for μ compared to the FBR.

The data also show that immobilisation within a FBR system using Siran beads increases Qs_{rAb}. This has also been seen for other supports, e.g. alginate (e.g. Lee *et al.*, 1993). So improvement in Qv_{rAb} in the FBR, compared to batch microcarrier cultures, was not due solely to an increase in cell numbers.

Conclusions

A systematic study of different recombinant cell lines, constructs, and culture systems was carried out in order to establish the important factors in obtaining high and stable expression of recombinant products in scaleable culture systems. The most significant factor was the culture type and not the cell line or construct, and a direct relationship between growth rate and productivity was established in non-immobilised cultures. The highest productivity was associated with systems which permitted formation of microenvironments (aggregates and porous carriers) and a rounded rather than flattened morphology. The effect of cell morphology on metabolism and regulatory pathways (e.g. integrin concentration and cytoskeleton polymerisation) is well recognised and further work is indicated on establishing its effect on productivity. The effect of all these parameters is qualitatively summarised in Table 5.

Using porous microcarrier systems (fixed and fluidised) the importance of a near-zero growth rate on increasing product expression was demonstrated (Table 4). High productivity was also consistently correlated to a high rate of glutamine catabolism, and the

Table 5. Qualitative comparison of rBHK culture parameters in various culture systems (0 – Zero; +++ Highest level)

	Fixed bed	Fluidised bed	Microcarrier	Aggregate	Free suspension
Productivity	+++	++	+	+++	++
Growth rate	0	+	++	0	++
Cell density	+++	+++	+	++	+
Morphology	SF	SF	F	S	S
Microenvironment/ Gradients	+++	++	0	++	0

F = flat; SF = semi-flat/spherical; S = spherical

physicochemical conditions were more important than the cell numbers *per se*. It was also demonstrated that stable expression can be obtained for over 100 days in a continuous culture.

Acknowledgements

This work was carried out as part of the European Community Bridge Programme (Contract No. BIOT-90-0185) in collaboration with Dr H. Hauser (GBF, Germany) and Professor M.J.T. Corrondo (IBET, Portugal). The authors thank Mrs J.E. Askey for the preparation of the manuscript.

References

Al-Rubeai M, Emery AN, Chalder S, Jan DC (1992) Specific monoclonal antibody productivity and the cell cycle – comparisons of batch, continuous and perfusion cultures. Cytotechnology 9: 85–97.

Benecke B-J, Ben-Ze'er A and Penman S (1978) The control of mRNA production, translation and turnover in suspended and reattached anchorage-dependent fibroblasts. Cell 14: 931–939.

Berger J, Hauber J, Hauber R, Geiger R, Cullen BR (1988) Secreted alkaline phosphatase: a powerful new quantitative indicator in eukaryotic cells. Gene 66: 1–10.

Cacan R, Labiau O, Mir A-M and Verbert A (1993) Effect of cell attachment and growth on the synthesis and fate of dolichol-linked oligosaccharides in Chinese hamster ovary cells. Eur. J. Biochem. 215: 873–881.

Cockett MI, Bebbington CR, Yarranton GT (1990) High level expression of tissue inhibitor of metalloproteinases in Chinese hamster ovary cells using glutamine synthetase gene amplification. Bio/Technol 8: 662–667.

Conradt HS, Nimtz M, Dittmar KEJ, Lindenmaier W, Hoppe J, Hauser H (1989) Expression of human interleukin-2 in recombinant baby hamster kidney, Ltk$^-$, and Chinese hamster ovary cells. Structure of O-linked carbohydrate chains and their location within the polypeptide. J. Biol. Chem. 264: 17368–17373.

Folkman J and Moscanna A (1978) Role of cell shape in growth control. Nature 273: 345–349.

Hayter PM, Curling EMA, Baines AJ, Jenkins N, Salmon I, Strange PG, Bull AT (1991) Chinese hamster ovary cell growth and interferon production kinetics in stirred batch cuiture. Appl. Microbiol. Biotechnol. 34: 559–564.

Hynes RO (1992) Integrins: versatility, modulation, signalling in cell adhesion. Cell 69: 11–25.

Kaluza B, Lenz H, Russmann E, Hock H, Rentrop O, Majdic O, Knapp W and Weidle UH (1991) Synthesis and functional characterisation of a recombinant monoclonal antibody directed against the α-chain of the human interleukin -2 receptor. Gene 107: 297–305.

Lee GM, Chuck AS, Palsson BO (1993) Cell culture conditions determine the enhancement of specific monoclonal antibody productivity of calcium alginate-entrapped S3H5/γ2bA2 hybridoma cells. Biotechnol. Bioeng. 41: 330–340.

Moreira JL, Alves PM, Aunins JG, Carrondo MJT (1992) Aggregate suspension cultures of BHK cells. In: Spier RE, Griffith JB, MacDonald C (eds) Animal Cell Technology: Developments, Processes and Products. Butterworth-Heinemann, Oxford, p411.

Park S and Stephanopoulos G (1993) Packed bed bioreactor with porous ceramic beads for animal cell culture. Biotechnol. Bioeng. 41: 25–34.

Pendse GJ, Karkare S, Bailey JE (1992) Effect of cloned gene dosage on cell growth and hepatitis B surface antigen in recombinant CHO cells. Biotechnol. Bioeng.40: 119–129.

Pollack R, Osborn M and Weber K (1975) Patterns of organisation of actin and myosin in normal and transformed cultured cells. Pro Natl. Acad. Sci., USA 72: 994–998.

Racher AJ, Looby D, Griffiths JB (1990) Studies on monoclonal antibody production by a hybridoma cell line (C1E3) immobilised in a fixed bed, porosphere culture system. J. Biotechnol. 15: 129–146.

Racher AJ, Griffiths JB (1993) Investigation of parameters affecting a fixed bed bioreactor process for recombinant cell lines. Cytotechnology 13: 125–131.

Racher AJ, Moreira JL, Alves PM, Wirth M, Weidle UH, Hauser H, Carrondo MJT, Griffiths JB (1994) Expression of recombinant antibody and secreted alkaline phosphatase in mammalian cells. Influence of cell line and culture system upon production kinetics. Appl. Microbiol. Biotechnol. 40: 851–856.

Robinson DK, Memmert KW (1991) Kinetics of recombinant immunoglobulin production by mammalian cells in continuous culture. Biotechnol. Bioeng. 38: 972–976.

Ryll T, Lucki-Lange M, Jager V, Wagner R (1990) Production of recombinant human interleukin-2 with BHK cells in a hollow

fibre and a stirred tank reactor with protein-free medium. J. Biotechnol. 14: 377-392.
Tucker RW, Sanford KK and Frankel FR (1978) Tubulin and actin in paired nonneoplastic and spontaneously transformed neoplastic cell lines *in vitro*: fluorescent antibody studies. Cell 13: 629–642.
Wagner R, Ryll T, Krafft H, Lehmann J (1988) Variation of amino acid concentrations in the medium of HU-β-IFN and HU Il-2 producing cell lines. Cytotechnology 1: 145–150.
Zaccharyl I and Rozengurt E (1992) A focal adhesion kinase ($p125^{FAK}$): a point of convergence in the action of neuropeptides, integrins, and oncogenes. Cell 71: 891–894.

Address for offprints: J. B. Griffiths, CAMR, Porton Down, Salisbury. Wiltshire SP4 OJG, U.K.

Cytotechnology **15**: 11–16, 1994.

Selecting and designing cell lines for improved physiological characteristics

J.R. Birch, R.C. Boraston, H. Metcalfe, M.E. Brown, C.R. Bebbington and R.P. Field[1]
Celltech Limited, 216 Bath Road, Slough, Berkshire, SL1 4EN, U.K.; [1] Present address: Zeneca Pharmaceuticals, Macclesfield, Cheshire, U.K.

Key words: Chemostat, cholesterol, choline, glutamine, glutamine synthetase, hybridoma.

Abstract

We have developed several approaches to create cell lines with improved characteristics in cell culture. In some cases it has been possible to isolate natural variants with useful properties. Cholesterol independent variants of the mouse NS0 myeloma cell line were isolated by cloning in a selective medium. A glutamine independent variant of a hyridoma was isolated by continuous (chemostat) culture under glutamine limited conditions in the presence of glutamate. Choline independent cells were isolated from a choline limited chemostat. In an alternative approach to modifying cell behaviour, we have used recombinant DNA techniques to introduce the glutamine synthetase (GS) gene to a hybridoma. This resulted in glutamine independence and increased productivity.

Introduction

In recent years there has been great progress in the development of genetic expression systems and of screening procedures for the isolation of cells expressing protein products at high levels. Less attention has been paid to the possibility of genetically manipulating the physiology of cells to improve their metabolism and growth characteristics in a production process. The potential exists in some cases to select natural variants with a desired property and in other cases it is possible to use recombinant DNA technology to introduce useful characteristics. These approaches are likely to be most useful for 'tailoring' cells which are widely applied, for example host cells used for expression of recombinant proteins and fusion partners for the creation of hybridomas.

Seaver (1992) pointed out that cell populations contain large numbers of genetic variants with a wide range of properties. It was possible to isolate hybridoma clones with, for example, enhanced stability to defrosting following cryopreservation. Screening in the presence of appropriate selection has been used to increase the probability of isolating a variant. Schumpp and Schlaeger (1992) were able to isolate hybridoma cell lines which were resistant to the toxic effects of ammonium and lactate by cloning in the presence of high concentrations of these substances. In this study we have used selective cloning techniques to isolate cholesterol independent variants of a mouse myeloma cell line. This makes it possible to grow the cells in media which are easier to prepare.

In situations where the desired variant is present with low frequency it is advantageous to grow the cells under appropriate selective conditions to enrich the variant population. The chemostat provides a particularly powerful method for applying a continuous selective pressure to a culture. Chemostat cultures have been used to study the selection of variants in microbial populations (Dykhuizen and Hartl, 1983) but the method has received little attention in animal cell culture. In the present paper we describe the use of the chemostat to isolate glutamine independent hybridoma cell lines, which could use glutamate, from a population which was absolutely dependent on glutamine. Some cell lines, including many hybridomas, have an absolute requirement for glutamine, whilst for other cell lines glutamine can be replaced by glutamate (Griffiths and Pirt, 1967). Glutamine independent variants of some myeloma and hybridoma cells lines have been isolated (Bebbington *et al.*, 1992) with a frequency that was cell line dependent. The ability to use glutamate seems to depend on the presence in the cell of glutamine synthetase. It is generally recognised that it is advantageous to use glutamate rather than glutamine in culture media since the latter is unstable and generates ammonia which can reach toxic levels in cultures (Glacken *et al.*, 1986).

There is increasing interest in the use of recombinant DNA techniques to usefully manipulate the physiology of cells. Stephanopoulos and Sinskey (1993) have discussed the potential for 'metabolic engineering'; the manipulation of cellular enzymatic, regulatory and transport processes for the purpose of enhancing specific product yield. There are already examples of recombinant DNA approaches being used to alter the properties of cells. Lee *et al.* (1989) for example describe the introduction of a sialyl transferase enzyme into CHO cells to produce glycoproteins having carbohydrate structures which more closely resemble naturally occurring human glycoproteins. In this paper we describe the introduction of the glutamine synthetase (GS) gene into a hybridoma cell line to confer glutamine independence.

Materials and methods

Selection of cholesterol independent cells

Mouse NSO myeloma cells were routinely cultured in a serum-free medium containing cholesterol. Cholesterol independent cells were isolated by dilution cloning in 96 well plates using medium lacking serum and cholesterol. Forty plates were seeded at an average distribution of one cell per well. Three cholesterol independent clones were isolated and subsequently grown in serum free shake flask culture. In some experiments, the cells were grown in shake flask culture in protein-free chemically defined medium.

Cell line selection using a chemostat

The cell line used for this study was a murine hybridoma derived from the NS-1 myeloma cell line. Cells were grown in an LH500 series reactor (LH Engineering Ltd) with a working volume of 700 to 800 ml. The vessel was top driven with a Rushton turbine impeller, and had automatic pH and dissolved oxygen control. A constant volume was maintained in the reactor by continuously pumping fresh medium into the vessel and by removing culture via an outflow pump activated by contact of the surface of the culture with a level sensor probe. The culture medium was DMEM supplemented with 3% foetal calf serum in the glutamine limited chemostat and a proprietary serum-free medium in the choline limited experiment. For the isolation of glutamine independent variants, the glutamine concentration in the feed was maintained at a limiting level (97 mg l^{-1}) and sodium glutamate (57 mg l^{-1}) was included in the medium. Glutamine, glutamate and ammonia were measured using commercially available assay kits (Boehringer Mannheim for glutamine and glutamate, Sigma for ammonia). For the isolation of the choline independent variant choline chloride (2 mg l^{-1}) was the limiting nutrient in the chemostat and for the experiment to select ammonia resistant cells, the feed contained ammonia at 100 mg l^{-1}, added as ammonium chloride. The principle of the chemostat is described by Pirt (1975). The dilution rate was changed once in both choline and glutamine limited chemostats to provide data for other studies.

Transfection of glutamine synthetase (GS) gene into hybridoma cell line

A murine, IgM secreting, hybridoma cell line was transfected with a vector containing the glutamine synthetase gene. Transfectants were identified by their ability to grow in glutamine-free medium. Transfectants were dilution cloned. One clone was chosen for further study. It was adapted to serum-free medium containing glutamate and asparagine and grown in shake flask culture. IgM was measured by ELISA.

Results

Isolation of cholesterol independent variants

Three cholesterol independent clones of the NSO myeloma cell line were isolated by dilution cloning in serum-free, cholesterol-free medium. One clone was chosen for further study. The cell line was capable of growth in protein-free and lipid-free medium in shake flask culture. In contrast the parent cell line died within 24 hours if cholesterol was omitted from the medium. The population doubling time (16 h) of the variant is comparable with the parent cell line.

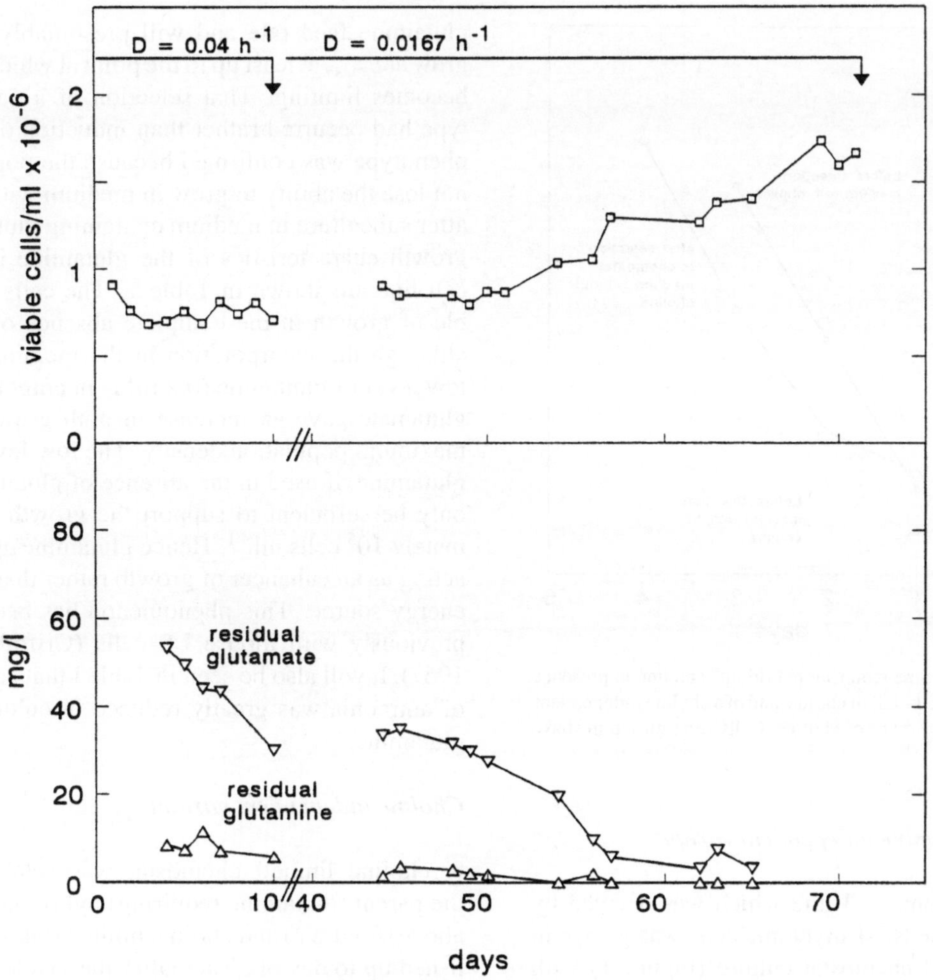

Fig. 1. Glutamine limited chemostat culture of a hybridoma cell line. The dilution rate was changed at the time indicated by the arrow.

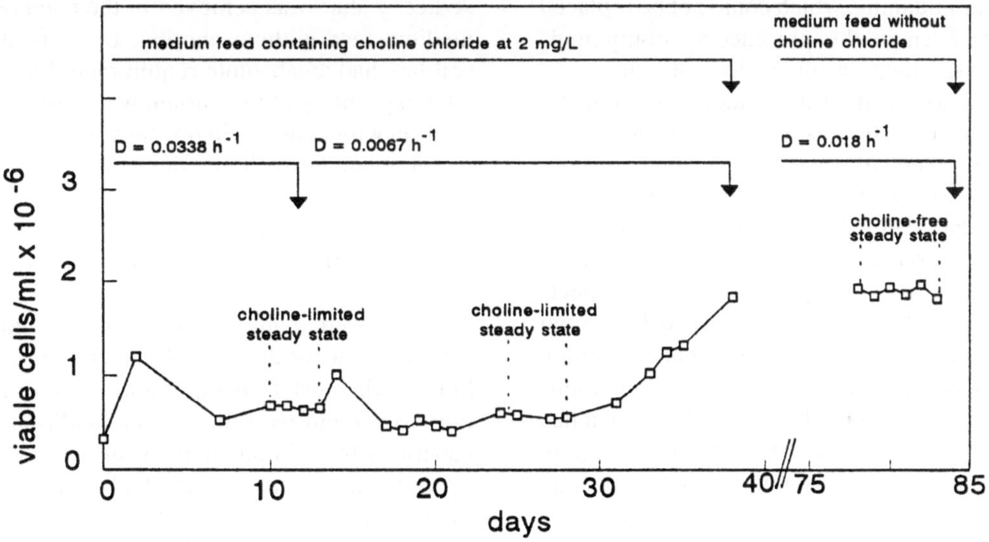

Fig. 2. Choline limited chemostat culture of a hybridoma cell line. Dilution rates were changed at the times indicated by the arrow.

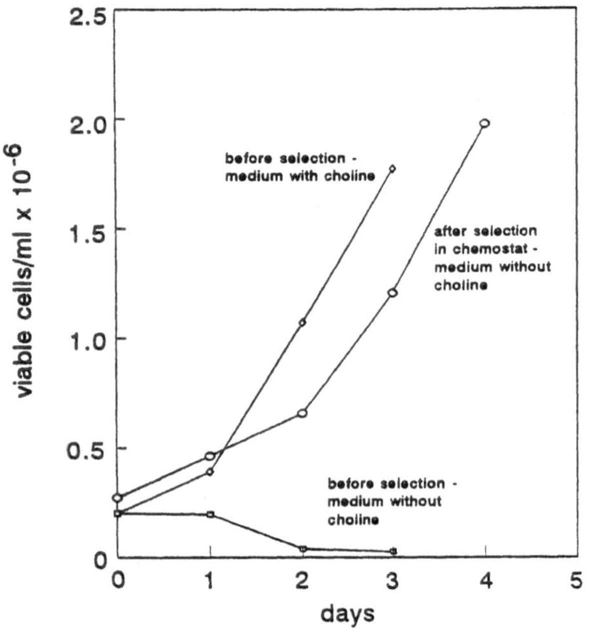

Fig. 3. Growth of choline requiring hybridoma cell line in presence (◇ – ◇) and absence (□ – □) of choline and of a choline independent variant (o – o) in the absence of choline. Cells were grown in shake flask culture.

Isolation of glutamine independent variant

A mouse hybridoma cell line which was derived by fusion with mouse NS-1 myeloma cells was grown in glutamine limited chemostat culture (Figure 1) with glutamate provided in the medium feed. The hybridoma had previously been shown to have an absolute requirement for glutamine which could not be replaced by glutamate, even in the presence of compounds whose synthesis might be dependent on glutamine (asparagine, aspartic acid, proline, nucleosides). In the chemostat a steady state was achieved during which glutamine was essentially exhausted. After approximately 55 days the viable cell concentration rose and this corresponded with a decrease in the residual glutamate concentration in the reactor. By this stage a glutamine independent subpopulation had been selected and cells taken from the reactor could be grown in medium containing glutamate rather than glutamine indicating that the variant population was now dominant. The variant is presumably selected because it has a growth rate advantage compared with the glutamine dependent population. The growth rate of cells requiring glutamine will be equal to the dilution date (D) which is less than μ_{max}. Any variant sub-population which can use glutamate will not be restricted by the glutamine feed rate and will presumably be able to grow at μ_{max}, at least up to the point at which glutamate becomes limiting. That selection of a variant genotype had occurred rather than induction of an altered phenotype was confirmed because the population did not lose the ability to grow in medium with glutamate after subculture in medium containing glutamine. The growth characteristics of the glutamine independent cell line are shown in Table 1. The cells were capable of growth in the complete absence of glutamine although the incorporation in the medium of a very low level of glutamine (0.2 mM) in combination with glutamate gave an increase in both growth rate and maximum population density. The low level of added glutamine, if used in the absence of glutamate, would only be sufficient to support the growth of approximately 10^5 cells ml^{-1}. Hence glutamine appears to be acting as an enhancer of growth rather than as a major energy source. This phenomenon has been observed previously with mouse LS cells (Griffiths and Pirt, 1967). It will also be seen in Table I that accumulation of ammonia was greatly reduced in cultures without glutamine.

Choline independent variant

A choline limited chemostat was established with the parent (glutamine requiring) hybridoma described above (Figure 2) and choline limited states were established up to day 30. Thereafter the viable cell density in the reactor increased and cells taken from the vessel were shown to no longer require choline, (Figure 3). A steady state was achieved in the chemostat using a medium feed without choline. In contrast the parent cell line had an absolute requirement for choline. The altered property of the variant was permanent; subcultivation with choline did not cause reversion to choline requirement. Growth kinetics of the variant were similar to the parent.

Attempt to establish ammonia tolerant variants

Attempts were made to isolate ammonia tolerant variants by growing the hybridoma cells in an ammonia limited chemostat. Ammonium chloride was added to the feed and we relied on the additional ammonia generated by cellular metabolism raising the ammonia concentration to a level which would limit growth. In practice we did not establish steady states and saw no evidence that ammonia tolerant cells were being selected.

Table 1. Growth of hybridoma variant with and without glutamine

Medium glutamine (mM)	Composition glutamate (mM)	Specific growth rate (h^{-1})	Maximum cell number ($\times 10^{-6}$ ml^{-1})	Ammonia accumulated (mM)
4.0	0	0.037	3.02	1.78
0	4.0	0.014	1.26	0.22
0.2	4.0	0.039	2.64	not tested

Table 2. Growth and productivity of GS transfected hybridoma

Medium	Specific growth rate (h^{-1})	Maximum viable cell number ($\times 10^{-6}$ ml^{-1})	Maximum IgM titre (μg ml^{-1})
− glutamine	0.034	5.5	517
+ glutamine (6 mM)	0.058	6.5	325

Results are the averages of duplicate shake flasks.

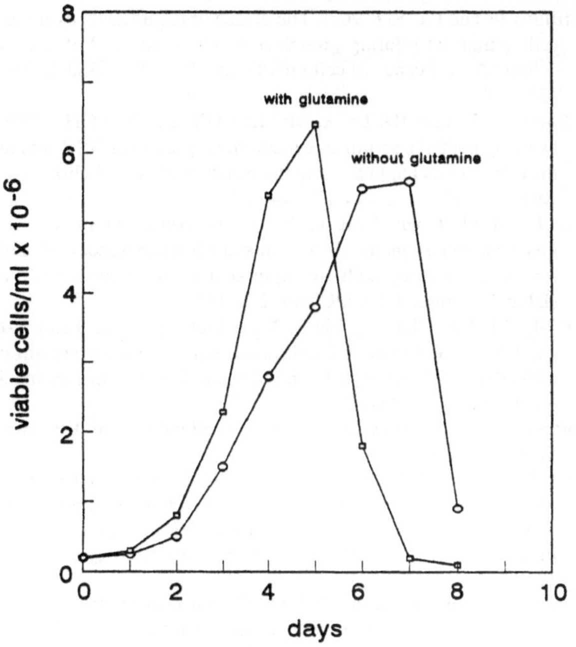

Fig. 4. Growth of a hybridoma cell line transfected with the glutamine synthetase (GS) gene. Cells were grown in medium with (□ – □) and without (o – o) glutamine in serum-free medium in shake flask culture.

Creation of glutamine independent cell line by introduction of glutamine synthetase

An IgM secreting hybridoma derived from a NS-1 myeloma was transfected with a vector (pCMGS, Bebbington *et al.*, 1992) carrying the GS gene. Transfectants were readily identified by their ability to grow in glutamine free medium. The growth of one such transfectant in shake flask culture is shown in Figure 4. The cells grew readily in serum-free medium in the absence of glutamine although the growth rate was increased if the identical medium was supplemented with glutamine (6 mM). It is interesting to note that antibody production was higher in the cultures without glutamine (Table 2). Whether this is a function of growth rate or a direct result of the altered nutrient supply is not clear. Similar growth profiles and productivity have been seen in airlift fermentations (5 litre scale) with the GS transfected cell line. Even in the presence of glutamine, productivity of the GS transfected cell line (325 mg l^{-1}) was higher than the non transfected parent (189 mg l^{-1}). It is not clear however whether this is a result of GS transfection or whether we simply isolated a particularly productive clone.

Discussion

We have demonstrated the value of exploiting naturally occurring genetic variants to obtain cell lines with useful physiological characteristics. It was possible to isolate cholesterol independent variants of the mouse NS0 cell line with high frequency. The NS0 cell is commonly used as a host for the production of recombinant proteins and particularly antibodies (Bebbington *et al.*, 1992). Removing the requirement for cholesterol in serum-free and in protein-free cultures significantly simplifies the preparation of culture media. Li *et al.* (1991) demonstrated that both rat myeloma cells (IR983F) and mouse myeloma cells (P3X63-Ag8-U1) required cholesterol even after six months growth in serum free medium. Kawamoto *et al.* (1983) were able to adapt mouse NS-1 myelomas to grow in a lipid deficient medium. After six months a clonal cell line was isolated which no longer required lipids. Sato *et al.* (1988) concluded that the cholesterol requirement of NS-1 myeloma cells resulted from a deficiency in 3-ketosteroid reductase activity.

Whilst some cell types can be adapted to utilise glutamic acid in place of glutamine (Griffiths and Pirt, 1967; Griffiths, 1973) several cell types, including some hybridomas, appear to have an absolute requirement for glutamine presumably as a result of glutamine synthetase deficiency. The disadvantages of using glutamine are well understood (Griffiths and Pirt, 1967; Glacken *et al.*, 1986) and various strategies have been proposed to overcome these disadvantages. It would appear from the present study that, at least in some cases, it is possible to isolate natural variants in which the glutamine synthetase deficiency is overcome, albeit by mechanisms which are unknown. The chemostat is an extremely powerful means of exposing cells to a continuous selection pressure. The isolation of choline independent hybridoma cells using the same technique, although of less practical value, is a further demonstration of the approach.

We have applied an alternative strategy using recombinant DNA technology to isolate hybridoma cells which no longer require glutamine. Transfection of the gene for glutamine synthetase (GS) into a hybridoma cell line gave rise to clones which grew readily in the absence of glutamine. GS has previousiy been used as a selective marker to confer glutamine independence in myeloma cell lines (Bebbington *et al.*, 1992). We and others have also transfected hybridomas with the GS gene to isolate glutamine independent cells (Bell *et al.*, 1991 and 1992). The present study gives the first indication that antibody productivity might be improved under glutamine-free conditions.

References

Bebbington CR, Renner G, Thomson S, King D, Abrams D and Yarranton GT (1992). High-level expression of a recombinant antibody from myeloma cells using a glutamine synthetase gene as an amplifiable selectable marker. Bio Technology 10: 169–175

Bell SL, Bebbington CR, Bushell ME, Sanders PG, Scott MF, Spier RE and Wardell JN (1991). Genetic engineering of cellular physiology. In: Spier RE, Griffiths JB and Meignier B (Eds.) Production of biologicals from animal cells in culture (pp 304–306). Butterworth-Heinemann.

Bell SL, Bushell ME, Scott MF, Wardell JN, Spier RE and Sanders PG (1992). Genetic modifications of hybridoma glutamine metabolism: physiological consequences. In: Spier RE, Griffiths JB and Macdonald C (Eds.) Animal cell technology: developments, processes and products (pp 180–182). Butterworth-Heinemann.

Dykhuizen DE and Hartl DL (1983). Selection in chemostats. Microbiological Reviews 47, 150–168.

Glacken MW, Fleischaker RJ and Sinskey AJ (1986). Reduction of waste product excretion via nutrient control: possible strategies for maximising product and cell yields on serum in cultures of mammalian cells. Biotech. Bioeng. XXVIII: 1376–1389.

Griffiths JB (1973). The effects of adapting human diploid cells to grow in glutamic acid media on cell morphology, growth and metabolism. J. Cell Sci. 12: 617–629.

Griffiths JB and Pirt SJ (1967). The uptake of amino acids by mouse cells (strain LS) during growth in batch culture and chemostat culture: the influence of cell growth rate. Proc. Roy. Soc. B. 168: 421–438.

Kawamoto T, Sato JD, Le A, McClure DB and Sato GH (1983). Development of a serum-free medium for growth of NS-1 myeloma cells and its application to the isolation of NS-1 hybridomas. Analytical Biochemistry 130: 445–453.

Lee EU, Roth J, and Paulson JC (1989). Alteration of terminal glycosylation sequences on N-linked oligosaccharides of Chinese hamster ovary cells by expression of β-galactoside α2, 6-sialyltransferase. J. Biol. Chem. 264: 13848–13855.

Li J-L, Li Y-J, Chao S, Lin L-X-M, Ouyang M-H, Pang Y-B and Chang W-S (1991). Cholesterol requirement for growth of 1R983F and P3X63-Ag8-U1 myeloma cells in serum-free medium. Cytobios 68: 15–22.

Pirt SJ (1975). Principles of microbe and cell cultivation. Blackwell Scientific Publications.

Sato JD, Cao H-T, Kayada Y, Cabot MC, Sato GH, Okamoto T and Welsh CJ (1988). Effects of proximate cholesterol precursors and steroid hormones on mouse myeloma growth in serum-free medium. In Vitro Cellular & Developmental Biology 24: 1223–1228.

Schumpp B and Schlaeger E-J (1992). Growth study of lactate and ammonia double-resistant clones of HL–60 cells. In: Spier RE, Griffiths JB and Macdonald C (Eds.). Animal cell technology: developments, processes and products (pp 183–185). Butterworth-Heinemann.

Seaver SS (1992). Enhancing monoclonal antibodies and hybridoma cell lines. Cytotechnology 9: 131–139.

Stephanopoulos G and Sinskey AJ (1993). Metabolic engineering – methodologies and future prospects. Tibtech 11: 392–396.

Address for offprints: J.R. Birch, Celltech Ltd., 216 Bath Road, Slough, Berkshire, SL1 4EN, U.K.

Cytotechnology **15**: 17–29, 1994.

Applications of improved stoichiometric model in medium design and fed-batch cultivation of animal cells in bioreactor

Liangzhi Xie and Daniel I.C. Wang
Department of Chemical Engineering, Biotechnology Process Engineering Center, Massachusetts Institute of Technology, Cambridge, MA 02139

Key words: Ammonia, animal cell, bioreactor, fed-batch culture, feeding strategy, lactate, medium design.

Abstract

In our previous work (Xie and Wang, 1994a), a simplified stoichiometric model on energy metabolism for animal cell cultivation was developed. Fed-batch experiments were performed in T-flasks using this model in supplemental medium design (Xie and Wang, 1994b). In this work, the major pathways of glucose and glutamine metabolism were incorporated into the stoichiometric model. Fed-batch culture was conducted in a 2-liter bioreactor with appropriate process control strategies. Nutrient concentrations, especially glucose and glutamine, were maintained at constant but low levels through the automated feeding of a supplemental medium formulated using the improved stoichiometric model. The formation of toxic byproducts, such as ammonia and lactate (Hassell *et al.*, 1991), was greatly reduced. The specific lactate production rate was decreased by 62-fold compared with batch culture in bioreactor and by 8-fold compared to fed-batch culture in T-flask using the previous stoichiometric model. Ammonia formation was also decreased compared with both the batch and fed-batch cultures. Most importantly, the monoclonal antibody concentration reached 900 mg l^{-1}, an increase of 17- and 1.6-fold compared with the batch and fed-batch cultures respectively.

Abbreviations: $C_{amm,\,n}$, $C_{amm,\,n-1}$: ammonia concentration in the n and (n-1)th samples respectively, mM; C^{s}_{amm}: ammonia concentration in the supplemental medium, mM; $C_{Ab,\,n}$, $C_{Ab,\,n-1}$: antibody concentration in the n and (n-1)th samples respectively, mg l^{-1}; C_i: concentration of the ith nutrient in the supplemental medium, mM; $C_{lac,\,n}$, $C_{lac,\,n-1}$: lactate concentration in the n and (n-1)th samples respectively, mM; C^{s}_{k}: concentration of glucose or glutamine in the supplemental medium, mM; $C_{k,\,n}$, $C_{k,\,n-1}$: concentration of glucose or glutamine in the n or(n-1)th sample respectively, mM; C_t: total concentration of glucose, amino acids, and vitamins in the supplemental medium, mM; F_1, F_2, F_3, F_4, F_5, F_6, F_7, F_8, F_9, F_{10}, F_{11}, F_{12}: fluxes described in Fig. 1, mmole cell^{-1}; m: total number of samples; N_t: total cell number in the reactor at culture time t, number of cells; $N_{t,\,0}$: total cell number in the reactor at the beginning of culture, number of cells; $N_{t,\,n}$, $N_{t,\,n-1}$: total cells in the reactor when the n and (n-1)th samples were taken, respectively, number of cells; $\Delta N_{t,\,n}$, $\Delta N_{t,\,m}$: total cells produced since the initiation when the nth sample and the mth sample were taken, respectively, number of cells; N_v: viable cell number in the reactor at culture time t, number of cells; $N_{v,\,n}$, $N_{v,\,n-1}$: viable cells in the reactor when the n and (n-1)th samples were taken, respectively, number of cells; $P_{Ab,\,n}$: monoclonal antibody produced between the n and (n-1) samples, mg; $P_{amm,\,n}$: ammonia produced between the n and (n-1) samples, mmole; $P_{i,\,n}$: amount of lactate or ammonia produced between the n and (n-1)th samples, mmole; $P_{lac,\,n}$: lactate produced between the n and (n-1) samples, mmole; P/O: number of ATP molecules generated per NADH molecule oxidized; $\bar{q}_i$: specific production rate of ammonia, lactate, or antibody respectively, mmole cell^{-1} h^{-1} (for ammonia and lactate), and mg cell^{-1} h^{-1} (for antibody); t:

culture time, h; t_m: culture time when the mth sample (last sample) was taken, h; t_n, t_{n-1}: culture time when the n and (n-1)th samples were taken respectively, h; ΔV_F: volume of supplemental medium fed to the reactor since the nth sample was taken, l; $\Delta V_{F,n}$: volume fed to the reactor between the (n-1)th and nth samples, l; V_n, V_{n-1}: total volume in the reactor after the nth or (n-1)th sample were taken respectively, l; $V_{s,j}$: volume of the jth sample, l; $V_{s,n}$, $V_{s,n-1}$: volume of the n and (n-1)th samples respectively, l; $X_{t,j}$: total cell density in the jth sample, cells l^{-1}; $X_{t,n}$: total cell density of the nth sample, cells l^{-1}; $X_{v,n}$, $X_{v,n-1}$: viable cell density in the n and (n-1)th samples respectively, cells l^{-1}; $Y_{amm/gln}$: ratio of ammonia production to glucose consumption, mmol $mmol^{-1}$; $Y_{i/cell}$: ratio of the total ammonia or lactate production to the total production of cells, mmole $cell^{-1}$; $Y_{lac/glc}$: ratio of lactate production to glucose consumption, mmol $mmol^{-1}$.

Greek letters: α: specific death rate, h^{-1}; β: total stoichiometric coefficient of glucose, amino acids, and vitamins, mmole $cell^{-1}$; $\delta_{glc,n}$, $\delta_{gln,n}$: Amount of glucose or glutamine, respectively, consumed between the n and the (n-1)th sample, mmole; $\delta_{k,n}$: Amount of glucose or glutamine consumed between the n and the (n-1)th sample, mmole; μ: specific growth rate, h^{-1}; θ_{ala}, θ_{asn}, θ_{asp}, θ_{glc}, θ_{glu}, θ_{gln}, θ_{gly}, θ_{pro}, θ_{ser}: stoichiometric coefficient of alanine, asparagine, aspartate, glucose, glutamate, glutamine, glycine, proline, and serine respectively including energy metabolism and biosynthesis of nonessential amino acids, mmole $cell^{-1}$; θ_{ala}^{cm}, θ_{asn}^{cm}, θ_{asp}^{cm}, θ_{glc}^{cm}, θ_{glu}^{cm}, θ_{gln}^{cm}, θ_{gly}^{cm}, θ_{pro}^{cm}, θ_{ser}^{cm}: stoichiometric coefficient of alanine, asparagine, aspartate, glucose, glutamate, glutamine, glycine, proline, and serine respectively in the biosyntheses of cell mass and product, excluding energy metabolism and synthesis from glutamine, mmole $cell^{-1}$; θ_{ATP}: stoichiometric coefficient for ATP in syntheses of cell mass and product, mmole $cell^{-1}$; θ_{glc}^{en}: stoichiometric coefficient of glucose in energy metabolism, mmol $cell^{-1}$; θ_i: stoichiometric coefficient of nutrient, mmol $cell^{-1}$; θ_i^{cm}: stoichiometric coefficient of nutrient in cell mass and product formation without consideration of consumption in energy formation and the production of nonessential amino acids from glutamine, mmol $cell^{-1}$; τ: integration of viable cells over culture time, cells h.

Superscripts: cm: cell mass (include product) syntheses; en: energy metabolism; s: supplemental medium.

Subscripts: Ab: antibody; ala: alanine; amm: ammonia; asn: asparagine; asp: aspartate; F: feeding; glc: glucose; gln: glutamine; glu: glutamate; gly: glycine; i: index for nutrients, index for ammonia or lactate; j: index of sample; k: stands for glucose or glutamine; lac: lactate; m: total number of sample; n: index of sample; p: product; pro: proline; ser: serine; t: time or total cells; v: vitamin or viable cells.

Introduction

Animal cell cultivation technology has shown great potential applications in manufacturing pharmaceutical biological products for diagnosis as well as for treating diseases such as cancer. For the past forty years, the application of this technology has faced some unsolved problems. These include low product concentration, low cell density, short culture span, toxic byproduct accumulation, and depletion of essential nutrients.

Recently, many efforts have been placed on increasing the product concentration as well as productivity (Adamson *et al.*, 1983; Fike *et al.*, 1993; Linardos *et al.*, 1992; Oh *et al.*,1993; Pendse and Bailey, 1990; Xie and Wang, 1994b). Growth factors were found to increase antibody secretion by 20% without affecting cell growth (Pendse and Bailey, 1990). Dialysis during cultivation has been employed in animal cell culture to remove waste products as well as to supply fresh nutrients (Adamson *et al.*, 1983; Linardos *et al.*, 1992). Although a significant increase in antibody concentration was reported, the efficiency of medium use in dialyzed culture was low. Hypertonic osmotic pressure was found to increase the specific antibody production rate (Oh, 1993; Ozturk, 1991). However, increase in antibody concentration was limited due to the negative effect of the hypertonic osmolality on cell growth. Fed-batch cultures have been widely used in animal cell cultivation (Lindell, 1992; Luan, 1987; Xie and Wang, 1994; Fike *et al.*, 1993). Concentrate supplementation was used in a hybridoma culture, where glucose and seven amino acids were fed to cells (Fike *et al.*, 1993). An empirical method was employed to determine the composition of the supplement. Antibody concentration was successfully increased to 132 mg l^{-1} compared with 36 mg l^{-1} in a batch culture.

In our previous work (Xie and Wang, 1994a and b), a stoichiometric model was developed, where glucose was assumed to be the sole energy source. Glutamine metabolism was simplified by neglecting its utilization for energy and nonessential amino acid production. These simplifications resulted in inaccuracies in supplemental medium composition, especially in the concentrations of glucose and glutamine. A feeding strategy was derived from this model and was subsequently employed to control nutrient concentrations at low levels in an effort to reduce ammonia and lactate formation. Fed batch experiments were performed in T-flasks by periodically feeding a complicated supplemental medium formulated with the stoichiometric model. Product concentration was increased 10-fold through a significant increase in cell density and a decrease in byproduct formation. However, data used to formulate the supplemental medium were obtained from literature which were different from the host cell line used in these experiments. This introduced errors in the composition of the supplemental medium. In addition, no process control could be performed in T-flask experiments. Hence, the performance of the T-flask experiments was limited by the above factors.

In this work, significant improvement has been achieved by incorporating more detailed energy metabolism pathways into the stoichiometric model. Cell composition from the host cell line was obtained and employed in the formulation of the supplemental medium. A fed-batch experiment conducted in a 2-liter bioreactor with proper computer process control was performed in this study.

Materials and Methods

Cell line and culture experiment

The host cell line used in fed-batch and batch experiment is a mouse-mouse hybridoma, CRL-1606, producing antifibronectin IgG monoclonal antibody. The cell line and its maintenance are described elsewhere (Xie and Wang, 1994b). For the preparation of inoculum, 50 ml cultures were conducted in T-flasks (750 ml, Falcon). After two days, cells from the T-flasks were transferred into two spinner flasks (200 ml, Bellco) and were cultivated for another two days.

In fed-batch experiment, cells from spinner flasks were centrifuged at 900 rpm in an IEC Centra-4B centrifuge (International Equipment Company) for 5 minutes and resuspended with an Initial medium (see Medium Design and Formulation) supplemented with 5% dialyzed fetal bovine serum (JRH Biosciences, CA). The experiment was performed in a 2-liter bioreactor (Braun Biostat M, B. Braun) with a starting volume of 900 ml and an initial cell density of 4×10^5 cells ml^{-1}. Feeding of a supplemental medium without serum (see Medium Design and Formulation) commenced right after the initiation. Culture process control was achieved through an ALERT 50 Mini dedicated real time computer and a control design program Alcom (Satt Control, Malmo, Sweden). The design program resided in a Compaq 286 monitor computer. Control strategies were developed through the program in the monitor computer and loaded into the ALERT control computer. Dissolved oxygen and pH were monitored and controlled at 60% of saturated air and 7.2, respectively, by adjusting the inlet gas composition. The inlet gas, composed of oxygen, carbon dioxide, and nitrogen, was passed through spiral silicone tubing (SIP Medical grade Silastic Q7-4750 Silicone tubing, Baxter, McGaw Park, IL.) immersed in the culture medium. Oxygen and carbon dioxide can diffuse through the tubing wall to provide oxygen and carbon dioxide.

In batch experiment, inoculum was prepared the same way as in the fed-batch experiment. Batch culture was initiated with Iscove's Modified Dulbecco's Medium supplemented with 5% dialyzed fetal bovine serum and 4 mM glutamine in the 2-liter reactor. Dissolved oxygen and pH were controlled the same way as in the fed-batch culture.

Samples of 4 ml were taken every 12 hours to measure the cell densities and osmolality, and followed by centrifugation at 1200 rpm for 10 minutes. Supernatant was then stored at –20 °C for later analysis. Cell enumeration has been described elsewhere (Xie and Wang, 1994b). The osmolality was measured via a μOSMETTE (model 5004, Precision Systems Inc. MA).

Analytical methods

Analyses for amino acids, glucose, lactate, ammonia, and monoclonal antibody were described previously (Xie and Wang, 1994b).

Dry Cell Weight

A 50 ml sample, containing about 10^8 cells, was placed into a 50 ml centrifuge tube (Corning, NY) and centrifuged at 1500 rpm for 10 min with an IEC Centra-4B centrifuge (International Equipment Co.). Total cell density was determined using four samples with a Neubauer hemacytometer (Reichert, Buffalo, NY). Cell pellet was then washed twice with 10 ml PBS and centrifuged. Supernatant was carefully discarded. The pellet was then transferred into a preweighed pan and dried at 60 °C to a constant weight.

Protein assay

Biuret assay (Read, 1984; Packer, 1967) was employed to determine the cellular protein content. Samples containing about 10^7 cells were placed into a 15 ml centrifuge (Corning, NY) tube and were prepared the same way as for dry cell weight measurement. Cells were disrupted in 0.5 ml lysis buffer (0.5% Triton X-100, 1 mM EDTA, 0.2 mM phenylmethylsulfonyl fluoride) at 4 °C for 40 min. After cell lysis, 1.5 ml Biuret reagent (1.5 g l^{-1} $CuSO_4 \cdot 5H_2O$, 6 g l^{-1}sodium potassium tartrate and 30 g l^{-1} NaOH) was added and mixed thoroughly. After a 30 min incubation at room temperature, absorbance was measured at 500 nm using a Perkin-Elmer Lambda-3 spectrophotometer (Norwalk, CT). A series of standards with concentrations from 1 to 10 mg ml^{-1} were prepared using bovine serum albumin (Sigma).

Carbohydrate assay

Total cellular carbohydrates were determined by the phenol reaction method (Hanson and Philips, 1981). The cell pellet was prepared the same way as in the protein assay. Cells were disrupted with 0.5 ml lysis buffer in a 15 ml centrifuge tube, and were transferred into a 20 ml thick-walled Pyrex test tube.The centrifuge tube was washed with another 0.5 ml lysis buffer and was combined with the liquid in the test tube. Then 1 ml phenol reagent (50 g L^{-1} aq.) was mixed rapidly and thoroughly with the sample. Finally, 5 ml concentrated sulfuric acid was added and rapidly mixed. After a 30 min of incubation at room temperature, absorbance was measured at 488 nm using a Perkin-Elmer Lambda-3 spectrophotometer (Norwalk, CT). A series of standards with concentrations from 10 to 100 μg ml^{-1} were prepared using glucose.

Lipid extraction and purification

The extraction of total cellular lipids was carried out using the methods of Folch *et al.*, (Bligh and Dyer, 1959; Folch *et al.*, 1956; Gerschenson *et al.*, 1967; Nelson, 1975; Packer, 1967; Stein and Smith, 1982). Samples containing about 3×10^8 cells (150 ml) were centrifuged in three 50 ml centrifuge tubes the same way as in the dry cell weight assay. Pellets were washed with 10 ml PBS and combined into one test tube. After centrifugation, the supernatant was discarded, then 0.5 ml Milli-Q water was added and mixed. In total 10 ml of methanol was mixed with the cells and the mixture was transferred quickly and completely into a 50 ml glass cylinder. Then 20 ml chloroform was added and mixed. After 40 min of extraction, the mixture was filtered through a 0.8 micron filter paper (MSI, Werstboro, MA). The filter paper was washed with 5 ml methanol and 10 ml chloroform. Filtrate was collected in a graduated cylinder and 8 ml NaCl (w/v 0.9% aq.) was added. The mixture was allowed to separate into two phases. After removing the upper phase, 13 ml of Chloroform/methanol/0.9% aq. NaCl (3:48:47) was added and mixed. The upper phase was carefully discarded. This step was repeated twice. The final lower phase contained lipids and was dried in a weighing pan at 40 °C in a vacuum desiccator until a constant weight.

DNA and RNA assay

The cellular DNA and RNA contents were determined by Diphenylamine and Orcinol reactions respectively. Samples containing about 2×10^7 cells were prepared and disrupted the same way as in the protein assay. The procedure for DNA and RNA extraction has been described elsewhere (Hanson and Philips, 1981). A DNA standard was prepared using deoxyribose with concentrations ranging from 10 to 50 μg ml^{-1}. An adenosine 5′-monophosphate solution with concentrations from 20 to 100 μg ml^{-1}, was used as a standard for the RNA assay.

Amino acid composition in cellular proteins

The amino acid composition of cellular proteins was determined by Analytical Biotechnology Services (Boston, MA). Hydrolysis was conducted with 6N HCl at 110 °C for 24 hours. Following hydrolysis, the sample was vacuum dried, taken up in a redrying solution (Ethanol:water:triethylamine; 2:2:1) and vacuum dried again. Finally, the hydrolysate was taken up in ethanol, water, and triothylamine (7:1:1) and derivatized by adding phenylisothiocyanate. The amino acids were analyzed using a Waters PicoTag HPLC system. Results for aspartate/asparagine and glutamate/glutamine were reported together. Average values from mammalian cell line for asparagine and glutarate were taken to determine the values for aspartate and glutamine. Samples containing 10^6 cells were centrifuged and washed twice with 1 ml PBS. Supernatant was completely discarded. Finally, 0.5 ml PBS was added to each sample, and they were shipped in dry ice.

Medium design and formulation

Two media were employed in the fed-batch experiment. An Initial medium was employed to initiate the cultivation and followed by the feeding of a supplemental medium according to a feeding strategy (see feeding strategy). The design of the Initial medium has been described in our previous article (Xie and Wang, 1994b). In order to decrease glutamine consumption and hence ammonia formation, the concentrations of nonessential amino acids in the Initial medium were increased based on results obtained from fed-batch experiments performed in T-flasks (unpublished). The composition of the Initial medium is shown in Table I.

The stoichiometric model (Xie and Wang, 1994 a and b) was modified by considering the major pathways of glucose and glutamine metabolism in energy and nonessential amino acid formation (Fig. 1). The fluxes (mmole cell^{-1}) in Fig. 1 were defined as the amount of substances converted per cell synthesized. The stoichiometric coefficients for nonessential amino acids defined in our previous work (Xie and Wang, 1994b) were actually the coefficients for cell mass and product formation without considering their production from glutamine, and they will be redefined here as θ_i^{cm} (i stands for gln, glc, and nonessential amino acids). The stoichiometric coefficients of nonessential amino acids, θ_i, will thus include the amount required for cell mass and product syntheses as well as the amount produced from glutamine. When the formation of non-essential amino acid (in most cases, alanine) from glutamine exceeds the amount required for cell mass and product synthesis, there is a net production and hence the stoichiometic coefficient is negative. In this case, the amino acid is not included in the supplemental medium. In order to correctly estimate the amount of energy derived from glutamine, glutamine consumption for the nonessential amino acid formation needs to be determined. Based on carbon balances, the following equations can be derived from Fig. 1:

$$2F_1 + F_{10} = F_2 + F_3 + F_4 \qquad (1)$$

$$F_{10} = F_5 - F_8 \qquad (2)$$

In the conversion of glutamate into α-ketoglutarate, the amino group in glutamate was assumed to be transferred to other nonessential amino acids. Ammonia

Table 1. Composition of culture media

Components	Initial medium (mM)	Supplemental medium (mM)
$CaCl_2$	2	
KCl	5	
$MgSO_4$	2	
NaCl	60	
$NaHCO_3$	60	
NaH_2PO_4	3	
$FeSO_4$	0.1	
Glucose	2	115.4
Alanine	0	0
Arginine	0.1	7.4
Asparagine	0.5	3.08
Aspartic acid	1	8.64
Cystine	0.1	3.49
Glutamic acid	0.5	5.01
Glutamine	0.2	41.9
Glycine	1	5.41
Histidine	0.1	2.69
Isoleucine	0.2	5.27
Leucine	0.2	10.1
Lysine	0.2	8.47
Methionine	0.1	2.71
Phenylalanine	0.1	3.98
Proline	0.2	2.0
Serine	0.5	3.19
Threonine	0.2	7.15
Tryptophan	0.1	1.37
Tyrosine	0.2	3.23
Valine	0.2	7.51
D-Biotin	5E-05	0.051
Choline Chloride	0.029	2.54
Folic Acid	0.009	0.09
Myo-Inositol	0.04	1.35
Niacinamide	0.033	0.37
Pyridoxal	0.02	0.51
Riboflavin	0.001	0.033
Thiamine	0.012	0.084
D-Ca Pantothenate	0.017	0.11
Vitamin B12	1E-05	0.007
Serum Percentage	5% (v/v)	0%
Osmolality	263 mOsm kg^{-1}	341 mOsm kg^{-1}

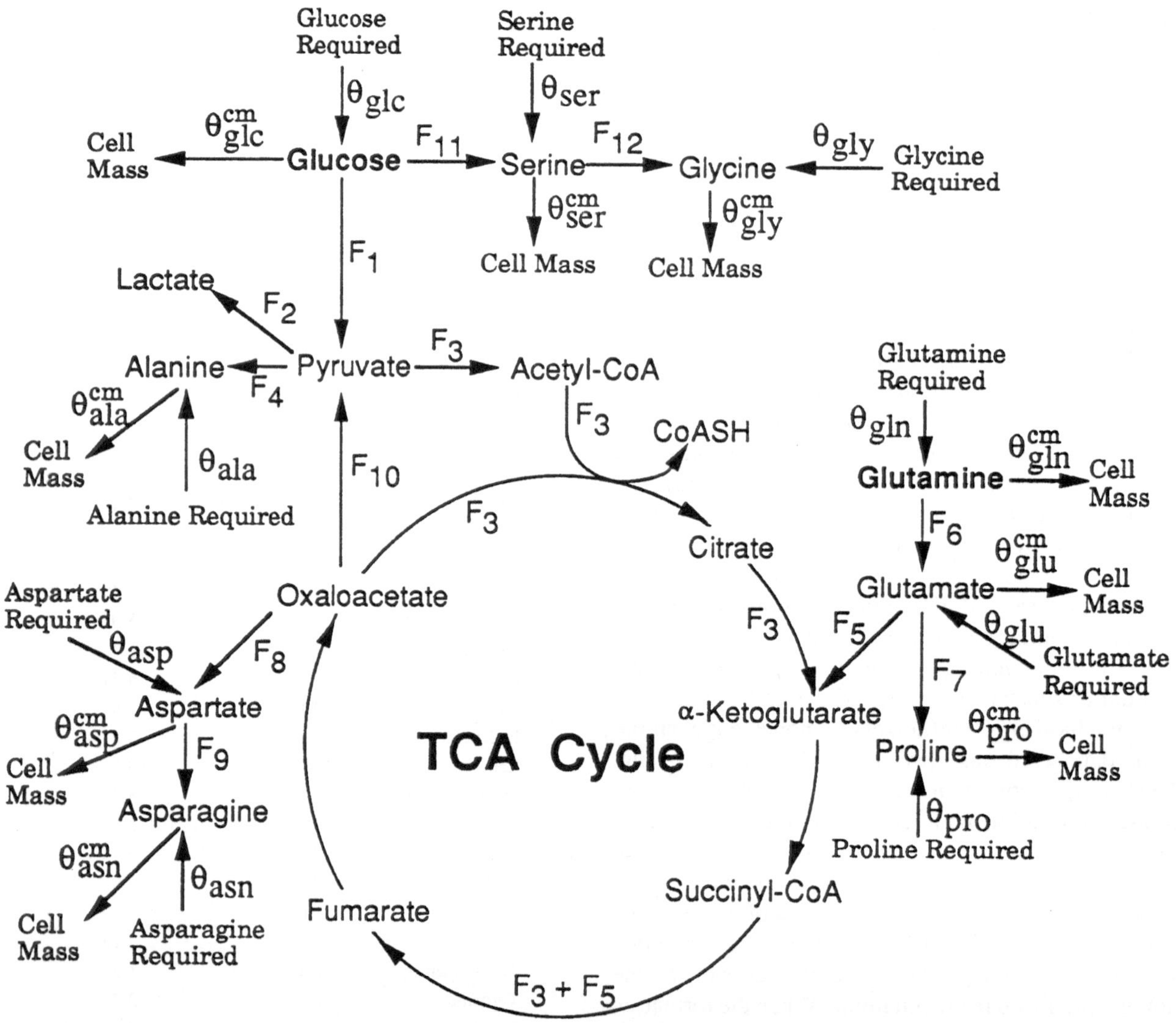

Fig. 1. Schematic figure of glucose and glutamine metabolism in energy production.

generated from this step was assumed to be negligible compared to the amount of glutamate converted into α-ketoglutarate. Equations (3) and (4) can be obtained using nitrogen and carbon balances respectively:

$$F_5 = F_4 + F_8 + 2F_{11} \tag{3}$$

$$F_6 = \theta_{glu}^{cm} - \theta_{glu} + F_5 + F_7 \tag{4}$$

Equation (5) was then derived from Equations (1), (2), and (3):

$$F_3 = 2F_1 - F_2 + 2F_{11} \tag{5}$$

According to an energy balance, the following equation was derived from biochemical reaction pathways:

$$\begin{aligned}\theta_{ATP} = {} & F_1(2ATP + 2NADH) - \\ & - F_2 \cdot NADH + F_3 \cdot 2NADH + \\ & + (F_3 + F_5)(2NADH + GTP + FADH_2) - \\ & - F_7 \cdot ATP - F_9 \cdot (ATP + 2NADH)\end{aligned} \tag{6}$$

By assuming a P/O ratio of 3, the above equation was simplified as:

$$\theta_{ATP} = 8F_1 - 3F_2 + 15F_3 + 9F_5 - F_7 - 7F_9 \tag{7}$$

Based on mass balances, the following equations were obtained:

$$\theta_{glc} = \theta_{glc}^{cm} + F_1 + F_{11} \quad (8)$$

$$F_{11} = \theta_{ser}^{cm} - \theta_{ser} + F_{12} \quad (9)$$

$$F_{12} = \theta_{gly}^{cm} - \theta_{gly} \quad (10)$$

$$F_4 = \theta_{ala}^{cm} - \theta_{ala} \quad (11)$$

$$\theta_{gln} = \theta_{gln}^{cm} + F_6 \quad (12)$$

$$F_7 = \theta_{pro}^{cm} - \theta_{pro} \quad (13)$$

$$F_8 = \theta_{asp}^{cm} - \theta_{asp} + F_9 \quad (14)$$

$$F_9 = \theta_{asn}^{cm} - \theta_{asn} \quad (15)$$

where

$$F_2 = \theta_{glc} \, Y_{lac/glc} \quad (16)$$

$$F_1 = \theta_{glc}^{en} \quad (17)$$

The stoichiometric coefficient for glucose in energy metabolism was then derived from Equations (5), (7), (16), and (17):

$$\theta_{glc}^{en} = (\theta_{ATP} + 18\,(F_{11} + \theta_{glc}^{en})\,Y_{lac/glc} - 9\,F_5 - - 30\,F_{11} + F_7 + 7\,F_9) \,/\, (38 - 18\,Y_{lac/glc}) \quad (18)$$

It is our assumption that the syntheses of nonessential amino acids from glutamine depends on the nutritional environment, especially the concentrations of glutamine, glucose, and nonessential amino acids. Hence there is no unique solution to the above model unless additional constraints are applied. A design criterion was employed to solve this problem. It is to minimize glutamine utilization in order to minimize toxic ammonia formation. In our previous work, it was concluded that when glutamine concentration was maintained at a constant but low level, ammonia formation was significantly reduced. Under that culture condition, the measured specific uptakes of nonessential amino acids could serve as the stoichiometric coefficients for the nonessential amino acids. The specific uptake values (mmole $cell^{-1}$) employed in the model were measured from our previous fed-batch experiment in which glutamine concentration was well controlled (hence minimum utilization), and are listed as the following: θ_{ala}, –0.17 × 10^{-9}, θ_{asn}, 0.046 × 10^{-9}, θ_{asp}, 0.13 × 10^{-9}, θ_{glu}, 0.076 × 10^{-9}, θ_{gly}, 0.082 × 10^{-9}, θ_{pro}, 0.048 × 10^{-9}.

The molar ratio of lactate to glucose was assumed to be 0.8, which was the average of the ratios measured from our previous fed-batch experiments (Xie and Wang, 1994b). The stoichiometric coefficients for cell mass, θ_i^{cm}, (i stands for glc, gln, and non-essential amino acids) and θ_{ATP} were also defined in our previous model (Xie and Wang, 1994b). The stoichiometric coefficients for glucose and glutamine were determined from Equations (18), (8), and (12).

Cell composition was measured for the host cell line and used to calculate the stoichiometric coefficients as the following (weight percentage): protein, 72.9; carbohydrates, 3.5; lipids, 13.5; DNA, 1.4; RNA, 3.8. Dry cell weight was measured as 25.0 × 10^{-8} mg $cell^{-1}$. The molar percentages of amino acids in cellular proteins were measured, and these were: ala, 8.25; arg, 5.93; asn, 4.4; asp, 4.69; cys, 2.8; gln, 6.2; glu, 4.96; gly, 8.54; his, 2.16; ile, 4.22; leu, 8.11; lys, 6.79; met, 2. 17; phe, 3. 19; pro, 5.26, ser, 6.89; thr, 5.73; trp, 1.1; tyr, 2.59; val, 6.02. Other information needed for the model was unchanged from our previous work (Xie and Wang, 1994b).

The stoichiometric coefficients of nutrients were determined according to our previous model (Xie and Wang, 1994b) except for glucose, glutamine, and nonessential amino acids. The total stoichiometric coefficient β was calculated by adding all of the positive stoichiometric coefficients together (nutrients with negative coefficients do not need to be included in supplemental medium due to net production in culture process). To determine the total concentration of nutrients in the supplemental medium (C_t), the osmolality of the supplemental medium and solubility of nutrients need to be considered (Xie and Wang, 1994b). A high C_t value was originally designed to lower the dilution effect, but it was finally lowered to 250 mM due to solubility problems. The composition of the supplemental medium was then determined by Equation (19) (Xie and Wang, 1994b), and is shown in Table I. The osmolality of the supplemental medium was measured as 341 mOsm kg^{-1}.

$$C_i = C_t \, \theta_i \,/\, \beta \quad (19)$$

Feeding strategy

The objective of the feeding strategy is to maintain a relatively constant nutritional environment. The concentrations of glucose and glutamine need to be controlled at a constant and low level in order to decrease toxic ammonia and lactate formation (Xie and Wang, 1994b). This can be achieved by feeding the nutrients to cells at the same rate as they are consumed. However, the accuracy will depend on prediction of cell growth. An off-line control strategy was employed to estimate the cell growth. Samples were taken every 12 hours and cell densities were determined. Specific growth and death rates were then calculated and input into the control program to predict cell growth before the next sample was taken.

The feeding strategy is described by the following equations:

$$V_n = V_{n-1} + \Delta V_{F,n} - V_{s,n} \qquad (19)$$

$$N_{v,n} = X_{v,n}(V_n + V_{s,n}) \qquad (20)$$

$$N_{t,n} = X_{t,n}(V_n + V_{s,n}) \qquad (21)$$

$$\alpha = \frac{\ln(N_{v,n}/N_{v,n-1})[(N_{t,n} - N_{v,n}) - (N_{t,n-1} - N_{v,n-1})]}{(t_n - t_{n-1})(N_{v,n} - N_{v,n-1})} \qquad (22)$$

$$\mu = \alpha + \frac{\ln(N_{v,n}/N_{v,n-1})}{(t_n - t_{n-1})} \qquad (23)$$

$$N_v = N_{v,n}\, e^{(\mu - \alpha)(t - t_n)} \qquad (24)$$

$$N_t = N_{t,n} + \frac{\mu N_{v,n}}{\mu - \alpha}[e^{(\mu - \alpha)(t - t_n)} - 1] \qquad (25)$$

$$\Delta V_F = \frac{\beta\,(N_t - N_{t,n})}{C_t} \qquad (26)$$

Calculations

The production of lactate, ammonia, and monoclonal antibody between the (n-1) and nth samples was calculated from the following equations:

$$P_{lac,n} = C_{lac,n}(V_n + V_{s,n}) - C_{lac,n-1} V_{n-1} \qquad (27)$$

$$P_{amm,n} = C_{amm,n}(V_n + V_{s,n}) - C_{amm,n-1} V_{n-1} - C^s_{amm}\Delta V_{F,n} \qquad (28)$$

$$P_{Ab,n} = C_{Ab,n}(V_n + V_{s,n}) - C_{Ab,n-1} V_{n-1} \qquad (29)$$

The average specific production rates of lactate, ammonia, and antibody were defined by Equation (30):

$$\bar{q}_i = \frac{1}{\tau}\sum_{n=1}^{m} P_{i,n} \quad (i = \text{lac, amm, or Ab}) \qquad (30)$$

where τ was the integral of viable cell number in the reactor over the culture time and was calculated by Equation (31):

$$\tau \int_0^{t_m} N_v\, dt = \sum_{n=1}^{m}(N_{v,n} + N_{v,n-1} - X_{v,n-1} V_{s,n-1})(t_n - t_{n-1})/2 \qquad (31)$$

The total amounts of glucose and glutamine consumed were calculated from mass balances using Equation (32):

$$\delta_{k,n} = C_{k,n-1} V_{n-1} + \Delta V_{F,n} C^s_k - C_{k,n} V_n \quad (k = \text{glc or gln}) \qquad (32)$$

The total number of cells produced when the nth sample was taken was calculated from Equation (33):

$$\Delta N_{t,n} = N_{t,n} + \sum_{j=1}^{n-1} X_{t,j} V_{s,j} - N_{t,0} \qquad (33)$$

The molar ratios of byproducts over total number of cells produced were determined from Equation (34):

$$Y_{i/cell} = \frac{1}{\Delta N_{t,m}}\sum_{n=1}^{m} P_{i,n} \quad (i = \text{lac or amm}) \qquad (34)$$

The ratios of byproducts over nutrients consumed were calculated from the following equations:

$$Y_{lac/glc} = \frac{\sum_{n=1}^{m} P_{lac,n}}{\sum_{n=1}^{m} \delta_{glc,n}} \qquad (35)$$

$$Y_{amm/gln} = \frac{\sum_{n=1}^{m} P_{amm,n}}{\sum_{n=1}^{m} \delta_{gln,n}} \qquad (36)$$

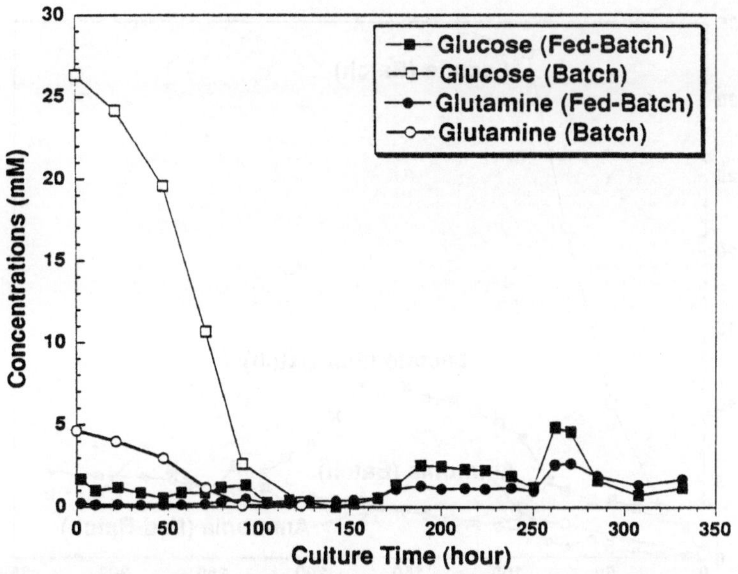

Fig. 2. Comparison of glucose and glutamine concentrations in fed-batch using rational medium design and conventional batch cultures.

Results and discussion

After the initiation of the fed-batch experiment and before the second sample was taken, the specific growth and death rates were assumed to be 0.02 and 0.0 h^{-1} respectively. After the second sample was taken, calculated values for growth from the experimental data were employed for the control of the nutrient feed. The nutrient concentrations in the reactor were well controlled as shown by the time profiles for glucose and glutamine during the entire culture process (Fig. 2). A batch experiment was also performed as a comparison in the 2-liter reactor with the same process control, and the results were also shown in Fig. 2. Both glucose and glutamine concentrations decreased rapidly during the culture process and almost reached zero at the end of the batch culture. However, both glucose and glutamine were maintained at low and constant levels in the entire culture process of the fed-batch culture (Fig. 2). This significant improvement in nutritional control compared with our previous study (Xie and Wang, 1994b) was contributed by improvement of the stoichiometric model, use of the actual cell composition data, and use of the automatic feeding strategy. The stoichiometric model relies on the cell composition to determine the composition of the supplemental medium. In our previous fed-batch experiment, cell composition data were the average values taken from literature for different mammalian cell lines. The total protein in our cell line is much higher than the value taken from literature. On the other hand, the total carbohydrates is much lower. The difference in cell composition resulted in inaccuracy of the supplemental medium composition and nutrient control in our previous fed-batch experiment. Results showed that the modified model was improved, especially with respect to the determination of stoichiometric demands for glucose and glutamine.

The formation of byproducts in the fed-batch bioreactor experiment (Fig. 3) was greatly reduced, compared with the batch experiment. Lactate concentration reached 11 mM at 180 hours and then decreased to 4 mM at the end of the fed-batch culture. The decrease in lactate concentration was not due to the dilution effect of supplemental medium feeding. This conclusion is supported by the fact that the total amount of lactate was decreasing after lactate concentration reached its maximum value. Ammonia concentration increased continuously during the entire culture process and reached 5.6 mM at the end of the experiment. A portion of the ammonia was introduced from the supplemental medium which had an ammonia concentration of 4 mM due to non-enzymatic degradation of glutamine and asparagine, The ammonia produced by cells was therefore actually less than 5.6 mM.

Fig. 4 shows the cell densities from the fed-batch and batch experiments, where both experiments were conducted in the 2-liter bioreactor. In the fed-batch

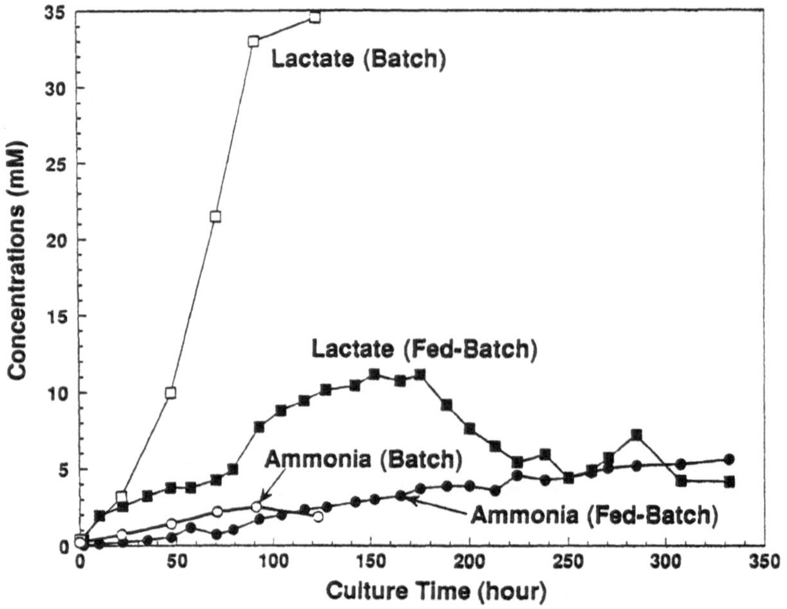

Fig. 3. Comparison of lactate and ammonia formation in fed-batch using rational medium design and conventional batch cultures.

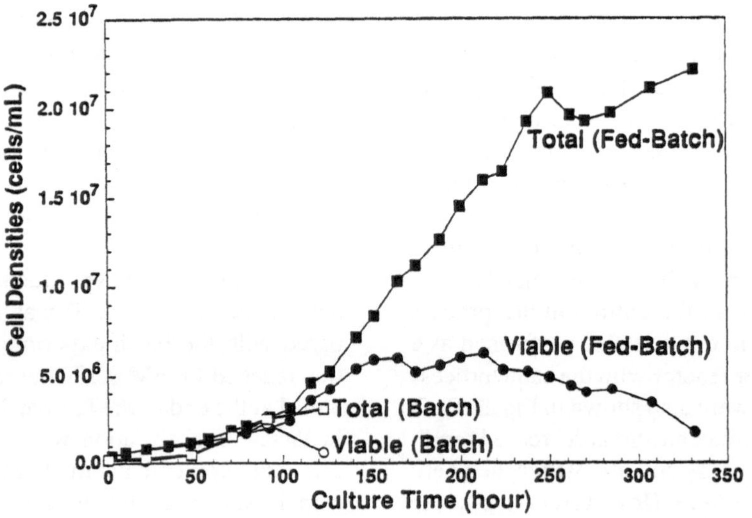

Fig. 4. Comparison of cell densities in fed-batch culture using rational medium design and conventional batch culture in bioreactor.

experiment, the specific growth rate was maintained at a low value through control of nutrient concentrations at low levels. This was preferred because it reduces toxic byproduct formation. Viable cell density reached a maximum of 6.3×10^6 cell ml^{-1}, and then remained constant for more than 100 hours while total cell density was increasing until the end of the culture. The same behavior was observed in our previous fed-batch experiments performed in T-flasks. Possible reasons for the decreasing viability were the following. The concentrations of glucose and glutamine at 100 hours were extremely low, which could result in cell death. In addition, the initial osmolality was intentionally designed to be low (263 mOsm kg^{-1} H_2O) based on the assumption that it would increase during the culture process. However, the byproduct formation was greatly reduced due to the nutritional control, which resulted in a decreased osmolality in the culture medium. After 160 hours, the osmolality decreased to less than 250 mOsm kg^{-1} H_2O which could have contributed to the decreased viability. The actual cause could be much more complicated, hence more detailed studies are needed to obtain a concrete explanation for this behavior.

The monoclonal antibody concentrations from the fed-batch and conventional batch experiments are shown in Fig. 5. In conventional batch culture, the antibody concentration increased very slowly during the first 50 hours due to the low cell density and reached only 50 mg l^{-1} when the cells began to die. In the fed-batch experiment, antibody concentration increased nearly exponentially at the beginning due to the higher cell density and viability. Unlike the batch experiment, the antibody concentration increased continuously until the end of the culture to 900 mg l^{-1}.

Table II shows the comparison of culture performances among batch and fed-batch (T-flask and 2-liter bioreactor) experiments. Toxic byproduct formation has been greatly reduced by controlling the nutrient concentrations at low levels. The final concentration of lactate was only 4.2 mM. At such a low concentration, the inhibitory effect of lactate was no longer a concern (Glacken, 1987). This low accumulation of lactate was resulted from the significant reduction in the specific lactate production rate, which was 62-fold lower than in batch culture and 8-fold lower than in our previous fed-batch culture performed in T-flasks. Consequently, the efficiency of glucose was significantly increased as supported by a very low lactate to glucose ratio. In total, only 5% of the glucose consumed was converted into lactate. Hence only a very small portion of energy was derived through the glycolytic pathways. These results confirmed our previous assumption that high glycolytic flux is not necessary for cell growth but a result of low efficient usage of glucose utilization caused by high residual glucose concentration. However, it has been found by other investigators that when glucose usage was reduced, ammonia formation increased as a result of increased glutamine utilization when glutamine concentration was not controlled (Hu, 1987). The effect of decreased glucose utilization on ammonia formation was not significant in our fed-batch experiment because glutamine concentration was also controlled at a very low level. The specific formation rate of ammonia was comparable to our previous fed-batch experiment but significantly reduced as compared with the batch culture. Hence, the efficiencies of both glucose and glutamine utilization were increased in both the fed-batch culture in the bioreactor and in our previous fed-batch experiment in T-flasks.

The concentration of monoclonal antibody reached 900 mg l^{-1}, an increase of 60% from our previous fed-batch experiment in T-flasks. This was an increase of 17-fold compared with conventional batch culture in the same bioreactor. The average specific antibody production rate was also significantly increased. This was probably due to the following factors. Lactate was found to inhibit antibody production (Glacken, 1989). Significantly decreased lactate concentration possibly contributed to the increase in specific antibody production rate. The specific production of antibody was found to be a linear function of death rate in a chemostat hybridoma culture (Linardos, 1991). However, no positive relationship between antibody production rate and cell death rate was observed in our experiment. It was found that total amount of antibody stored in hybridoma CRL-1606 cells was less than 0.025% of the total amount of antibody in the culture medium (Lindell, 1992). Hence, the contribution of possible cell lysis in the late culture stage to the increased specific antibody production rate was insignificant for this cell line. The improvement in the bioreactor environment through process control could also contribute to the increase.

Conclusions

Our previous simplified stoichiometric model for animal cell cultivation was improved by considering more detailed cellular energy metabolism pathways for glucose and glutamine as well as pathways for non-essential amino acid synthesis. This modified model was employed in a supplemental medium formulation, together with experimental cell composition data of the host cell line. Fed-batch culture was performed in a 2-liter bioreactor with process and feeding control. Nutritional environment was well controlled, especially the concentrations of glucose and glutamine. Culture performance was significantly improved from our previous fed-batch experiment, which had already been greatly improved from conventional batch and fed-batch experiments. This was supported by an 8-fold reduction in specific lactate production rate and a 60% increase of antibody concentration compared with our previous fed-batch experiments. The reason for high average specific antibody production rate needs to be further investigated.

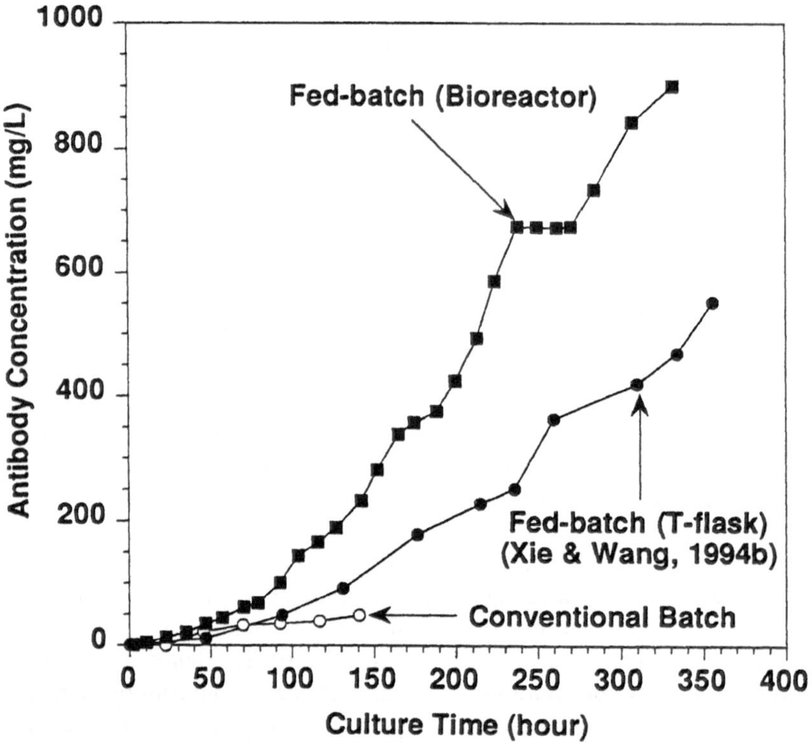

Fig. 5. Comparison of product concentrations in fed-batch using rational medium design and conventional batch cultures.

Table 2. Comparison of culture performance between batch and fed-batch experiments

Parameters	Units	Batch (reactor)	Fed-batch (reactor)	Fed-batch (T-flask)[1]
Maximum viable cell density	10^6 cells ml^{-1}	2.28	6.3	7.5
Maximum total cell density	10^6 cells ml^{-1}	3.1	22.2	14.5
Culture span	h	123	332	356
Final lactate concentration	mM	34.5	4.2	30.9
Final ammonia concentration	mM	1.85	5.6	5.4
Ratio of lactate to glucose	mole $mole^{-1}$	1.33	0.10	1.42
Ratio of lactate to cell	10^{-9} mmole $cell^{-1}$	11.9	0.21	3.1
Average specific formation rate of lactate	10^{-9} mmole $cell^{-1}$ h^{-1}	0.30	0.0048	0.038
Ratio of ammonia to glutamine	mole/mole	0.40	0.30	0.45
Ratio of ammonia to cell	10^{-9} mmole $cell^{-1}$	0.64	0.20	0.38
Average specific formation rate of ammonia	10^{-9} mmole $cell^{-1}$ h^{-1}	0.016	0.0045	0.005
Antibody concentration	mg l^{-1}	52	900	551
Average specific antibody production rate	10^{-9} mmole $cell^{-1}$ h^{-1}	0.45	0.93	0.47

[1] Experiment performed in T-flask with stoichiometric feeding (Xie and Wang, 1994b)

Acknowledgement

The authors acknowledge the financial support from the National Science Foundation (EEC 80-03014) through their Engineering Research Center initiative. We are very grateful to Prof. Charles L. Cooney for his kindness in letting us use the reactor and control system. We thank Dr. Per I. Lindell and Mr. Gregg Nyberg for their help in setting up the bioreactor, and Ms. Kit Yue Wong for performing some of the assays and preparing media. Mr. Steve Meier's editorial effort for this manuscript is also appreciated.

References

Adamson SR, Fitzpatrick SL and Behie LA (1983) In vitro production of high titer monoclonal antibody by hybridoma cells in dialysis culture. Biotechnol. Lett. 5: 573–578.

Bligh EG and Dyer WJ (1959) A rapid method of total lipid extraction and purification. Can. J. Biochem. Physiol. 37: 911–917.

Fike R, Kubiak J, Price P and Jayme D (1993) Feeding strategies for enhanced hybridoma productivity: automated concentrate supplementation. BioPharm. 6: 49–54.

Folch J, Lees M and Sloane Stanley GH (1957) A simple method for the isolation and purification of total lipids from animal tissues. J. Biol. Chem. 226: 497–509.

Gerschenson LE, Mead JF, Harary I and Haggerty Jr. DF (1967) Studies on the essential fatty acids on growth rate, fatty acid composition, oxidative phosphorylation and respiratory control of HeLa cells in culture. Biochim. Biophys. Acta 131: 42–49.

Glasken MW, Huang C and Sinskey AJ (1989) Mathematical descriptions of hybridoma culture kinetics. III. simulation of fed-batch bioreactors. J. Biotechnol. 10: 39–66.

Glacken MW (1987) Development of mathematical descriptions of mammalian cell culture kinetics for the optimization of fed-batch bioreactors. Ph. D. thesis, MIT, Cambridge, MA, USA

Hanson RS and Philips JA (1981) Chemical Composition. In: Gerhardt P et al. (eds.). Manual of Methods for General Bacteriology (pp. 329–364) American Society for Microbiology, Washington, D.C.

Hassell T, Gleave S and Butler M (1991) Growth inhibition in animal cell culture: The effect of lactate and ammonia. Applied Biochem. Biotechnol. 30: 29–41.

Hu WS, Dodge TC, Frame KK and Himes VB (1987) Effect of glucose on the cultivation of mammalian cells. Develop. Biol. Stand. 66: 279–290.

Linardos TI, Kalogerakis N and Behie LA (1991) The effect of specific growth rate and death rate on monoclonal antibody production in hybridoma chemostat cultures. Can. J. Chem. Eng. 69: 429–438.

Linardos TI, Kalogerakis N and Behie LA (1992) Monoclonal antibody production in dialyzed continuous suspension culture. Biotechnol. Bioeng. 39: 504–510.

Lindell PI (1992) Dynamic operation of mammalian cell fed-batch bioreactors. Ph.D. thesis, MIT, Cambridge, MA, USA.

Luan YT, Mutharasan R and Magee WE (1987) Strategies to extend the longevity of hybridomas in culture and promote yield of monoclonal antibodies. Biotechnol. Lett. 9: 691–696.

Nelson GJ (1975) Isolation and purification of lipids from animal tissues. In: Perkins EG (ed.) Analysis of lipids and lipoproteins (pp. 1–22) American Oil Chemists Society, Champaign, Illinois.

Oh SKW, Vig P, Chua F, Teo WK Yap MGS (1993) Substantial overproduction of antibodies by applying osmotic pressure and sodium butyrate. Biotechnol. Bioeng. 42: 601–610.

Ozturk S.S and Palsson BO (1991) Effect of medium osmolarity on hybridoma growth, metabolism, and antibody production. Biotechnol. Bioeng. 37: 989–993.

Packer L (1967) Experiments in cell physiology. Academic press, New York.

Pendse GJ and Bailey JE (1990) Effects of growth factors on cell proliferation and monoclonal antibody production of batch hybridoma cultures. Biotechnol. Lett. 12: 487–492.

Read SM (1984) Techniques in proteins and enzyme biochemistry, part I supplement: techniques for determining protein concentration. In: Tipton KF (ed.) Techniques in the life sciences. (pp. 1–34) Elsevier Scientific, New York.

Stein J and Smith G (1982) Techniques in lipid and membrane biochemistry, part I: extraction methods. In: Hesketh TR et al. (eds.) Techniques in the life sciences. (pp. 1–10) Elsevier/North-Holland Scientific, New York.

Xie L and Wang DIC (1994a) Stoichiometric analysis of animal cell growth and its application of medium design. Biotechnol. Bioeng. 43: 1164–1174.

Xie L and Wang DIC (1994b) Fed-batch cultivation of animal cells using different medium design concepts and feeding strategies. Biotechnol. Bioeng. 43: 1175–1189.

Address for offprints: Daniel I.C. Wang, Room 20A-207, MIT, 18 Vassar St., Cambridge, MA 02139, U.S.A.

Cytotechnology **15:** 31–50, 1994.

Diverse effects of essential (*n*-6 and *n*-3) fatty acids on cultured cells

Stephanos I. Grammatikos[1], Papasani V. Subbaiah[2], Thomas A. Victor[3] and William M. Miller[1]

[1] *Department of Chemical Engineering, Northwestern University, 2145 Sheridan Rd., Evanston, Illinois 60208–3120;* [2] *Departments of Biochemistry and Medicine, Rush Medical College, 1653 W. Congress Pkwy, Chicago, Illinois 60612;* [3] *Department of Pathology, Northwestern University Medical School, Evanston Hospital, Evanston, Illinois 60201, USA*

Key words: Desaturation; eicosanoids; lipid peroxidation; membrane fluidity; protein kinase C; serum-free medium

Abstract

Fatty acids (FAs) have long been recognized for their nutritional value in the absence of glucose, and as necessary components of cell membranes. However, FAs have other effects on cells that may be less familiar. Polyunsaturated FAs of dietary origin (*n*-6 and *n*-3) cannot be synthesized by mammals, and are termed 'essential' because they are required for the optimal biologic function of specialized cells and tissues. However, they do not appear to be necessary for normal growth and metabolism of a variety of cells in culture. The essential fatty acids (EFAs) have received increased attention in recent years due to their presumed involvement in cardiovascular disorders and in cancers of the breast, pancreas, colon and prostate. Many *in vitro* systems have emerged which either examine the role of EFAs in human disease directly, or utilize EFAs to mimic the *in vivo* cellular environment. The effects of EFAs on cells are both direct and indirect. As components of membrane phospholipids, and due to their varying structural and physical properties, EFAs can alter membrane fluidity, at least in the local environment, and affect any process that is mediated via the membrane. EFAs containing 20 carbons and at least three double bonds can be enzymatically converted to eicosanoid hormones, which play important roles in a variety of physiological and pathological processes. Alternatively, EFAs released into cells from phospholipids can act as second messengers that activate protein kinase C. Furthermore, susceptibility to oxidative damage increases with the degree of unsaturation, a complication that merits consideration because lipid peroxidation can lead to a variety of substances with toxic and mutagenic properties. The effects of EFAs on cultured cells are illustrated using the responses of normal and tumor human mammary epithelial cells. A thorough evaluation of EFA effects on commercially important cells could be used to advantage in the biotechnology industry by identifying EFA supplements that lead to improved cell growth and/or productivity.

Abbreviations: AA: arachidonic acid (20 carbons: 4 double bonds, *n*-6); BHA: butylated hydroxyanisole; BHT: butylated hydroxytoluene; cAMP: cyclic adenosine monophosphate; CHO: Chinese hamster ovary; DAG: diacylglycerol; DGLNA: dihomo-γ-linolenic acid (20:3, *n*-6); DHA: docosahexaenoic acid (22:6, *n*-3); EFA: essential fatty acid; EGF: epidermal growth factor; EGFR: epidermal growth factor receptor; EPA: eicosapentaenoic acid (20:5, *n*-3); FA: fatty acid; FBS: fetal bovine serum; GLNA: γ-linolenic acid (18:3, *n*-6); LA: linoleic acid (18:2, *n*-6); LNA: α-linolenic acid (18:3, *n*-3); LT: leukotriene; MDA: malondialdehyde; NAD: nicotinamide adenine dinucleotide; NDGA: nordihydroguaiaretic acid; OA: oleic acid (18: 1, *n*-9); PG: prostaglandin; PKC: protein kinase C; PUFA: polyunsaturated fatty acid; SFM: serum-free medium; TX: thromboxane

Introduction

Polyunsaturated fatty acids (PUFAs) are common in higher plants and animals, but they are rare in bacteria. The different PUFA families or series are distinguished by the position of the double bond furthest from the carboxyl group. They are commonly named according to the position of this double bond by counting from the methyl terminal, with the methyl carbon (*n* or ω carbon) as number 1 (Fig. 1). The four PUFA families can be identified by the precursor fatty acids (FAs) from which all other PUFAs in that family can be derived, namely oleic (OA, 18 carbons: 1 double bond, *n*-9 or ω-9), palmitoleic (16:1, *n*-7), linoleic (LA, 18:2, *n*-6) and α-linolenic (LNA, 18:3, *n*-3) acids. All mammalian species lack the enzymes necessary for introducing double bonds above C-9 (from the carboxyl end, Fig. 1) in the fatty acid (FA) chain; therefore, mammals cannot synthesize LA or LNA (Stryer, 1988). Because a lack of *n*-6 or *n*-3 FAs results in various health disorders, these two PUFA families are termed *essential* and must be derived from the diet. For example, corn and sunflower oils are rich in LA (*n*-6), linseed oil is rich in LNA (*n*-3) and various oils of marine origin are rich in long-chain *n*-3 FAs such as eicosapentaenoic (EPA, 20:5) and docosahexaenoic (DHA, 22:6) acids. Dietary intake of high amounts of essential fatty acids (EFAs), however, does not always have a beneficial effect. Although they are believed to protect against some human diseases, such as cardiovascular and central nervous system disorders, high amounts of certain EFAs have been associated with increased risk of breast, colon, pancreatic and prostate cancer. For breast cancer in particular, experiments with laboratory animals have linked LA to increased tumor incidence, growth and metastasis, while the *n*-3 FAs have a protective effect against the disease (Carroll and Hopkins, 1979; Katz and Boylan, 1987; Hubbard and Erickson, 1987; Cave, 1991; Pritchard *et al.*, 1989).

The EFAs can be metabolized by a combination of desaturation/saturation and chain elongation/shortening reactions to give a variety of PUFAs *of the same family* (Fig. 2) (Stryer, 1988; Sprecher, 1981). The major *n*-6 and *n*-3 PUFAs that accumulate in most tissues in the form of complex, neutral lipids and phospholipids are LA (*n*-6), arachidonic acid (AA, 20:4, *n*-6), docosapentaenoic acid (22:5, *n*-6), LNA (*n*-3), EPA (*n*-3) and DHA (*n*-3) (Cook *et al.*, 1991). All major 20- and 22-carbon PUFAs are considered necessary for the optimal biological function of specialized

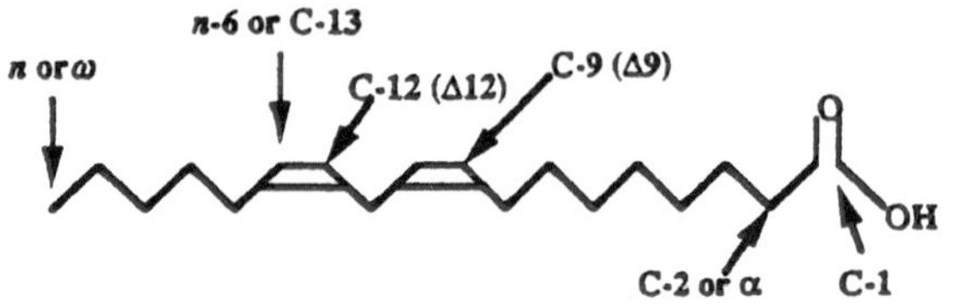

Linoleic Acid (18 carbons : 2 double bonds, *n*-6)

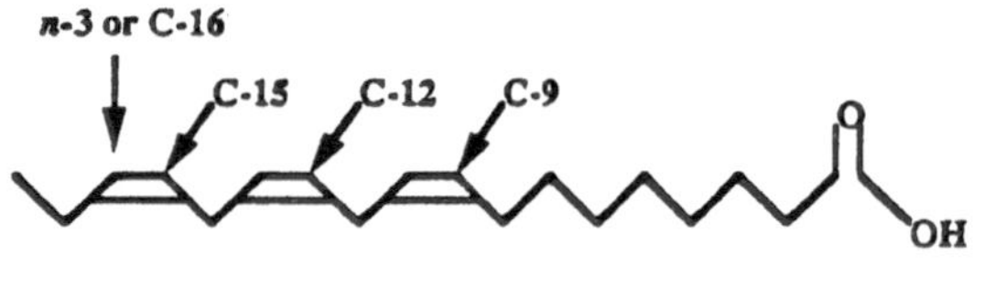

α-Linolenic Acid (18:3, *n*-3)

Fig. 1. Structure of the parent essential fatty acids linoleic (*n*-6) and α-linolenic (*n*-3).

cells and tissues such as brain, retina, testes, heart, liver and kidney (Cook *et al.*, 1991; Tinoco, 1982). Although the bulk of FA metabolism is thought to take place in the liver, cells from a variety of tissues are capable of processing EFAs to varying extents.

Numerous studies with a variety of cultured cells from different species suggest that cells in culture are also capable of at least partially processing exogenous EFAs (Grammatikos *et al.*, 1994a; Spector *et al.*, 1981; Rosenthal, 1987). Conversion of the parent EFAs LA (*n*-6) and LNA (*n*-3) to the 20-carbon PUFAs AA (*n*-6) and EPA (*n*-3), respectively, requires chain elongation and Δ^6 (or Δ^8) and Δ^5 desaturations (Fig. 2). All cultured cells are able to elongate exogenous LA and LNA, and most can perform Δ^5 desaturation, so Δ^6 and Δ^8 desaturation are typically the limiting steps in AA and EPA production (Grammatikos *et al.*, 1994a). It should be noted that production of AA and EPA from LA and LNA, respectively, is thought to proceed preferentially by Δ^6 desaturation followed by a two-carbon elongation and Δ^5 desaturation (Fig. 2) (Rosenthal, 1987). Longer FAs that have more double bonds than AA or EPA are less frequently produced because cultured cells are typically deficient in Δ^4 desaturating ability. The process of retroconversion (chain shortening) is less extensively studied, but evidence from a variety of cells suggests that this type of metabolic conversion is normally active (Grammatikos *et al.*, 1994a; Rosenthal, 1987). Inherent characteristics of different cell types, the length of time in culture, and

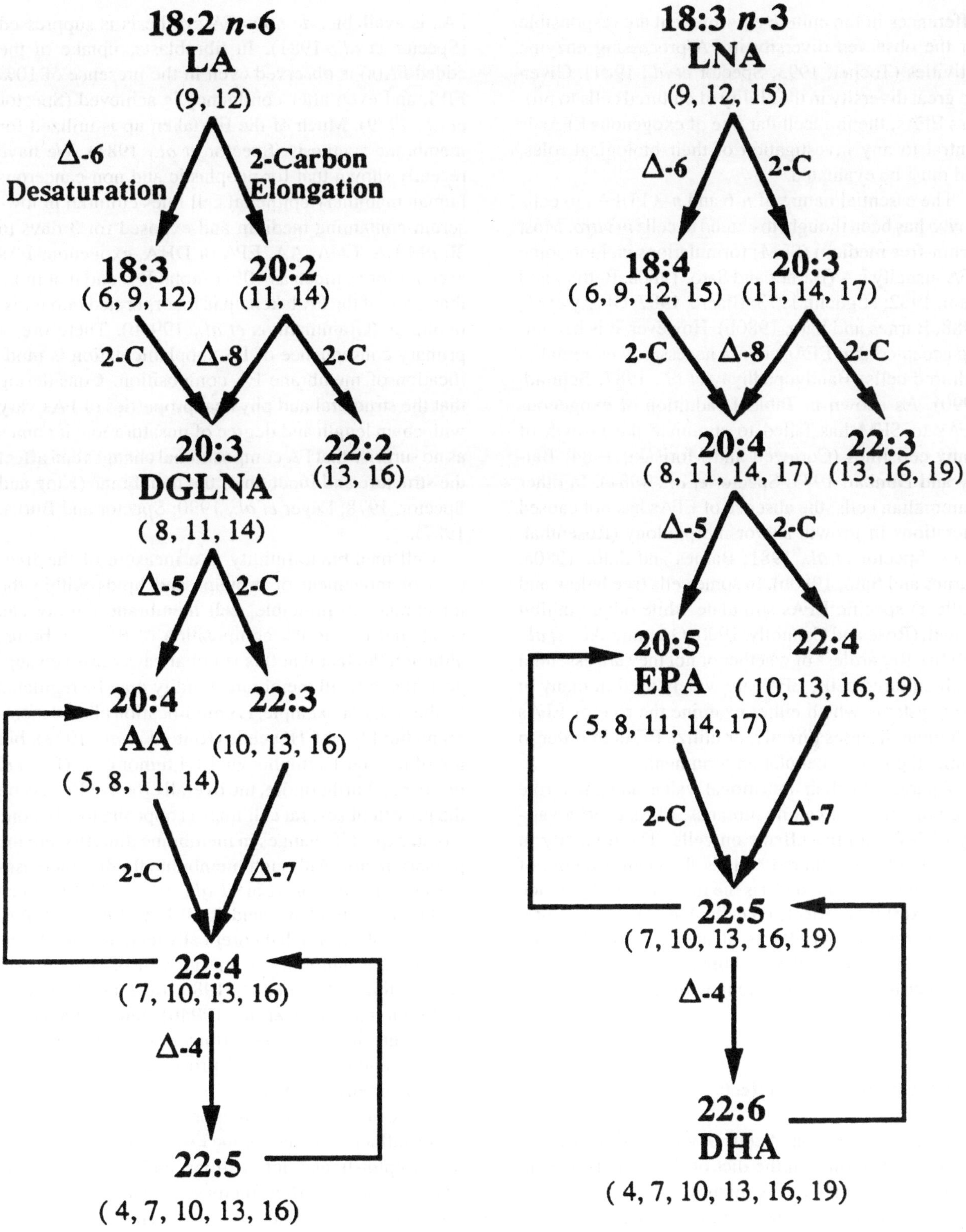

Fig. 2. Metabolism of the parent essential fatty acids linoleic (*n*-6) and α-linolenic (*n*-3) to other fatty acids of the same family. The positions of the double bonds, relative to the carboxyl end, are shown in parentheses for each fatty acid.

differences in the culture environment are responsible for the observed diversity in FA-processing enzyme activities (Tocher, 1993; Spector *et al.*, 1981). Given the great diversity in the ability of cultured cells to process EFAs, the intracellular fate of exogenous EFAs is central to any investigation of their biological roles, and must be evaluated.

The essential nature of *n*-6 and *n*-3 PUFAs to cells *in vivo* has been thought to extend to cells *in vitro*. Most serum-free medium (SFM) formulations include some EFA, usually LA (Barnes and Sato, 1980a; Bettger and Ham, 1982; Higuchi, 1973; Bjaere, 1987; Glassy *et al.*, 1988; Barnes and Sato, 1980b). However, it is becoming accepted that EFAs are not necessarily essential to cultured cells (Bandyopadhyay *et al.*, 1987; Schmid, 1990). As shown in Table 1, addition of exogenous EFAs to SFM has failed to stimulate the growth of many cell lines (Cornwell and Morisaki, 1984; Bailey and Dunbar, 1973; Spector *et al.*, 1981). In other mammalian cells, the absence of EFAs has not caused alterations in growth and/or morphology (Rosenthal, 1987; Spector *et al.*, 1981; Barnes and Sato, 1980a; Barnes and Sato, 1980b). In some cells (see below and Table 1) specific EFAs stimulate while others inhibit growth (Rose and Connolly, 1990; Grammatikos *et al.*, 1994b). Regardless of whether or not they are essential to all cultured cells, EFAs are widely used in many *in vitro* systems which either examine the role of EFAs in human diseases directly, or utilize EFAs in order to mimic the *in vivo* cellular environment.

Apart from their nutritional value and their role as components of cell membranes, EFAs exert a variety of less familiar effects on cells. The diversity in EFA effects on cultured cells is the central theme of this review. We first discuss the mechanisms by which EFAs exert their effects on cells. We conclude by presenting the growth effects of a variety of EFAs on two mammary epithelial cell lines, one neoplastic and the other non-cancerous, along with an analysis of the likely mechanisms.

Membrane-associated effects

The EFA composition of cells *in vivo* reflects the prevailing EFA family in the diet of the organism. Supply of specific FAs or FA mixtures to cells in culture invariably leads to changes in the cellular FA composition that reflect the exogenous additions (Spector and Yorek, 1985). This is because cells preferentially utilize exogenous FAs. If an adequate supply of FAs is available, *de novo* FA synthesis is suppressed (Spector *et al.*, 1981). In fibroblasts, uptake of the added FA(s) is observed even in the presence of 10% FBS, and even after confluency is achieved (Spector *et al.*, 1979). Much of the FA taken up is utilized for membrane synthesis (Spector *et al.*, 1981). We have recently shown that for neoplastic and non-cancerous human mammary epithelial cell lines cultured in low-serum-containing medium and exposed for 3 days to 30 μM LA, LNA, AA, EPA or DHA, exogenous FAs account for as much as 40% of total FA, and that more than 85% of the FA taken up is incorporated into phospholipids (Grammatikos *et al.*, 1994b). Therefore, a primary consequence of FA supplementation is modification of membrane FA composition. Considering that the structural and physical properties of FAs vary with chain length and degree of unsaturation, it comes as no surprise that FA compositional changes can affect the structure and function of the membrane (King and Spector, 1978; Léger *et al.*, 1990; Spector and Burns, 1987).

Cell membrane fluidity is a measure of the freedom of movement of proteins and lipids within the membrane. In principle, cell membrane fluidity can be altered by the FA composition of the membrane, although the fact that this is not always observed suggests that overall membrane fluidity may be regulated by the cell. For example, FA modification altered membrane fluidity of CHO cells (Rintoul *et al.*, 1978), but not of murine T lymphocyte EL4 tumor cells (Poon *et al.*, 1981). Furthermore, the overall effect of PUFAs on the growth of several cell lines is opposite to what one would expect if changes in membrane fluidity were the primary factor. Although membrane fluidity increases during cell division (Lai *et al.*, 1980), PUFAs such as dihomo-γ-linolenic acid (DGLNA, 20:3, *n*-6), AA, EPA and DHA, which theoretically increase membrane fluidity, are inhibitory to the growth of several cancer cell lines (Begin *et al.*, 1986; Rose and Connolly, 1990; Grammatikos *et al.*, 1994b). Small changes in membrane fluidity are nevertheless possible and could affect membrane-mediated processes at least in the local microenvironment.

More important than membrane fluidity effects may be the influence of membrane FA composition changes on receptor-ligand interactions and receptor signal transduction properties. Increased levels of unsaturated FAs in cell membranes have been shown to enhance the prostaglandin E_2-stimulated activity of adenylate cyclase in murine T lymphocyte EL4 tumor cells (Poon *et al.*, 1981), and to reduce the binding affinity of the

Table 1. Effect of various essential fatty acids on the growth of cells in culture

Cell Type	Fatty Acid(s)	Growth Effect	Reference
mammary epithelial(r)[1]	LA, LNA	+	Wicha *et al.*, 1979
MCF-10A mammary epith. (h)	LA, AA, LNA, EPA, DHA	+	Grammatikos *et al.*, 1994b
mammary carc. (r)	LA	+	Wicha *et al.*, 1979
MDA-MB-231 breast carc. (h)	LA	+	Rose and Connolly, 1990
rBHK (hm)	LA	+	Schmid *et al.*, 1991
LM fibroblasts (m)	LNA, AA	no effect	Doi *et al.*, 1978
MCF-7 breast carc. (h)	LA	no effect	Grammatikos *et al.*, 1994b
liver (r)	lipid depletion	no effect	Takaoka and Katsuta, 1971
liver (m)	lipid depletion	no effect	Evans *et al.*, 1964
thymus (r)	lipid depletion	no effect	Takaoka and Katsuta, 1971
hepatoma (r)	lipid depletion	no effect	Takaoka and Katsuta, 1971
Hela-S3 ovarian carc. (h)	lipid depletion	no effect	Evans *et al.*, 1964
skin epithelial (h)	lipid depletion	no effect	Evans *et al.*, 1964
B14-FAF28 fibroblasts (hm)	lipid depletion	no effect	Evans *et al.*, 1964
MDA-MB-231 breast carc. (h)	EPA, DHA	–	Rose and Connolly, 1990
MCF-7 breast carc. (h)	AA, LNA, EPA, DHA	–	Grammatikos *et al.*, 1994b
ZR-75-1 breast carc. (h)	GLNA, DGLNA, AA, EPA	–[2]	Begin *et al.*, 1986
A-549 lung carc. (h)	GLNA, DGLNA, AA, EPA	–[2]	Begin *et al.*, 1986
PC-3 prostatic carc. (h)	GLNA, DGLNA, AA, EPA	–[2]	Begin *et al.*, 1986
CCD-41SK fibroblasts (h)	GLNA, DGLNA, AA, EPA	–	Begin *et al.*, 1986
MDCK kidney (d)	GLNA, DGLNA, AA, EPA	–	Begin *et al.*, 1986
CV-1 kidney (mk)	GLNA, DGLNA, AA, EPA	–	Begin *et al.*, 1986
BSC-1 kidney (mk)	GLNA, DGLNA, AA, EPA	–	Begin *et al.*, 1986
skin fibroblasts (h)	AA, LNA	–	Spector *et al.*, 1979

[1] Species are abbreviated as follows: r: rat; h: human; hm: hamster; m: mouse; d: dog; mk: monkey
[2] Cell killing was induced by the fatty acids studied

insulin receptor of Friend erythroleukemia cells (Ginsberg *et al.*, 1981). Although it has not been demonstrated yet, different PUFAs may alter epidermal growth factor (EGF) binding to its receptor (EGFR) and affect EGFR phosphorylation. Binding of EGF is known to be affected by the membrane phospholipid environment and requires phosphatidyl ethanolamine (Kano-Sueoka *et al.*, 1990). Furthermore, the proliferation of mouse mammary epithelial cells in response to EGF is enhanced by LA (Bandyopadhyay *et al.*, 1987).

In addition to receptor-ligand interactions, FA modifications have been shown to affect other aspects of membrane-mediated cellular function such as carrier-mediated transport, ion channels, the activity and properties of membrane-bound enzymes, and such processes as phagocytosis, endocytosis and exocytosis (Huang *et al.*, 1992; Spector and Yorek, 1985; Spector and Burns, 1987; Field *et al.*, 1990). It is therefore likely that PUFAs play an important role in altering the local microenvironment of receptors and transporters, and in this way affect cellular proliferation and function. However, it is not possible to generalize these effects to all cells because no consistent pattern can be identified. From this and the discussion in the sections that follow it is important to recognize that EFA effects can be very different for each cell type.

Eicosanoids

Eicosanoids are oxygenated metabolites of the 20-carbon PUFAs DGLNA (20:3, *n*-6), AA (20:4, *n*-6), and EPA (20:5, *n*-3). Prostaglandins (PG), thromboxanes (TX), prostacyclin (PGI_2), leukotrienes (LT), lipoxins and other hydroxy FAs belong to this family of compounds. They are derived enzymatically by the action of a membrane-bound cyclooxygenase (Fig. 3) or specific lipoxygenases (Fig. 4). The eicosanoids are biologically very active in an autocrine or paracrine fashion and play an important role in many physio-

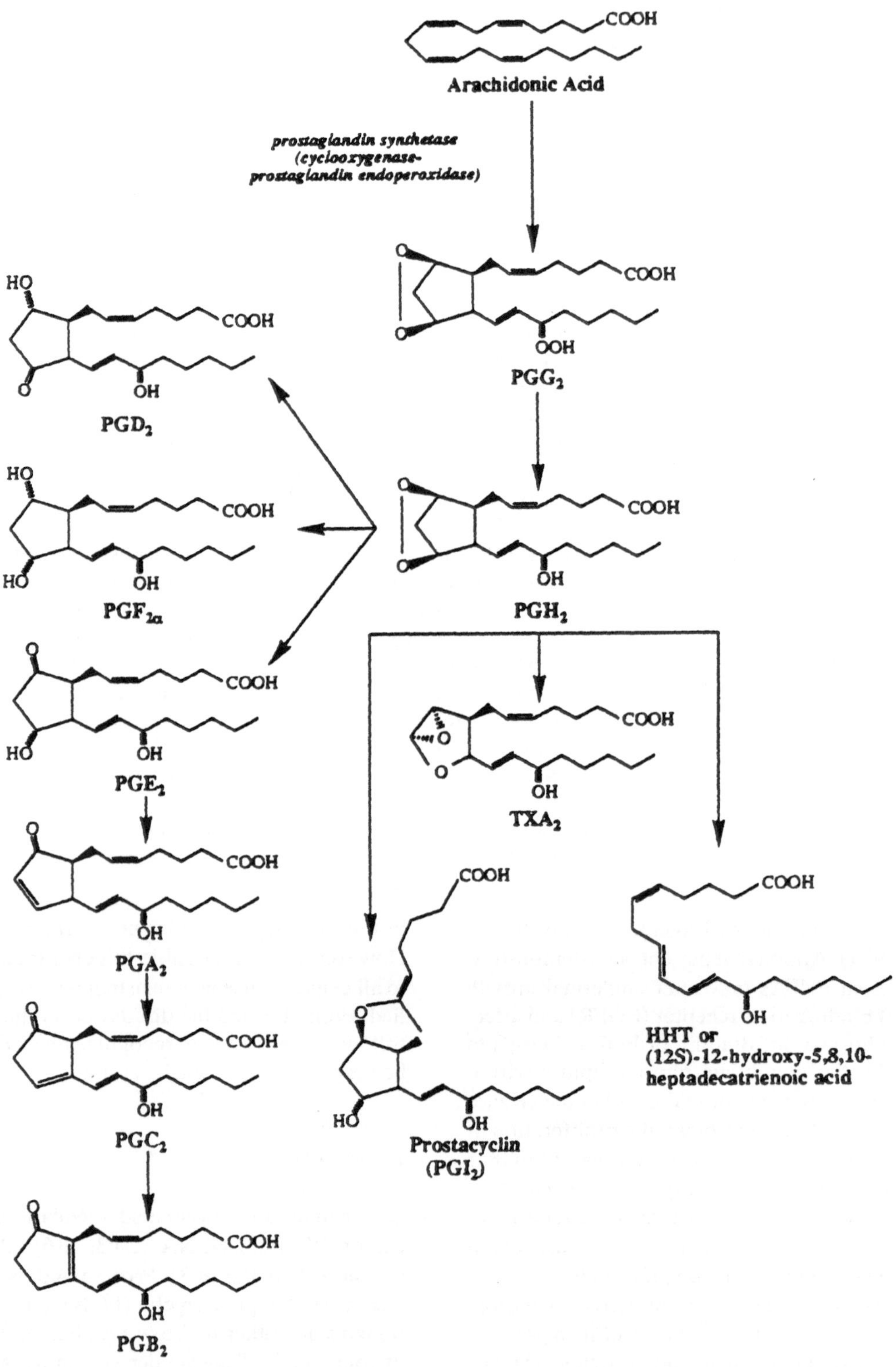

Fig. 3. Major eicosanoids produced from arachidonic acid by the action of cyclooxygenase (adapted from Fischer *et al.*, 1989). Similar compounds are produced from dihomo-γ-linolenic acid (1-series thromboxane and prostaglandins lacking the double bond at C-5 from the carboxyl end, no prostacyclin, see Willis and Smith, 1989 and references therein) and eicosapentaenoic acid (3-series thromboxane and prostaglandins containing an additional double bond at C-3 from the methyl end, see Salem Jr, 1989 and references therein).

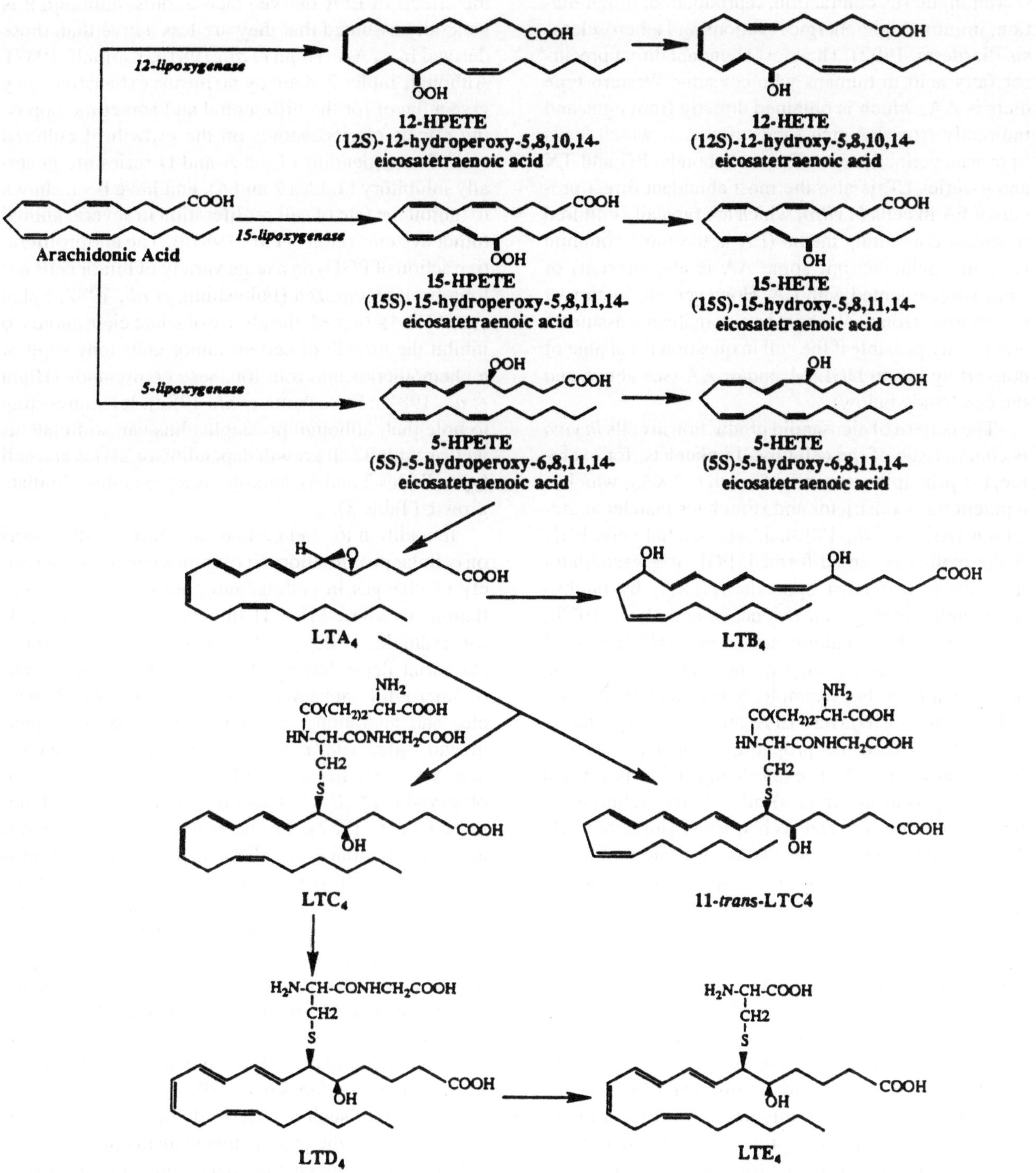

Fig. 4. Major eicosanoids produced from arachidonic acid by the action of various lipoxygenases (adapted from Fischer *et al.*, 1989). Similar compounds are produced from dihomo-γ-linolenic acid (corresponding HPETEs and HETEs, and 3-series leukotrienes lacking the double bond at C-14 from the carboxyl end, see Willis and Smith, 1989 and references therein) and eicosapentaenoic acid (corresponding HPETEs and HETEs, and 5-series leukotrienes containing an additional double bond at C-3 from the methyl end, see Salem Jr, 1989 and references therein).

logical and pathological processes including stomach secretion, uterus contraction, reproduction, inflammation, immunologic/allergic reactions and atherosclerosis (Sardesai, 1992). The most abundant direct precursor fatty acid in humans who consume Western-type diets is AA, which is oblained directly from eggs and indirectly from LA-rich foodstuffs. AA, which leads to prostacyclin, 2-series (2 double bonds) PG and TX and 4-series LT, is also the most abundant direct precursor FA in cells *in vitro*, which are typically cultured in serum-containing media (LA is the most common EFA in bovine serum; some AA is also present) or SFM supplemented with LA. However, production of eicosanoids from LA cannot be automatically assumed, and is only possible if the cell in question is capable of converting LA to DGLNA and/or AA (see above and the case study below).

The pattern of eicosanoid production in cells *in vivo* is charcteristic of the cell type. In platelets, for example, the principle eicosanoid fonned is TXA_2, which is a potent vasoconstrictor and stimulates platelet aggregation (Arita *et al.*, 1989). In endothelial cells PGI_2 is the main eicosanoid formed. PGI_2 is a vasodilator that acts in a manner opposite to TXA_2 by inhibiting platelet aggregation (Moncada and Vane, 1979). Although cells in culture are in general capable of forming eicosanoids, they do not always follow the *in vivo* patterns. For example, human endothelial cell cultures do not convert EPA into PGI_3 (Spector *et al.*, 1983), which has properties similar to those of PGI_2 (Sardesai, 1992), even though PGI_3 is formed in humans (but not in rats) after dietary administration of EPA (Fischer and Weber, 1984; Hornstra *et al.*, 1981). Such discrepancies can complicate the design and interpretation of *in vitro* experiments aimed at understanding *in vivo* phenomena.

During the past fifteen years and mainly during the 1980s, much attention has been directed towards a possible role of eicosanoids, most notably PGs, in carcinogenesis (Honn *et al.*, 1981a). Eicosanoids are believed to be involved in all stages of the carcinogenic process, from tumor initiation to metastasis. While the involvement of eicosanoids in immune reactions and platelet aggregation certainly implies that eicosanoids may play a role in tumor progression and metastasis, no solid evidence for such a role has been obtained to date. Experiments with a variety of cells in culture have produced conflicting results and have exposed a multitude of effects of eicosanoids on cells. Most frequently studied are the effects of *n*-6 FA-derived eicosanoids on cellular proliferation, which are presented in Tables 2–4 for a variety of cells. Much less is known about the effects of EPA-derived eicosanoids, although it is generally assumed that they are less active than those derived from AA (Honn *et al.*, 1981a; Karmali, 1987). Although Tables 2–4 are by no means exhaustive, they give a flavor for the differential and sometime opposing effects of eicosanoids on the growth of cultured cells. Prostaglandins of the A and D series are generally inhibitory (Tables 2 and 4), and have been shown to inhibit the rate of cell proliferation in several animal tumor systems (Honn *et al.*, 1981a). The antiproliferative action of PGD_2 on a large variety of tumor cells has long been recognized (Fukushima *et al.*, 1982; Sakai *et al.*, 1984). Indeed, the ability of some eicosanoids to inhibit the growth of certain tumor cells may suggest a chemotherapeutic role for these compounds (Honn *et al.*, 1981b; Fukushima *et al.*, 1982). It is interesting to note that, although prostaglandins can stimulate as well as inhibit cell growth depending on series and cell type (Tables 2 and 4), leukotrienes generally stimulate growth (Table 3).

In addition to, and perhaps in concert with effects on cellular proliferation, eicosanoids can induce a variety of changes in cellular morphology and differentiation. In MDA-MB-231 human breast cancer cells for example, PGE_1, $PGF_{2\alpha}$ and PGA_1 cause mitochondrial deformation, shortening or disappearance of microvilli, appearance of many lysosomal vesicles and alterations in nuclear morphology (Chegini and Safa, 1987). With the exception of nuclear alterations, similar morphological changes were also observed in MCF-7 human breast cancer cells (Chegini and Safa, 1987). Furthermore, some eicosanoids are able to induce or inhibit cellular differentiation (Honn *et al.*, 1981a). Prostaglandins of the A, D and E series for example, induce differentiation of several animal and human leukemic cell lines (Marini *et al.*, 1990). Leukotriene B_4 is thought to participate in myelopoiesis either directly or via the induction of cytokines and growth factors (Rola-Pleszczynski and Stankova, 1992). Nanomolar quantities of LTB_4 induce production of IL-6 by human monocytes (Rola-Pleszczynski and Stankova, 1992).

Because of such diverse biological effects on various cell types, the precise mechanisms of eicosanoid action are not well characterized. One mode of action, at least for PGA_1, $PGF_{2\alpha}$, and prostaglandins of the E series may involve the elevation of intracellular cyclic AMP (cAMP) levels (Chegini and Safa, 1987; Johnson and Pastan, 1971; Eisenbarth *et al.*, 1974; McCumbee *et al.*, 1983; Samuelsson, 1983). It is well known

Table 2. Effects of arachidonic acid-derived prostacyclin and prostaglandins on the growth of a variety of cells in culture

	Stimulates	No effect	Inhibits	Reference
PGA_2	VX_2-L carcinoma (rb)[1]			Yoneda *et al.*, 1985
		RPMI 7932 melanoma (h)		Bhuyan *et al.*, 1986
		SK Mel 28 melanoma (h)		Bhuyan *et al.*, 1986
			Friend erythroleukemia[2] (m)	Marini *et al.*, 1990
			L-1210 leukemia (m)	Bhuyan *et al.*, 1986
			HeLa ovarian carc. (h)	Suzuki *et al.*, 1988
			Ca 9–22 gingival carc. (h)	Suzuki *et al.*, 1988
			B-16 melanoma (m)	Bhuyan *et al.*, 1986
PGD_2			C6 glioma[3] (r)	Conde *et al.*, 1991
			glioma (m)	Keyaki *et al.*, 1984
			Ca 9–22 gingival carc. (h)	Suzuki *et al.*, 1988
			HeLa ovarian carc. (h)	Suzuki *et al.*, 1988
			VX_2-L carcinoma (rb)	Yoneda *et al.*, 1985
			B-16 melanoma (m)	Bhuyan *et al.*, 1986; Simmet and Jaffe, 1983
			K562 leukemia[3] (h)	Santoro *et al.*, 1986
			L-1210 leukemia (m)	Bhuyan *et al.*, 1986
PGE_2	hepatocytes (r)			Skouteris and McMenamin, 1992
	VX_2-L carcinoma (rb)			Yoneda *et al.*, 1985
	embryo fibroblasts (m)			Durant *et al.*, 1989
	mammary epithelial (m)			Bandyopadhyay *et al.*, 1987
		glioma (m)		Keyaki *et al.*, 1984
			aortic smooth muscle (gp)	Huttner *et al.*, 1977
			S-91 melanoma (m)	Abdel-Malek *et al.*, 1987
			B-16 melanoma (m)	Santoro *et al.*, 1986; Santoro *et al.*, 1976
$PGF_{2\alpha}$	3T3 fibroblasts (m)			Jimenez de Asua *et al.*, 1975
	embryo fibroblasts (m)			Durant *et al.*, 1989
	hepatocytes (r)			Skouteris and McMenamin, 1992
	VX_2-L carcinoma (rb)			Yoneda *et al.*, 1985
	MCF-7 breast carc. (h)			Barnes and Sato, 1980b
			embryo fibroblasts (m)	Johnson and Pastan, 1971
			L-929 fibroblasts (m)	Johnson and Pastan, 1971
PGI_2	VX_2-L carcinoma (rb)			Yoneda *et al.*, 1985

[1] Species are abbreviated as follows: rb, rabbit; m, mouse; h, human; r, rat; gp, guinea pig;
[2] Differentiation was assessed and was induced in these cells
[3] Inhibition is due to a cytotoxic effect on these cells

that elevated levels of cAMP can either enhance or inhibit cell proliferation depending on cell type: fibroblasts are inhibited (Johnson and Pastan, 1971; Taylor-Papadimitriou *et al.*, 1980), whereas epithelial (Taylor-Papadimitriou *et al.*, 1980; Stampfer, 1982; Silberstein *et al.*, 1984; Yang *et al.*, 1980) and endothelial (Davison and Karasek, 1981) cells are stimulated. In this regard, it has been recently reported that cAMP modulates EGF-induced signal transduction pathways (Wu *et al.*, 1993; Cook and McCormick, 1993). However, the effect of prostaglandins on cell proliferation can

Table 3. Effects of arachidonic acid-derived leukotrienes on the growth of a variety of cells in culture

	Stimulates	No effect	Reference
LTB_4	B lymphocytes[1] (h)[2]		Dugas *et al.*, 1990
		aortic endothelial (b)	Modat *et al.*, 1987
LTC_4	adult melanocytes (h)		Medrano *et al.*, 1993
	myeloid progenitors (h)		Miller *et al.*, 1989
	HL-60 leukemia (h)		Miller *et al.*, 1989
	K-562 leukemia (h)		Miller *et al.*, 1989
	KG-1 leukemia (h)		Miller *et al.*, 1989
	aortic endothelial (b)		Modat *et al.*, 1987
	glomerular epithelial (h)		Baud *et al.*, 1985
LTD_4	glomerular epithelial (h)		Baud *et al.*, 1985

[1] Differentiation was assessed and was induced in these cells
[2] Species are abbreviated as follows: h, human; b, bovine

Table 4. Effects of prostaglandins derived from dihomo-γ-linolenic acid on the growth of a variety of cells in culture

	Stimulates	Inhibits	Reference
PGA_1		L-1210 leukemia (m)[1]	Bhuyan *et al.*, 1986
		K-562 leukemia (h)	Santoro *et al.*, 1986
		Rel chondrosarcoma (m)	Eisenbarth *et al.*, 1974
		chondrosarcoma (h)	McCumbee *et al.*, 1983
		B-16 melanoma (m)	Bhuyan *et al.*, 1986
PGE_1	mammary epithelial (m)		Bandyopadhyay *et al.*, 1987
		aortic smooth muscle (gp)	Huttner *et al.*, 1977
		embryo fibroblasts (m)	Johnson and Pastan, 1971
		transformed fibroblasts (m)	Johnson and Pastan, 1971
		S-91 melanoma (m)	Abdel-Malek *et al.*, 1987
$PGF_{1\alpha}$	VX_2-L carcinoma (rb)		Yoneda *et al.*, 1985
	K-562 leukemia (h)		Santoro *et al.*, 1986

[1] Species abbreviations are as follows: m, mouse; h, human; gp, guinea pig; rb, rabbit

not be explained in terms of their effects on cAMP levels for all cell types studied (Honn *et al.*, 1981a).

From the above examples it is evident that eicosanoids exert no single effect on cell growth and that the effects of individual eicosanoids are very much cell type specific. It must be emphasized, however, that these effects were delineated by the exogenous addition of individual eicosanoids to the culture medium. Given the multitude of eicosanoids that can be produced from precursor FAs and the different and sometimes opposing effects that different eicosanoids exert on cells, the net effect of all eicosanoids produced by a particular cell from a precursor FA is what eventually contributes to the overall effect of exogenous EFA addition. The balance among various eicosanoids may represent a form of cellular control of proliferation *versus* differentiation. This is illustrated by the case of human adult melanocytes in which LTC_4 induced growth stimulation, loss of contact inhibition and formation of structures resembling tumor spheroids (Medrano *et al.*, 1993). In the same cells, cholera toxin, a cAMP stimulator, induced features of terminal differentiation. As mentioned above, some prostaglandins, such as PGE_2, are able to stimulate cAMP production. It is conceivable, therefore, that the production of LTC_4 relative to PGE_2 from AA could determine the fate of these cells.

Pharmacological as well as dietary approaches can be used to modulate the eicosanoid cascade either in culture or *in vivo*. The pharmacological approach is centered around the use of cyclooxygenase or lipoxygenase inhibitors. Prostaglandin inhibitors include indomethacin (Shen and Winter, 1977), aspirin (Vane, 1971) and piroxicam [I, 4-hydroxy-2-methyl-N-(2-pyridyl)-2H-1,2-benzothiazine-3-carboxamide-1,1-dioxide] (Carty *et al.*, 1980a; Carty *et al.*, 1980b). Leukotriene inhibitors include ETYA (5,8,11,14-eicosatetraynoic acid) (Tobias and Hamilton, 1979), nordihydroguaiaretic acid (4,4(2,3-dimethyl-1,4-butanediyl)bis(1,2-benzenediol), NDGA) (Bach, 1984), and esculetin (6,7-dihydroxycoumarin) (Neichi *et al.*, 1983). The most serious drawback of using these inhibitors arises from the fact that they also inhibit enzymes of other biochemical pathways and can therefore induce potentially undesireable side-effects. Indomethacin, for example, partially inhibits a cAMP-dependent protein kinase (Kantor and Hampton, 1978) and 3-hydroxysteroid dehydrogenase (Penning and Talalay, 1983) at 100 nM. At 1 μM it also inhibits the activity of phospholipase A_2 in rabbit polymorphonuclear leukocytes (Kaplan *et al.*, 1978). In addition to being an effective lipoxygenase inhibitor, NDGA also inhibits nonspecific lipid peroxidation and directly blocks calcium channels in pituitary cells and ventricular myocytes (Huang *et al.*, 1992).

The dietary approach to eicosanoid modulation attempts to alter the formation of one eicosanoid series in favor of another by supplying the appropriate precursor FA. In this regard, EPA, itself a precursor of 3-series prostaglandins and 5-series leukotrienes, competitively inhibits the action of cyclooxygenase on AA (Rosenthal, 1987). Some PUFAs which are not substrates for eicosanoid synthesis enzymes have been found to interfere with the eicosanoid cascade. DHA, for example is a strong inhibitor of prostaglandin, but not leukotriene, formation from AA (Corey *et al.*, 1983). Based on the observation that leukotrienes are generally stimulatory to cell growth, we would expect DHA and EPA to exert an overall growth stimulatory effect on a variety of cells. That the exact opposite is generally observed emphasizes the diversity and complexity of EFA effects on cultured cells.

Non-esterified fatty acids as second messengers

The potential second messenger role of unsaturated FAs released from phospholipids by the action of specific phospholipases (A_1 or more often A_2) has only in recent years become apparent. AA (*n*-6), for example, inhibits isolated Ca^{2+}/calmodulin-dependent kinases such as smooth muscle myosin light chain kinase (Kigoshi *et al.*, 1990), which enables the activation of myosin and is thus involved in smooth muscle and non-muscle cell contraction, or brain protein kinase II (Piomelli *et al.*, 1989), which enhances neurotransmitter release by nerve cells. EPA (*n*-3) modulates the expression of the oncogene HER-2/*neu* in intact MCF-7 breast cancer cells transfected with v-H-ras (Tiwari *et al.*, 1991). Both LNA (*n*-3) and EPA induce increased cytoplasmic mRNA levels of interferon-inducible genes in intact MCF-7 cells (Tiwari *et al.*, 1991).

Activation of protein kinase C

Protein kinase C (PKC) consists of a family of closely related isoenzymes that plays an important role in signal transduction and cell regulation (Nishizuka, 1988; Nishizuka, 1989). Enzyme activity depends on phospholipids and (for some members of the family) on Ca^{2+} (Nishizuka, 1989). Diacylglycerols (DAGs), which are formed by phospholipase C-induced breakdown of membrane phospholipids, activate PKC, as do phorbol esters and related tumor promoters (Nishizuka, 1989; Rhee *et al.*, 1989; Castagna *et al.*, 1982). When activated, PKC phosphorylates various target proteins at specific serine or threonine residues. In many animal cells, for example, PKC is thought to activate the Na^+-H^+ exchanger that controls intracellular pH and thereby increase the intracellular pH. Apart from possible effects on proliferation, increased intracellular pH may affect such processes as recombinant protein expression and glycosylation (Borys *et al.*, 1993). Unsaturated fatty acyl moieties within DAG are not required for activation of membrane-bound PKC (Conn *et al.*, 1985). Furthermore, the specificity of PKC activation is directed at the glycerol backbone rather than at the FA chain (Boni and Rando, 1985). It has been recently suggested, however, that PKC regulation may occur by two pathways: (a) via DAG generated by phospholipase C action, and (b) via an unsaturated FA released from phospholipids by phospholipase A_2 (Khan *et al.*, 1992). The former pathway operates on membrane-bound PKC, whereas the latter operates on soluble enzyme in the cytosol. One study estimates that 60–70% of PKC is in the cytosolic fraction in the presence of 50–500 nM calcium (Khan *et al.*, 1992). Consistent with the soluble PKC activation mode, AA and other unsaturated FAs were found to activate isolated human neutrophil PKC in a dose- and calcium-dependent fashion (McPhail *et al.*, 1984). By contrast, DHA was recently found to inhibit dramatically the stimulation by phosphatidyl serine and diolein of purified rat colon PKC (Holian and Nelson, 1992).

Modulation of steroid receptor characteristics

While EFAs esterified to the glycerol backbone of membrane phospholipids can affect the function of membrane-bound hormone and growth factor receptors and free (non-esterified) unsaturated FAs may activate soluble PKC (see above), it is now apparent that free FAs released from phospholipids can also affect cytosolic steroid hormone receptors. In various tissues, for example, non-esterified unsaturated FAs negatively modulate the concentration of receptors for estrogens, progestins, androgens, glucocorticoids, 1, 25-dihydroxyvitamin D_3 and thyroid hormones (Borrás and Leclercq, 1992). In MCF-7 breast cancer cells in particular, unsaturated FAs were found to modulate the structure and binding characteristics of the estrogen receptor (Borrás and Leclercq, 1992). Activated steroid receptors bind to specific genes in the nucleus and regulate their transrription. By altering steroid receptor binding, EFAs can play an indirect role in protein production by cells in culture.

Fatty acid peroxidation

PUFAs with methylene-interrupted double bonds are highly susceptible to enzymatic and non-enzymatic peroxidation. Enzymatic peroxidation is brought about by the action of cell-derived peroxidizing enzymes such as cyclooxygenase and the various lipoxygenases involved in the eicosanoid hormone cascade described above. Non-enzymatic peroxidation is initiated by free radicals, requires oxygen, and has the characteristics of a chain reaction (Esterbauer *et al.*, 1990). The process begins when a free radical abstracts a hydrogen atom from a PUFA, thus forming a lipid radical (Fig. 5) (Pryor and Stanley, 1975). The lipid radical reacts quickly with oxygen to form a lipid peroxy radical, which in turn either forms a lipid endoperoxide or abstracts a hydrogen atom from another PUFA to form a new lipid radical and a lipid hydroperoxide. As before, the new lipid radical reacts with O_2 and the reaction enters a self-propagating phase that could convert all PUFAs to lipid hydroperoxides (Buege and Aust, 1978; Esterbauer *et al.*, 1990). Traces of transition metals such as Fe^{2+} or Cu^{+} can easily induce the decomposition of lipid hydroperoxides to give lipid alkoxy radicals, which initiate new chain reactions (Esterbauer *et al.*, 1990).

Decomposition of the lipid hydroperoxides and endoperoxides, including those formed enzymatically, eventually gives rise to a variety of end products including alcohols, ketones, ethers and aldehydes. Some end products, such as malondialdehyde (MDA, fragment of the interior part of the PUFA chain) and 4-hydroxy-2-nonenal (fragment of the methyl end of the PUFA chain) are biologically very active and possess toxic and mutagenic properties (Esterbauer *et al.*, 1990). In primary cultures of rat hepatocytes, for example, 4-hydroxy-2-nonenal was highly cytotoxic at concentrations exceeding 100 μM, and at subcytotoxic concentrations (0.1–10 μM) it produced genotoxic effects such as an increase in micronuclei, chromosomal aberrations and sister-chromatid exchange (Esterbauer *et al.*, 1990). In another study (Bird and Draper, 1980), MDA at 1–100 μM had little or no effect on the morphology and growth of rat skin fibroblasts, but induced severe morphological alterations and growth inhibition at 1 mM.

Lipid peroxidation is a complication that must be considered in EFA-supplemented cultures, especially because of the presence of oxygen and iron in the cellular environment. However, lipid peroxidation in cells *in vitro* and *in vivo* may not be as uncontrolled as it appears in experiments with non-living systems, and may not be as harmful as suggested by experiments in which the cytotoxicity/mutagenicity of peroxidation products is determined via their exogenous addition to cells in high amounts. Efficient enzymatic detoxification systems have been described in cells (Galeotti *et al.*, 1991; Esterbauer *et al.*, 1990). In rat hepatocytes, for example, metabolism of 4-hydroxy-2-nonenal to less toxic compounds is possible by (a) cytosolic NADH-dependent alcohol dehydrogenase (reduction of the aldehyde to the corresponding alcohol) (Esterbauer *et al.*, 1985), (b) mitochondrial NAD(P)-dependent aldehyde dehydrogenases (oxidation to the the carboxylic acid) (Lamé and Segall, 1986), and (c) glutathione transferases (conjugation to glutathione) (Danielson *et al.*, 1987). These systems prevent the intracellular accumulation of 4-hydroxy-

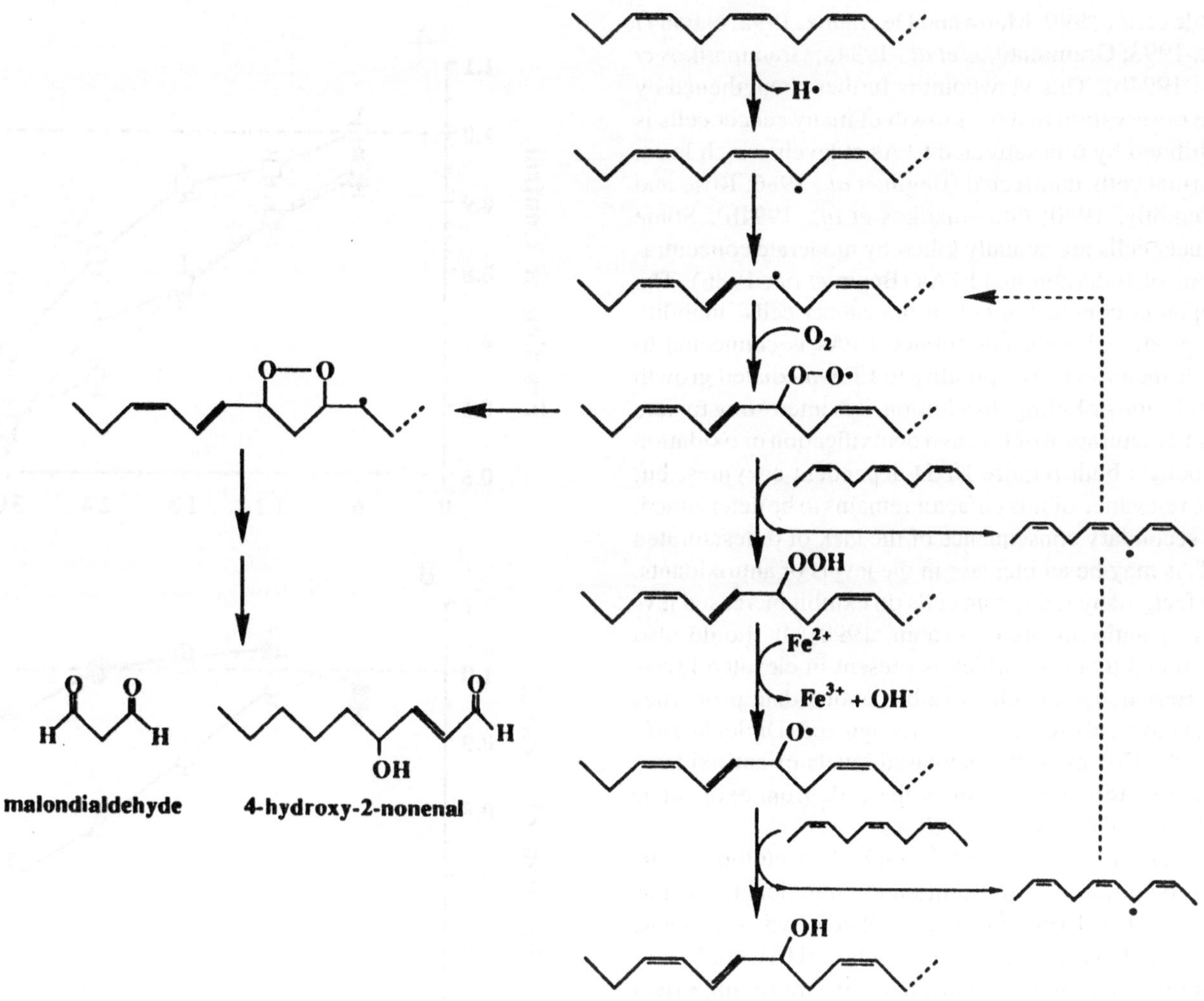

Fig. 5. The process of non-enzymatic polyunsaturated fatty acid peroxidation and subsequent degradation. The cascade begins by oxygen (or hydroxyl) free radical abstraction of a hydrogen atom, and is catalyzed by transition metals.

2-nonenal to cytotoxic levels. While alcohol and aldehyde dehydrogenases are present mainly in liver cells, glutathione transferases are present in cells from a variety of tissues.

Lipid peroxidation may even play a role in regulating cell proliferation. Rapidly dividing cells, for example, show low levels of lipid peroxidation as compared to slowly dividing or non-dividing cells (Cheeseman *et al.*, 1984; Horrobin, 1990; Galeotti *et al.*, 1991). Much can be learned about the role of lipid peroxidation in cellular regulation by comparing cancer and normal cells in terms of their lipid peroxide content. Consistent with a regulatory role for lipid peroxidation in cellular proliferation, cancer cells, which are capable of rapid and unlimited proliferation, have low levels of lipid peroxides. Horrobin (1990) suggests three reasons for this phenomenon. Cancer cells could (a) lack substrate, (b) contain excess antioxidants, or (c) contain mechanisms for rapid removal of lipid peroxides. The first reason is supported by the fact that cancer cells have low levels of 6-desaturated EFAs (double bond introduced at C-6 of LA or LNA; DGLNA, AA, EPA and DHA are all 6-desaturated) (Utsumi *et al.*, 1965; Bartoli *et al.*, 1980; Cheeseman *et al.*, 1984). This may be due to increased EFA utilization by the more rapidly dividing tumor cells, but it is also a consequence of reduced capacity to desaturate LA and LNA. Indeed, many but not all transformed and malignant cells *in vitro* have reduced desaturating ability (Dunbar and Bailey, 1975; Maeda *et al.*, 1978; Iturralde *et al.*, 1990; Marra and De Alaniz, 1992; Naval *et al.*, 1993; Grammatikos *et al.*, 1994a; Grammatikos *et al.*, 1994b). This viewpoint is further strengthened by the observation that the growth of many cancer cells is inhibited by 6-desaturated EFAs at levels which leave normal cells unaffected (Begin *et al.*, 1986; Rose and Connolly, 1990; Grammatikos *et al.*, 1994b). Some cancer cells are actually killed by moderate concentrations of 6-desaturated EFAs (Begin *et al.*, 1986). The apparent conclusion is that the cancer cells' inability to produce highly unsaturated EFAs is connected to their increased susceptibility to EFA-mediated growth inhibition or killing. In addition, it is interesting to note that desaturation of FAs and detoxification of oxidation products both require NAD-dependent enzymes, but the relevance of this cofactor remains to be determined. A secondary consequence of the lack of 6-desaturated EFAs may be an increase in the levels of antioxidants. In fact, many malignant cells do exhibit increased levels of antioxidants (Horrobin, 1990). It should also be noted that OA, which is present in elevated levels in EFA-deficient cells, exhibits antioxidant properties that have only recently been recognized (Diplock *et al.*, 1988). However, the increased levels of antioxidants are not sufficient to protect the cells from exogenous addition of 6-desaturated EFAs.

Peroxidation of EFAs supplied to cultured cells can be minimized by addition of antioxidants such as α-tocopherol (vitamin E), butylated hydroxytoluene (BHT), butylated hydroxyanisole (BHA), and selenium. Vitamin E is considered the most important and most potent inhibitor of fatty acid peroxidation (Cheeseman *et al.*, 1984; Gower, 1988) because it is an effective free radical scavenger and is incorporated into cellular membranes. Selenium, which is a universal supplement to serum-free media, is an important component of the enzyme glutathione peroxidase, which converts FA hydroperoxides to hydroxy acids. BHT and BHA do not occur naturally, but are biocompatible. All of the above antioxidants have been reported to interfere with EFA-mediated tumor cell killing *in vitro* (Begin *et al.*, 1986; Begin, 1987; Begin *et al.*, 1988).

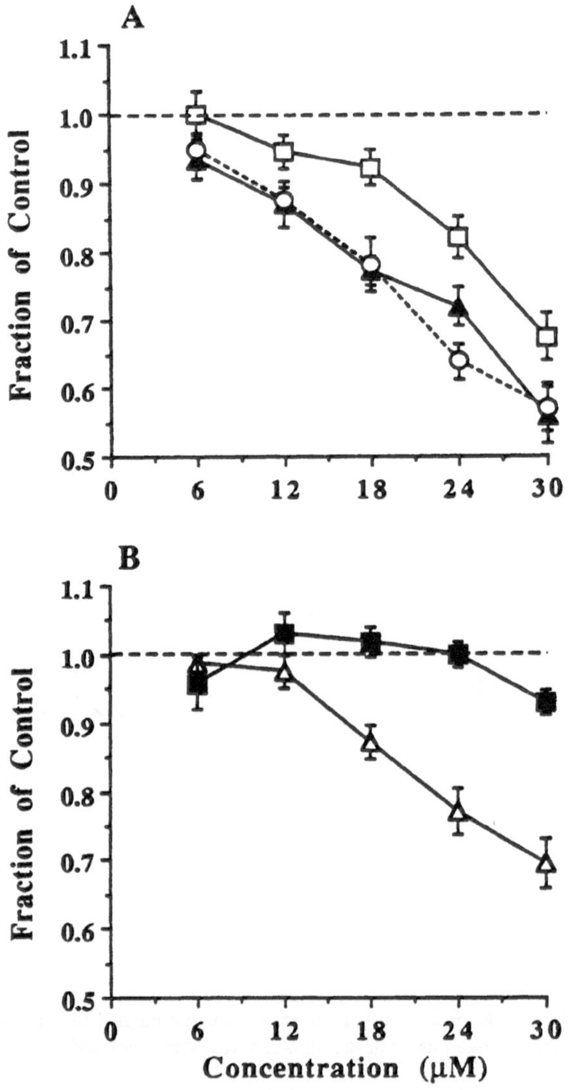

Fig. 6. Effect of (A) the *n*-3 FAs LNA (open squares), EPA (filled triangles) and DHA (open circles), and (B) the *n*-6 FAs LA (filled squares) and AA (open triangles) on the growth of MCF-7 breast cancer cells cultured in a low-serum containing medium (0.5% FBS v/v) in order to avoid interference from serum FAs. MCF-7 cells inoculated at 70,000 cells per well in fibronectin-coated 6-well plates were cultured for 48 h before the FAs were introduced. Oleic acid (OA, *n*-9) at 30 μM was used as control. Total exogenous FA was kept constant at 30 μM by supplementing with OA. Supplementation with 30 μM total FA also provided 12 μM cholesterol and 100 μM vitamin E. The cells from each well were collected after 7 days. Control cells exhibited a 5-fold increase in cell number. *Points*, mean values relative to control for 3 separate experiments each set up in triplicate wells; *bars*, SE. Statistical significance was assessed by a paired Student's t test on the raw data from all three experiments. Points that are significantly different from control with $P<0.05$: LNA 18 μM, EPA 6 μM, LA 30 μM; with $P<0.01$: LNA 24-30 μM, EPA 12–30 μM, DHA 12–30 μM, AA 18–30 μM. Reproduced from Grammatikos *et al.* (1994b) with permission.

Case study: growth effects of essential fatty acids on neoplastic and non-cancerous human mammary epithelial cell lines

The involvement of dietary fat in breast cancer initiation, promotion and metastasis is controversial.

Although numerous animal and some epidemiological studies suggest a role for EFA type in breast cancer (Rose, 1986; Cave, 1991; Pritchard *et al.*, 1989), there have been very few attempts to study this problem at the cellular level (Rose and Connolly, 1990; Wicha *et al.*, 1979). We have examined the influence of three *n*-3 (LNA, EPA and DHA) and two *n*-6 (LA and AA) FAs, presented as the free FA bound to albumin, on the growth of the breast cancer cell line MCF-7 and the non-cancerous mammary epithelial cell line MCF-10A (Grammatikos *et al.*, 1994b).

All of the *n*-3 FAs inhibited MCF-7 cell growth in a dose-dependent manner, with EPA and DHA being most effective (Fig. 6A). For the *n*-6 FAs, LA had no effect, while AA was as inhibitory as LNA (Fig. 6B). Inhibition by *n*-3 EFAs, but not by LA is consistent with animal models showing tumor inhibition by fish oils (rich in *n*-3 EFAs) relative to corn oil (rich in LA) (Cave, 1991; Pritchard *et al.*, 1989). The differing effect of LA compared to AA and the less extensive inhibition by LNA compared to EPA and DHA indicate that differences exist between EFAs of the same family, and suggest that LA and LNA are not processed to AA and EPA, respectively, by MCF-7 cells. Indeed, analysis of total cellular lipid extracts by gas chromatography revealed that, although the exogenous EFAs were extensively incorporated into MCF-7 cells, their metabolic conversion was limited to two-carbon elongations, very limited Δ^8 and/or Δ^6 desaturation, and retroconversion (mainly DHA to EPA) (Grammatikos *et al.*, 1994b). MCF-7 cells are unable to perform appreciable Δ^6, Δ^5 and Δ^4 desaturations, and are therefore unable to produce AA from LA or EPA and DHA from LNA. Despite the limited metabolic conversions, 90% of the EFA taken up by MCF-7 cells was incorporated into phospholipids.

In contrast to MCF-7 cells, MCF-10A cells were not inhibited by any of the *n*-3 or *n*-6 FAs at levels below 24 μM, and were even stimulated by as much as 50% at the lower concentrations (Fig. 7) (Grammatikos *et al.*, 1994b). At 30 μM, however, the cells were inhibited by all of the FAs. Unlike MCF-7 cells, the effects of LA and AA on MCF-10A cell growth were qualitatively the same, suggesting differences in the EFA processing patterns of the two cell lines. Evaluation of the fate of the exogenous *n*-3 and *n*-6 FAs revealed that MCF-10A cells desaturated and elongated the exogenous EFAs via all the known pathways. This observation is of particular interest because the MCF-7 and MCF-10A cell lines are both derived from

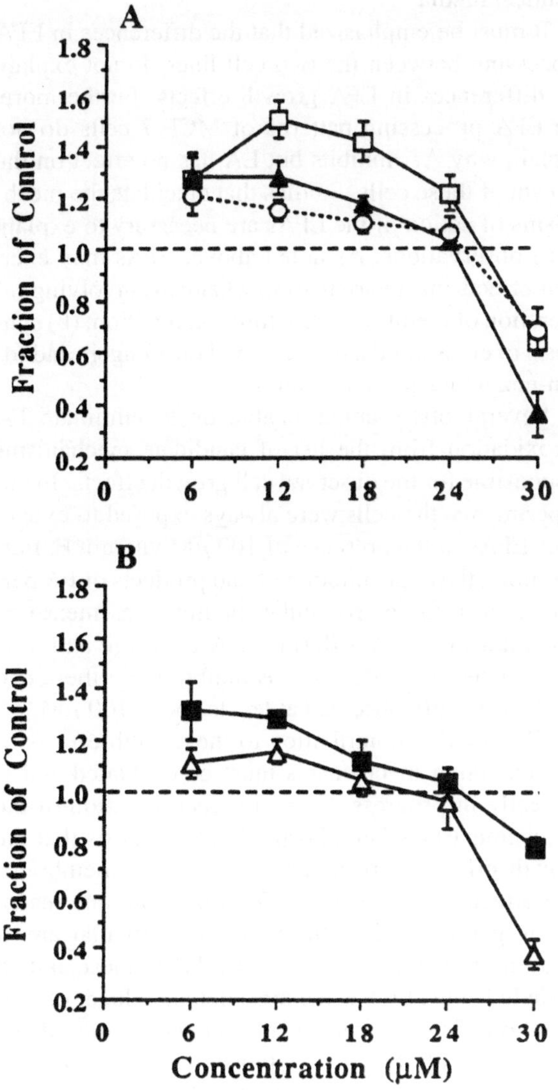

Fig. 7. Effect of (A) the *n*-3 FAs LNA (open squares), EPA (filled triangles) and DHA (open circles), and (B) the *n*-6 FAs LA (filled squares) and AA (open triangles) on the growth of MCF-10A non-cancerous human mammary epithelial cells. Experimental conditions were as described in the legend of Figure 6 with the following modifications to account for the faster growth rate of these cells: (i) 40,000 cells per well were cultured for 24 hours prior to FA addition, and (ii) the cells from each well were collected and counted after 4 days. Control cells exhibited an 8-fold increase in cell number. *Points*, mean values relative to control for 3 separate experiments each set up in triplicate wells; *bars*, SE. Statistical significance was assessed by a paired Student's t test on the raw data from all three experiments. Points that are significantly different from control with $P<0.05$: LNA 18 μM, EPA 6 μM, LA 30 μM; with $P < 0.01$: LNA 24–30 μM, EPA 12–30 μM, DHA 12–30 μM, AA 18–30 μM. Reproduced from Grammatikos *et al.* (1994b) with permission.

human mammary epithelial cells, and were cultured in identical media.

It must be emphasized that the differences in EFA processing between the two cell lines do not explain the differences in EFA growth effects. Furthermore, the EFA processing patterns of MCF-7 cells do not explain why AA inhibits but LA has no effect on the growth of these cells. Studies that elucidate the mechanisms of action of the EFAs are necessary to explain these observations. As noted above, EFAs may exert their effects via a variety of mechanisms involving (a) alteration of membrane structure and function, (b) synthesis of eicosanoids, (c) as second messengers and (d) non-enzymatic peroxidation.

Several observations enable us to eliminate FA peroxidation from the list of candidate mechanisms responsible for the observed cell growth effects. In our experiments, the cells were always exposed to exogenous EFAs in the presence of 100 μM vitamin E. Furthermore, the concentration of end products of FA peroxidation in the extracellular medium was measured at several time points during EFA exposure and never exceeded 0.5 μM, whereas inhibition by these end products is only observed at levels above 100 μM.

The relative contribution to the growth effects of the remaining mechanisms must be evaluated and is currently in progress. Extensive incorporation of the exogenous EFAs into phospholipids suggests that the growth effects were initiated at the cell membrane. The effects of exogenous EFAs on membrane receptors in general and on the EGFR in particular are of great interest considering the fact that in these and all epithelial cells, EGF is the most potent inducer of proliferation. Possible second messenger actions of the EFAs, particularly their relationship to PKC, are also under investigation. The differential ability of MCF-7 and MCF-10A cells to produce eicosanoid hormone precursor FAs from parent EFAs may imply a role for eicosanoids in the observed growth effects. The fact that MCF-7 cells do not produce AA from LA and are inhibited by it contradicts the notion that the breast cancer promoting effect of LA is mediated by eicosanoid hormones. However, production of eicosanoids was not measured in our experiments. Until the patterns of eicosanoid production in both cell lines becomes known, the relative contribution of these compounds to the growth effects cannot be clearly assessed.

Conclusions

Essential (*n*-6 and *n*-3) FAs exert a multitude of effects on cultured cells that extend beyond their roles in cellular nutrition and membrane integrity. Incorporation of exogenous EFAs into cells in culture leads to alterations in membrane structure and function, second messenger effects on a variety of signal transduction pathways, modulation of gene expression, and enzymatic and non-enzymatic production of a variety of peroxidized compounds which are potent physiological modulators (eicosanoids), or exhibit toxic and mutagenic properties (malondialdehyde, 4-hydroxy-2-nonenal, etc). Cells *in vitro* from a variety of tissues are capable of metabolically converting exogenous EFAs to varying extents. In principle, therefore, it is possible that supply of one EFA leads to intracellular enrichment of a variety of EFAs of the same family, including ones that may have more profound effects than the supplied EFA. The net effect of these competing and/or cooperating mechanisms varies greatly with cell type and supplied fatty acid.

Because of such diverse EFA-mediated effects on a variety of cells *in vitro*, it is conceivable that altering a cell's EFA composition and/or eicosanoid production could lead to improved growth and productivity of commercially important cells. It has recently been demostrated (Schmid *et al.*, 1991) that antithrombin III formation by recombinant baby hamster kidney and Chinese hamster ovary cells is a function of combinations and concentrations of FAs added to the culture medium. The usefulness of EFAs and EFA metabolites in this regard has been largely overlooked or ignored by the biotechnology industry. The majority of studies focus on EFA involvement in human diseases, and little or no data exist regarding EFA effects on commercially used cells.

Acknowledgements

This work was supported in part by NIH grant R01CA49564-04 (TAV), NSF grant BCS-9058416 (WMM) and contributions to WMM from Eli Lilly and Co., Schering Plough Research, and Abbott Laboratories.

References

Abdel-Malek ZA, Swope YB, Amornsiripanitch N and Nordlund JJ (1987) *In vitro* modulation of proliferation and melanization of S91 melanoma cells by prostaglandins. Cancer Res 47: 3141–3146.

Arita H, Nahano T and Hanasaki K (1989) Thrombaxane A_2: its generation and role in platelet activation. Prog Lipid Res 28: 273–301.

Bach MK (1984) The Leukotrienes. Academic, New York.

Bailey JM and Dunbar LM (1973) Essential fatty acid requirements of cells in tissue culture: A review. Exp Mol Pathol 18: 142–161.

Bandyopadhyay GK, Imagawa W, Wallace D and Nandi S (1987) Linoleate metabolites enhance the *in vitro* proliferative response of mouse mammary epithelial cells to epidermal growth factor. J Biol Chem 262: 2750–2756.

Barnes D and Sato GH (1980a) Methods for growth of cultured cells in serum free-medium. Anal Biochem 102: 255–270.

Barnes D and Sato GH (1980b) Serum-free cell culture: A unifying approach. Cell 22: 649-655.

Bartoli GM, Bartoli S, Galeotti T and Bertoli E (1980) Superoxide dismutase content and microsomal lipid composition of tumours with different growth rates. Biochim Biophys Acta 620: 205-211.

Baud L, Sraer J, Perez J, Nivez M-P and Ardaillou R (1985) Leukotriene C_4 binds to human glomerular epithelial cells and promotes their proliferation *in vitro*. J Clin Invest 76: 374–377.

Begin ME, Ells G, Das UN and Horrobin DF (1986) Differential killing of human carcinoma cells supplemented with *n*-3 and *n*-6 polyunsaturated fatty acids. J Natl Cancer Inst 77: 1053-1062.

Begin ME (1987) Effects of polyunsaturated fatty acids and of their oxidation products on cell survival. Chem Phys Lipids 45: 269–313.

Begin ME, Ells G and Horrobin DF (1988) Polyunsaturated fatty acid-induced cytotoxicity against tumor cells and its relationship to lipid peroxidation, J Natl Cancer Inst 80: 188–194.

Bettger WJ and Ham RG (1982) The nutrient requirements of cultured mammalian cells. Adv Nutr Res 4: 249–286.

Bhuyan BK, Adams EC, Badiner CJ, Li LH and Barden K (1986) Cell cycle effects of prostaglandins A_1, A_2, and D_2 in human murine melanoma cells in culture. Cancer Res. 46: 1688–1693.

Bird RP and Draper HH (1980) Effect of malonaldehyde and acetaldehyde on cultured mammalian cells: Growth, morphology, and synthesis of macromolecules. J Toxicol Env Health 6: 811–823.

Bjaere U (1987) Serum-free cultivation of lymphoid cells. In: Fiechter A. (ed.) Advances in Biochemical Engineering/Biotechnology. (pp. 95–109) Springer-Verlag, Berlin.

Boni LT and Rando RR (1985) The nature of protein kinase C activation by physically defined phospholipid vesicles and diacylglycerols. J Biol Chem 260: 10819–10825.

Borrás M and Leclercq G (1992) Modulatory effect of nonesterified fatty acids on structure and binding characteristics of estrogen receptor from MCF-7 human breast cancer cells. J Receptor Res 12: 463–484.

Borys MC, Linzer DIH and Papoutsakis ET (1993) Culture pH affects expression rates and glycosylation of recombinant mouse placental lactogen proteins by Chinese hamster ovary (CHO) cells. Biotechnology 11: 720–724.

Buege JA and Aust SD (1978) Microsomal lipid peroxidation. Meth Enzymol 52: 302–310.

Carroll KK and Hopkins GJ (1979) Dietary polyunsaturated fat versus saturated fat in relation to mammary carcinogenesis. Lipids 14: 155–158.

Carty TJ, Eskra JD, Lomberdino JG and Hoffman WW (1980a) Piroxicam, a potent inhibitor of prostaglandin production in cell culture. Structure-activity study. Prostaglandins 19: 51–59.

Carty TJ, Stevens JS, Lombardino JG, Pany MJ and Randall MJ (1980b) Piroxicam, a structurally novel anti-inflammatory compound. Mode of prostaglandin synthesis inhibition. Prostaglandins 19: 671-682.

Castagna M, Takai Y, Kaibuchi K, Sano K, Kikkawa U and Nishizuka Y (1982) Direct activation of calcium-activated, phospholipid-dependent protein kinase by tumor-promoting phorbol esters. J Biol Chem 257: 7847–7851.

Cave Jr, WT (1991) Dietary *n*-3 (ω-3) polyunsaturated fatty acid effects on animal tumorigenesis. FASEB J 5: 2160–2166.

Cheeseman KH, Burton GW, Ingold KU and Slater TF (1984) Lipid peroxidation and lipid antioxidants in normal and tumor cells. Toxicol Pathol 22: 235–239.

Chegini N and Safa AR (1987) Morphological alterations induced by prostaglandins E_1,F_{2a} and A_1 in MDA-MB-231 and MCF-7 human breast cancer cell lines. Cancer Lett 37: 189–197.

Conde B, Tejedor M, Sinues E and Alcala A (1991) Modulation of cell growth and differentiation induced by prostaglandin D_2 in the glioma cell line C6. Anticancer Res 11: 289–296.

Conn PM, Ganong BR, Ebeling J, Staley D, Neidel JE and Bell RM (1985) Diacylglycerols release LH: Structure-activity relations reveal a role for protein kinase C. Biochem Biophys Res Commun 126: 532–539.

Cook HW, Byers DM, Palmer FBStC, Spence MW, Rakoff H, Duvall SM and Emken EA (1991) Alternate pathways in the desaturation and chain elongation of linolenic acid, 18:3(*n*-3), in cultured glioma cells. J Lipid Res 32: 1265–1273.

Cook SJ and McCormick F (1993) Inhibition by cAMP of Ras-dependent activation of Raf. Science 262: 1069–1072.

Corey EJ, Shih C and Cashman JR (1983) Docosahexaenoic acid is a strong inhibitor of prostaglandin but not leukotriene biosynthesis. Proc Natl Acad Sci USA 80: 3581–3584.

Cornwell DG and Morisaki N (1984) Fatty acid paradoxes in the control of cell proliferation: Prostaglandins, lipid peroxides and cooxidation reactions. In: Pryor WA. (ed.) Free Radicals in Biology. (pp.95-148) Academic Press, Orlando.

Danielson UH, Esterbauer H and Mannervik B (1987) Structure-activity relationships of 4-hydroxyalkenals in the conjugation catalysed by mammalian glutathione transferases. Biochem J 247: 707–713.

Davison PM and Karasek MA (1981) Human dermal microvascular endothelial cells *in vitro*: Effect of cyclic AMP on cellular morphology and proliferation rate. J Cell Phys 106: 253–258.

Diplock AT, Balasubramanian KA, Manohar M, Mathan VI and Ashton D (1988) Purification and chemical characterisation of the inhibitor of lipid peroxidation from intestinal mucosa. Biochim Biophys Acta 962: 42–50.

Doi O, Doi F, Schroeder F, Alberts AW and Vagelos PR (1978) Manipulation of fatty acid composition of membrane phospholipid and its effects on cell growth in mouse LM cells. Biochim Biophys Acta 509: 239–250.

Dugas B, Paul-Eugene N, Cairns J, Gordon J, Calenda A, Mencia-Huerta JM and Braquet P (1990) Leukotriene B_4 potentiates the expression and release of FCeRII/CD23, and proliferation and differentiation of human B lymphocytes induced by IL4. J Immunol 145: 3406–3411.

Dunbar LM and Bailey JM (1975) Enzyme deletions and essential fatty acid metabolism in cultured cells. J Biol Chem 250: 1152–1153.

Durant S, Duval D and Homo-Delarche F (1989) Effect of exogenous prostaglandins and nonsteroidal anti-inflammatory agents on prostaglandin secretion and proliferation of mouse embryo fibroblasts in culture. Prostaglandins Leukotrienes Essential Fatty Acids 38: 1–8.

Eisenbarth GS, Wellman DK and Lebovitz HE (1974) Prostagndin A_1 inhibition of chondrosarcoma growth. Biochem Biophys Res Commun 60: 1302–1308.

Esterbauer H, Zollner H and Lang J (1985) Metabolism of the lipid peroxidation product 4-hydroxynonenal by isolated hepatocytes and by liver cytosolic fractions. Biochem J 228: 363–373.

Esterbauer H, Eckl P and Ortner A (1990) Possible mutagens derived from lipids and lipid precursors. Mutation Res 238: 223–233.

Evans VJ, Bryant JC, Kerr HA and Schilling EL (1964) Chemically defined media for cultivation of long-term cell strains from four mammalian species. Exp Cell Res 36: 439–474.

Field CJ, Ryan EA, Thomson ABR and Clandinin MJ (1990) Diet fat composition alters membrane phospholipid composition, insulin binding, and glucose metabolism in adipocytes from control and diabetic animals. J Biol Chem 265: 11143-11150.

Fischer S and Weber PC (1984) Prostaglandn I_3 is formed *in vivo* in man after dietary eicosapentaenoic acid. Nature 307: 165–168.

Fischer SM, Cameron GS, Baldwin JK, Jasheway DW, Patrick KE and Belury MA (1989) The arachidonic acid cascade and multistage carcinogenesis in mouse skin. In: Slaga TJ (ed.) Skin Carcinogenesis: Mechanisms and Human Relevance. (pp. 249–264) A.R. Liss, New York.

Fukushima M, Kato T, Ueda R, Ota K, Narumiya S and Hayaishi O (1982) Prorlaglandin D_2, a potential antineoplastic agent. Biochem Biophys Res Commun 105: 956–964.

Galeotti T, Masotti L, Borello S and Casali E (1991) Oxy-radical metabolism and control of tumour growth. Xenobiotica 21: 1041–1051.

Ginsberg BH, Brown TJ, Simon I and Spector AA (1981) Effect of the membrane lipid environment on the properties of insulin receptors. Diabetes 30: 773–780.

Glassy MC, Tharakan JP and Chau PC (1988) Serum-free media in hybridoma culture and monoclonal antibody production. Biotechnol Bioeng 32: 1015–1028.

Gower JD (1988) A role for dietary lipids and antioxidants in the activation of carcinogens. Free Radical Biol Med 5: 95–111.

Grammatikos SI, Subbaiah PV, Victor TA and Miller WM (1994a) Diversity in the ability of cultured cells to elongate and desaturate essential (*n*-6 and *n*-3) fatty acids. Ann NY Acad Sci 745: 92–105.

Grammatikos SI, Subbaiah PV, Victor TA and Miller WM (1994b) (*n*-3 and *n*-6) fatty acid processing and growth effects in neoplastic and non-cancerous human mammary epithelial cell lines. Br J Cancer 70: 219–227.

Higuchi K (1973) Cultivation of animal cells in chemically defined media: A review. Adv Appl Microbiol 16: 111–136.

Holian O and Nelson R (1992) Action of long-chain fatty acids on protein kinase C activity: Comparison of omega-6 and omega-3 fatty acids. Anticancer Res 12: 975–980.

Honn KV, Bockman RS and Marnett LJ (1981a) Prostaglandins and cancer: A review of tumor initiation through tumor metastasis. Prostaglandins 21: 833–864.

Honn KV, Cicone B and Skoff A (1981b) Prostacyclin: A potent antimetastatic agent. Science 212: 1270–1272.

Hornstra G, Christ-Hazelhof E, Haddeman E, ten Hoor F and Nugteren DH (1981) Fish oil feeding lowers thromboxane and prostacyclin production by rat platelets and aorta and does not result in the formation of prostaglandin I_3. Prostaglandins 21: 727–738.

Horrobin DF (1990) Essential fatty acids, lipid peroxidation, and cancer. In: Horrobin DF. (ed.) Omega-6 Fatty Acids: Pathophysiology and Roles in Clinical Medicine (pp. 351-377) A.R. Liss, New York.

Huang JM-C, Xian H and Bacaner M (1992) Long-chain fatty acids activate calcium channels in ventricular myocytes. Proc Natl Acad Sci USA 89: 6452–6456.

Hubbard NE and Erickson KL (1987) Enhancement of metastasis from a transplantable mouse mammary tumor by dietary linoleic acid. Cancer Res 47: 6171–6175.

Huttner JJ, Gwebu ET, Panganamala RV, Milo GE and Cornwell DG (1977) Fatty acids and their prostaglandin derivatives: Inhibitors of proliferation in aortic smooth muscle cells. Science 197: 289–291.

Iturralde M, González B and Pineiro A (1990) Linoleate and linolenate desaturation by rat hepatoma cells. Biochem Intl 20: 37–43.

Jimenez de Asua L, Clingan D and Rudland PS (1975) Initiation of cell proliferation in cultured mouse fibroblasts by prostaglandin $F_{2\alpha}$. Proc Natl Acad Sci USA 72: 2724–2728.

Johnson GS and Pastan I (1971) Change in growth and morphology of fibroblasts by prostaglandins. J Natl Cancer Inst 47: 1357–1364.

Kano-Sueoka T, King DM, Fisk HA and Klug SJ (1990) Binding of epidermal growth factor to its receptor is affected by membrane phospholipid environment. J Cell Physiol 145: 543–548.

Kantor HS and Hampton M (1978) Indomethacin in submicromolar concentrations inhibits cyclic AMP-dependent protein kinase. Nature 276: 841–842.

Kaplan L, Weiss J and Elsbach P (1978) Low concentrations of indomethacin inhibit phospholipase A_2 of rabbit polymorphonuclear leukocytes. Proc Natl Acad Sci USA 75: 2955–2958.

Karmali RA (1987) Fatty acids: inhibition. Am J Clin Nutr 45: 225–229.

Katz EB and Boylan ES (1987) Stimulatory effect of high polyunsaturated fat diet on lung metastasis from the 13762 mammary adenocarcinoma in female retired breeder rats. J Natl Cancer Inst 79: 351–358.

Keyaki A, Handa H, Yamashita J, Tokuriki Y, Otsuka S-I, Yamasaki T and Gi H (1984) Growth-inhibitory effect of prostaglandin D_2 on mouse glioma cells. J Neurosurg 61: 912–917.

Khan WA, Blobe GC and Hannun YA (1992) Activation of protein kinase C by oleic acid. Determination and analysis of inhibition by detergent micelles and physiologic membranes: Requirement for free oleate. J Biol Chem 267: 3605–3612.

Kigoshi T, Uchida K, Kaneko M, Iwasaki R, Nakano S, Azukizawa S and Morimoto S (1990) Direct inhibition of smooth muscle light chain kinase by arachidonic acid in a purified system. Biochem Biophys Res Commun. 171: 369–374.

King ME and Spector AA (1978) Effects of specific fatty acyl enrichments on membrane physical properties detected with a spin label probe. J Biol Chem 253: 6493–6501.

Lai CS, Hopwood LE and Swartz HM (1980) Electron spin resonance studies of changes in membrane fluidity of Chinese hamster ovary cells during the cell cycle. Biochim Biophys Acta 602: 117–126.

Lamé MW and Segall HJ (1986) Metabolism of the pyrrolizidine alkaloid metabolite trans-4-hydroxy-2-hexanal by mouse liver aldehyde dehydrogenases. Toxicol Appl Pharmaol 82: 94-103.

Léger CL, Daveloose D, Christon R and Viret J (1990) Evidence for a structurally specific role of essential polyunsaturated fatty acids depending on their peculiar double-bond distribution in biomembranes. Biochemistry 29: 7269–7275.

Maeda M, Doi O and Akamatsu Y (1978) Metabolic conversion of polyunsaturated fatty acids in mammalian cultured cells. Biochim Biophys Acta 530: 153–164.

Marini S, Palamara AT, Garaci E and Santoro MG (1990) Growth inhibition of Friend erythroleukaemia cell tumours *in vivo* by a synthetic analogue of prostaglandin A: an action independent of natural killer-activity. Br J Cancer 61: 394–399.

Marra CA and De Alaniz MJT (1992) Incorporation and metabolic conversion of saturated and unsaturated fatty acids in SK-Hep_1 human hepatoma cells in culture. Mol Cell Biochem 117: 107–118.

McCumbee WD, Harrelson JM and Lebovitz HE (1983) Hormonal and metabolic regulation of human chondrosarcoma *in vitro*. Cancer Res 43: 513–516.

McPhail LC, Clayton C and Snyderman R (1984) A potential second messenger role for unsaturated fatty acids: Activation of Ca^{2+}-dependent protein kinase. Science 224: 622-624.

Medrano EE, Farooqui JZ, Boissy RE, Boissy YL, Akadiri B and Nordlund JJ (1993) Chronic growth stimulation of human adult melanocytes by inflammatory mediators *in vitro*: Implications for nevus formation and initial steps in melanocyte oncogenesis. Proc Natl Acad Sci USA 90: 1790–1794.

Miller AM, Cullen MK, Kobb SM and Weiner RS (1989) Effects of lipoxygenase and glutathione pathway inhibitors on leukemic cell line growth. J Lab Clin Med 113: 355–361.

Modat G, Muller A, Mary A, Grégoire C and Bonne C (1987) Differential effects of leukotrienes B_4 and C_4 on bovine aortic endothelial cell proliferation *in vitro*. Prostaglandins 33: 531-538.

Moncada S and Vane JR (1979) Pharmacology and endogenous roles of prostaglandin endoperoxides, thromboxane A_2 and prostacyclin. Pharmacol Rev 30: 293–331.

Naval J, Martinez-Lorenzo MJ, Marzo I, Desportes P and Pineiro A (1993) Alternative route for the biosynthesis of polyunsaturated fatty acids in K562 cells. Biochem J 291: 841–845.

Neichi T, Koshihara Y and Murota S (1983) Inhibitory effect of esculetin on 5-lipoxygenase and leukotriene biosynthesis. Biochim Biophys Acta 753: 130–132.

Nishizuka Y (1988) The molecular heterogeneity of protein kinase C and its implications for cellular regulation. Nature 334: 661–665.

Nishizuka Y (1989) Studies and prospectives of the protein kinase C family for cellular regulation. Cancer 63: 1892–1903.

Penning TM and Talalay P (1983) Inhibition of a major NAD(P)-linked oxidoreductase from rat liver cytosol by steroidal and non-steroidal anti-inflammatory agents and by prostaglandins. Proc Natl Acad Sci USA 80: 4504–4508.

Piomelli D, Wang JKT, Sihra TS, Nairn AC, Czernik AJ and Greengard P (1989) Inhibition of Ca^{2+}/calmodulin-dependent protein kinase II by arachidonic acid and its metabolites. Proc Natl Acad Sci USA 86: 8550–8554.

Poon R, Richards JM and Clark WR (1981) The relationship between plasma membrane lipid composition and physical-chemical properties. II. Effect of phospholipid fatty acid modulation on plasma membrane physical properties and enzymatic activities. Biochim Biophys Acta 649: 58–66.

Pritchard GA, Jones DL and Mansel RE (1989) Lipids in breast carcinogenesis. Br J Surg 76: 1069–1073.

Pryor WA and Stanley JP (1975) A suggested mechanism for the production of malonaldehyde during the autoxidation of polyunsaturated fatty acids. Nonenzymatic production of prostaglandin endoperoxides during autoxidation. J Org Chem 40: 3615–3617.

Rhee SG, Suh P-G, Ryu S-H and Lee SY (1989) Studies of inositol phospholipid-specific phospholipase C. Science 244: 546–550.

Rintoul DA, Sklar LA and Simoni RD (1978) Membrane lipid modification of Chinese hamster ovary cells. Thermal properties of membrane phospholipids. J Biol Chem 253: 7447–7452.

Rola-Pleszczynski M and Stankova J (1992) Leukotriene B_4 enhances interleukin-6 (IL-6) production and IL-6 messenger RNA accumulation in human monocytes *in vitro*: transcriptional and posttranscriptional mechanisms. Blood 80: 1004–1011.

Rose DP (1986) Dietary factors and breast cancer. Cancer Surveys 5: 671–687.

Rose DP and Conolly JM (1990) Effects of fatty acids and inhibitors of eicosanoid synthesis on the growth of a human breast cancer cell line. Cancer Res 50: 7139–7144.

Rosenthal MD (1987) Fatty acid metabolism of isolated mammalian cells. Prog Lipid Res 26: 87–124.

Sakai T, Yamaguchi N, Kawai K, Nishino H and Iwashima A (1984) Prostaglandin D_2 inhibits the proliferation of human malignant tumor cells. Prostaglandins 27: 17–26.

Salem Jr N (1989) Omega-3 fatty acids: Molecular and biochemical aspects. In: Scala J and Spiller GA (eds.) New Protective Roles for Selected Nutrients. (pp. 109–228) A. R. Liss, New York.

Samuelsson B (1983) Leukotrienes: Mediators of immediate hypersensitivity reactions and inflammation. Science 220: 568–575.

Santoro MG, Philpott GW and Jaffe BM (1976) Inhibition of tumour growth *in vivo* and *in vitro* by prostaglandin E. Nature 263: 777–779.

Santoro MG, Crisari A, Benedetto A and Amici C (1986) Modulation of the growth of a human erythroleukemic cell line (K562) by prostaglandins: Antiproliferative action of prostaglandin A. Cancer Res 46: 6073–6077.

Sardesai VM (1992) The essential fatty acids. Nutr Clin Practice 7: 179-186.

Schmid G (1990) Lipid metabolism of animal cells in culture-A review. In: Spier RE, Griffiths JB and Meignier B. (eds.) Production of Biologicals from Animal Cells in Culture. (pp. 61–66) Butterworth-Heinemann, Stoneham, MA.

Schmid G, Zilg H, Eberhard U and Johannsen R (1991) Effect of free fatty acids and phospholipids on growth of and product formation by recombinant baby hamster kidney (rBHK) and Chinese hamster ovary (rCHO) cells in culture. J Biotechnol 17: 155–167.

Shen TY and Winter CA (1977) Chemical and biological studies on indomethacin, sulindac and their analogs. Adv Drug Res: 12: 90–245.

Silberstein GB, Strickland P, Trumpbour V, Coleman S and Daniel CW (1984) *In vivo*, cAMP stimulates growth and morphogenesis of mouse mammary ducts. Proc Natl Acad Sci USA 81: 4950-4954.

Simmet T and Jaffe BM (1983) Inhibition of B-16 melanoma growth *in vitro* by prostaglandin D_2. Prostaglandins 25: 47–54.

Skouteris GG and McMenamin M (1992) Transforming growth factor-α-induced DNA synthesis and c-*myc* expression in primary rat hepatocyte cultures is modulated by indomethacin. Biochem J 281: 729–733.

Spector AA, Kiser RE, Denning GM, Koh S-WM and DeBault LE (1979) Modification of the fatty acid composition of cultured human fibroblasts. J Lipid Res 20: 536–547.

Spector AA, Mathur SN, Kaduce TL and Hyman BT (1981) Lipid nutrition and metabolism of cultured mammalian cells. Prog Lipid Res 19: 155–186.

Spector AA, Kaduce TL, Figard PH, Norton KC, Hoak JC and Czervionke RL (1983) Eicosapentaenoic acid and prostacyclin production by cultured human endothelial cells. J Lipid Res 24: 1595–1604.

Spector AA and Yorek M (1985) Membrane lipid composition and cellular function. J Lipid Res 26: 1015–1035.

Spector AA and Burns CP (1987) Biological and therapeutic potential of membrane lipid modification in tumors. Cancer Res 47: 4529–4537.

Sprecher H (1981) Biochemistry of essential fatty acids. Prog Lipid Res 20: 13–22.

Stampfer MR (1982) Cholera toxin stimulation of human mammary epithelial cells in culture. In Vitro 18: 531–537.

Stryer L (1988) Biochemistry. Freeman and Company, New York.

Suzuki K, Kobayashi N, Moriya Y, Abiko Y, and Suzuki H (1988) Inhibition of human gingival carcinoma cell growth by prostaglandins. Gen Pharmac 19: 273–276.

Takaoka T and Katsuta H (1971) Long-term cultivation of mammalian cell strains in protein- and lipid-free chemically defined synthetic media. Exp Cell Res 67: 295–304.

Taylor-Papadimitriou J, Purkis P and Fentiman IS (1980) Cholera toxin and analogues of cyclic AMP stimulate the growth of cultured human mammary epithelial cells. J Cell Phys 102: 317–321.

Tinoco J (1982) Dietary requirement and functions of α-linolenic acid in animals. Prog Lipid Res 21: 1–45.

Tiwari RK, Mukhopadhyay B, Telang NT and Osborne MP (1991) Modulation of gene expression by selected fatty acids in human breast cancer cells. Anticancer Res 11: 1383-1388.

Tobias LD and Hamilton JG (1979) The effect of 5,8,11,14-eicosatetraynoic acid on lipid metabolism. Lipids 14: 181–193.

Tocher DR (1993) Elongation predominates over desaturation in the metabolism of 18:3n-3 and 20:5n-3 in turbot (*Scopthalmus maximus*) brain astroglial cells in primary culture. Lipids 28: 267–272.

Utsumi K, Yamamoto G and Inaba K (1965) Failure of Fe^{2+}-induced lipid peroxidation and swelling in the mitochondria isolated from ascites tumor cells. Biochim Biophys Acta 105: 368–371.

Vane JR (1971) Inhibition of prostaglandin synthesis as a mechanism of action for aspirin-like drugs. Nature 278: 456–459.

Wicha MS, Liotta LA and Kidwell WR (1979) Effects of free fatty acids on the growth of normal and neoplastic rat mammary epithelial cells. Cancer Res 39: 426–435.

Willis AL and Smith DL (1989) Dihomo-gamma-linolenic and gamma-linolenic acids in health and disease. In: Spiller GA and Scala J (eds.) New Protective Roles for Selected Nutrients. (pp. 39–108) A.R. Liss, New York.

Wu J, Dent P, Jelinek T, Wolfman A, Weber MJ and Sturgill TW (1993) Inhibition of the EGF-activated MAP kinase signaling pathway by adenosine 3',5'-monophosphate. Science 262: 1065–1069.

Yang I, Guzman R, Richarda J, Imagawa W, McCormick K and Nandi S (1980) Growth factor- and cyclic nucleotide-induced proliferation of normal and malignant mammary epithelial cells in primary culture. Endocrinology 107: 35–41.

Yoneda T, Kitamura M, Ogawa T, Aya S and Sakuda M (1985) Control of VX_2 carcinoma cell growth in culture by calcium, calmodulin, and prostaglandins. Cancer Res 45: 398–405.

Address for offprints: W.M. Miller, Northwestern University, Department of Chemical Engineering, 2145 Sheridan Rd., 60208–3120, Evanston, Illinois, U.S.A.

Cytotechnology **15:** 51–56, 1994.

Enhancement effects of BSA and linoleic acid on hybridoma cell growth and antibody production

Masaki Kobayashi, Satoru Kato, Takeshi Omasa, Suteaki Shioya and Ken-ichi Suga
Department of Biotechnology, Faculty of Engineering, Osaka University, Yamadaoka 2-1, Suita, Osaka 565, Japan.

Key words: Hybridoma, bovine serum albumin, fatty acid, linoleic acid, antibody production rate

Abstract

The effects of linoleic acid and bovine serum albumin on hybridoma cell growth and antibody production were investigated. In dish cultivation, linoleic acid on its own promoted cell growth when used at concentrations below 50 mg L^{-1}, but strongly inhibited growth at a concentration of 100 mg L^{-1} on more. However, linoleic acid bound to bovine serum albumin did not inhibit cell growth, even at a concentration as high as 100 mg L^{-1}. Also, linoleic acid did not affect the specific antibody production rate, with or without bovine serum albumin. In order to elucidate the enhancement of antibody production by bovine serum albumin, fractions were prepared by ultrafiltration (98% molecular weight cut-offs, 50,000 and 17,000) and the effects of the fractionation on antibody production were studied in batch cultivation. The high-molecular-weight fraction ($\geqq$ 50,000) promoted antibody production whereas the low-molecular-weight fraction ($\leqq$ 17,000) inhibited it. In continuous cultivation, the high-molecular-weight fraction was also found to enhance antibody production.

Introduction

Improvements in culture techniques have enabled many cell lines to be cultured in serum-free media, which are supplemented with known growth factors instead of serum. Serum-free media used for hybridoma have generally been supplemented with insulin, transferrin, ethanolamine, selenite and bovine serum albumin (BSA) (Murakami *et al.*, 1982).

BSA consists of a single polypeptide chain of molecular weight 67,000 ± 2000 (Spector, 1975), and is normally supplemented in serum-free media as a protein source. The main roles of BSA in serum-free culture are as an antioxidant and as a carrier of long-chain fatty acids to the cells (Halliwell, 1985). Many researchers have reported on the roles of fatty acids and BSA in cell growth. Albumin was observed to protect cells against the toxic effects of fatty acids in free solution (Halliwell, 1988). All unsaturated fatty acids were found to be growth-promoting, while the saturated acids were growth inhibiting (Nilausen, 1978: Spiker-Polet and Polet, 1981). Though fatty acid-free albumin failed to support cell growth, albumin promoted growth when it was combined with a fatty acid (Yamane *et al.*, 1978). There have, however, been few quantitative investigations into the effects of BSA and fatty acids on cell growth and antibody production.

In a previous study (Omasa *et al.*, 1991), we found that BSA acted as a growth factor, enhancing the antibody production in hollow-fiber and perfusion cultures. Here we investigate the effects of BSA and a fatty acid (linoleic acid) on cell growth and antibody production quantitatively, and examine the reason for the enhancement of antibody production by BSA in hollow-fiber and perfusion cultures.

Materials and methods

Cell line, culture medium and sample analysis

The cell line employed in the experiments was the mouse-mouse hybridoma 3A21, which produces an anti-RNase A monoclonal antibody (IgG) (Omasa *et al.*, 1992a). The tissue culture medium was the serum-free, RDF-ITES with bovine serum albumin (BSA) (Murakami *et al.*, 1982). Cohn fraction V BSA (Sigma Chemical Co., St Louis, MO; A-4503) was used as the BSA-fatty acid combination. A BSA (Sigma A-6003) prepared from the A-4503, was used as the fatty acid-free BSA (Chen, 1967). Linoleic acid (Sigma L-1012) the most growth-promoting of the fatty acids (Rockwell *et al.*, 1980), was selected as the fatty acid to be used in the study. However, as the linoleic acid was not sufficiently soluble in a water at a high concentration, a stock solution was prepared by dissolving the acid in ethanol at 1.0 mg mL^{-1}.

The viable and total cell numbers, and the glucose, lactate and antibody concentrations were analyzed as previously described (Omasa *et al.*, 1992a, 1992b). The linoleic acid concentration was measured by gas chromatography (Hitachi Co. Ltd., Tokyo; 163) with FID using Shinchrom A (Chromato Packing Center Co. Ltd, Kyoto) coated with 5% Advance-DS (80–100 mesh, Chromato Packing Center) in a tubular glass column (length, 1m; inner diameter, 3 mm) (Johnson and Stocks, 1971). Nitrogen was employed as the carrier. Other operation conditions were as follows: inlet pressure of nitrogen, 0.8 kg cm^2; column temperature, 200 °C; injection-port temperature, 250 °C. Palmitic acid (Sigma P-5917) solution (0.333 mg L^{-1}) dissolved in ethanol was used as an internal standard, mixed at a ratio of 1:10 with the sample solution.

Dish and spinner-flask cultivation

A glass spinner flask with jacket (Shibata Haria Co. Ltd, Tokyo; Culstar 1609–500) was used for spinner culture. The working volume of the flask was 600 mL. The temperature was maintained at 37 °C and the agitation speed was 60 rpm. The medium was aerated at a flow rate of 40 mL min^{-1} using a permeable Teflon tube. The pH was controlled at 7.4 by on-off control of CO_2 gas.

Fractionation of BSA

The Cohn fraction V BSA (Sigma A-4503) was fractionated by thin-channel ultrafiltration equipment (Amicon, inc. Beverly, MA; TCF10) or hollow-fiber unit. The filter membrane of the thin-channel equipment was Advantec UK-20, with a 98% molecular weight cut-off of 50,000. The hollow-fiber unit, which was made of cellulose acetate, had a 98% molecular weight cut-off of 17,000. BSA solution (500 mL) was concentrated by ultrafiltration or the hollow-fiber unit, respectively, to give volumes of 100 mL. Each concentrated BSA solution was diluted 5-fold by distilled water and concentrated 5-fold again by ultrafiltration or the hollow-fiber unit, respectively. These concentrated solutions were used as BSA fractions with molecular weights of $\geqq$ 50,000 and $\geqq$ 17,000. The filtrate from the hollow-fiber unit was concentrated by a rotary evaporator and used as fractionated BSA (molecular weight $\leqq$ 17,000).

Results and discussion

Effect of linoleic acid on hybridoma cell growth

Many researchers have studied the effects of fatty acids on the growth of mammalian cells. Nilausen (1978) reported that saturated fatty acids had an inhibitory effect on Chinese hamster CHEF cell growth, but that unsaturated fatty acids, such as oleic and linoleic acid, stimulated the cell growth. Rockwell *et al.* (1980) showed that a combination of albumin and linoleic acid was most effective for the growth of Balb/c 3T3 cells, and Schmid *et al.* (1991) obtained a similar result with BHK and CHO cells. In a previous study, the serum-free RDF medium was developed for hybridoma cultivation (Omasa *et al.*, 1991, 1992a, 1992b). Here, we employed a fatty acid, linoleic acid

$$(CH_3(CH_2)_4CH = CHCH_2CH = $$
$$= CH(CH_2)_7 - COOH).$$

Using dish cultivation, we first investigated the effect of the linoleic acid concentration (0.02–200 mg L^{-1}) on cell growth. Fig. 1 shows the initial specific growth and death rates under various linoleic acid concentrations without BSA. At a linoleic acid concentration of about 0.5 mg L^{-1} the specific growth rate showed the maximum value; maximum cell number was also obtained at 10 mg L^{-1} concentration. In the case of

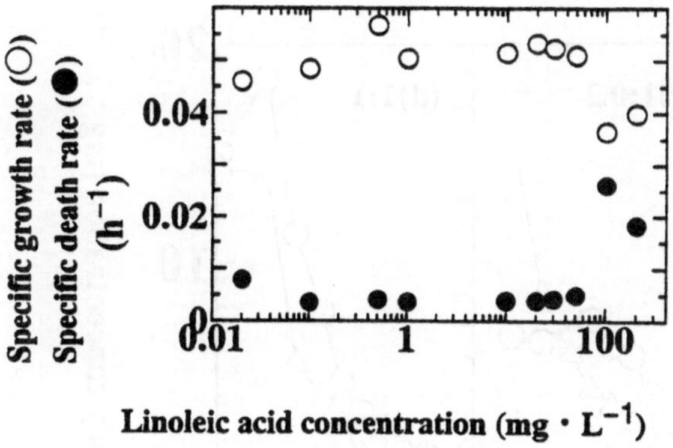

Fig. 1. Effect of linoleic acid on specific growth and death rates (without BSA).

CHEF cells, the maximum cell number was obtained at about 2 mg L^{-1} (Nilausen, 1978). Thus, the optimal concentration of linoleic acid with the 3A21 cells used here was higher than that with CHEF cells. At linoleic acid concentrations of 100 mg L^{-1} on more, the specific growth rate decreased while there was a rise in the specific death rate. A similar result was obtained with the CHEF cell line using linoleic and oleic acids (Nilausen, 1978). The reason for this inhibitory effect is not clear, but it may be due to the oxidation or surfactant activity of linoleic acid or the effect of ethanol concentration. Fatty acids are mainly utilized as energy sources and building blocks for cells (Schmid, 1991). In this experiment, since neither the specific glucose consumption nor the antibody production rates were affected by the linoleic acid concentration (data not shown), the linoleic acid is thought not to have affected the glucose metabolic pathway, i.e., it was not mainly used as an energy source for the cells. Bandyopadhyay *et al.* (1987) reported that linoleic acid was metabolized to arachidonic acid and prostaglandin E2, enhancing the proliferation of mammary epithelial cells. Since the relationship between linoleic acid and prostaglandin E2 in hybridoma 3A21 is not clear, further experiments are necessary to elucidate this point. BSA is an antioxidant and acts as a carrier of long-chain fatty acids to the cells in a culture (Halliwell, 1988). We therefore investigated the effect of linoleic acid and BSA on hybridoma cell growth.

Effect of linoleic acid bound to BSA on cell growth

It has been reported that fatty acid bound to BSA was stably maintained during cultivation (Spector, 1975). First, we investigated cell growth under various ratios of fatty acid-free BSA and linoleic acid. The time courses of batch cultivations are shown in Fig. 2. In this experiment, the linoleic acid concentration was fixed at 100 mg L^{-1} and the BSA concentration was varied in order to change the BSA to linoleic acid mole ratio. Without the addition of BSA cell growth was strongly inhibited as shown in Fig. 1, but in the medium containing 24 g L^{-1} fatty acid-free BSA and 100 mg L^{-1} linoleic acid (mole ratio, 1:1), cell growth was not inhibited. However, cell growth was inhibited at concentrations of 100 mg L^{-1} linoleic acid and less than 4.8 g L^{-1} fatty acid-free BSA (BSA/linoleic acid $\leqq 0.2$ (mole/mole)). In the case of human plasma albumin, five strong fatty acid binding sites exist in one HSA molecule (HSA/maximum acid = 0.2 mole/mole) (Spector, 1975). In the case of the higher linoleic acid to BSA mole ratio, all the linoleic acid may be unable to bind to the BSA, and the free linoleic acid may inhibit cell growth. BSA may have the role of detoxification of linoleic acid.

To elucidate the effect, if any, of BSA and fatty acid on hybridoma cultivation, the cell growth and antibody concentration were investigated under much less ratio of linoleic acid to BSA. In order to investigate the higher ratio of BSA, the linoleic acid concentration should be decreased compared with that in Fig. 2. Where, we fixed the linoleic acid concentration to 1.02 mg L^{-1} Fig. 3 shows the effects of mole ratio of linoleic acid to BSA on cell growth; The maximum cell concentration increased with increase of BSA concentration. The specific growth rate hardly changed with the different mole ratio of linoleic acid to fatty acid-free BSA. The specific antibody production rate also didn't change. Therefore, linoleic acid did not affect the antibody production.

Enhancement of antibody production by fractionated BSA

Linoleic acid did not affect antibody production. However, in a previous study (Omasa *et al.*, 1991) growth factor BSA enhanced the antibody production rate in the hollow-fiber which system have the two types of hollow-fiber unit and could keep the cell and high molecular weight growth factor inside, and in the perfusion cultures. BSA was directly fed into EC side of the hollow-fiber system. In these hollow-fiber and perfusion systems, a membrane (98% molecular weight cut-off, 17,000) was used for the separation of low-molecular-weight materials. From the results, we supposed that BSA may contain constituents which pro-

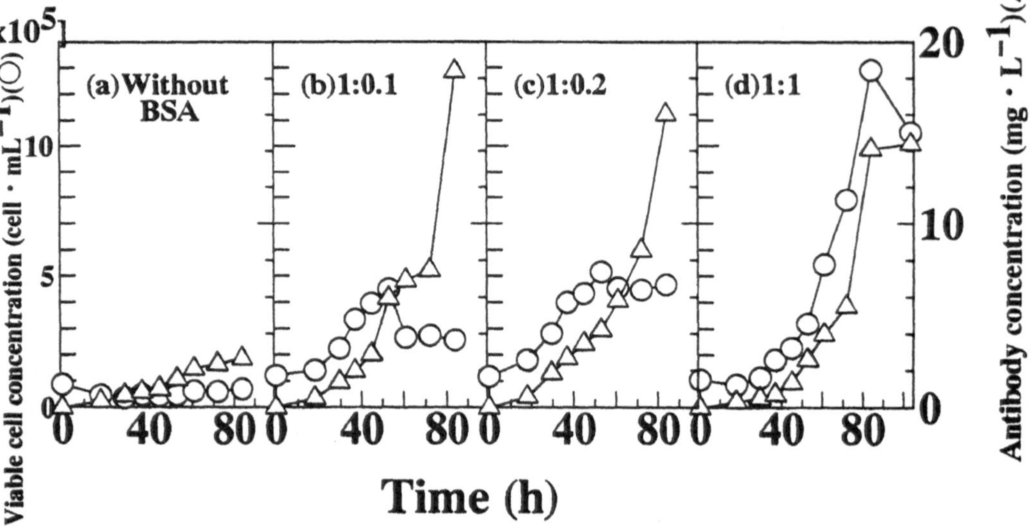

Fig. 2. Time courses of batch cultures using various ratios of linoleic acid and fatty acid-free BSA (No. 1). (The added linoleic acid concentration was fixed at 100 mg L^{-1}), (a) without BSA (b) 1:0.1 (c) 1:0.2 (d) 1:1 (linoleic acid: BSA mole ratio).

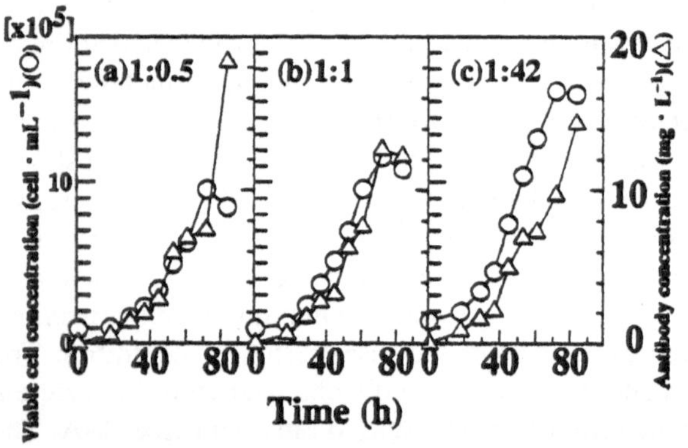

Fig. 3. Time courses of batch cultures using various ratios of linoleic acid and fatty-acid free BSA (No. 2). (The added linoleic acid concentration was fixed at 1.02 m gL^{-1}), (a) 1:0.5 (b) 1:1 (c) 1:42 (linoleic acid: BSA mole ratio).

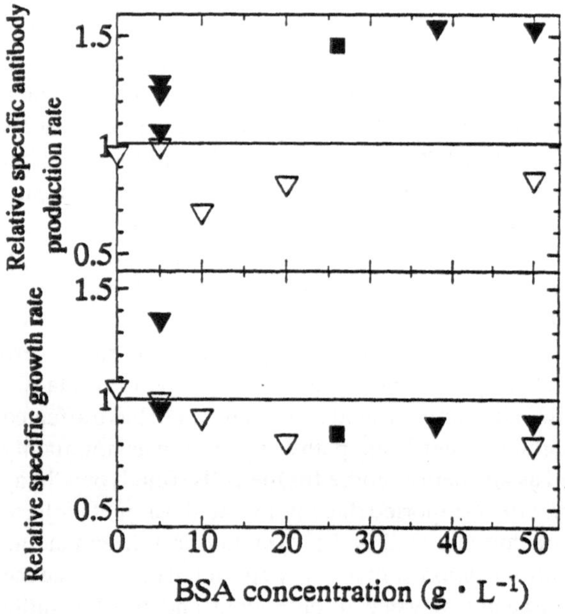

Fig. 4. Effect of BSA concentration on relative specific growth and antibody production rate (standard BSA 5 g L^{-1} = 1.0), △ standard BSA (Cohn fraction V), ■ Fraction 2 BSA (M.W. ≧ 17,000), ▲ Fraction 4 BSA (M.W. ≧ 50,000). The specific growth and antibody production rates were normalized using the standard cultivation value (BSA concentration = 5 g L^{-1}).

mote and/or reduce antibody production and that these reducible constituents could be removed from the culture medium by a molecular weigth cut-off membrane. Here, we investigated the effects of fractionation on antibody production in batch cultivation by separating the Cohn fraction V BSA into four fractions, Fractions 1 and 2 with molecular weights of ≦ 17,000 and ≧ 17,000, respectively, using the hollow-fiber unit, and Fractions 3 and 4 with molecular weights of ≦ 50,000 and ≧ 50,000, respectively, using an ultrafiltration unit.

The effects of the concentration of the Cohn fraction V BSA and Fractions 2 and 4 on the specific growth and antibody production rates are shown in Fig. 4. In the case of 5 g L^{-1} of fraction 4 BSA concentration, the data of specific growth rate was scattered. However, it can been seen that there were no differences with respect to the effect of the BSA concentration

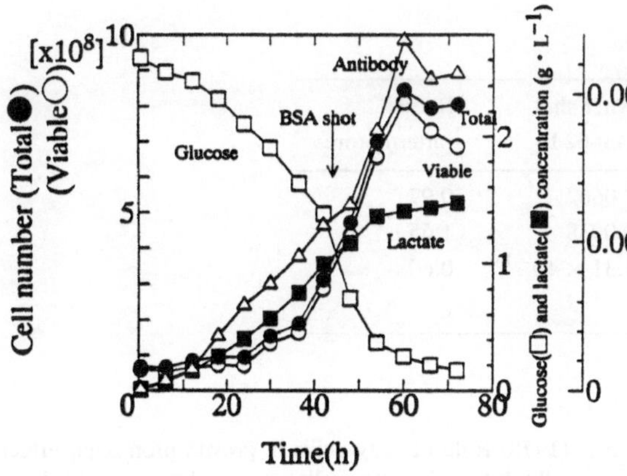

Fig. 5. Time courses of spinner batch culture with fractionated BSA supplement (Fraction 1 BSA (M.W. ≦ 17,000) was added at 42 h).

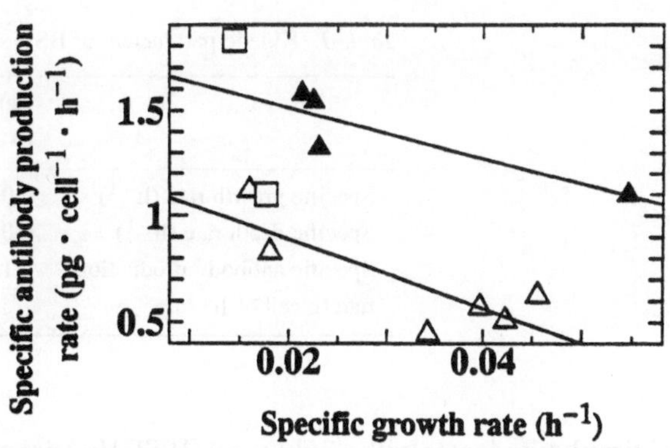

Fig. 6. Relationship between specific growth rate and specific antibody production rate in glucose-limited continuous culture, □ without BSA, △ standard BSA (Cohn fraction V), ▲ Fraction 4 BSA (M.W. ≧ 50,000)

on the specific growth rate among the Cohn fraction V BSA, Fraction 2 (molecular weigth ≧ 17,000) and Fraction 4 (molecular weigth ≧ 50,000). While, the specific antibody production rate gradually decreased with increasing BSA concentration in the case of the Cohn fraction V BSA, whereas it increased 50% in the case of Fractions 2 and 4 of BSA. These results indicate that Cohn fraction V BSA might contain a substance or substances that promote and/or reduce antibody production.

Batch cultivation with shot addition of low-molecular-weight fractionated BSA

The BSA fractionation experiments indicated that a low-molecular-weight (≦ 17,000) fraction may reduce antibody production. To check this assumption, BSA Fraction 1 (molecular weight ≦ 17,000) was prepared by concentration of the filtrate using a rotary evaporator. Using spinner-flask batch cultivation, this fraction was then fed into the culture medium in the mid-growth phase and the specific antibody production rates were compared before and after the addition. Fig. 5 shows the time course of spinner-flask batch cultivation with Fraction 1; the cell growth seemed not to be affected by the shot addition of the fraction. The kinetic parameters of this cultivation are shown in Table 1. As can be seen from the table, the specific growth rate did not change after the addition, whereas the specific antibody production rate decreased 60%. In the normal batch cultivation without fraction 1 BSA, the specific growth and antibody production rates are kept constant over 70 hours. This suggests that Fraction 1 might contain a substance(s) that reduces antibody production.

Enhancement of antibody production in glucose-limited continuous cultivation with addition of high-molecular-weight fraction

Using batch cultivation, we also showed the enhancement effect of BSA on antibody production. Fraction 4 (molecular weight ≧ 50,000) improved the specific antibody production rate. However, the specific antibody production rate is closely related to the specific growth rate (Omasa *et al.*, 1992a). Therefore, in order to confirm the enhancement of antibody production, the effect must be investigated taking into consideration the relationship between the specific growth and antibody production rates. As the specific growth rate can be controlled by the dilution rate in glucose-limited continuous culture, the specific antibody production rate could be investigated under various specific growth rates. We investigated the effect of the concentration of Fraction 4 on the specific antibody production rate.

Fig. 6 shows the relationship between the specific growth and antibody production rates in glucose-limited continuous cultivation. In our earlier work (Omasa *et al.*, 1992a), we found a 'trade-off' relationship existed between the specific antibody production rate and specific growth rate in fed-batch cultivation. A similar trade-off relationship between the two rates was also found to exist in the continuous cultivation conducted in the present experiment (Fig. 6). When the high-molecular-weight BSA fraction (≧ 50,000) was added, the specific antibody production rate increased. Fraction 4 enhanced only the specific antibody production rate without affecting the specific growth rate. Sugahara *et al.* (1991a) reported that an immunoglobu-

Table 1. Kinetic parameters of BSA shot batch culture

	Before shot (0–42 h)	After shot (43–62 h)	Ratio (after/before)
Specific growth rate (h^{-1})	0.0619	0.0602	0.97
Specific death rate (h^{-1})	0.0023	0.0038	1.65
Specific antibody production rate (g $cell^{-1}$ h^{-1})	2.15×10^{-13}	1.31×10^{-13}	0.60

lin stimulation factor with a 36kD unit (IPSF IIα) exists in Namalwa cells. Yamada *et al.* (1991) also reported that various types of caseins and their protease digests stimulated the immunoglobulin production of human-human hybridoma. It was not clear whether the antibody production enhancement was actually due to the BSA or to an other factor in the Cohn fraction V BSA. Since the immunoglobulin-stimulating factor IPSF IIα had glyceraldehyde-3-phosphate dehydrogenase activity (Sugahara *et al.*, 1991b), further investigation of the glyceraldehyde-3-phosphate dehydrogenase activity in fractionated BSA should be conducted.

In conclusion, linoleic acid was found to be growth promotive, and its binding to the BSA was an important factor for cell growth. Linoleic acid alone did not affect antibody production, but Cohn fraction V BSA is believed to contain constituents that promote or reduce it. In continuous cultivation, the addition of a high-molecular-weight BSA fraction was effective for antibody production. As a next step, it will be of benefit to investigate the enhancement factor in Cohn fraction V BSA.

References

Bandyopadhyay GK, Imagawa W, Wallace D and Nandi S (1987) Linoleate metabolites enhance the *in vitro* proliferative response of mouse mammary epithelial cells to epidermal growth factor. J. Biol. Chem. 262: 2750–2756.

Chen RF (1967) Removal of fatty acids from serum albumin by charcoal treatment J. Biol. Chem. 242: 173–181.

Halliwell B (1988) Albumin – An important extracellular antioxidant? Biochem. Pharmacol. 37: 569–571.

Johnson AR and Stocks RB (1971) Gas-liquid chromatography of lipids. In: Biochemistry and Methodology of Lipids, pp. 195–218.

Murakami H, Masui H, Sato GH, Sueoka N, Chow TP and Sueoka TK (1982) Growth of hybridoma cells in serum-free medium: Ethanolamine is an essential component. Proc. Natl. Acad. Sci. USA 79: 1158–1162.

Nilausen K (1978) Role of fatty acids in growth-promoting effect of serum albumin on hamster cells *in vitro*. Journal of Cellular Physiology 96: 1–14.

Omasa T, Kobayashi M, Nishikawa T, Shioya S, Suga K, Uemura S, Kitani Y, and Imamura Y (1991) Hybridoma culture in hollow-fiber system – The effects of growth factors –. In: R. Sasaki and K. Ikura (eds.) Animal cell culture and production of biologicals. (pp. 229–236) Kluwer Academic Publishers, Dordrecht.

Omasa T, Higashiyama K, Shioya S, and Suga K (1992a) Effects of lactate concentrations on hybridoma culture in lactate-controlled fed-batch operation Biotechnol.Bioeng. 39: 556–564.

Omasa T, Ishimoto M, Higashiyama K, Shioya S, and Suga K (1992b) The enhancement of specific antibody production rate in glucose- and glutamine-controlled fed-batch culture. Cytotechnology 8: 75–84.

Rockwell GA, Sato GH, and McClure DB (1980) The growth requirements of SV40 virus transformed Balb/c-3T3 cells in serum-free monolayer culture. Journal of Cellular Physiology 103: 323–331.

Schmid G, Zilg H, Eberhard U, and Johannsen R (1991) Effect of free fatty acids and phospholipids on growth of and product formation by recombinant baby hamster kidney (rBHK) and Chinese hamster ovary (rCHO) cells in culture. J. of Biotechnology 17: 155–167.

Spector AA (1975) Fatty acid binding to plasma albumin. J. Lipid Res. 16: 165–179

Spieker-Polet H, and Polet H (1981) Requirement of a combination of a saturated and an unsaturated free fatty acid and a fatty acid carrier protein for *in vitro* growth of lymphocytes. J. Immunol. 126; 949–954.

Sugahara T, Shirahata S, Yamada K and Murakami H (1991a) Purification of immunoglobulin production stimulating factor IIα derived from Namalwa cells. Cytotechnology 5: 255–263.

Sugahara T, Shirahata S, Akiyoshi K, Isobe T, Okuyama T and Murakami H (1991b) Immunoglobulin production stimulating factor-IIα (TPSF-IIα) is glyceraldehyde-3-phosphate dehydrogenase like protein. Cytotechnology 6: 115-120.

Yamada K, Ikeda I, Nakajima H, Shirahata S and Murakami H (1991) Stimulation of proliferation and immunoglobulin production of human-human hybridoma by various types of caseins and their protease digests. Cytotechnology 5: 279–285.

Yamane I, Murakami 0, and Kato M (1978) Role of bovine serum albumin in a serum-free suspension cell culture medium (38823). Proc. Soc. Exp. Biol. Med. 149: 439-442.

Address for offprints: T. Omasa, Department of Biotechnology, Faculty of Engineering, Osaka University, Yamadaoka 2-1, Suita, Osaka 565, Japan.

Cytotechnology **15**: 57–64, 1994.

Growth rate suppression of cultured mammalian cells enhances protein productivity

Kazunari Takahashi, Satoshi Tereda, Hiroshi Ueda, Fusao Makishima and Eiji Suzuki
Department of Chemistry and Biotechnology, Faculty of Engineering, The University of Tokyo, Hongo, Bunkyo-ku, Tokyo, 113 Japan

Key words: Enhancement of protein production, growth suppression, cell differentiation, mammalian cell culture, mRNA stability

Abstract

Suppression of proliferation of cells which contain stable or stabilized mRNA coded for a protein to be produced, a partial mimic of cell differentiation, was examined for enhancing protein production by cultured mammalian cells. Hybridoma 2E3 cells which were adapted to be interleukin-6 sensitively growth-suppressed accumulated the mRNA of IgG_1 which is reported stable, and IgG_1 production rate increased as a result when their growth was suppressed with interleukin-6. A myeloma cell line was similarly adapted; the obtained myeloma cells can be used as host cells for enhancing production of exogenous proteins by suppressing growth with interleukin-6. Temperature-sensitively growth-suppressible mutants of mouse mammary carcinoma FM3A were transfected with cDNA of IgM λ_1 chain and cultured at nonpermissive temperature to enhance production of λ_1. Addition of various growth-suppressive reagents to culture medium was studied for finding methods suitable for suppressing growth while maintaining high cell viability. Caffeine yielded the best results among these reagents. Deprivation of various growth-supporting components in culture medium was also tested; simultaneous deprivation of insulin and transferrin viably suppressed growth of hybridoma 2E3 cells, resulting in enhanced antibody productivity.

Abbreviations: IL6 – recombinant human interleukin-6; TGF-β – recombinant human TGF-β_1; X63.653-P3X63 – Ag8.653 myeloma

Introduction

Production rate q of a specific protein by a mammalian cell was represented by eq. (1) as a first approximation (Savinell *et al.*, 1989; Suzuki and Ollis, 1990)

$$q = \frac{\alpha \cdot \beta \cdot D \cdot E}{\mu + k} \qquad (1)$$

Here, α is the coefficient of mRNA synthesis rate equation assumed as the first order of the number of the corresponding gene molecule; β is the coefficient of mRNA translation rate equation assumed as the first order of the number of the mRNA molecule; D is the copy number of the gene in a cell; E is the efficiency for splicing; μ is the specific growth rate of mammalian cells; and k is the coefficient of the mRNA decomposition rate equation assumed as the first order of the number of the mRNA molecule. The first order equation for mRNA decomposition rate was proposed by Kafatos as a frst approximation (Kafatos, 1972), and supported by several measured rates of mRNA decay (Koeller *et al.*, 1991; Brock and Shapiro, 1983). Equation (1) was modeled for simulating production rate per mammalian cell of a protein that is stable in cells and constitutively secreted out; immunoglobulin produced by hybridomas and many of exogenous proteins designed to be secreted out by adding signal peptides are examples of such proteins. The equation is applicable only when ribosomes and the other components required for protein synthesis are abundant in cells and accordingly their concentrations are not rate limiting factors. The equation indicates that the protein produc-

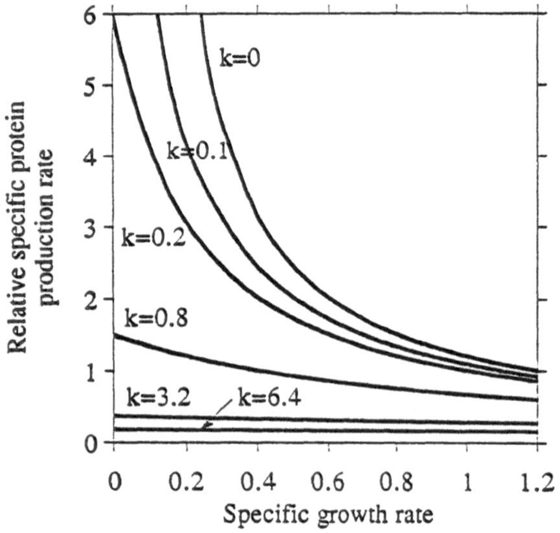

Fig. 1. Effect of the growth rate of cells and the stability of mRNA on production rate per cell of the protein coded in the mRNA. Production rates of the protein per cell were estimated by eq. (1) at various values of μ(day^{-1}) and k(day^{-1}), the specific growth rate and the coefficient for mRNA decomposition rate equation, and were plotted as ratio to that at μ=1.2 and k=0.

tion rate increases with decrease in μ and k, the specific growth rate and the coefficient of mRNA decomposition rate equation, as shown in Fig. 1. In other words, cells can be specialized to produce a specific protein preferably when the cell deivision is suppressed if the mRNA coded for the protein is stable, or stabilized. The minimization of $\mu + k$ may be a simplified, partial mimic of mechanism of cell differentiation (Carnerio and Schibler, 1984), although the actual mechanism is reportedly much more complicated and diverse.

We tried to establish methods for minimizing μ+k to maximize the protein production rate. First, aiming at minimization of μ, we examined about twenty methods for suppressing cell-growth viably and reversibly; the desirable methods should be inexpensive, easy to apply, and not inhibitory for protein synthesis. Such growth-suppression methods are useful not only for enhancing protein production rate, but also for prolonging effective production period of a batch culture by preventing total cell death due to over-growth. The methods we examined are categorized into the following four groups. The first is cytokine-sensitive growth-suppression using IL6. We adapted a mouse myeloma X63.653, which is widely used as fusion partners of hybridomas and as host cells to produce exogenous proteins, to be IL6- sensitively growth-suppressed. The second is temperature-sensitive growth-suppression using ts mutants of mouse mammary carcinoma cell line FM3A, tsFT210 and tsFT20 which grow fast at 33 °C and not at 39 °C (Murakami *et al.*, 1985). We transfected tsFT210 cells with cDNA of immunoglobulin λ_1 chain, cultured them at 33 °C, suppressed their growth at 39 °C and examined production rate of the protein. The third is addition of growth-suppressible reagents such as caffeine. The fourth is limitation of nutrients or growth factors.

Next, we examined the effect of minimization of k, in other words, the effect of stability of mRNA. Equation (1) represents the following notion: stable mRNAs accumulate in cells when the cells do not at all or quite slowly divide, because the mRNAs are incessantly formed, less frequently distributed to newly divided cells, and stable enough to decompose slowly; as a result, the production rate per cell of the protein coded in the mRNAs significantly increases as μ decreases, if k is about or less than 1.0 day^{-1} as shown in Fig. 1. For confirming this notion, we determined the concentration of immuniglobulin λ_1 chain mRNA, which was stable enough to satisfy the above conditions (Kuehl, 1977; Storb, 1973), in growth-suprpessed hybridoma cells.

Materials and methods

Reagents and culture medium

Reagents used were recombinant interleukin-6 (5 × 10^6 unit mg^{-1}, purity over 98%, Wako Junyaka, Tokyo), recombinant human TGF-β_1 (King Brewing, Kakogawa, Japan), caffeine (Sigma), and actinomycin D (Sigma). Culture media used were serum-free media ASF103 (Ajinomoto, Tokyo) and SF-O (Sanko Junyaku Co., Tokyo), RPMI-1640 select amine kit (Sigma), and a serum replacement, CPSR-3 (Sigma).

Cells and adaptation

Mouse myeloma X63.653 cells were provided from Japanese Cancer Research Resources Bank (JCRB). Mouse mammary carcinoma FM3A cells and their temperature-sensitive mutants, tsFT20 and tsFT210 cells, were kind gifts from Dr. F. Hanaoka (The Institute of Physical and Chemical Research, Wako, Japan) (Murakami *et al.*, 1985). Cell line 2E3 was a hybridoma established in the authors laboratory by fusing myeloma P3X63 AG8U.1 with BALB/c mouse spleen cells secreting IgG_1 specific to a trinitrophenyl (TNP)-hapten (Makishima *et al.*, 1992).

X63.653 cells were cultured in SF-O medium for eight weeks until they became growth-suppressible by IL6. During the adaptation, 0 to 100 U ml^{-1} of IL6 was added to parts of the culture, and the growth rates of the cells were determined by cell counting. Hybridoma 2E3 cells were similarly adapted to be growth-suppressible by IL6 in ASF 103 medium.

Hybridoma 2E3 cells were adapted for three days in glucose-deficient RPMI-1640 medium supplemented with CPSR-3 at 10%; the adapted 2E3 cells could grow at a normal rate (μ=1.0 day^{-1}) in the glucose-deficient medium.

Construction of plasmids

The cDNA of mouse IgM light chain λ_1 was prepared by reverse transcription followed by PCR amplification from the mRNAs which were extracted from mouse myeloma J558L cells as described elsewhere. The SalI-NotI fragment of λ_1 cDNA was inserted into pBCMGSNeo (Karasuyama *et al.*, 1990) to construct BCMG$\lambda_1$neo. The plasmid carrying chloramphenicol resistant gene, pHSG397 (JCRB-*Gene*), was digested with HindIII and inserted into pBCMG$\lambda_1$neo replacing its HindIII fragment containing BPV69% to construct pCMVλ_1.

DNA transfection

FM3A, tsFT20 and tsFT210 cells were transfected with pCMVλ_1 by electroporation, and then selected at 1 mg ml^{-1} of G418 (Sigma).

Determination of mRNA and protein

Total RNA was extracted by the guanidine isothiocyanate/hot phenol method from 5 × 10^6 of 2E3 cells, and Northern-blotted; λ_1 mRNA was determined by using ^{32}P-labeled DNA fragments of C region of λ_1 chain. IgG_1 and λ_1 in culture supernatants were determined by ELISA.

Flow cytometry

Cells were washed with PBS twice, fixed in 70% ethanol at 4 °C for 2 h, washed with PBS twice, treated with RNase A at 500 mg ml^{-1} at 37 °C for 1 h, washed with PBS three times, stained with propidium iodide at 50 mg ml^{-1}1 for 10 min, and then analyzed with the flow cytometer (EPICS-CS, Coulter, Florida) using 488 nm laser beam.

Results and discussion

Cytokine-sensitive growth regulation

X63.653 cells were usually not sensitive to IL6 when they were cultured in medium supplemented with serum. They, however, became IL6-sensitively growth-suppressible after eight week adaptation in serum-free SF-O medium; then single clone of the IL6-sensitively growth-suppressible X63.653 cells was selected from the culture by limiting dilution. Addition of IL6 at 10 U ml^{-1} prolonged their doubling time to 50 h from 20 h of control (Fig. 2a) while the cell viability was maintained at 80–90% as high as that of IL6-free control culture (Fig. 2b). The adapted cell line can be used as fusion partners for obtaining IL6-sensitively growth-suppressible hybridomas or as host cells for producing exogenous proteins. 2E3 cells were similarly adapted to be IL6-sensitively growth-regulated (Makishima *et al.*, 1992). The cell cycle analysis of 2E3 cells by flow cytometry showed that the cell population in G1/G0 phase increased to 48% from 28% of control by growth-suppression with IL6. 2E3 cells were also TGF-β-sensitively growth-regulated in SF-O medium without adaptation (Table 1).

Since the mRNA of IgG_1 is stable enough to have k between 0.2 and 1.8 day^{-1} (Kuehl, 1977; Storb, 1973), IgG_1 production rate of 2E3 cells was enhanced fivefold by the growth-suppression induced by IL6 as eq. (1) predicted (Table 1). However, the production was not enhanced when the growth was suppressed by TGF-β (Table 1). Althopugh the reason why the growth-suppression induced by TGF-β did not enhance IgG_1 production rate was unknown, the growth-suppression through slowed protein synthesis such as that induced by limitation of amino acids or addition of cycloheximide also did not increase the specific antibody production rates as reported later in this work. We, therefore, presumed that TGF-β simultaneously suppressed IgG_1 synthesis as well as the cell growth. In other words, the enhancing effect of reduced μ in the denominator of eq. (1) was approximately cancelled by that of reduced β in the numerator.

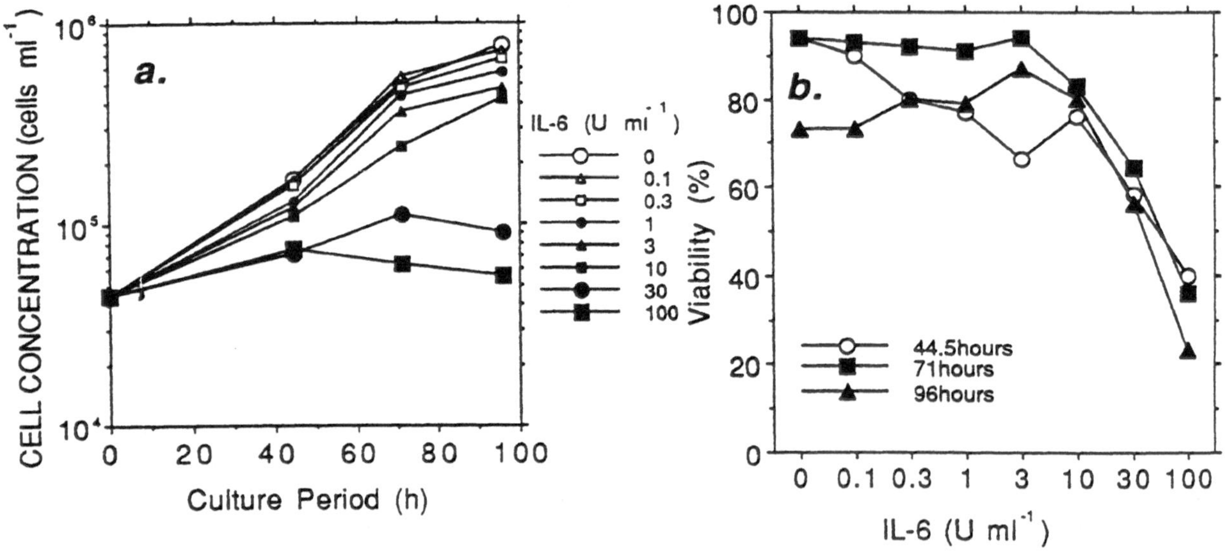

Fig. 2. Culture of myeloma X63.653 adapted to be IL6-sensitively growth-suppressible. a) growth, b) cell viability.

Table 1. Antibody productivity of hybridomas growth-suppressed by reagents

Reagents Concentrations for optimum growth suppression	Hybridomas	Optimum growth suppression (μ/μ normal)	Enhancement of antibody productivity (q/q normal)
Caffeine 1.92mM	2E3	0.22	2.8
Sodium n-butyrate[a] 1.2mM	VII H-8	0.39	2.4
Potassium acetate[a] 20mM	VII H-8	0.50	2.0
Thymidine 15mM	2E3 VII H-8[a]	0.30	2.0
Hydroxyurea[a] 0.1mM	VII H-8 VII H-8[a]	0.22 0.22	2.3 2.3
c-AMP[a] 0.3mM	VII H-8	0.82	1.3
Cycloheximide[a]	VII H-8	no-optimum	no-enhance
Actinomycin D	2E3, VII H-8	no-optimum	no-enhance
Interleukin-6	2E3	0.4	5.0
TGF-β	2E3	no-optimum	no-enhance

[a] Suzuki, 1990

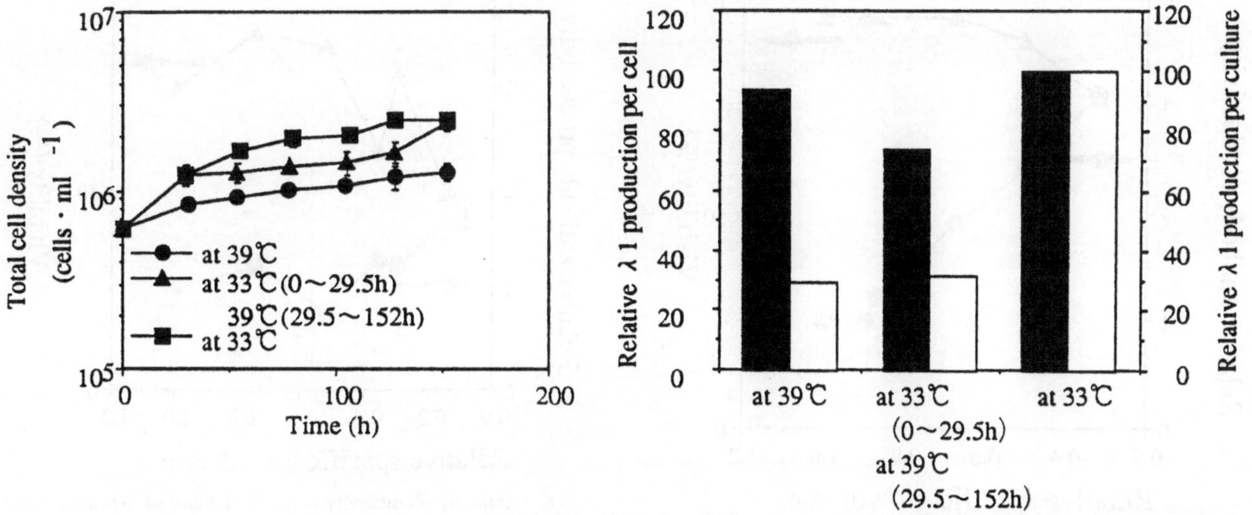

Fig. 3. Culture of tsFT210, temperature-sensitive mutant of FM3A, transfected with λ DNA. Cultured at nonpermissive (39 °C) or permissive (33 °C) temperature, or at the combination of the two temperatures. a) growth, b) λ_1 production rate per cell (solid bar) and total production per batch of culture (open bar).

Temperature-sensitive growth-regulation

The temperature-sensitive mutants, tsFT210 and tsFT20, grow at 33 °C but not at 39 °C. Since the cdc2 kinase of tsFT210 cells is temperature-sensitively deactivated at 39 °C, the cells are arrested at the G2 phase of the cell cycle at 39 °C (Yasuda *et al.*, 1991). On the other hand, tsFT20 cells are arrested at the border of the G1 and S phases at 39 °C, since DNA polymerase α of the cells is temperature-sensitively deactivated at 39 °C (Murakami *et al.*, 1985).

tsFT210 cells which had been transfected with λ_1 cDNA were cultured at 33 °C to about 10^6 cells ml^{-1} ands then growth-suppressed by raising culture temeprature to 39 °C; the λ_1 production was determined by ELISA. Fig. 3a shows the growth curves, and Fig. 3b does the production rate per cell and the product amount of the culture batch in comparison with two controls, one cultured always at 33 °C and the other at 39 °C. No enhancement of the λ_1 protein production was observed in the culture which was growth-suppressed at 39 °C. This seemed to be due to the fact that tsFT210 cells immediately started dying when their growth was suppressed at 39 °C; tsFT20 cells died further rapidly at 39 °C.

Addition of growth-suppressive reagents

Table 1 summarized the antibody productivities of hybridoma cells which were growth-suppressed by adding various reagents or cytokines to medium. Seven of the ten reagents or cytokines enhanced antibody productivity while suppressing the hybridoma growth. Among them, caffeine, a relatively inexpensive reagent, suppressed the growth of hybridoma 2E3 cells without significant decrease of cell viability (viability is over 90% between μ=0.55 and 1.0 day^{-1}, about 80% between μ=0.2 and 0.55, and 55% at μ=0), and enhanced antibody production rate per cell threefold (Fig. 4). The IgG_1 productivity of the cells which were completely growth-suppressed (μ=0) was lower than the maximum productivity observed

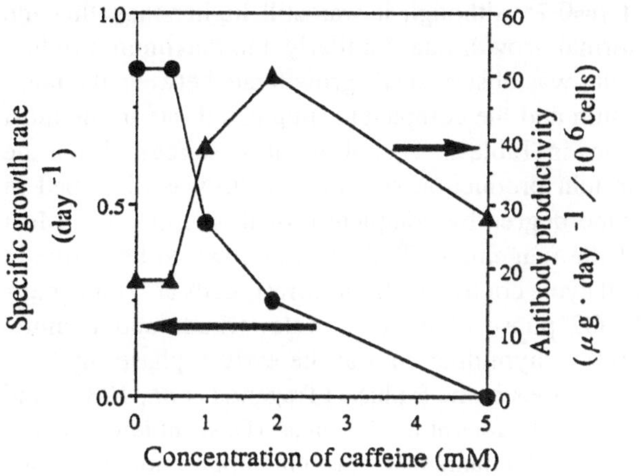

Fig. 4. Antibody productivity of hybridoma 2E3 whose growth was suppressed by caffeine.

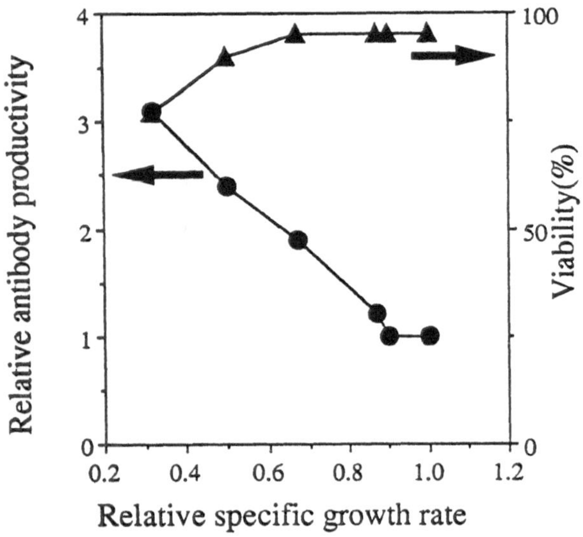

Fig. 5. Antibody productivity and cell viability of hybridoma 2E3 whose growth was suppressed by partial deprivation of insulin and transferrin.

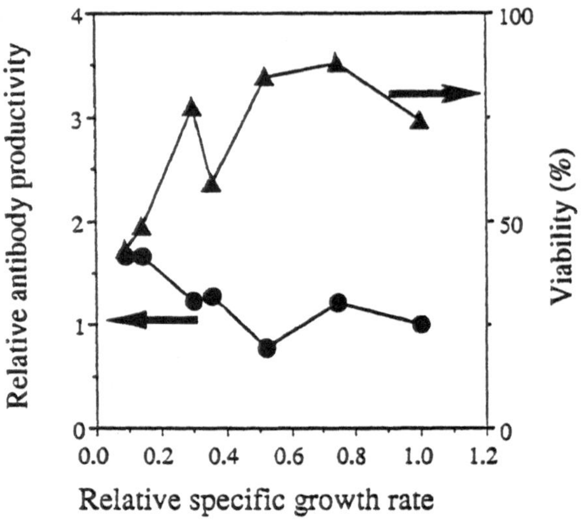

Fig. 6. Antibody productivity and cell viability of hybridoma 2E3 whose growth was suppressed by partial deprivation of glutamine in glucose-free medium.

at μ=0.25 although it was still higher than that the normal growth rate. Similarly, the maximum productivity was observed at a growth rate between the maximum and the completely suppressed rate in the most cases in Table 1. This observation suggested that the protein production system of cells was disturbed in some degree by complete growth-suppression. Most of the reagents in Table 1 are known to be active in cell cycle control. Caffeine arrests cells at G1 and partly G2 phase of the cell cycle (Okuda and Kimura, 1988), thymidine does at the early S phase, hydroxyurea does at the S phase (Traganos *et al.*, 1982), and n-butyrate does at the G1 phase (Darzynkiewicz *et al.*, 1981). Mechanisms of growth-suppression by potassium acetate and cyclic AMP are unclear. Cycloheximide and actinomycin D were examined as examples of undesriable reagents for enhancing protein production because they suppress cell-growth by inhibiting protein synthesis. Actinomycin D is an inhibitor of RNA synthesis (Darzynkiewicz *et al.*, 1981), and cycloheximide is an inhibitor of peptide chain elongation (Sonenshein and Brawerman, 1976). Both of the reagetns did not enhance IgG production when they suppressed the growth (Table 1).

Deprivation of growth-supporting components

Growth factors. Growth of hybridoma 2E3 cells was suppressed by partial deprivation of growth factors, insulin and transferrin, from the serum-free medium ASF103 which contained each of the proteins at 5 mg l^{-1} for standard use; the concentrations of the two growth factors were evenly reduced. The cell viability was higher than 75% at all the growth rate above 30% of the normal rate (μ=1.0 day^{-1}). The IgG_1 productivity increased threefold while the growth rate decreased from 100% to 30% of the normal rate as the concentrations of the growth factors were reduced from 5 mg l^{-1} to 1.5 mg l^{-1} each as shown in Fig. 5.

Glutamine. Since glutamine is a principal energy source for cell growth in a glucose-deficient culture, it was anticipated that mild glutamine limitation in glucose-deficient medium suppresses growth while enhancing protein productivity (Suzuki *et al.*, 1992), To realize this anticipation, 2E3 cells which had been adapted to grow at a normal rate (μ=1.0 day^{-1}) in glucose-deficient medium containing glutamine at 2.05 mM were cultured at various glutamine concentrations from 0.04 mM to 2.05 mM in the glucose-deficient medium; the cells grew at the rate decreasing with glutamine concentration. The IgG_1 productivity increased by 70% while the growth rate decreased to 10% of the normal growth rate, as the concentration of glutamine was reduced from 2.05 mM to 0.04 mM (Fig. 6). The optimum concentration of glutamine for suppressing growth while enhancing antibody productivity seemed to be close to the minimum concentration needed for maintaining protein synthesis system and much below that required for normal growth as an energy source. The achieved enhancement, 70%, of antibody productivity was less than that obtained by addition of caffeine

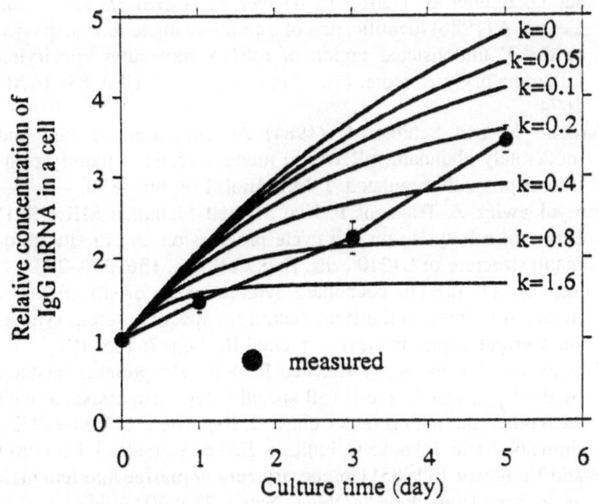

Fig. 7. Accumulation of λ_1 mRNA in hybridoma 2E3 cells whose growth was suppressed at day 0 by 20 U ml^{-1} of IL6. Solid circle: measured by Northern blotting, curve: estimated by eq. (4).

or deprivation of the growth-factors; a possible explanation given by the semi-quantitative model (Suzuki *et al.*, 1992) is that the positive effect of the mRNA accumulation in the growth-suppressed cells mentioned later was partially cancelled by the reduced protein synthesis rate per mRNA molecule due to the limited glutamine concentration. *Other amino acids.* Since essential amino acids other than glutamine such as leucine and isoleucine are not crucial energy source but building blocks of proteins, it is presumable that limitation of these amino acids suppresses growth by means of suppression of protein synthesis and, therefore, the enhancing effect of smaller μ in eq. (1) is cancelled by reduced β. For examples, growth-suppression of 2E3 cells by leucine limitation in RPMI1640 supplemented with 10% CPSR-3 did not enhance antibody productivity (data not shown); isoleucine limitation yielded the similar result (Suzuki and Ollis, 1990).

Accumulation of mRNA

2E3 cells growing at the specific growth rate μ of 1.4 day^{-1} were growth-suppressed at day 0 by adding IL6 at 20 units ml^{-1}; μ decreased gradually to 0 at day 5: the averasged μ through day 0 to day 5 was 0.2 day^{-1}. The concentration of λ_1 mRNA in the cells determined by Northern blotting was plotted in Fig. 7. The mRNA concentration increased 3.5-fold; the antibody productivity was enhanced 2- to 3.5-fold in a different run of experiment under the same condition (Makishima *et al.*, 1992). This increase of protein productivity paralleling the accumulation of the mRNA supported the notion which was explained as the basis of eq. (1) in Introduction. Time change of the averaged number M of molecules per cell of a specific mRNA is represented by differential equation (2).

$$\frac{dM}{dt} = \varepsilon - kM - \mu M \tag{2}$$

Here ε is the rate of the mRNA synthesis ($\varepsilon = \alpha\, D\, E$), and t is culture time (day).

In the following, we deal with the case that the cells grow exponentially for long period until time t_0 at a normal specific growth rate μ_0, and then they are growth-suppressed and grow exponentially at a reduced specific growth rate μ after time t_0. Integrating eq. (2) and assigning I to the initial number of the molecules per cell of the mRNA gives eq. (3).

$$M = \varepsilon/(k+\mu) - \{\varepsilon - (k+\mu)I\}$$
$$EXP\{-(k+\mu)t\}/(k+\mu) \tag{3}$$

When t_0 is assumed large enough, 4 days as an example, to diminish the second term of the right hand side of eq. (3), M_0(M at t=t_0) is approximated by $\varepsilon/(k+\mu_0)$. Integrating eq. (2) between t_0 and t substituting ε by $M_0(k+\mu_0)$ and arranging gives eq. (4).

$$\frac{M}{M_0} = \frac{k+\mu_0}{k+\mu} - \frac{\mu_0-\mu}{k+\mu} exp\{-(k+\mu)(t-t_0)\} \tag{4}$$

The ratio of the number of the mRNA molecules M/M_0 estimated by eq. (4) using 1.4 and 0.2 day^{-1} for μ_0 and μ and various values for k is shown in Fig. 7 in comparison with the experimental results. Since μ was assumed for simplification to be constant at 0.2 day^{-1} after the growth surpression, the estimated curves do not simulate excatly the observed accumulation of the mRNA in the culture of which the growth gradually slowed down between day 0 and 3. Another explanation of the discord between the measured and the simulated accumulation of the mRNA was that the mRNA decomposition rate in the cells might have decreased or its transcription rate might have increased due to some unknown mechanism as the growth rate decreased. Fig. 7 indicates that the coefficient k of the λ_1 mRNA decomposition rate equation is about or less than 0.2 day^{-1} corresponding to the half life of 3.4 day. This semi-quantitatively estimated half life is close to the reported maximum half life, 3.3 day, of immunoglobulin mRNAs (Kuehl, 1977; Storb, 1973).

Conclusions

The simulation by using eq. (1) shown in Fig. 1 represents the hypothesis that mRNA which is stable enough to have k about or less than 1 day^{-1} accumulate in cells when the cells grow slowly, resulting in enhancement of specific production rate of the protein coded in the accumulated mRNA. This hypothesis was experimentally supported: the mRNA of IgG which was reportedly stable with k between 0.2 and 1.8 day^{-1} (Kuehl, 1977; Storb, 1973) accumulated in the hybridoma cells and the production rate per cell of the coded protein, IgG, was enhanced when growth of the cells was suppressed by the various methods. However, other explanations for the enhanced protein productivity at the slowed growth could not be discarded yet: for examples, the mRNA might have been stabilized or synthesized faster due to some unknown mechanism when the growth was suppressed. To support the hypothesis more firmly, it should be experimentally confirmed that some stable mRNAs other than IgG mRNA accumulates in growth-suppressed cells, resulting in the enhancement of production rate of the coded protein.

Among the methods we examined to suppress growth of the cells, addition of caffeine or sodium n-butyrate into the culture medium was most effective to enhance the protein production rate of the cells. These methods were inexpensive, easy to apply, and more generally applicable to other cell lines. An IL6-sensitively growth-suppressible myeloma cell line was established by adapting myeloma X63.653. This cell line can be utilized as host cells to be transfected with some DNA coded for useful protein and produce the protein at a rate enhanced by growth-suppression. Although IL6 is expensive for the time being, the amount needed is minute 1 ng ml^{-1}.

Acknowledgements

We thank Dr. F. Hanaoka for offering us FM3A, tsFT210, and tsFT20 cells. This work was supported by a grant for the 'Biodesign Research Program' from RIKEN to F. Makishima.

References

Brock ML and Shapiro DJ (1983) Estrogen stabilizes vitellogenin mRNA against cytoplasmic degradation. Cell 34: 207–214.

Caput D, Beutler B, Hartog K, Thayer R, Brown-Shimer S and Cerami A (1986) Identification of a common nucleotide sequence in the 3' untranslated region of mRNA molecules specifying inflammatory mediators. Proc. Natl. Acad. Sci. USA 83: 1670–1674.

Carneiro M and Schibler U (1984) Accumulation of rare and moderately abundant mRNAs in mouse L-cells is mainly post-transcriptionally regulated. J. Mol. Biol. 178: 869–880.

Darzynkiewicz Z, Traganos F, Xue SH and Melamed MR (1981) Effect of n-butyrate on cell cycle progression and in situ chromatin structure of L1210 cells. Exp. Cell Res. 136: 279–293.

Kafatos FC (1972) The cocoonase zymogen cells of silk moths: a model of terminal cell differentiation for specific protein synthesis. Current Topics in Developmental Biology 7: 125–191.

Karasuyama H, Kudo A and Melchers F (1990) The proteins encoded by the V_{preB} and λ_5 pre-B cell-specific genes can associate with each other and with μ heavy chain. J. Exp. Med. 172: 969–972.

Kashima N, Nishi-Takaoka C, Fujita T, Taki S, Yamada G, Hamuro J and Taniguchi T (1985) Unique structure of murine interleukin-2 as deduced from cloned cDNAs. Nature 313: 402–404.

Koeller, DM, Horowitz JA, Casey JL, Klausner RD and Harford JB (1991) Translational and the stability of mRNAs encoding the transferrin receptor and c-*fos*. Proc. Natl. Acad. Sci. USA 88: 7778–7782.

Kuehl WM (1977) Synthesis of immunoglobulin in myeloma cells. Curr. Top. Microbiol. Immunol. 76: 1–46.

Makishima F, Terada S, Mikami T and Suzuki E (1992) Interleukin-6 is antiproliferative to a mouse hybridoma cell line and promotive for its antibody productivity. Cytotechnol. 10: 15–23.

Murakami F, Yasuda H, Miyazawa H, Hanaoka F and Yamada M (1985) Characterization of a temperature-sensitive mutant of mouse FM3A cells defective in DNA replication. Proc. Natl. Acad. Sci. USA 82: 1761–1765.

Okuda A and Kimura G (1988) Elongation of G1 phase by transient exposure of rat 3Y1 fibroblasts to caffeine during the previous and present generations. J. Cell Sci. 89: 379–386.

Savinell JM, Lee GM and Palsson BO (1989) On the orders of magnitude of epigenic dynamics and monoclonal antibody production. Bioprocess Eng. 4: 231–234.

Suzuki E (1990) Kinetics of monoclonbal antibody production. Doctoral dissertation, North Carolina State University, Raleigh, North Carolina.

Suzuki E and Ollis DF (1990) Enhanced antibody production at slowed growth rates: experimental demonstration and a simple structured model. Biotechnol. Prog. 6: 231–236.

Suzuki E, Takahashi K and Ollis DF (1992) A simple structured model predicted positively-, negatively, or non-growth associated antibody production rate depending on culture conditions. In: Murakami H, Shirahata S and Tachibana H (eds.) Animal Cell Technology: Basic and Applied Aspects vol. 4 (pp. 375–381).

Storb U (1973) Turnover of myeloma messenger RNA. Biochem. Biophys. Res. Commun. 52: 1483–1491.

Traganos F, Darzynkiewicz Z and Melamed MR (1982) The ratio of RNA to total nucleic acid content as a quantitative measure of unbalanced cell growth. Cytometry 2: 212–218.

Yasuda H, Kamijo M, Honda R, Nakamura M, Hanaoka F and Ohba Y (1991) A point mutation in C-terminal region of cdc2 kinase causes a G2-phase arrest in a mouse temperature-sensitive FM3A cell mutant. Cell Struct. Funct. 16: 105–112.

Address for offprints: E. Suzuki, Department of Chemistry and Biotechnology, Faculty of Engineering, The University of Tokyo, Hongo, Bunkyo-ku, Tokio, 113 Japan.

Cytotechnology **15**: 65–71, 1994.
© 1994 *Kluwer Academic Publishers.*

Growth and interferon-γ production in batch culture of CHO cells

V. Leelavatcharamas, A.N. Emery and M. Al-Rubeai
BBSRC Centre for Biochemical Engineering, School of Chemical Engineering, The University of Birmingham, Edgbaston, Birmingham B15 2TT, UK

Key words: Animal cell culture, interferon, CHO, growth rate, batch culture, cell cycle

Abstract

The relationship between growth and interferon-γ (IFN-γ) production in the recombinant cell line CHO 320 was studied by varying the foetal calf serum (FCS) concentration. The specific growth rate varied with the initial FCS concentration in a manner which could be well fitted by the Monod model. The K_s and μ_{max} values were found to be 0.771% (v/v) serum and 0.031 h^{-1} respectively. The average specific IFN-γ production rates during the exponential phase increased with increasing FCS concentration. A good correlation between specific production rate and specific growth rate was found in all phases of the culture except the lag phase and it was clearly demonstrated that IFN-γ production was growth associated. Specific glucose and glutamine utilisation rates were inversely related to specific growth rates.

Introduction

To optimise a bioreaction process in animall cell culture, the process has to be designed in such a way that the biological potential of the cell is optimally exploited. In simple batch culture, a seemingly preferred operation for the production of pharmaceuticals from animal cells, maximisation of product formation is achieved through two main strategies viz.: maximisation of viable cell number and improvement of medium formulation. Manipulations of the chemical and physical environment are among the basic methods used to improve cell number. A significant increase in productivity has been reported using these two approaches for a number of cell lines (Luan *et al.*, 1987; Field *et al.*, 1991; Jan *et al.*, 1991; Reid *et al.*, 1992; Buntemeyer *et al.*, 1992).

Understanding the relationship between growth and productivity is also important for effective design and operation of production processes. Such relationships have mainly been studied in hybridomas but no really consistent picture has emerged, product production kinetics being observed to depend both on cell line and on cell culture method (for review see Al-Rubeai *et al.*, 1992). In that work it was suggested that the observable relationship between growth rate and monoclonal antibody (MAb) productivity is affected both by the events which regulate MAb synthesis and secretion during the various phases of the cell cycle and by the amount of intracellular MAb released during cell death.

The Chinese Hamster Ovary (CHO) cell line CHO 320, producing interferon-γ, has been extensively studied by Hayter *et al.* (1991, 1992, 1993), who determined the effects of the physiological state of CHO cells on both the production and the quality of the heterologous protein. One important and relevant finding was the positive association seen between growth rate (as determined by dilution rate in 'chemostat' culture) and interferon-γ specific production, a result which suggests that CHO cells are unlike those hybridoma cells for which the data shows a negative association. In this study such relationships was examined in batch cultures in which the growth rate is determined by variation of the serum concentration. This approach has allowed the separation of the effect of growth phase from the effect of growth rate. The dependency of peak cell number and growth rate on the concentration of foetal calf serum as well as the utilisation of glutamine and glucose at different rates were also investigated.

Materials and methods

The recombinant CHO 320 cell line was kindly provided by the Wellcome Foundation, Ltd. (Beckenham, Kent, U.K.). The cells expressed IFN-γ which was co-amplified with dihydrofolate reductase by methotrexate selection.

The inoculum, which was taken from the late exponential phase of 5% FCS cultures was spun down and resuspended in 255 ml RPMI with 1 μM methotrexate. 50 ml of this suspension culture was inoculated into each batch of 1.0, 2.5, 5.0, 7.5 or 10.0% FCS culture to give the initial cell number of 2×10^5 cells/ml in 200 ml working volume. All cultures, in stirred bottles, were incubated at 37 °C in an incubator and magnetically stirred at an agitation rate of 150 rpm. Samples were taken every 6 h in the first 2 days and every 12 h for another 2 days after incubation. After this, samples were taken every 24 h until the end of the batch. Samples were cell counted and centrifuged at 1000 rpm for 5 min. The supernatants were kept at –20 °C for analysis at the end of the batch and cells were fixed with 70% ethanol and kept at the same temperaure for flow cytometry analysis.

Glucose concentrations were measured using a glucose strip test kit REFLOLUX 2 (Boehringer-Manheim). Glutamine concentrations were measured using the Sigma glutamine/glutamate determination kit. Lactic acid was analysed by conversion to pyruvate by the enzyme lactate dehydrogenase and the coenzyme NAD^+. The product, NADH, was detected spectrophotometrically at 340 nm. The $NH_4{}^+$ ion concentrations were detected by conversion to NH_3 which reacted with sodium phenate and hupochlorite to produce a blue colour which could be quantified spectrophotometrically. The reaction was catalysed by nitroprusside (Fawcett and Scott, 1960). IFN-γ concentrations were measured by a sandwich enzyme linked immunosorbent assay (ELISA) (P. Hayter, personal communication) using anti-IFN-γ polyclonal antibody coated plates. 20B8 monoclonal antibody raised against IFN-γ from *E. coli* was used as the second antibody. The optical density at 490 nm was recorded after treating the plates with anti-mouse polyvalent immunoglobulins, peroxidase comjugate. The polyclonal and monoclonal antibodies used were kindly provided by Drs. P. M. Hayter and N. Jenkins (University of Kent, UK).

Results and discussion

Growth of CHO 320 cells

Generally, substrate limitation of growth and production can be regarded as being of two types, stoichiometric and kinetic. A stoichiometric limitation refers to the limitation on product (cell mass and/or IFN-γ) yield achieved by complete utilisation of a certain nutrient (substrate). A kinetic limitation, on the other hand, implies the diminution of a rate, i.e., cell growth rate or production rate, due to a reduction in the amount of the kinetically limiting substrate. While many substrates can stoichiometrically limit growth or product formation, they may not necessarily control the kinetics of growth and production formation.

The growth curves for each batch culture at the various FCS concentrations are presented in Fig. 1. Both cell growth rates and cell yields increased with increasing FCS concentration, showing both stoichiometric and kinetic limitation of substrate. At 10% FCS the cells entered a stationary phase at an earlier point than for the lower FCS concentrations and the ultimate cell yields was reduced slightly. Both cell growth rate and cell yield were severely reduced at 1% FCS.

Taking the time period of the early exponential growth phase (during 30–60 h in Fig. 1) for each batch culture, one may calculate a value for μ using the log-linear plot which essentially represents a 'maximum' specific growth rate for that condition. Fig. 2 shows the dependency of this specific growth rate on the initial amount of serum in each batch (S_0). The nature of this curve suggests that a saturation model of the Monod type can be used to represent this relationship.

The Monod model:

$$\mu = \frac{\mu_{max} \times S_0}{K_s + S_0} \tag{1}$$

where S_0 is the starting substrate concentration, has been used to describe the effect of serum concentration on specific growth rate by Dalili and Ollis, (1989) and Hild *et al.* (1992). The values for the maximum specific growth rate, μ_{max}, and Monod constant K_s were obtained via the double reciprocal plot (μ_{max}=0.031 h^{-1} and K_s=0.771% (v/v) serum). The specific growth rate of CHO 320 cells, therefore, is controlled by the level of certain (unidentified) serum component(s).

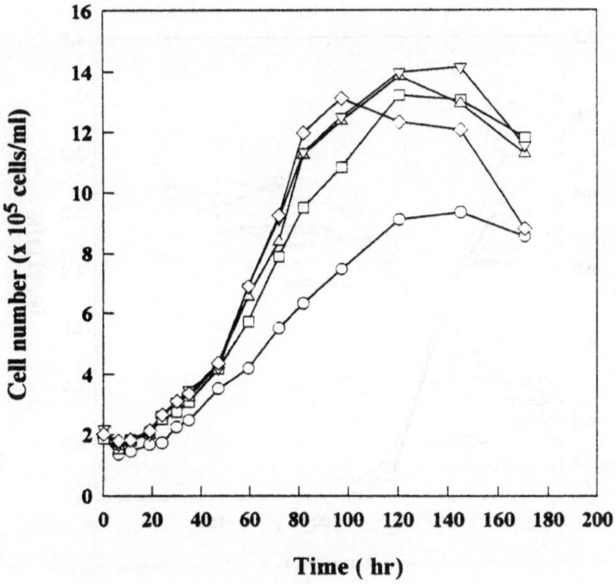

Fig. 1. Growths of CHO 320 cells at various % FCS. ○ = 1%, □ = 2.5%, △ = 5%, ▽ = 7.5% and ◇ = 10% FCS.

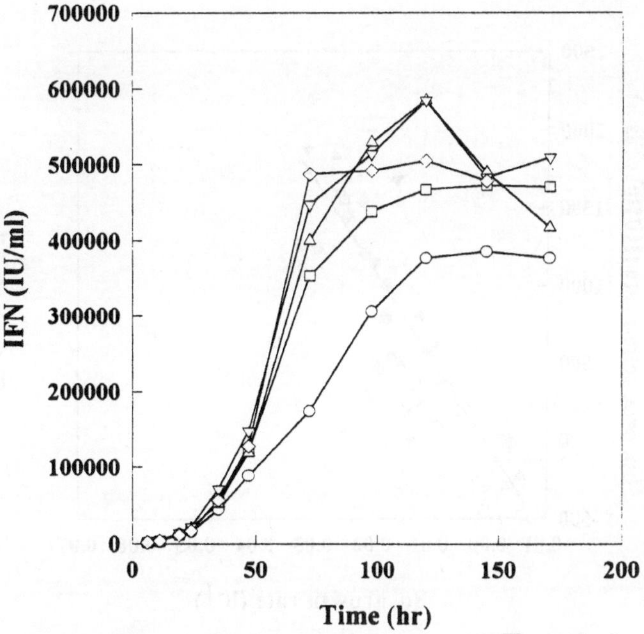

Fig. 3. IFN production at various % FCS. ○ = 1%, □ = 2.5%, △ = 5%, ▽ = 7.5% and ◇ = 10% FCS.

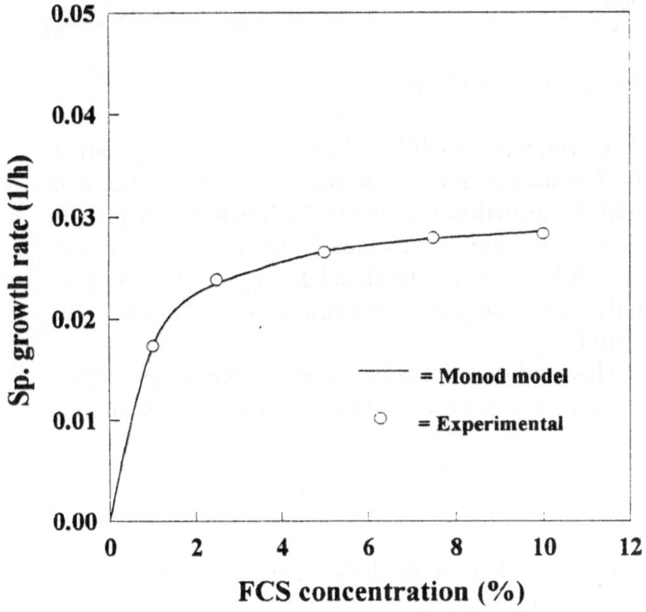

Fig. 2. The dependency of specific growth rate on % FGCS.

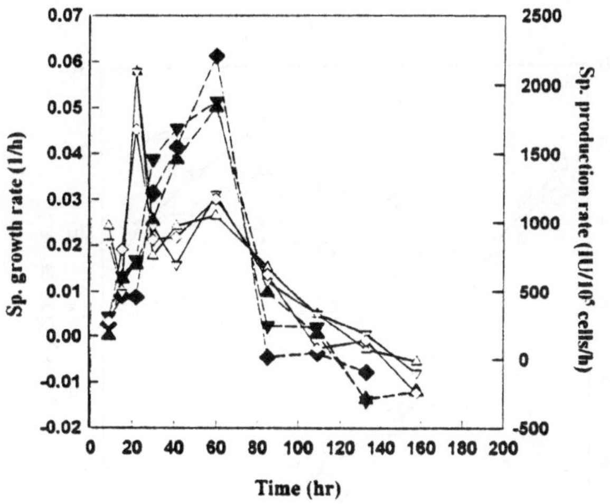

Fig. 4. Graphs of specific growth rate (open symbols) and specific production rate (solid symbols) over time, △ = 5% FCS, ▽ = 7.5% and ◇ = 10% FCS.

Comparison with the non-recombinant cell line CHO K1 grown under similar conditions (Leelavatcharamas, 1994), shows that the μmax of CHO K1 (0.04 h^{-1}) was a little higher than the μ_{max} of CHO 320 cells. The μ_{max} of CHO 320 cells in a stirred batch of serum-free medium was also reported as 0.031 h^{-1} by Hayter *et al.* (1991). CHO K1 is the nonproducing parent of CHO 320 and its higher growth rate indicates that production of foreign proteins by cells is imposing burden on biomass production and cell division. The contribution made by serum (above 2.5%) to cell growth is negligible in relation to the added cost of doing so and it is therefore reasonable to suggest that for a cost effective operation 2.5% serum should be

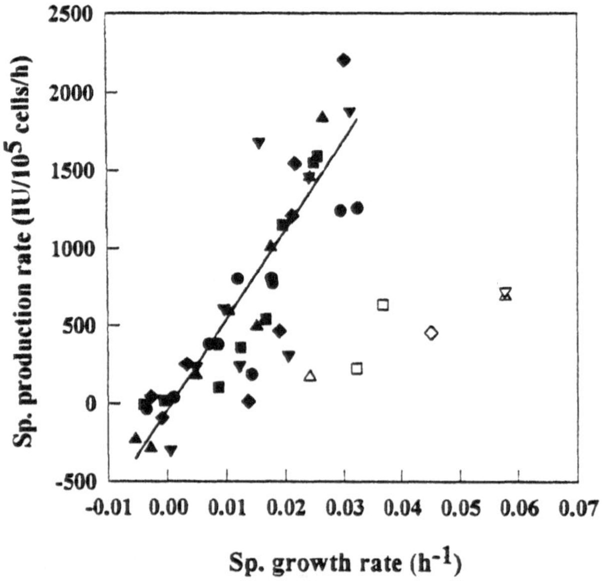

Fig. 5. The plot of specific growth rate againstt specific production rate during the growth phase of batch cultures. (Note that only the open symbols, representing data points from the lag phase of the growth curves can be identified as significantly deviating from the regression line). ● = 1%, ■ = 2.5%, ▲ = 5%, ▼ = 7.5% and ◆ = 10% FCS.

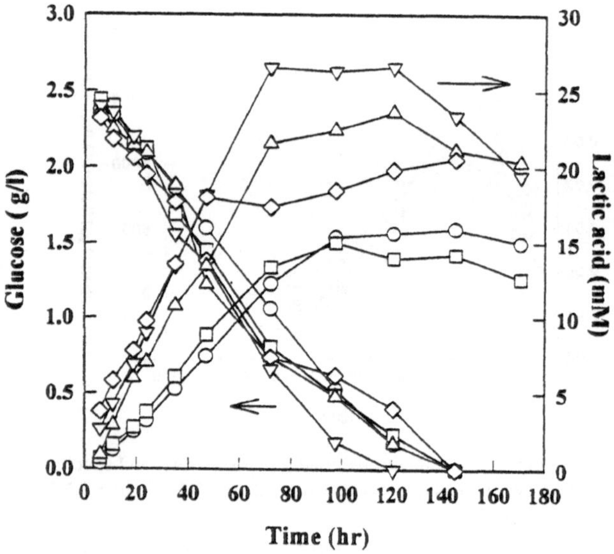

Fig. 6. Glucose utilization and lactic acid accumulation at various % FCS. ○ = 1%, □ = 2.5%, △ = 5%, ▽ = 7.5% and ◇ = 10% FCS.

used while attempts are made to improve basal medium composition.

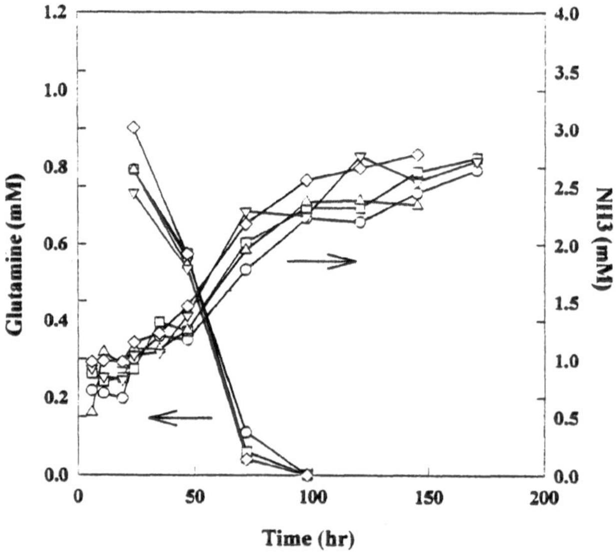

Fig. 7. Glutamine utilization and ammonia accumulation at various % FCS. ○ = 1%, □ = 2.5%, △ = 5%, ▽ = 7.5% and ◇ = 10% FCS.

Interferon-γ production

The production of IFN-γ by CHO 320 cells in all % FCS concentrations is shown in Fig. 3. Comparison with the growth curves in Fig. 1, shows that the growth and production curves clearly follow the same pattern. IFN-γ was accumulated during the growth phase only and the highest concentration was about 6×10^5 IU/ml.

If c and q_{IFN} are IFN-γ concentration rate respectively then a specific production rate can be determined as:

$$q_{IFN} = \frac{(c_2 - c_1)(\ln x_2 - \ln x_1)}{(x_2 - x_1)(t_2 - t_1)} \qquad (2)$$

where c_2 and c_1 are the IFN-γ concentrations at times t_2 and t_1 respectively.

As compared with these averaged figures over the early exponential phase, Fig. 4 shows the instant specific growth rates and instant specific production rates measured between successive time intervals over the whole incubation time. The averaging disguises sharp increases in the specific growth rates during the early growth phase to values well above those recorded in Fig. 2. After 60 h of incubation, both specific growth rates and specific production rates declined sustantially and remained so to the end of the batch. When the instant specific growth rate is plotted against the instant

Table 1. The specific nutrient consumption and product accumulation rates of cells in various % FCS cultures

	Specific consumption rate		Specific production rate		
FCS (%)	Glucose (mM/10^5 cells/h)	Glutamine (mM/10^5 cells/h)	IFN-γ (IU/10^5 cells/h)	Lactic (mM/10^5 cells/h)	Ammonia (mM/10^5 cells/h)
1.0	0.030	0.0027	1014	0.042	0.0051
2.5	0.022	0.0019	1484	0.027	0.0033
5.0	0.019	0.0017	1585	0.033	0.0029
7.5	0.021	0.0016	1508	0.037	0.0029
10.0	0.016	0.0019	1848	0.018	0.0031

specific production rate (Fig. 5) a good correlation is obtained apart from some obvious outliers (open symbols) which represent the high growth rates referred to above for which correspondingly high IFN-γ production rates were not recorded.

By studying more carefully the results shown in Fig. 4 and 5, it is found that the points represented with open symbols originate from the time period between 12 and 24 h in all cases. These high apparent growth rates at the start of batch cultures can, however, be demonstrated to be directly influenced by the physiological and synthetic state of the inoculated cells. By analyses of the cell cycle (Leelavatcharamas, 1994) a large proportion (~70%) of cells in the inoculum is found to be in the G1 phase. The inoculation of this cell population into fresh and highly nutritious medium led to a partial synchronisation of cell division with cells reaching the G2/M phase within 20 hr from inoculation. Consequently the dividing fraction was then at a maximum, thus considerably increasing the apparent growth rate as measured by cell numbers. Of course specific rates based on cell mass may show a quite different behaviour.

The averaged specific IFN-γ production rates over the exponential growth phases for all the FCS concentrations are shown in Table 1. A clear relationship between % FCS and specific INF-γ productivity based on cell number is demonstrated suggesting that serum concentration has a direct kinetic effect on the synthesis and secretion of IFN-γ at least during the exponential phase (24–72 hr). This may not however be the whole story since Hayter *et al.* (1992, 1993) have shown a similar effect in serum free medium. What may be being observed in each case is a differential effect on the distribution of cells in the more productive phases of the cell cycle. By estimation of the specific IFN-γ productivity based on cell biomass it is expected that this relationship would be even further pronounced due to the larger cell size at the lower serum concentrations. This interesting effect remains to be more rigorously examined using absolute (as opposed to relative) values of cell size. The kinetic effect of serum is on biomass production and on the specific and volumetric IFN-γ productivity.

Metabolic trends

Glucose metabolism

Glucose utilisation and lactate production were followed for all the cultures already described and these are shown in Fig. 6. At all FCS concentrations the reduction of the glucose concentration was essentially linear up to the mid-exponential period (~70 hours). Only at the lowest serum concentration (1%) was the rate at which the glucose concentration declined significantly reduced. As growth rates slowed in the late exponential phase so the rate at which the glucose was being depleted also declined, but the behaviour at 1% and 10% FCS concentrations was interesting. In 1% FCS the total glucose utilisation was comparable to that at higher FCS concentrations, and therefore, given the much reduced cell numbers at this FCS concentration, the specific glucose use was substantially higher (see Table 1). At 10% FCS the lower rates of glucose depletion reflected the earlier entry of cells at this concentration into a longer 'stationary' phase. In all cases glucose was completely exhausted at the end of this phase (~144 hours). Dalili and Ollis (1989) reported an inverse relationship between glucose consumption and growth rate for hybridoma cells suggesting an inefficient consumption of glucose by cells at low sapecific growth rates. The effect is not so distinctly growth rate related for these CHO cells but clearly the low growth rate seen at the lowest FCS concentration has led to a qualitatively similar effect. The specific lactic acid production rates were generally consistent with the glucose utilisation. The conversion rate (mole/mole) of glucose to lactic acid was slightly erratic (1.2–1.8) but the averaged figures again disguise the growth-phase related effects referred to above.

Glutamine metabolism

Glutamine functions as an oxidizable source of energy for many cells cultured *in vitro*, and the metabolic end products of glutamine metabolism are similar for a variety of cells (Thomas, 1990). The major end products of glutamine metabolism are generally CO2, ammonia, lactate, glutamate, aspartate and alanine.

The utilisation of glutamine by cells at each FCS concentration, together with the corresponding ammonia accumulation are presented in Fig. 7. What is striking is that the glutamine utilisation is not only rapid, leading to complete exhaustion in the late exponential phase, but it also appears to be essentially independent of the growth rate, except at the lowest FCS concentration again. Limitation of growth by glutamine exhaustion has been commonly observed, for example in CHO cells by Hayter *et al.* (1991) and in hybridoma cells by Geaugey *et al.* 61989) and Dalili and Ollis (1989) during batch cultivation. There was no significant difference on the amount of ammonia produced at 1–10% FCS. The maximum ammonia concentration was about 2.7 mM.

The specific glutamine utilisation and ammonia production rates (Table 1) were generally similar for the 2.5–10% FCS cultures (about 0.002 mM/10^5 cells/h and 0.003 mM/10^5 cells/h, respectively). Again, the 1% FCS culture showed higher specific glutamine utilisation and ammonia production rates. Such higher rates are consistent with higher maintenance energy requirements by cells at lower growth rates resulting from either a less nutritional medium or a stressful environment. Such patterns of glutamine utilisation and ammonia production are again similar to those reported in hybridoma cells by Dalili and Ollis (1989).

As the exhaustion of glucose and glutamine coincided with the end of the growth phase, it is reasonable to assume that both or either of these nutrients were limiting cell yield. However, Hayter *et al.* (1991) showed that increasing the concentrations of glucose and glutamine in the medium did not improve the cell yield and interferon production in this cell line. They also reported that the 6 mM ammonia and 22 mM lactate concentrations that had assumulated by the end of their batch runs were not inhibitory to growth. It is still necessary therefore to identify the inhibitory metabolites that do arise with such medium supplementation and for this purpose it is necessary to adopt a control techniques using multi-nutrient feeds and to apply rational medium design protocols.

Conclusions

In batch cultures of IFN-γ producing CHO cells the apparent growth-associated production kinetics reflect a cell cycle related phenomenon. The cause of growth reduction appears to be prolongation of the cell cycle and an increase in the relative frequency of the G1 cells rather than an increase in cell death, resulting from stoichiometric and kinetic limitation by serum. It is plausible to suggest that the G1 phase is the least productive phase of the cell cycle in CHO cells.

References

Al-Rubeai M, Emery AN, Chalder S, Jan DC (1992) Specific monoclonal antibody productivity and the cell cycle- comparisons of batch, continuous and perfusion cultures. Cytotechnology 9: 85–97.

Buntemeyer H, Wallerius C, Lehmann J (1992) Optimal medium use for continuous high density perfusion processes. Cytotechnology 9: 59–67.

Dalili M and Ollis DF (1989). Transient kinetics of hybridoma growth and monoclonal antobody production in serum-limited cultures. Biotechnol. Bioeng. 33: 984–990.

Fawcett JK, and Scott JE (1960) A rapid and precise method for the determination of urea J. Clin. Path. 13: 156–159.

Field RP, Brand H, Renner GL, Robertson HA, Boraston R (1991) Production of a chimeric antibody for tumour imaging and therapy from chinese hamster overy (CHO) and myeloma cells. In: Production of biologicals from animal cells in culture. (Eds.) Spier RE, Griffiths JB, meignier B, Butterworth-Heinemann, Oxford pp. 742–744.

Geaugey V, Duval D, Geahel I, Marc A, Engasser JM (1989) Infulence of amino acids on hybridoma cell viability and antibody secretion. Cytotechnology 2: 119–129.

Hayter PM, Curling EMA, Baines AJ, Jenkins N, Salmon I, Strange PG, Bull AT (1991) Chinese hamster ovary cell growth and interferon production kinetics in stirred batch culture. Appl. Microbiol; 34: 559–564.

Hayter PM, Curling EMA, Baines AJ, jenkins N, Salmon I, Strange PG, Tong JM, Bull AT, (1992) Glucose-limited chemostat culture of chinese hamster overy cells producing recombinant human interferon-γ. Biotechnology and Bioengineering 29: 327–335.

Hayter PM, Curling EMA, Gould ML, Baines AJ, Jenkins N, Salmon I, Strange PG, Bull AT, (1993) The effect of the dilution rate on the CHO cell physiology and recombinant interferon-γ production in glucose-limited chemostat culture. Biotechnology and Bioengineering 42: 1077–1085.

Hild HM, Emery AN, and Al-Rubeai M, (1992) The effect of pH, temperature, serum concentration and media composition on the growth of insect cells In: Workshop on baculovirus and recombinant protein production processes. Edited by JM Vlak, EJ, Schlaeger and AR, Bernard, March 29- April 1, Interlaken, Switzerland.

Jan DC, Al-Rubeai M, Emery AN, (1991) Towards low-cost medium formulation for intensive animal cell culture. In: (eds.) Huyghebaert A, and Vandamme E, Upstream and downstream processing in biotechnology III (pp. 1.17- 1.26) Royal Flemish Society of Engineers, Antwerpen.

Leelavatcharamas V, (1994) Growth, cell cycle and interferon-γ production of CHO cells in batch culture M. Phil. thesis. School of chemical engineering. University of Birmingham.

Luan YT, Mutharasan R and Magee WE, (1987) Strategies to extend longevity of hybridomas in culture and promote yield of monoclonal antibodies Biotechn. Lett., 9: 691–696.

Reid S, Saxena V, greenfield PF, Weiss SA, (1992) Applicability of fed batch processes for improving recombinant protein production by the baculovirus expression system. In: Spier RE, Griffiths JB, and MacDonald C, Animal cell Technology: developments, processes and products (eds.) pp. 276–181. Butterworth-Heinemann, Oxford.

Thomas JN, (1990) Mammalian cell physiology in: (eds.) Lubiniecki AS, p 119, Large-scale mammalian cell culture technology, Marcel Dekker Inc. New York.

Address for offprints: M. Al-Rubeai, BBSRC Centre for Biochemical Engineering, School of Chemical Engineering, The University of Birmingham, Edgbaston, Birmingham B15 2TT, UK.

Cytotechnology **15:** 73–86, 1994.

Metabolic and kinetic studies of hybridomas in exponentially fed-batch cultures using T-flasks

Ana E. Higareda[1], Lourival D. Possani[2] and Octavio T. Ramírez[1] *
Departments of Bioengineering[1] and Biochemistry[2], Instituto de Biotecnología, Universidad Nacional Autónoma de México, Avenida Universidad 2001, Apdo. Postal 510-3, Cuernavaca, Morelos 62271, México

Key words: Hybridoma cell culture, monoclonal antibody, exponentially fed-batch, culture kinetics, metabolism, chemostat.

Abstract

Exponentially fed-batch cultures (EFBC) of a murine hybridoma in T-flasks were explored as a simple alternative experimental tool to chemostats for the study of metabolism, growth and monoclonal antibody (MAb) production kinetics. EFBC were operated in the variable volume mode using an exponentially increasing and predetermined stepwise feeding profile of fresh complete medium. The dynamic and steady-state behaviors of the EFBC coincided with those reported for chemostats at dilution rates below the maximum growth rate. In particular, steady-state for growth rate and concentration of viable cells, glucose, and lactate was attained at different dilution rates between 0.005 and 0.05 h^{-1}. For such a range, the glucose and lactate metabolic quotients and the steady-state glucose concentration increased, whereas total MAb, volumetric, and specific MAb production rates decreased 65-, 6-, and 3-fold, respectively, with increasing dilution rates. The lactate from glucose yield remained relatively constant for dilution rates up to 0.03 h^{-1}, where it started to decrease. In contrast, viability remained above 80% at high dilution rates but rapidly decreased at dilution rates below 0.02 h^{-1}. No true washout occurred during operation above the maximum growth, as concluded from the constant viable cell number. However, growth rate decreased to as low as 0.01 h^{-1}, suggesting the requirement of a minimum cell density, and concomitant autocrine growth factors, for growth. Chemostat operation drawbacks were avoided by EFBC in T-flasks. Namely, simple and stable operation was obtained at dilution rates ranging from very low to above the maximum growth rate. Furthermore, simultaneous operation of multiple experiments in reduced size was possible, minimizing start-up time, media and equipment costs.

Abbreviations: EFBC – exponentially-fed batch culture; CSC – continuous suspended culture; MAb – monoclonal antibody; D – dilution rate; q_i – metabolic quotient or specific rate of consumption or production of i.

Introduction

Complete kinetic characterization of hybridoma cultures not only provides an important insight into cell physiology and metabolism, but is also a necessary prerequisite for successful optimization and scaling-up of monoclonal antibody (MAb) production. The reported dissimilar behaviors between different hybridoma cell lines emphasize the importance of conducting kinetic studies in every case. For instance, there exists contradictory information as to the effect of growth rate on the specific MAb production rate (q_{MAb}). While many authors have shown that q_{MAb} increases with decreasing growth rate (Miller *et al.*, 1988; Ramírez and Mutharasan, 1990; Martens *et al.*, 1993), others have shown, either the opposite behavior (Low *et al.*, 1987), an optimum at intermediate growth rates (Ray

* To whom correspondence should be addressed.

et al., 1989), or no effect at all (Flickinger *et al.*, 1990). A broad range in q_{MAb} has also been reported; from as low as 0.2×10^{-10} mg/cell-h (Leno *et al.*, 1992) to as high as 33×10^{-10} mg/cell-h (Kurokawa *et al.*, 1993). Stability of Mab production (Frame and Hu, 1990) and sensitivity to ammonia and lactate (Omasa *et al.*, Ozturk *et al.*, 1992) also varies widely among cell lines. In particular, Reuveny *et al.* (Reuveny *et al.*, 1987) have found that the commonly toxic lactate can stimulate growth of an hybridoma cell line at concenvations below 22 mM. While most data indicate that the yield of lactate from glucose increases with increasing growth rate (Miller *et al.*, 1988; Ramírez and Mutharasan, 1990) other studies show the opposite behavior (Ray *et al.*, 1989). Furthermore, environmental factors, such as hydrodynamic (Papoutsakis, 1991) and osmotic stress (Reddy *et al.*, 1994) have also been shown to be cell line and growth rate dependent.

The effect of environmental parameters on cell growth, metabolism and product formation of various hybridoma cell lines has been determined in continuous suspension culture (CSC) (Miller *et al.*, 1988; Frame and Hu, 1990; Hiller *et al.*, 1991; Kurokawa *et al.*, Martens *et al.*, 1993). CSC is usually considered the most appropriate operation mode for such studies since homogeneous and constant environmental conditions can be maintained. Nevertheless, CSC can be plagued by significant drawbacks. Namely, CSC operation is expensive due to the inefficient use of sophisticated media and the need of costly equipment and instrumentation, especially if multiple simultaneous experiments are performed. CSC operation at low dilution rates can be problematic due to the higher death rates and lower viability of hybridomas, possibly due to increased shear sensitivity at reduced growth (Boraston *et al.*, 1984; Miller *et al.*, 1988; Ramírez and Mutharasan, 1990). An accurate control of volume and flow rates is also difficult in such regions (Yamané and Shimizu, 1984). Likewise, CSC presents increased risk of contamination during long operating times at low dilution rates, and due to cell line instability a true steady state might not be attainable (Frame and Hu, 1990, and Martens *et al.*, 1993). Furthermore, washout of the culture can occur at high dilution rates. Such disadvantages have drawn cell culturists to use batch operation as a simpler tool for hybridoma characterization. Nevertheless, batch culture data can provide misleading information (Miller *et al.*, 1988).

Fed-batch culture has proven to be a powerful experimental tool for microbial energetics and kinetic studies (Keller and Dunn, 1978; Esener *et al.*, 1981). Accordingly, such mode of operation can represent an attractive alternative to CSC. Nonetheless, application of fed-batch culture has been mainly restricted to its conventional form of operation in which only small volume variations are allowed and where the main goal is the achievement of high cell densities and productivities (Glacken *et al.*, 1989; Ramírez and Mutharasan, 1990; Hu and Piret, 1992; Jo *et al.*, 1993a). In conventional fed-batch operation, medium containing a highly concentrated carbon or energy source, but usually deficient in other nutrients, is fed in several modes and a quasi-steady state is attained only for a single limiting substrate. In contrast, Ramírez *et al.* (Ramírez *et al.*, 1994a, 1994b) have shown for microbial systems that a steady state condition can be attained for most parameters if a variable volume exponentially fed-batch culture (EFBC) is employed. The dynamic and steady state behavior of EFBC is the same as CSC. However, EFBC are technically simpler to run and permit stable operation in regions where continuous cultivation is not feasible, i.e. very low or high dilution rates. Furthermore, equipment, medium utilization, and start-up time are minimized in fed-batch culture of hybridomas in T-flasks, and multiple experiments can be performed simultaneously with reduced risk of contamination (Truskey *et al.*, 1990).

Fed-batch operation using various modes of feedback flow rate control has been used for different mammalian cells (Glacken *et al.*, 1989; Truskey *et al.*, 1990; Omasa *et al.*, 1992; Chevalot and Marc, 1993). In such studies, the pseudo-steady state condition has been attained only for a limited number of parameters. Semicontinuous or repeated fed-batch operation has also been used as a means of approximating continuous operation in plant and animal cell culture (Tharakan and Chau, 1986; Ramírez and Mutharasan, 1990; Lee *et al.*, Westgate *et al.*, 1991; Leno *et al.*, Schmid *et al.*, 1992; Jo *et al.*, 1993b; Salazar-Kish and Heath, 1993). In this work, we report for the first time the use of predetermined variable-volume EFBC in T-flasks as an alternative to CSC for hybridoma kinetic characterization studies. The effect of dilution rate on hybridoma growth, glucose consumption, and lactate and MAb production is presented.

Materials and methods

Cell line and culture medium

A murine hybridoma, designated BCF2, that secretes a neutralizing monoclonal antibody (MAb) specific to toxin 2 of the scorpion *Centruroides noxius* Hoffmann was generated and used in this study. Generation of the hybridoma cell line and the *in vivo* neutralizing activity of MAb has been described elsewhere (Zamudio *et al.*, 1992). The medium used was Dulbecco's modified Eagle's medium prepared from powdered formulation (Sigma Chemical Co., St. Louis, MO). The final medium contained 4 g l^{-1} glucose, 4 mM L-glutamine, 2.2 g l^{-1} $NaHCO_3$, 13.2 mg l^{-1} oxaloacetic acid, 0.8 mg l^{-1} crystalline insulin, 5.5 mg l^{-1} sodium pyruvate, 1% (v/v) of lOOX nonessential amino acid solution (Sigma), 1% (v/v) of antibiotic antimycotic solution [penicillin (100,000 U l^{-1}), streptomycin (100 mg l^{-1}), amphotericin B (250 μg l^{-1})] (Sigma), and 10% (v/v) fetal bovine serum (Sigma). The concentration of antibiotics was much lower than those known to be cytotoxic.

Exponentially fed-batch cultures

All fed-batch experiments were performed in 75-cm^2 T-flasks maintained in a CO_2 incubator (37 °C, 6% CO_2). Seven dilution rates, divided in two groups, were tested. In the first group of cultures, operated at dilution rates of 0.005, 0.01, 0.02 and 0.07 h^{-1}, the flasks initially contained a 4-ml cell suspension consisting of hybridomas from mid-exponential phase of batch cultures and sufficient fresh culture medium to give a concentration below 0.5 $\times$ 10^6 cel ml^{-1}. For the second group, operated at dilution rates of 0.03, 0.04 and 0.05 h^{-1}, a 4-ml cell suspension of late-exponential phase hybridomas was placed directly in the flasks without diluting in fresh medium in order to give an initial concentration of 0.9 $\times$ 10^6 cell ml^{-1}. After allowing the cells to grow in batch mode for a predetermined period of time, a known volume of fresh culture medium was added and a sample was removed for cell number determination and other assays. The sample volume never exceeded 1-ml and was typically 0.65-ml. Homogeneity of T-flasks before sampling and after medium addition was achieved by gentle mixing; the remaining time, the T-flasks remained static. This procedure was repeated successively until a final volume of 40-ml was reached. Only the cultures operated at a dilution rate of 0.03, 0.04, and 0.05 h^{-1} were allowed to grow in batch mode after the final volume was reached and until cell viability decreased to less than 30%. The time between medium additions and the volume added in every instance was adjusted in order to obtain a volume profile that increased exponentially with time. The time between medium additions never exceeded 24 h, and the volume added in every occasion remained between 1 and 4-ml. A fraction of the medium added corresponded to the volume removed during sampling. Such a procedure resulted in a stepwise profile that closely traced the ideal exponential profile that would be generated if a continuous feeding had been established. As described below, a constant and predetermined dilution rate can be fixed through the use of exponentially-fed batch cultures. In addition to the fed-batch experiments, batch control cultures in 30-ml T-flasks were also performed.

Analytical methods

Cell number was determined by trypan blue exclusion method. Samples of supernatant taken from the cultures were frozen at –20 °C for later determination of glucose, lactate, and MAb concentration. Glucose and lactic acid were determined enzymatically with a YSI model 2700 analyzer (Yellow Springs Instruments Co., Yellow Springs, OH). MAb concentration was determined, as described before (Ramírez and Mutharasan 1990), by an alkaline phosphatase 'sandwich' ELISA technique. ELISA plates were read at 405 nm with a Titertek Multiscan MCC/340 reader (Flow Laboratories, McLean, VA) and reported MAb titers were obtained from the average of 12 wells. The values of MAb concentration reported correspond only to the fed-batch phase.

Mathematical considerations

For a well-mixed fed-batch culture with sterile feed and no outflow, a cell balance over the T-flask is given by:

$$\frac{d(xV)}{dt} = \mu(xV) \tag{1}$$

where x, V, t, and μ are viable cell concentration, culture volume, time, and apparent specific growth rate, respectively. Thus, upon integration of Equation (1), μ, can be readily obtained from the slope of a curve of Ln[xV] versus time. As described elsewhere (Ramírez *et al.*, 1994a), a constant and predetermined dilution rate, D, can be obtained in a fed-batch culture if an exponentially increasing feeding flow rate, F(t), is set as:

$$\frac{dV}{dt} = F(t) = V_0 D\exp(Dt) \qquad (2)$$

where the subindex 0 refers to the time of feeding initiation. Combining Equations (1) and (2), and rearranging, yields:

$$\frac{d(Lnx)}{dt} = \mu - D \qquad (3)$$

Similarly, a substrate balance yields:

$$\frac{dS}{dt} = D(S_r - S) - \frac{\mu x}{Y} \qquad (4)$$

where S and S_r are the substrate concentrations in the T-flask and in the feed, respectively, and Y is the yield of biomass on substrate. As shown by Equations (3) and (4), the dynamic and steady state behavior of an EFBC is identical to that of a CSC, underscoring the utility of the former as an alternative to the latter.

The specific glucose consumption, q_G, and lactate production, q_L, rates were determined as the average of the individual rates between medium additions, which in turn were determined as:

$$q_{G_{i+1}} = \frac{[(G_i V_i) - (G_{i+1}V_{i+1}) + G_r(V_{i+1} - V_i)]}{\left[(t_{i+1} - t_i)\left(\frac{x_{i+1}V_{i+1} + x_i V_i}{2}\right)\right]} \qquad (5)$$

$$q_{L_{i+1}} = \frac{[(L_{i+1}V_{i+1}) - (L_i V_i) - L_r(V_{i+1} - V_i)]}{\left[(t_{i+1} - t_i)\left(\frac{x_{i+1}V_{i+1} + x_i V_i}{2}\right)\right]} \qquad (6)$$

where G_i and L_i are residual glucose and lactate concentrations, respectively, at the *i*th time interval. The subindex r refers to the concentration in the feeding medium. The apparent molar yield of lactate from glucose was determined by dividing the average lactate and glucose metabolic molar quotients. The total MAb produced, ΔMAb; the overall volumetric MAb production rate, r_{MAb}; and the average specific MAb production rate were determined from Equations 7, 8 and 9, respectively.

$$\Delta MAb = MAb_f V_f - MAb_0 V_0 \qquad (7)$$

$$r_{MAb} = \frac{\Delta MAb}{(t_f - t_0)V_f} \qquad (8)$$

$$q_{MAb} = \frac{\Delta MAb}{(t_f - t_0)\tilde{x}} \qquad (9)$$

where [MAb] is the antibody concentration, the subindex f refers to the concentration at the end of the fed-batch phase, and $\tilde{x}$ is the average viable cell number defined as:

$$\tilde{x} = \frac{\int_{t_0}^{t_f}(xV)dt}{(t_f - t_0)} \approx \frac{\sum_{i=0}^{f-1}\left[\frac{x_{i+1}V_{i+1} + x_i V_i}{2}(t_{i+1} - t_i)\right]}{(t_f - t_0)} \qquad (10)$$

Results and discussion

Batch culture

Fig. 1 shows typical behavior of BCF2 hybridoma batch culture in 75-cm^2 T-flasks. After a maximum viable cell concentration of 1.3 × 10^6 cell ml^{-1} was attained, a rapid decrease in viability was observed. We have previously shown (Higareda *et al.*, 1993) that for BCF2 hybridomas in serum containing media, glucose and glutamine are the main carbon and energy source, respectively, and that depletion of either one limits growth. Therefore, the observed growth cessation in batch is possibly due to accumulation of toxic metabolic byproducts, such as lactate or ammonia, since glutamine (data not shown) and glucose were still present at non-limiting concentrations. The maximum growth rate, calculated for the exponential growth phase, was 0.046 h^{-1}. As shown in Fig. 1B, glucose consumption and lactate production ceased a few hours after the maximum viable cell concentration was attained, whereas MAb continued to increase until the end of the culture where a maximum of 36 mg l^{-1} was reached (Fig. 1C). The batch metabolic quotients for glucose and lactate, calculated as the ratio of maximum growth rate to the corresponding cell yield, were 2.9 × 10^{-13} and 5.7 × 10^{-13} mol/cell-h, respectively. The apparent lactate from glucose yield was 1.96 mol mol^{-1}, suggesting that during batch growth glucose is converted almost entirely into lactate and that little is catabolized through the TCA cycle. Such a high yield has also been reported for continuous or pseudocontinuous culture at high growth rates (Miller *et al.*, 1988; Ramírez and Mutharasan, 1990). The total MAb produced and the volumetric and specific MAb production rates were 1.02 mg (1.4 mg if corrected for 40-ml), 0.28 mg/l-h, and 4.2 × 10^{-10} mg/cell-h, respectively.

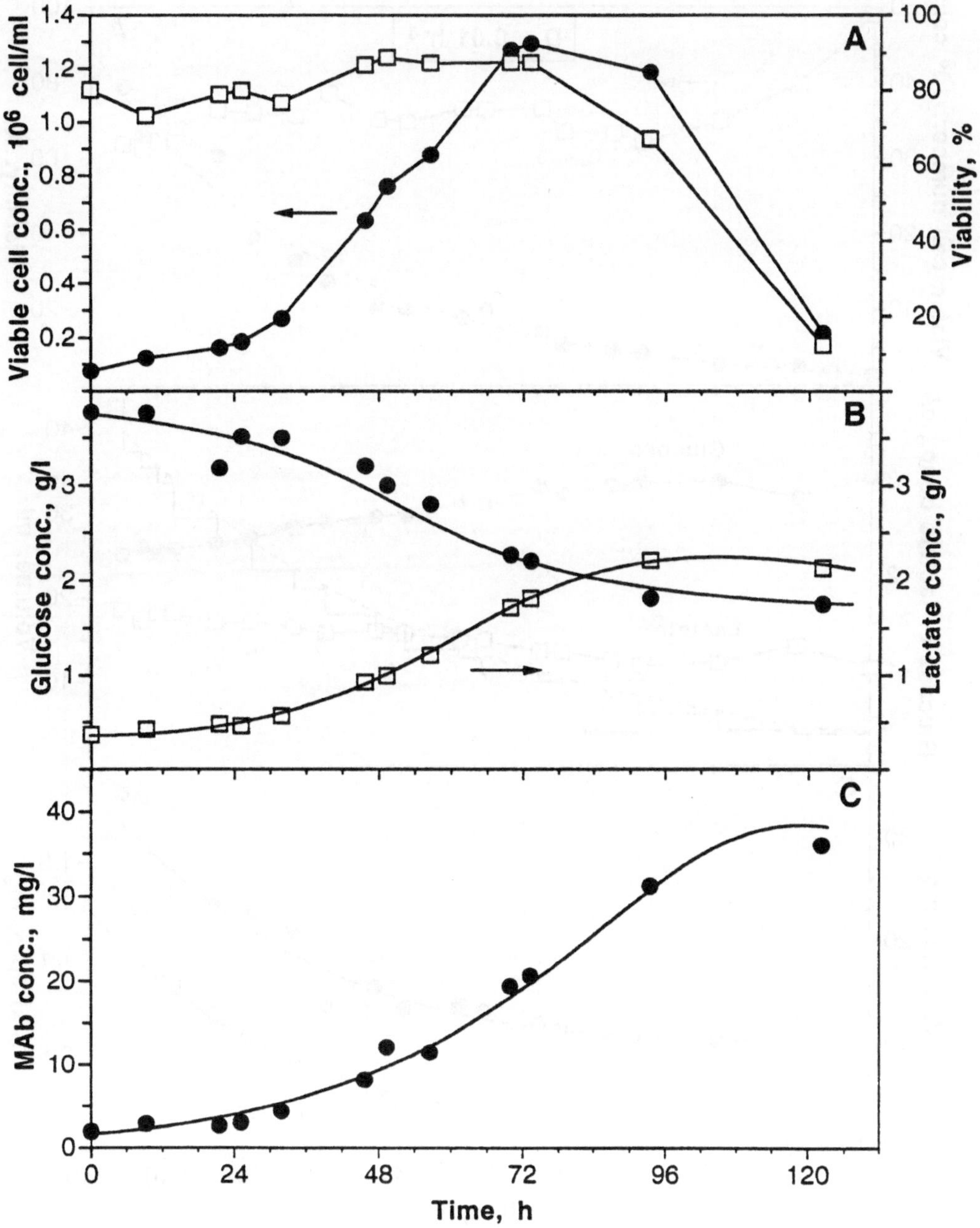

Fig. 1. Batch culture of BCF2 hybridoma in 75-cm^2 T-flask. A; viable cell concentration (closed circles) and viability (open squares). B; glucose (closed circles) and lactate (open squares) concentration. C; MAb concentration (closed circles).

Exponentially fed-batch cultures

Fig. 2 shows the results of BCF2 cells during EFBC in T-flasks at a dilution rate of 0.01 h^{-1}. Such results illustrate the typical behavior observed at intermediate dilution rates. As shown by the dotted-line in Fig. 2A, viable cell number increased exponentially, indicating that a constant growth rate was attained. Cell viability rapidly decreased from 98% to around 70% during the first 2 days of culture, but then remained at a relatively constant value of about 65% (Fig. 2A). Cell concentration increased from 0.38 × 10^6 cell ml^{-1} at inoculation, to a steady-state value of 0.98 × 10^6 cell ml^{-1} after 8 days of operation. Residual glucose

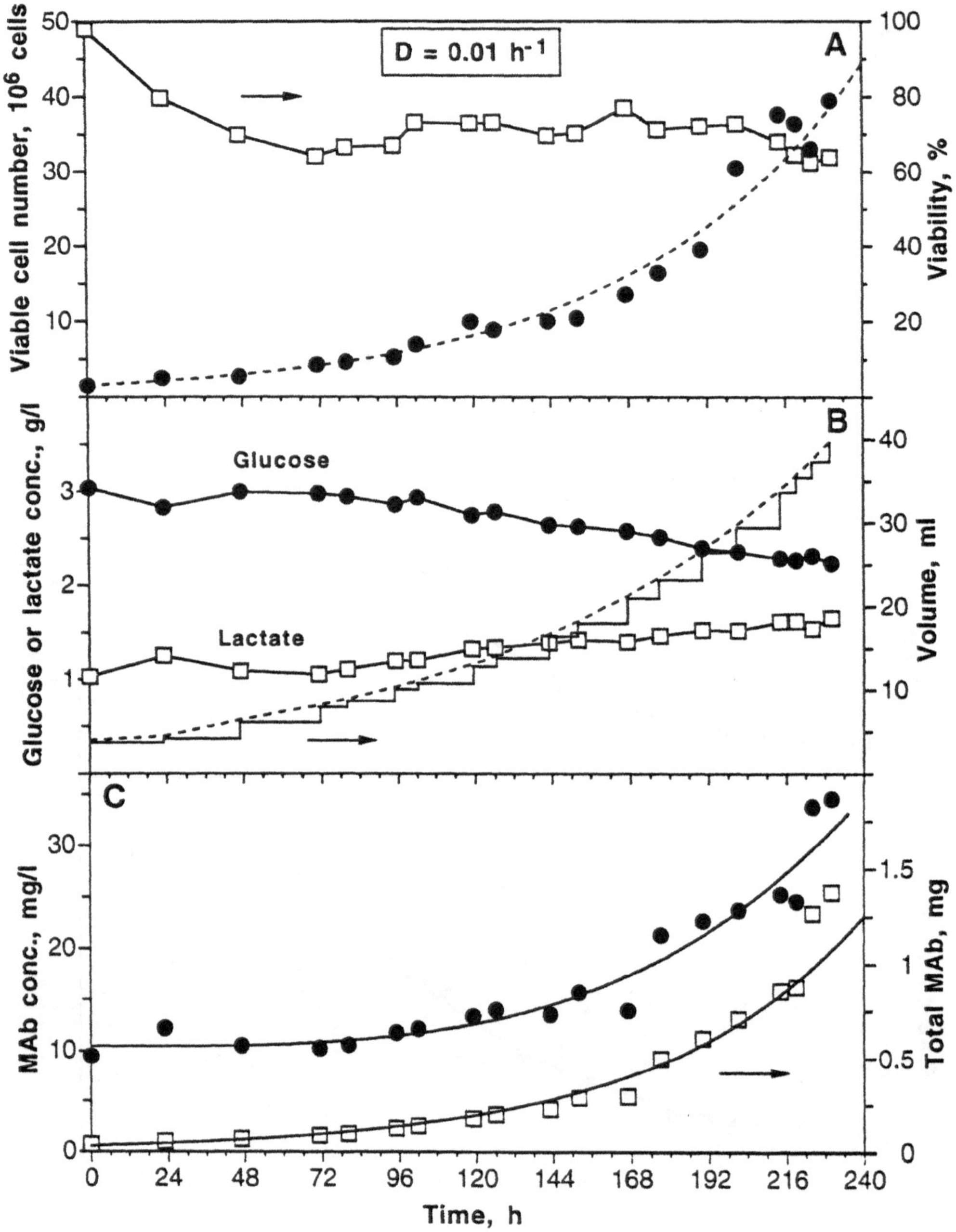

Fig. 2. Exponentially fed-batch culture of BCF2 hybridoma in 75-cm^2 T-flask at a dilution rate of 0.01 h^{-1}. A; viable cell number and viability. B; glucose and lactate concentration. C; MAb concentration and total MAb (open squares). Symbols as in Fig. 1. The continuous and dotted-lines in A and C correspond to an exponential fit. In panel B, the dotted-line corresponds to the ideal exponentially increasing volume profile that would have been generated if a continuous feeding had been established, and the stepwise profile corresponds to the actual volume.

and lactate concentrations also approached steady-state values after the same time period (Fig. 2B). No pH determinations were made, however from the lactate data and the color of the indicator phenol red in the medium, it can be inferred that pH also reached a steady-state. It should be noted that steady-state was attained using only a stepwise predetermined feeding profile, which is in contrast to other more complicated systems where some means of feedback control is required for maintaining a desired condition (Glacken *et al.*, 1989; Truskey *et al.*, 1990; Omasa *et al.*, 1992). Furthermore, in other types of fed-batch operation, steady-state condition is usually attained only for a limited number of parameters, whereas large variations occur in the remaining (Glacken *et al.*, 1989; Chevalot and Marc, 1993; Jo *et al.*, 1993b). As seen in Fig. 2C, MAb concentration did not reach a steady-state but continuously increased, with accumulated MAb exponentially increasing. The complete MAb production profile was determined just for this culture, whereas only the MAb concentrations at the beginning and end of the fed-batch phase were determined for the other dilution rates tested.

Figs. 3 and 4 present the results of the EFBC operated at dilution rates of 0.005 and 0.07 h^{-1}, respectively, the lowest and highest dilution rates tested. Such a broad range in dilution rates has not been, to our knowledge, studied before for an hybridoma cell line. Most CSC reports have only studied a 1.5- to 5.3-fold increase in dilution rate (Miller *et al.*, 1988; Ray, 1989; Frame and Hu, 1990; Hiller *et al.*, 1991; Martens *et al.*, 1993), being the range between 0.008 to 0.055 h^{-1} among the broadest one described (Leno *et al.*, 1992). Small scale bioreactor operation in continuous mode at very low dilution rates can present significant problems, such as difficulties in maintaining a constant volume due to inaccurate control of low flow rates and losses by evaporation, increased risk of contamination, and increased shear sensitivity. On the other hand, stable continuous or pseudocontinuous operation at growth rates close to the maximum is difficult and culture washout occurs if the critical dilution rate is exceeded. Such difficulties may explain the limited range of dilution rates studied in CSC. In contrast, EFBC permitted simple operation at a dilution rate almost one order of magnitude lower than the maximum growth rate (Fig. 3). Furthermore, since no cell removal exists in the EFBC, a true washout will not occur. This is shown in Fig. 4 where even at a dilution rate considerably higher than the batch maximum growth rate, the viable cell number slightly increased from 1.5×10^6 to 2×10^6 cells, even though the cell concentration decreased. Operation at such a high dilution rate resulted in the plentiful supply of all nutrients at non-limiting concentrations and very low concentration of toxic metabolic wastes, as illustrated in Fig. 4B for glucose and lactate concentration.

As exemplified in Figs. 2 to 4, the trajectory followed by the various parameters towards steady-state corresponds to the expected behavior of a first order system following Monod-type kinetics (see Equations (3) and (4)). Namely, a steep initial response succeeded by a slower direct, i.e. non oscillating, approach to steady-state. Such a transient behavior is the same as the one traced during start-up and step perturbations of CSC, with each particular system requiring a characteristic response time to reach steady-state. For instance, more than 13 residence times were needed to reach steady-state glucose concentration during CSC start-up and dilution rate step increases (Frame and Hu, 1990; Hiller *et al.*, 1991). In contrast, less than 3 residence times were needed for lactate concentration to reach steady-state after a step reduction in the feed medium glucose concentration (Ray *et al.*, 1989). The observed response times in the EFBC for the various parameters ranged between those cited above (see Figs. 2 to 4), underscoring the similarities between EFBC and CSC.

Effect of dilution rate

Fig. 5 to 7 summarize the results of the EFBC at all dilution rates (D) tested. As seen in Fig. 5, the steady-state viable cell concentration showed a maximum around 1×10^6 cell ml^{-1} at a D of 0.01 h^{-1}, but rapidly decreased at lower or higher D. A similar behavior has been reported for hybridomas in CSC or pseudo-continuous culture where maximum steady-state cell concentration has been found to occur between D of 0.015 to 0.03 h^{-1} (Miller *et al.*, 1988; Ray *et al.*, 1989; Ramírez and Mutharasan, 1990; Hiller *et al.*, 1991; Leno *et al.*, 1992). Residual glucose approached the concentration of feeding medium as D increased. Glucose did not limit growth as it remained above 1.9 g l^{-1} for all the conditions tested. Cell viability remained relatively constant at 80% for D above 0.02 h^{-1}, but rapidly decreased to less than 30% as D decreased to 0.005 h^{-1}. Such a high sensitivity of the viability at low dilution rates has been described previously for pseudocontinuous (Ramírez and Mutharasan. 1990) and CSC (Boraston *et al.*, 1984; Miller *et al.*, 1988), and is probably caused, at least partially, by the higher levels of byproducts and lower concentration of essential nutrients. For instance, as seen in Figs. 2 to 4, the steady-state lactate concentration decreased from 1.8 to 0.2 g l^{-1} as D increased from 0.005 to 0.07 h^{-1}. Such a change in lactate concentration corresponds to a change in pH of only 0.06, as measured in cell-free control medium. Thus, differences in viability and other parameters discussed below, such as MAb production, between the various dilution rates cannot be attributed to the differences in pH. The rate at which the viability decreased as D decreased was found to be lower in the EFBC than in the CSC reports. This sug-

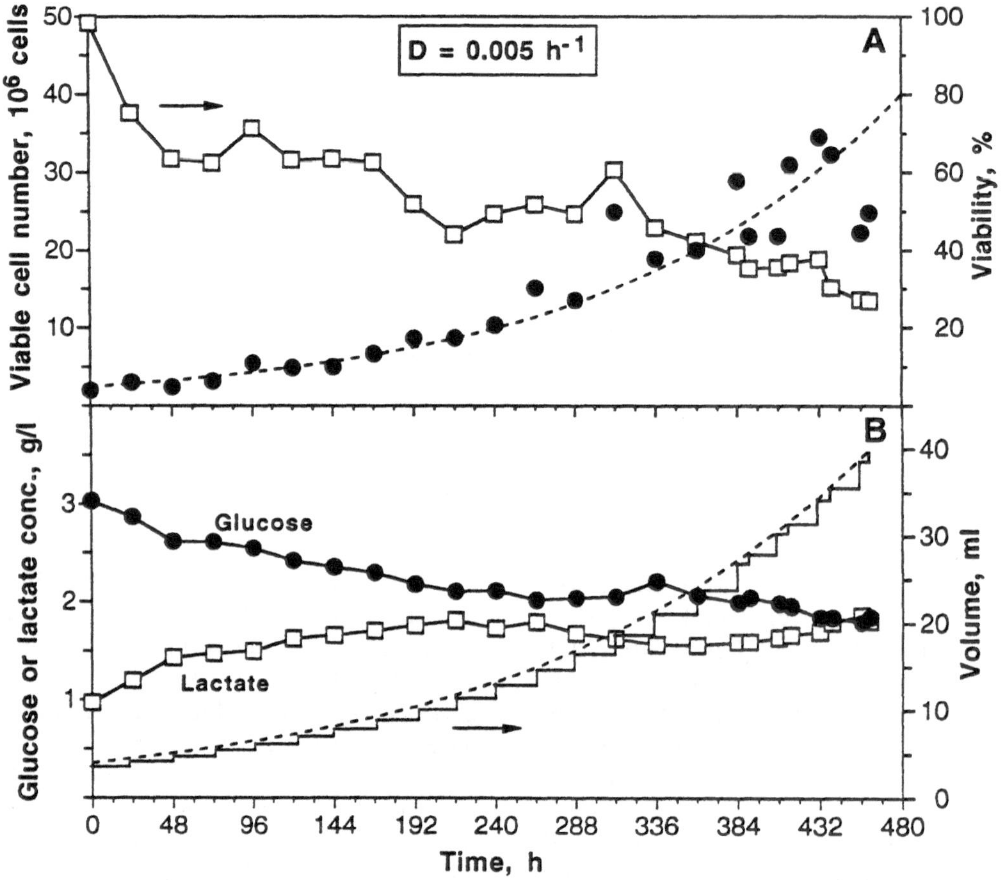

Fig. 3. Exponentially fed-batch culture of BCF2 hybridoma in 75-cm^2 T-flask at a dilution rate of 0.005 h^{-1}. Rest of legend as in Fig. 2.

gests that at low growth rates the death rates in the EFBC are lower than in the agitated CSC, which can be explained by the absence of hydrodynamic damage in the static culture conditions.

Figure 5B shows that the growth rate at steady-state followed the same behavior of an ideal CSC (indicated in the graph by the dotted-diagonal line) for growth rates below the maximum observed in batch cultures (indicated by the dotted-horizontal line). That is, a predetermined growth rate equal to the dilution rate can be attained. Surprisingly, for D above the maximum growth rate, the apparent specific growth rate did not remain at its maximum, but rather decreased rapidly to very low values. Such a behavior contrasts with that of microbial EFBC where the specific growth rate has been found to remain at its maximum value as D increased above the maximum specific growth rate (Ramírez *et al.*, 1994a). As mentioned before, even though a net increase of 0.5×10^6 cells was observed in the EFBC at a D of 0.07 h^{-1}, the high rate of medium addition caused a decrease in cell concentration from an initial value of 0.4×10^6 cel ml^{-1} to less than 0.07×10^6 cel ml^{-1} after 24 h of operation. Therefore, the low growth rate observed at such a high dilution indicates that a minimum cell density is needed for active growth to occur. Such a requirement also strongly suggests the existence of autocrine growth factors necessary for hybridoma proliferation. Similar requirements, in medium with or without serum, and evidence of the existence of autrocrine growth factors have been reported before (Tharakan and Chau, 1986; Lee *et al.*, 1991). The work of Leno *et al.* (Leno *et al.*, 1992) was found to be the only report of a culture operated at D above the batch maximum specific growth rate. In contrast to this work, they found growth rates as much as 25% higher than the maximum batch specific growth rate, which may be explained by the much higher cell concentration present when the high D were initiated, as well as the possibility of substantial growth on walls due to the particular system employed.

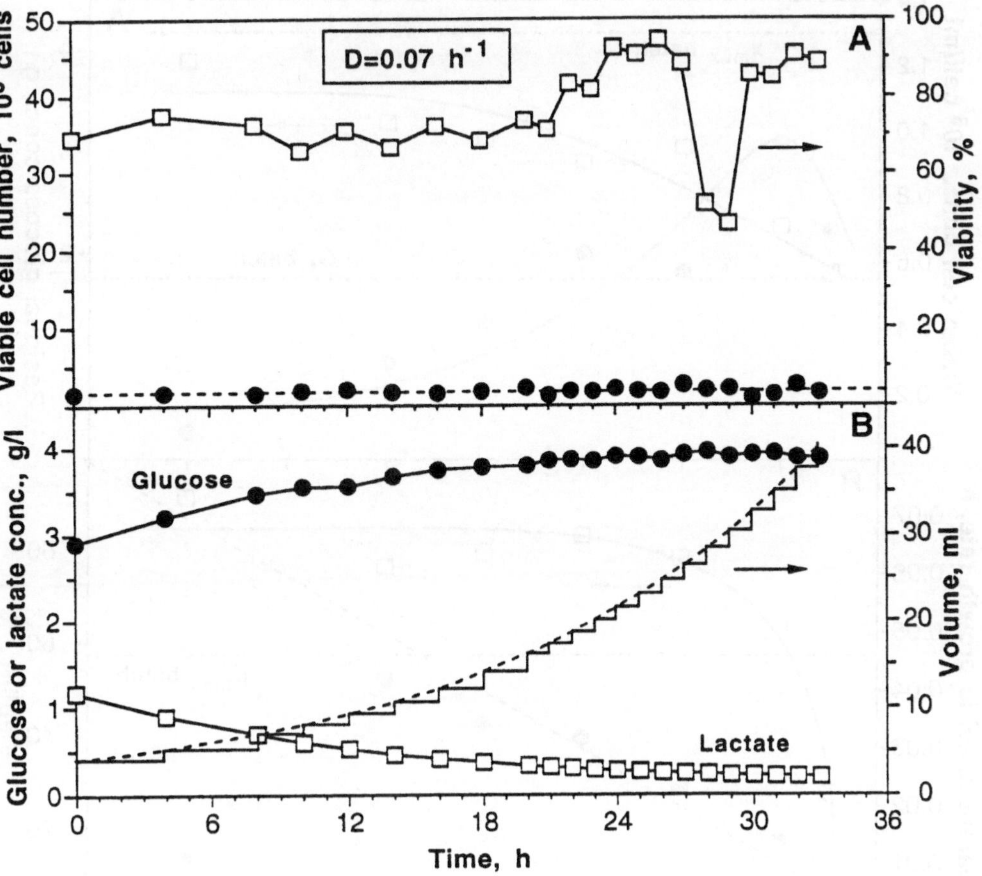

Fig. 4. Exponentially fed-batch culture of BCF2 hybridoma in 75-cm^2 T-flask at a dilution rate of 0.07 h^{-1}. Rest of legend as in Fig. 2. In A, the same scale as in Figs. 2 and 3 was maintained to facilitate comparison.

The average specific glucose uptake rate (q_G) and the average specific lactate production rate (q_L) as a function of D are shown in Fig. 6A. As observed by others (Miller *et al.*, 1988; Ramírez and Mutharasan, 1990; Hiller *et al.*, 1991) q_G and q_L increased linearly with D for growth rates below the maximum. For D above 0.045 h^{-1}, q_G and q_L decreased due to the lower apparent specific growth rates that occurred during such high D. The apparent yield of lactate from glucose ($Y_{L/G}$) remained relatively constant at around 1.7 mol/mol up to a D of 0.03 h^{-1} where it started to decrease (Fig. 6B). Hiller *et al.* (Hiller *et al.*, 1991) have also observed that $Y_{L/G}$ remains relatively constant in CSC between D of 0.015 to 0.03 h^{-1}, but in contrast to this work, they observed an increasing $Y_{L/G}$ as D increased above 0.03 h^{-1}. The $Y_{L/G}$ at all the D tested in the EFBC were lower than in the batch culture, which indicates a higher efficiency of glucose utilization under the controlled growth conditions.

The effect of D on MAb production during EFBC is summarized in Fig. 7. The data points represent the overall parameters determined from the MAb concentrations at the initiation and end of the fed-batch phase (Equations (7) to (9)). It should be noted that the culture volume increased from 4 to 40-ml during all the EFBC. Therefore, in every condition tested, the same mean number of 3.33 cell generations was allowed. Accordingly, such a protocol results in a more reliable comparison between experiments and avoids interpretation problems due to generation-associated phenomena, such as loss of MAb productivity due to cell line instability (Frame and Hu, 1990; Hiller *et al.*, 1991). For instance, Martens *et al.* (Martens *et al.*, 1993) have observed in CSC the occurrence of two distinctive hybridoma populations and a drop in specific MAb productivity after about 30 generations. Such behavior should caution on the use of CSC for extended periods of time during hybridoma kinet-

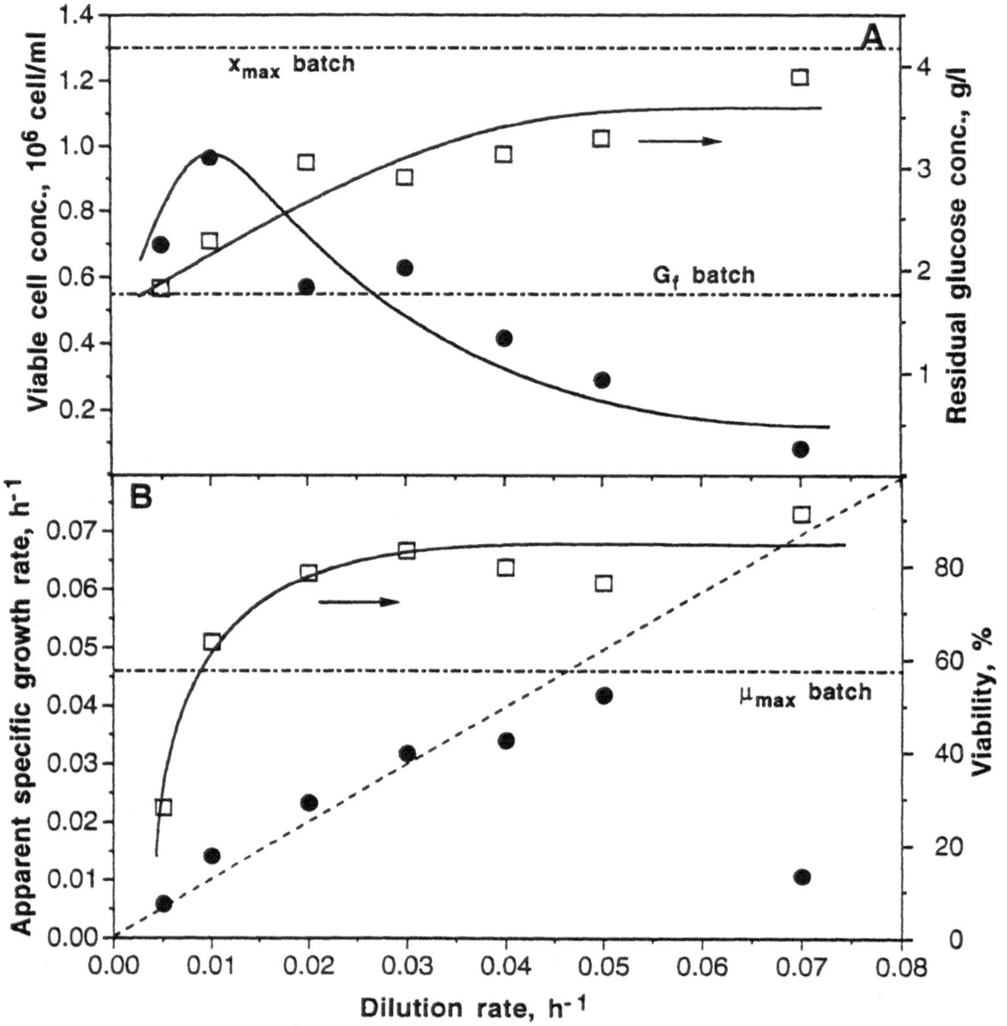

Fig. 5. Summary of exponentially fed-batch cultures. A; viable cell (closed circles) and residual glucose (open squares) conc. B; apparent specific growth rate (closed circles) and viability (open squares). All data points represent the average of the values at steady-state. The horizontal dotted-lines represent the values obtained during batch culture shown in Fig. 1. The diagonal dotted-line in B represents the ideal case where μ = D. Continuous lines drawn to show trend.

ic studies. As seen in Fig. 7A, the volumetric MAb production rate and the total MAb produced during the fed-batch phase decreased 6- and 65-fold, respectively, with increasing D. Likewise, the specific MAb production rate decreased from 6.5×10^{-10} mg/cell-h at a D of 0.005 h^{-1} to around 2×10^{-10} mg/cell-h at a D of 0.05 h^{-1} (Fig. 7B). Such a negative correlation between MAb secretion rate and growth rate has been commonly reported for different hybridoma cell lines (Boraston *et al.*, 1984; Miller *et al.*, 1988; Ramírez and Mutharasan, 1990; Leno *et al.*, 1992; Martens *et al.*, 1993). For instance, Ramírez and Mutharasan (Ramírez and Mutkarasan, 1990) observed that q_{MAb} decreased from 2×10^{-10} to 0.5×10^{-10} mg/cell-h as dilution rate increased from 0.007 to 0.042 h^{-1}, and attributed such a behavior to a higher MAb production rate of G_1 cells compared to cells traversing through mitosis and other stages of interphase. Likewise, Miller *et al.* (Miller *et al.*, 1988) found an approximately 16- and 2-fold increase in antibody concentration and q_{MAb}, respectively, for a 2.5-fold decrease in dilution rate. Such an increase was proposed to be either due to a stressed condition that favored protein production or to differential MAb production rates during the various phases of the cell cycle. In addition to such possibilities, it has also been discussed (Martens *et*

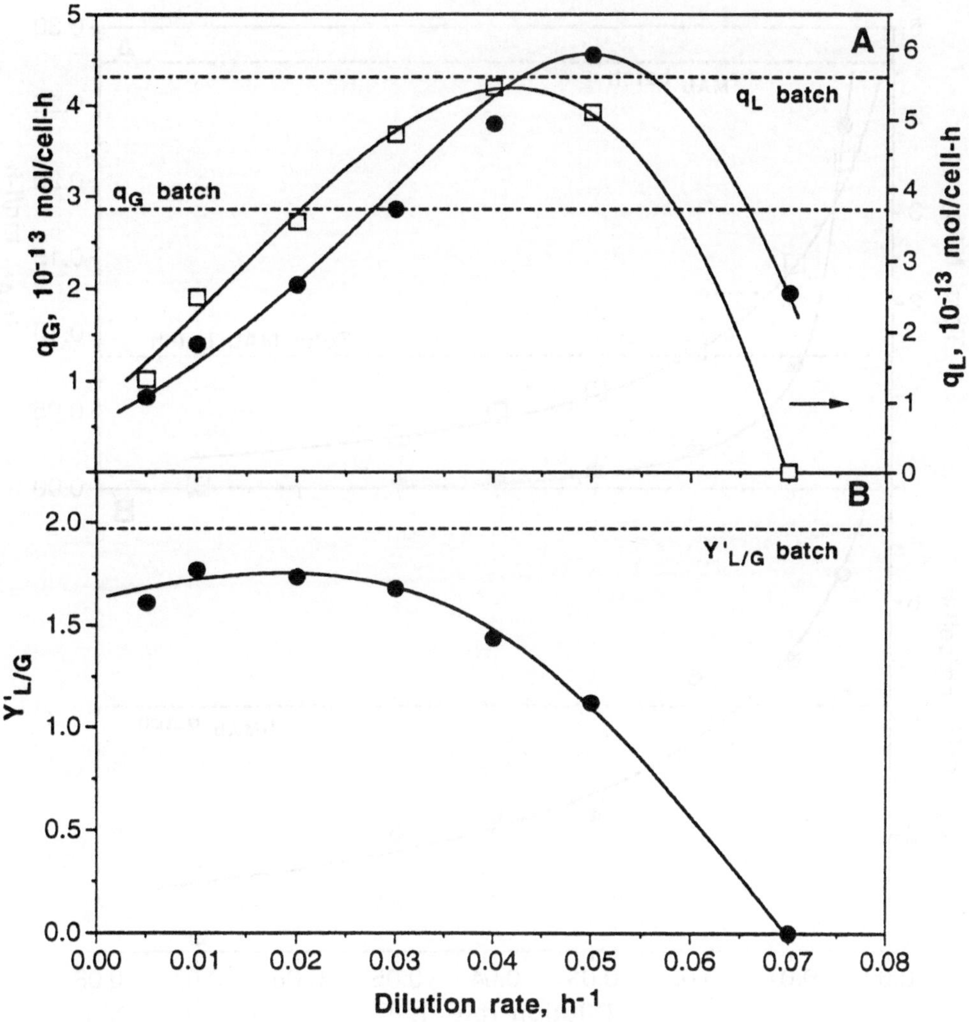

Fig. 6. Summary of exponentially fed-batch cultures. A; metabolic quotients for glucose (closed circles) and lactate (open squares). B; apparent molar yield of lactate from glucose. The horizontal dotted-lines represent the values obtained during batch culture shown in Fig. 1. Continuous lines drawn to show trend.

al., 1993) that the increased MAb production at low dilution rates could be due to an increased passive release of antibodies or its fractions by the larger concentration of dead cells under such conditions. Nevertheless, independently of the explanation given for the observed behavior, it can be seen that the EFBC can be used as an important experimental tool to determine such trends, especially since dissimilar behaviors exist between different hybridoma cells. For instance, increasing q_{MAb} with increasing D (Low *et al.*, 1987), an optimum q_{MAb} at intermediate growth rates (Ray *et al.*, 1989), or no effect at all (Flickinger *et al.*, 1990) have also been reported. Any of these behaviors can be explained from simulation analysis of the assembly rate of the heavy and light chains of the antibody molecule (Flickinger *et al.*, 1992).

Finally, results of the batch culture, corrected for 40-ml culture volume, have also been included in Fig. 7 for comparison and are represented as the horizontal dotted lines. As seen, the overall volumetric MAb productivity obtained in batch culture was higher than in EFBC at any of the dilution rates tested. This results from the higher cell concentration maintained during a longer period of time in batch compared to the EFBC. In contrast, the total MAb and q_{MAb} in batch culture only reached the values obtained at intermediate dilution rates in the EFBC. Since in batch culture cells grow from very low to very high rates as they progress

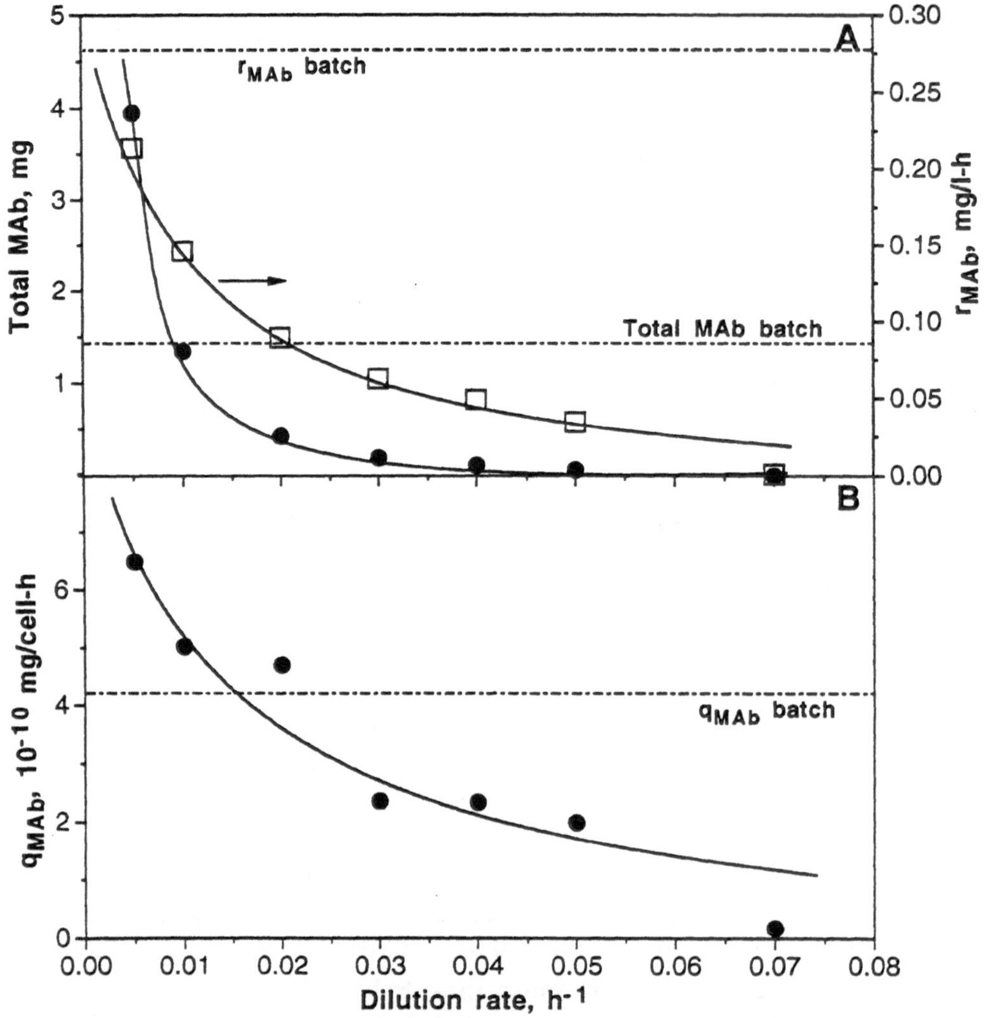

Fig. 7. Summary of exponentially fed-batch cultures. A; total MAb (closed circles) and volumetric rate of MAb production (open squares). B; specific MAb production rate. The data points represent overall parameters for the fed-batch phase. The horizontal dotted-lines represent the values obtained during batch culture shown in Fig. 1, and have been corrected for 40-mL culture volume to allow comparison. Continuous lines drawn to show trend.

through the various phases, it is reasonable that the total and specific MAb values correspond to those of an average intermediate growth rate.

Conclusions

The results of this study show that exponentially fed-batch cultures of hybridomas in T-flasks can be used as a simple alternative experimental tool to chemostats for the study of metabolism, growth, and MAb production kinetics. EFBC allowed simple and stable operation of simultaneous multiple experiments at either very low or very high dilution rates, minimizing the start-up time, equipment cost, media utilization, and risk of contamination. Furthermore, the dynamic and steady-state behaviors of the EFBC were found to be similar to those reported for CSC. For the hybridoma cell line generated and studied here, it was found that with increasing growth rate the viability, metabolic quotients for glucose and lactate, and residual glucose concentration increased; whereas viable cell concentration, total MAb, MAb specific and volumetric production rates, and residual lactate decreased. Finally, low growth rates observed at dilution rates above the batch maximum growth rate suggest the requirement of a minimum cell density, and concomitant autocrine growth factors, for growth. Production studies, underway, on the implementation of EFBC in bioreactors should overcome interpretation difficulties due to the lack of an accurate control of pH and dissolved oxygen; such limitations are always associated with any study performed in T-flasks.

Acknowledgements

This research was supported by grants DGAPA-National University of Mexico No. IN300991 and National Council of Science and Technology Mexico No. 1105-N9201. We wish to express our gratitude to Mrs. Maribel Flores Martínez and Mr. Jesus Martínez Dorantes for their technical assistance in this work.

References

Boraston R, Thompson PW, Garland S and Birch JR (1984) Growth and oxygen requirements of antibody producing mouse hybridoma cells in suspension culture. Dev. Biol. Stand. 55: 103–101

Chevalot I and Marc A (1993) Interest of fed-batch culture for the production of a membrane-bound protein by an adherent animal cell. Biotech. Letts. 15: 791–796.

Esener AA, Roels JA and Kossen NWF (1981) Fed-batch culture: modeling and applications in the study of microbial energetics. Biotech. Bioeng. 23: 1851–1871.

Flickinger MC, Goebel NK and Bohn MA (1990) Determination of specific monoclonal antibody secretion rate during very slow hybridoma growth. Bioproc. Eng. 5: 155–164.

Flickinger MC, Bibila TA and Kitchin K (1992) Prediction and experimental investigation of the effect of altered peptide chain assembly rate on antibody secretion using a structured model of antibody synthesis and secretion. In: Ladisch MR and Bose A (ed.) Harnessing Biotechnology for the 21st Century. (pp. 345–350) American Chemical Society.

Frame KK and Hu WS (1990) The loss of antibody productivity in continuous culture of hybridoma cells. Biotech. Bioeng. 35: 469–476.

Glacken MW, Huang C and Sinskey AJ (1989) Mathematical descriptions of hybridoma culture kinetics. III. Simulation of fed-batch bioreactors. J. Biotech. 10: 39–66.

Higareda AE, Possani LD and Ramírez OT (1993) Use of on-line culture redox potential and oxygen uptake rate measurements in advanced computerized nutrient feeding strategies for optimization of MAb production. Paper 104g American Institute of Chemical Engineers Annual Meeting. St. Louis MO, U.S.A.

Hiller GW, Aeschlimann, AD, Clark DS and Blanch HW (1991) A kinetic analysis of hybridoma growth and metabolism in continuous suspension culture on serum-free medium. Biotech. Bioeng. 38: 733–741.

Hu WS and Piret JM (1992) Mammalian cell culture processes. Curr. Opinion Biotech. 3: 110–114.

Jo ECh, Kim DI and Moon HM (1993a) Step-fortifications of nutrients in mammalian cell culture. Biotech. Bioeng. 42: 1218–1228.

Jo ECh, Park HJ, Kim DI and Moon HM (1993b) Repeated fed-batch culture of hybridoma cells in nutrient-fortified high-density medium. Biotech. Bioeng. 42: 1229-1237.

Keller R and Dunn IJ (1978) Fed-batch microbial culture: models, errors and applications. J. Appl. Chem. Biotech. 28: 508–514.

Kurokawa H, Ogawa T, Kamihira M, Park YS, Iijima S and Kobayashi T (1993) Kinetic study of hybridoma metabolism and antibody production in continuous culture using serum-free medium. J. Ferm. Bioeng. 76: 128–133.

Lee GM, Varma A and Palsson BO (1991) Production of monoclonal antibody using free suspended and immobilized hybridoma cells: effect of serum. Biotech. Bioeng. 38: 821–830.

Leno M, Merten OW and Hache J (1992) Kinetic analysis of hybridoma growth and monoclonal antibody production in semicontinuous culture. Biotech. Bioeng. 39: 596–606.

Low KS, Harbour C, Barford JP (1987) A study of hybridoma cell growth and antibody production kinetics in continuous culture. Biotech. Tech. 1: 239–244.

Martens DE, de Gooijer CD, Velden-de Groot CAM Beuvery, EC and Tramper J (1993) Effect of dilution rate on growth, productivity, cell cycle and size and shear sensitivity of a hybridoma cell in a continuous culture. Biotech. Bioeng. 41: 429–439.

Miller WM, Blanch HW and Wilke CR (1988) A kinetic analysis of hybridoma growth and metabolism in batch and continuous suspension culture: effect of nutrient concentration, dilution rate and pH. Biotech. Bioeng. 32: 947–965.

Omasa T, Higashiyama K, Shioya S and Suga Ki (1992) Effects of lactate concentration on hybridoma culture in lactate-controlled fed-batch operation. Biotech. Bioeng. 39: 556–564.

Ozturk SS, Riley MR and Palsson BO (1992) Effects of ammonia and lactate on hybridoma growth, metabolism and antibody production. Biotech. Bioeng. 39: 418–431.

Papoutsakis ET (1991) Media additives for protecting freely suspended animal cells agitation and aeration damage. TIBTECH. 9: 316–324.

Ramírez OT and Mutharasan R (1990) Cell cycle and growth phase-dependent variations in size distribution, antibody productivity and oxygen demand in hybridoma cultures. Biotech. Bioeng. 36: 839–848.

Ramírez OT, Zamora R, Quintero R and López-Munguía A (1994a) Exponentially fed-batch cultures as an alternative to chemostats: the case of penicillin acylase production by recombinant E. coli Enzyme Microb. Tech. 16: 895–903.

Ramírez OT, Aguilar-Aguila A and Quintero R (1994b) Dynamic behavior of activated-sludge in exponentially fed-batch cultures subjected to step perturbations. In: Galindo E and Ramírez OT (eds.) Advances in Bioprocess Engineering. Kluwer Academic Publishers, The Netherlands, pp. 345-353.

Ray NG, Karkare SB and Runstadler PW (1989) Cultivation of hybridoma cells in continuous cultures: kinetics of growth and product formation. Biotech. Bioeng. 33: 724–730.

Reddy S, Bauer KD and Miller, WM (1992) Determination of antibody content in live versus dead hybridoma cells: analysis of antibody production in osmotically-stressed cultures. Biotech. Bioeng. 40: 947–964.

Reuveny S, Velez D, Macmillan JD and Miller L (1987) Factors affecting monoclonal antibody production in culture. Dev. Biol. Stand. 66: 169–174.

Salazar-Kish JM and Heath CA (1993) Comparison of a quadroma and its parent hybridomas in fed batch culture. J. Biotech. 30: 351–365.

Schmid G, Zilg H and Johannsen R (1992) Repeated batch cultivation of rBHK cells on cytodex 3 microcarriers: antithrombin III, amino acid, and fatty acid metabolic quotients. Appl. Microb. Biotech. 38: 328–333.

Tharakan JP and Chau PC (1986) Serum free fed batch production of IgM. Biotech. Letts. 8: 457–462.

Truskey GA, Nicolakis DP, DiMasi D, Haberman A and Swartz RW (1990) Kinetic studies and unstructured models of lymphocyte metabolism in fed-batch culture. Biotech. Bioeng. 36: 797–807.

Westgate PJ, Wayne CR, Emery AH, Hasegawa PM and Heinstein PF (1991) Approximation of continuous growth of *Cephalotaxus harringtonia* plant cell cultures using fed-batch operation. Biotech. Bioeng. 38: 241–246.

Yamané T and Shimizu S (1984) Fed-batch techniques in microbial processes. Adv. Biochem. Eng. Biotech. 30: 145–194.

Zamudio F, Saavedra R, Martin BM, Gurrola-Briones G, Herion P and Possani LD (1992) Amino acid sequence and immunological characterization with monoclonal antibodies of two toxins from the venom of the scorpion *Centruroides noxius* Hoffmann. J. Biochem. 204: 281–292.

Address for offprints: O.T. Ramírez, Department of Bioengineering, Instituto de Biotecnología, Universidad Nacional Autónoma de México, Avenida Universidad 2001, Apdo. Postal 510-3, Cuernavaca, Morelos 62271, México.

Cytotechnology **15:** 87–94, 1994.

Adaptation of mammalian cells to non-ammoniagenic media

Michael Butler and Andrew Christie
Department of Microbiology, University of Manitoba, Winnipeg, Manitoba, Canada R3T 2N2

Key words: Adaptation, ammonia, cell culture, glutamine, glutamate, dipeptides

Abstract

Although glutamine is used as a major substrate for the growth of mammalian cells in culture, it suffers from some disadvantages. Glutamine is deaminated through storage or by cellular metabolism, leading to the formation of ammonia which can result in growth inhibition. Non-ammoniagenic alternatives to glutamine have been investigated in an attempt to develop strategies for obtaining improved cell yields for ammonia sensitive cell lines.

Glutamate is a suitable substitute for glutamine in some culture systems. A period of adaptation to glutamate is required during which the activity of glutamine synthetase and the rate of transport of glutamate both increase. The cell yield increases when the ammonia accumulation is decreased following culture supplementation with glutamate rather than glutamine. However some cell lines fail to adapt to growth in glutamate and this may be due to a low efficiency transport system.

The glutamine-based dipeptides, ala-gln and gly-gln can substitute for glutamine in cultures of antibody-secreting hybridomas. The accumulation of ammonia in these cultures is less and cell yields in dipeptide-based media may be improved compared to glutamine-based controls. In murine hybridomas, a higher concentration of gly-gln is required to obtain comparable cell growth to ala-gln or gln-based cultures. This is attributed to a requirement for dipeptide hydrolysis catalyzed by an enzyme with higher affinity for ala-gln than gly-gln.

Introduction

High cell yields are required in culture processes designed for the large-scale production of biologicals. In such cultures, there are three major factors that may reduce cell growth: (i) The depletion of an essential nutrient; (ii) The accumulation of a growth inhibitory product; (iii) The complete utilisation of the growth surface required by anchorage-dependent cells.

In some culture systems the accumulation of ammonia has been identified as the likely limiting factor which determines the final cell yield (Butler *et al.*, 1983; Butler and Spier, 1984). The ammonia can arise by glutamine degradation in the culture medium (Tritsch and Moore, 1962) or by metabolic deamination in the cells (Reitzer *et al.*, 1979). Glutamine is normally provided in standard culture medium at a concentration of 2–6 mM, which can degrade spontaneously at 37 °C to form ammonia at about 0.2–0.6 mM day^{-1}. This rate may be enhanced by the presence of glutaminase from cell lysis during the later stages of culture.

Glutamine is required by cells as a precursor for protein and nucleotide synthesis as well as a major energy source for cellular metabolism (Zielke *et al.*, 1984; Wice *et al.*, 1981). Its catabolic breakdown (glutaminolysis) involves a two stage deamination (or transamination), resulting in the formation of free ammonia which is released into the medium.

A further disadvantage to the use of glutamine in culture medium is that its chemical instability limits the shelf life of liquid medium even when stored at 4 °C. This also prohibits the use of autoclaving as a means of sterilisation.

In this paper, the importance of the growth inhibition of ammonia is reviewed in various culture systems. Substitutes for glutamine are considered in an attempt to alleviate the problem of ammonia accumulation in cultures of sensitive cell lines. In particular, the potential of using glutamate or glutamine-based dipeptides as non-ammoniagenic substitutes for glutamine in culture medium is explored.

Results and discussion

The extent of growth inhibition by ammonia

A combination of cellular consumption and chemical breakdown causes the rapid depletion of glutamine from cultures in 3/4 days under standard batch conditions. This results in an accumulation of ammonia ($NH_3 + NH_4^+$), typically to a concentration of 1–4 mM.

Growth limitations caused by the accumulation of ammonia have been widely recognised (Ryan and Cardin, 1966; Butler and Spier, 1984; Reuveny *et al.*, 1986; Glacken *et al.*, 1986). Estimates of the concentration of ammonia to inhibit cell growth by 50% (IC-50) have varied from 0.5 mM for mouse L cells (Ryan and Cardin, 1966), 3.5 mM for a murine hybridoma (Reuveny *et al.*, 1986) to 9.6 mM for human myelomas (Taya *et al.*, 1986). However, it is difficult to draw conclusions from such comparisons because of differences in culture conditions. Growth inhibition by ammonia inhibition is dependent on a variety of culture parameters including pH (Doyle and Butler, 1990) and lactate concentration (Hassell *et al.*, 1991).

The effect of added ammonia on the growth of a number of cell lines under standardised culture conditions was determined by Hassell *et al.* (1991). The cultures were supplemented with ammonium chloride at a concentration range up to 2.5 mM (Table I). In the case of the hybridoma, the concentration range was extended to 20 mM. Cell densities of these cultures were determined after 3 days growth.

The cell lines were classified into three levels of sensitivity to ammonia, based upon the degree of growth inhibition in the presence of 2 mM NH_4Cl. A low level of sensitivity was shown by four cell lines (293, HDF, Vero and PQXB1/2). For two of these cell lines (HDF and Vero), no growth inhibition was observed even at the highest concentration of added NH_4Cl (2.5 mM). In the case of the 293 and PQXB1/2 cells, the growth inhibition observed at 2 mM NH_4Cl was low (< 15%). MDCK and McCoy cells were classified as intermediate in terms of their sensitivity to ammonia with a 50–60% decrease in cell yield in the presence of 2 mM NH_4Cl. A high level of growth inhibition was observed in the third group (HeLa and BHK) in the presence of 2 mM NH_4Cl (> 75%).

Table 1. The effect of added ammonium chloride on the growth of several cell lines

Cell line	Final $[NH_4^+]$ in control cultures (mM)	% decrease in cell yield at 2 mM NH_4Cl	IC-50 (mM)
HDF	1.6	0	> 2.5
Vero	2.9	0	> 2.5
293	2.4	8	> 2.5
PQXB1/2	1.3	14	5.1
MDCK	1.8	50	1.8
McCoy	2.3	60	1.7
HeLa	2.7	75	0.8
BHK	2.3	80	1.3

The anchorage-dependent cells (all except PQXB1/2) were inoculated at 3.5-5.5 $\times$ 10^5 cells into 10 ml GMEM and grown for 3 d. Final cell numbers of control cultures (no added ammonia) were between 0.75 and 3.0 $\times$ 10^6 cells. The PQXB1/2 hybridomas were inoculated at 1.4 $\times$ 10^5 cells and grown in RPMI 1640 (2 ml) for 3 d. Control cultures increased to a cell density of 1.06 $\times$ 10^6 cells/ml (from Hassell *et al.*, 1991).

This experiment confirms the varying sensitivity of cell lines to the ammonia concentration (1–3 mM) that typically accumulates in culture. At 2 mM NH_4Cl, the growth of some cell lines such as Vero are unaffected whereas other cell lines such as BHK show extensive (80%) inhibition.

Adaptation to glutamate-based media

Glutamine is the major energy source for most mammalian cell culture systems and derives from the original media formulations which were based upon observed cellular requirements (Eagle, 1955). The preferred route of glutamine catabolism appears to be via phosphate-activated glutaminase followed by transamination. The resulting carbon skeleton can enter the TCA cycle as 2-oxoglutarate for further catabolism and generation of ATP (Jenkins *et al.*, 1992). The amino groups stripped from the glutamine are released into the culture medium as free ammonia or as extracellular amino acids, usually alanine or aspartate.

The use of a substitute for glutamine in culture medium is an attractive proposition for two main reasons. Firstly, because of the potential for reducing the concentration of accumulated ammonia during cell growth and secondly in order to improve the thermostability of media.

Glutamate would appear to be a suitable candidate as it has one less amino group than glutamine and is relatively thermostable. However, cell lines vary in their ability to adapt. HeLa cells were adapted to growth in a medium containing a high concentration of glutamate (20 mM) by Eagle *et al.* (1956) but mouse L fibroblasts failed to adapt using the same protocol. The human diploid fibroblasts, WI-38 and MRC-5 were reported to have an increased growth potential in a medium in which glutamine (4 mM) was substituted by glutamate on a mole to mole basis (Griffiths, 1973).

The varying ability of a range of anchorage-dependent cell lines to adapt to glutamate-based media was reported by Hassell and Butler (1990). McCoy cells adapted most readily within 10 days of continuous growth in a glutamate-based medium (GMEM+gmate). After this initial period, cell growth assumed the same rate in GMEM+gmate as in GMEM+gmine. For other cell lines (eg: BHK and Vero) the process of adaptation took longer (15–30 d) and involved the use of an initially higher glutamate concentration before normal growth rates were established in the glutamate-based media. Other cell lines (eg: MDCK and a hybridoma, CC9C10) failed to adapt to glutamate-based media using this protocol.

The effect of growth in glutamate-based media

The growth of McCoy cells was monitored in microcarrier cultures containing glutamine (GMEM+gmine) or glutamate (GMEM+gmate) (Hassell and Butler, 1990). The growth rate was equivalent in the two media but the final cell yield attained in GMEM+gmate was approximately 15% higher than in GMEM+gmine (Figure 1). Daily supplementation of the cultures with glutamine or glutamate (1 mM) from day 2 enhanced cell yields but the relative difference between cultures containing the two media remained.

The rates of utilisation of either glutamine or glutamate were comparable (Figure 2) with complete depletion of either amino acid occurring on day 4. However, the rates of ammonia accumulation in the cultures were quite different. The ammonia concentration increased to a maximum of 3.7 mM in GMEM+gmine cultures,

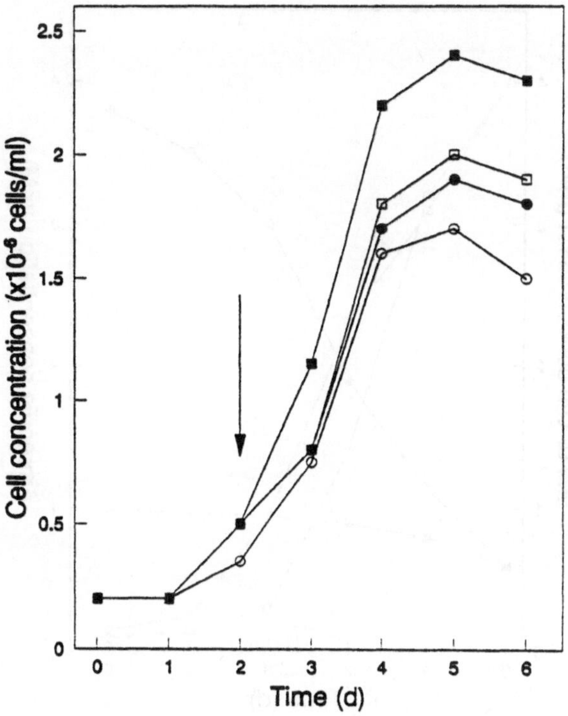

Fig. 1. Growth of McCoy cells in microcarrier cultures (100 ml). Selected cultures were fed with daily supplements of single amino acids (from the arrow). Cell concentrations are shown for cultures in GMEM+gmine with (●) and without (○) feeding of glutamine, and GMEM+gmate with (■) and without (□) feeding of glutamate, (n = 2) (from Hassell and Butler, 1990).

whereas the ammonia concentration of GMEM+gmate remained less than 1.1 mM.

Changes in the glucose and lactate concentrations were also monitored over the culture period (Figure 3). Glucose utilisation was significantly greater in the GMEM+gmine cultures with a decrease in concentration of 17 mM compared to a much lower value (5 mM) in GMEM+gmate. The values for the metabolic coefficient, lactate/glucose were also significantly different between the two cultures at 1.15 in GMEM+gmine and 0.59 in GMEM+gmate.

Factors governing adaptability of cells to glutamate-based media

The reason for differences in the ability of cell lines to adapt to glutamate-based media was investigated by a comparison of two cell lines – MDCK and McCoy (McDermott and Butler, 1993). Both cell lines had a normal growth rate with a doubling time of 18–20 h in GMEM+gmine. However, McCoy cells adapted to

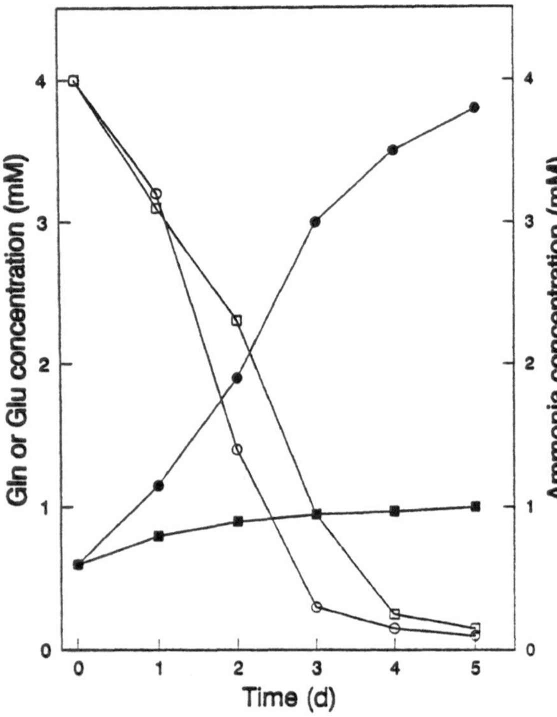

Fig. 2. Glutamine/glutamate utilisation and ammonia production for microcarrier cultures of McCoy cells. Glutamine (○) and ammonia (●) were measured in GMEM+gmine cultures. Glutamate (□) and ammonia (■) were measured in GMEM+gmate cultures, (n = 2) (from Hassell and Butler, 1990).

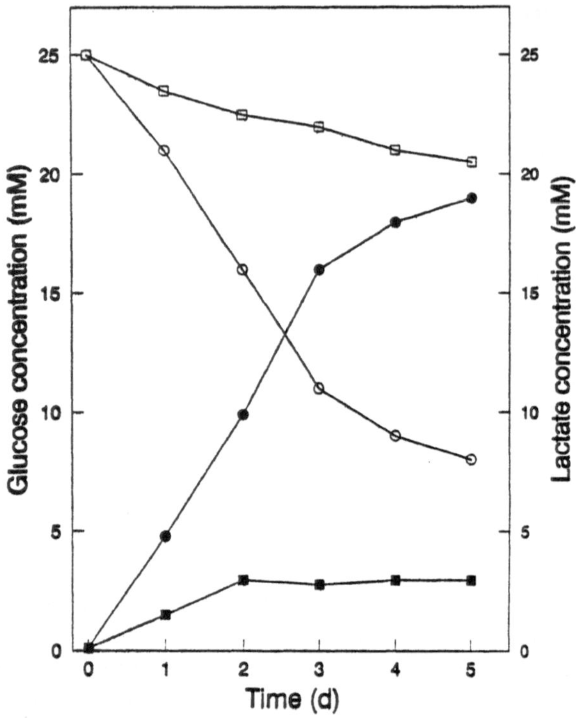

Fig. 3. Glucose utilisation and lactate production for microcarrier cultures of McCoy cells. Glucose (open symbols) and lactate (filled) were measured daily in cultures of GMEM+gmine (○, ●) and GMEM+gmate (□, ■), (n = 2) (from Hassell and Butler, 1990).

Table 2. The rate of uptake of L-[-^{14}C]-glutamine and L-[-^{14}C]-glutamate into cells

Cell line	Medium	Initial Uptake Rate (nmol/min per mg cell protein)	
		Glutamine	Glutamate
McCoy	GMEM+gmine	3.22 ± 0.59	0.250 ± 0.044*
McCoy	GMEM+gmate	4.50 ± 0.41	0.540 ± 0.095*
MDCK	GMEM+gmine	2.63 ± 0.50*	0.042 ± 0.003*

The values are means (± standard deviation) of three independent experiments. The significance of differences between values was determined by the Student t-test. * Denotes values which are significantly different from the other two values for that amino acid ($P < 0.01$) (from McDermott and Butler, 1993).

GMEM+gmate and assumed a normal growth rate following 10 days continuous culture, whereas MDCK cells ceased growth immediately following transfer to the glutamate-based medium.

(a) Glutamine synthetase

Changes in the activities of specific enzymes would be expected during the adaptation of cells to alternative substrates. Part of the adaptation process to glutamate as a substrate is the induction of the activity of glutamine synthetase (GS) which is known to respond to changes in glutamine concentration (Feng *et al.*, 1990). Glutamine synthetase enables the synthesis of glutamine from glutamate. In glutamine-free cultures this metabolic conversion is essential to ensure the formation of at least the minimum quantity of intracellular glutamine required as a precursor for the synthesis of proteins or nucleic acids.

Changes in the specific activity of GS were determined during growth of McCoy and MDCK cells in GMEM+gmine (Figure 4). A rapid increase in GS activity of up to x9 was shown following the depletion of glutamine from the medium for both cell lines. For McCoy cells, a high level of specific GS activity was maintained during growth in GMEM+gmate. However, no significant difference was observed between cell lines in terms of the response to glutamine depletion during growth in GMEM+gmine. Therefore the

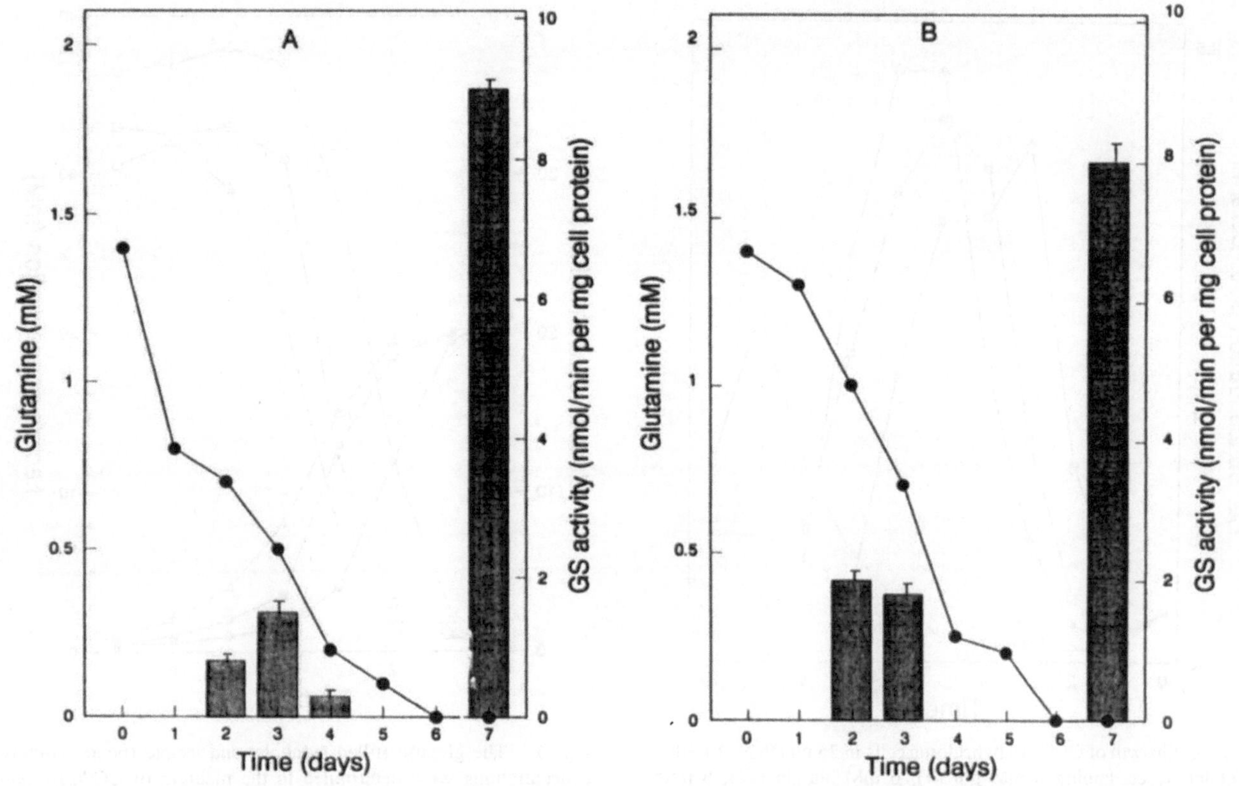

Fig. 4. **Glutamine synthetase activity of McCoy cells and MDCK cells during growth in the presence of glutamine. McCoy cells (A) and MDCK cells (B) were grown in GMEM+gmine (50 ml). Glutamine concentrations (•) were determined in the culture medium and intracellular glutamine synthetase activity (columns) was assayed during the culture period, (n = 3) (from McDermott and Butler, 1993).**

GS response to change in glutamine concentration is unlikely to be the sole factor in determining adaptability of cells to the glutamate-based medium.

(b) Glutamate transport

Glutamine synthetase activity and glutamate transport have been shown to be co-ordinately regulated in some cells by the availability of glutamine (Low *et al.*, 1994). Glutamine and glutamate transport were measured by a radioactive incorporation assay in the MDCK and McCoy cells, which differed in their adaptability to glutamate-based media (McDermott and Butler, 1993).

The initial rate of transport of glutamine by McCoy cells was significantly higher than in MDCK cells in either media by a factor of approximately 1.5 (Table 2). The rate of uptake of glutamate was lower than that of glutamine in all cases. However, the adaptation of McCoy cells to GMEM+gmate was associated with a substantial increase (x2) in the capacity to transport glutamate from the medium. Of particular significance in this experiment was the observation that the rate of glutamate uptake in MDCK cells (0.04 nmol min^{-1} per mg protein) was an order of magnitude less than for McCoy cells (0.54 nmol min^{-1} per mg protein) in GMEM+gmate. This suggests that glutamate may not be transported into the MDCK cells at a sufficient rate to support the energy metabolism necessary for growth and may account for the differences in ability of the cell lines to adapt to glutamate-based media.

Cell growth in glutamine-based dipeptides

Glutamine-based dipeptides can also be used as substitutes for glutamine (Roth *et al.*, 1988; Minamoto *et al*; 1991; Holmlund et al., 1992). They have the advantage of thermostability, which allows the media to be autoclaved and also increases the shelf life.

The murine hybridoma, CC9C10 was grown in media supplemented with either glutamine, alanyl-glutamine or glycyl-glutamine (Christie and Butler, 1994). These experiments suggested that one dipeptide (gly-gln) should be added at a higher concentration (20 mM) than glutamine (6 mM) in order to obtain equivalent cell growth. To allow a period of adaptation,

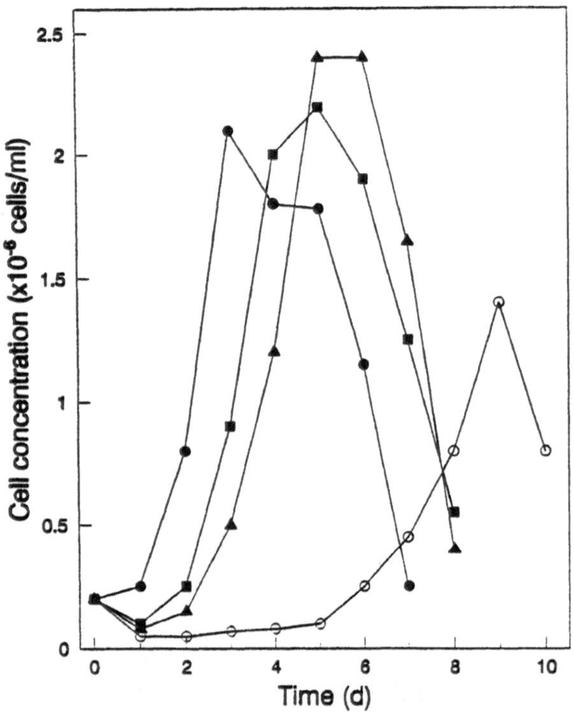

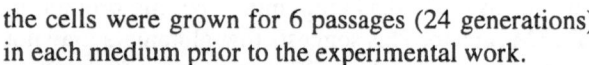

Fig. 5. Growth of CC9C10 hybridoma cells in 25 ml DMEM + 10% Fetalclone containing 6 mM gln (•), 6 mM ala-gln (■), 6 mM gly-gln (○) or 20 mM gly-gln (▲). To allow adaptation, the cells were grown for 6 passages in each medium prior to this experiment (from Christie and Butler, 1994).

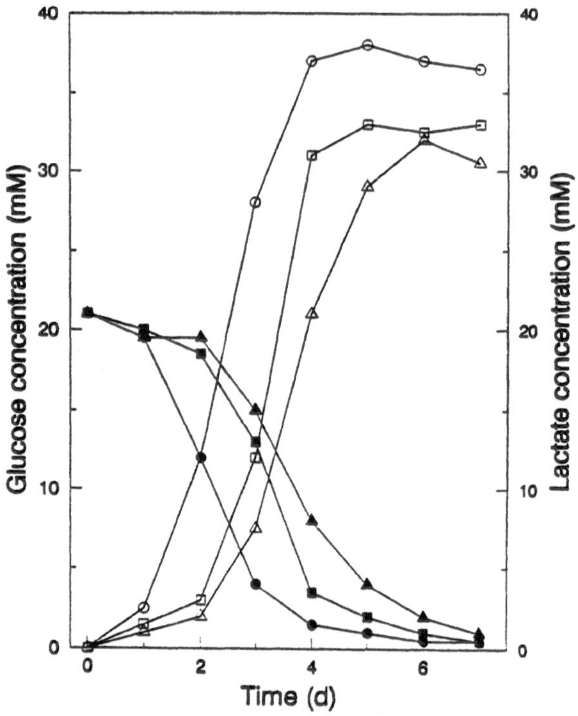

Fig. 6. The glucose (filled symbols) and lactate (open symbols) concentrations were determined in the medium of CC9C10 cells grown in 25 ml DMEM + 10% Fetalclone containing 6 mM gln (•, ○), 6 mM ala-gln (■, □), or 20 mM gly-gln (▲, △) (from Christie and Butler, 1994).

the cells were grown for 6 passages (24 generations) in each medium prior to the experimental work.

Figure 5 shows the growth characteristics of the cells in 4 media formulations. High yields were obtained from 3 media in order of gly-gln >ala-gln >gln with a maximum cell yield of 2.4×10^6 cells ml^{-1} obtained in 20 mM gly-gln. Cell growth in the lower concentration of gly-gln was poor and was not considered for further analysis. The antibody production of the hybridoma was not significantly different in the 3 cultures that attained maximum yields above 2×10^6 cells ml^{-1}.

Human leukaemia (Roth *et al.*, 1988), human lymphoma (Minamoto *et al.*, 1991) and Chinese hamster ovary (CHO) cells (Holmlund et al., 1992) have all been grown successfully in ala-gln or gly-gln at a concentration of 2-4 mM. However, Minamoto *et al.* (1991) found poor cell yields when a murine hybridoma was grown in gly-gln at a concentration up to 3.5 mM. This was probably due to the low concentration of dipeptide and confirms that a higher concentration (20 mM) of gly-gln is required to obtain acceptable growth in murine cells.

The rates of utilisation of glutamine and glucose appear to be co-ordinately regulated as shown originally for lymphocytes (Ardawi and Newsholme, 1983). Thus, it is significant that a lower rate of glucose utilisation occurred in glutamate or dipeptide-supplemented CC9C10 cultures (Christie and Butler, 1994). The specific rates of glucose consumption and lactate production during the exponential phase of growth of the CC9C10 cells were significantly lower (40-50%) in the dipeptide-based cultures compared to the glutamine-based culture (Figure 6). The metabolic coefficient (lactate/glucose) was also significantly lower in the dipeptide-based cultures. The most plausible explanation for this is that under conditions that generate ammonia, some glycolytic intermediates are utilised to sequestrate the ammonia. This idea is supported by the observed release of alanine in the presence of ammonia (Butler *et al.*, 1991) and also the effect of lactate in reducing ammonia toxicity (Hassell *et al.*, 1991). Thus, in the presence of non-ammoniagenic

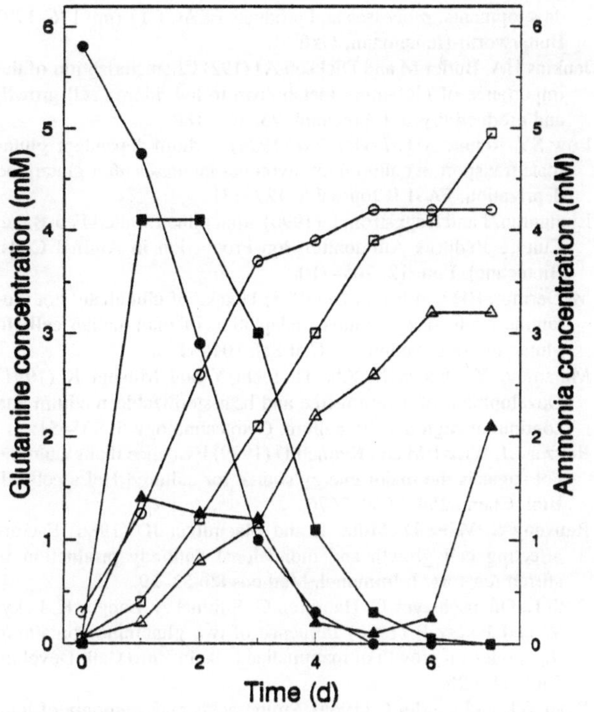

Fig. 7. The glutamine (filled symbols) and ammonia (open symbols) concentrations were determined in the medium of CC9C10 cells grown in 25 ml DMEM + 10% Fetalclone containing 6 mM gln (●, ○), 6 mM ala-gln (■, □), or 20 mM gly-gln (▲, △) (from Christie and Butler, 1994).

substrates such as glutamate or glutamine-dipeptides, the requirement for glycolytic products is reduced.

Dipeptide hydrolysis

The increase in the length of the lag phase in the dipeptide cultures is attributed to a period of extracellular release of a peptidase and subsequent dipeptide hydrolysis. This allows a gradual increase in availability of glutamine to the cells as shown in Figure 7. Peaks of glutamine concentration were observed in the dipeptide cultures at 24 h after which time the rate of utilisation of glutamine by the cells becomes greater than the rate of dipeptide hydrolysis. The rate of accumulation of ammonia in the cultures during cell growth was significantly greater in the glutamine-based medium compared to either the dipeptide-based cultures. In the case of the gly-gln culture, this difference was substantial (x2).

The slow release of glutamine by peptide hydrolysis can be compared with the kinetics of a fed-batch culture in which glutamine is slowly fed into the culture at

Table 3. Kinetic parameters of peptidase activity

Substrate	Km (mM)	Specific activity μmol min^{-1} per mg protein)
gly-gln	14.02	0.820
ala-gln	1.21	0.635

a low concentration (Ljunggren and Haggstrom, 1990). The advantage of maintaining a low concentration of glutamine is that the rate of accumulation of ammonia is reduced. This effect is substantial in the gly-gln supplemented cultures reported here, which maintain a lower concentration of ammonia and a higher cell yield (15%) compared to the control.

The requirement for a higher concentration of gly-gln in the CC9C10 cultures was investigated by determination of the specificity and activity of the peptidase enzyme necessary for peptide hydrolysis. The kinetic parameters of the enzyme located in the cytosol of the cells were determined from curve fitting to a Michaelis-Menten plot (Table 3). Although the maximum specific rate of hydrolysis of gly-gln was slightly higher (> 20%) with respect to ala-gln, the Km values differed by an order of magnitude between the two dipeptides. This indicates a higher affinity of the peptidase enzyme for ala-gln. This explains why a considerably higher gly-gln concentration would be required to obtain equivalent rates of glutamine release compared to ala-gln.

Conclusion

In conclusion, the use of glutamine-free media for selected cell lines can lead to higher cell yields in batch culture. This is coincident with a significant reduction of accumulated ammonia. A further observed effect of the glutamine-free cultures is a significant reduction in glucose utilisation and associated lactate production.

Acknowledgements

Part of the work reported here was supported by an NSERC/Apotex Industrial Research grant.

References

Ardawi MSM and Newsholme EA (1983) Glutamine metabolism in lymphocytes of the rat. Biochem. J. 212: 835–842.

Butler M, Imamura T, Thomas J and Thilly WG (1983) High yields from microcarrier cultures by medium perfusion. J. Cell Sci. 61: 351–363.

Butler M and Spier RE (1984) The effects of glutamine utilisation and ammonia production on the growth of BHK cells in microcarrier cultures. J. Biotech. 1: 187–196.

Butler M, Hassell T, Doyle C, Gleave S and Jennings P. (1991) The effect on metabolic by products on animal cells in culture. In: Spier RE, Griffiths JB and Meigner B (eds.) Production of biologicals from animal cells in Culture, ESACT 10 (pp. 226–228) Butterworth-Heinemann, Oxford.

Christie A and Butler M (1994) Growth and metabolism of a murine hybridoma in cultures containing glutamine-based dipeptides. Focus 16: 9–13.

Doyle C and Butler M (1990) The effect of pH on the toxicity of ammonia to a murine hybridoma. J. Biotech. 15: 91–100.

Eagle H (1955) Nutrition needs of mammalian cells in tissue culture. Science 122: 501–504.

Eagle H, Oyama VI, Levy M, Horton CL and Fleischman R (1956) The growth response of mammalian cells in tissue culture to L-glutamine and L-glutamic acid. J. Biol. Chem. 218: 607–616.

Feng B, Shiber SK and Max SR (1990) Glutamine regulates glutamine synthetase expression in skeletal muscle cells in culture. J. Cell Physiol. 145: 376–380.

Glacken MW, Fleischaker, RJ and Sinskey, AJ (1986) Reduction of waste product excretion via nutrient control: possible strategies for maximising product and cell yields in cultures of mammalian cells. Biotech. Bioeng. 28: 1376–1389.

Griffiths JB (1973) The effects of adapting human diploid cells to grow in glutamic acid media on cell morphology, growth and metabolism. J. Cell Sci. 12: 617–629.

Hassell T and Butler M (1990) Adaptation to non-ammoniagenic medium and selective substrate feeding lead to enhanced yields in animal cell cultures. J. Cell Sci. 96: 501–508.

Hassell T, Gleave S and Butler M (1991) Growth inhibition in animal cell culture: the effect of lactate and ammonia. Appl. Biochem. Biotech. 30: 29–41.

Holmlund A-C, Chatzisavido N, Bell SL and Lindner-Olsson E (1992) Growth and metabolism of recombinant CHO cell-lines in serum-free medium containing derivatives of glutamine. In: Spier RE, Griffiths JB and MacDonald C (eds) Animal cell technology: developments, processes and products ESACT 11 (pp. 176–179) Butterworth-Heinemann, Oxford.

Jenkins HA, Butler M and Dickson AJ (192) Characterization of the importance of glutamine metabolism to hybridoma cell growth and productivity. J. Biotechnol. 23: 167–182.

Low SY, Rennie MJ, Taylor PM (1994) Sodium-dependent glutamate transport in cultured rat myotubes increases after glutamine deprivation. FASEB Journal 8: 127–131.

Ljunggren, J and Haggstrom L (1990) Glutamine Limited Fed-Batch Culture Reduces Ammonium Ion Production in Animal Cells. Biotechnol. Lett. 12: 705–710.

McDermott RH and Butler M (1993) Uptake of glutamate, not glutamine synthetase, regulates adaptation of mammalian cells to glutamine-free medium. J. Cell Sci. 104: 51–58.

Minamoto Y, Ogawa K, Abe H, Iochi Y and Mitsugi K (1991) Development of a serum-free and heat-sterilizable medium and continuous high-density culture. Cytotechnology 5, S35–51.

Reitzer LJ, Wice BM and Kennell D (1979) Evidence that glutamine not sugar is the major energy source for cultured HeLa cells. J. Biol. Chem. 254: 2669–2676.

Reuveny S, Velez D, Miller L and Macmillan JD (1986) Factors affecting cell growth and monoclonal antibody production in stirred reactors. J. Immunol. Methods 86: 53–59.

Roth E, Ollenschlager G, Hamilton G, Simmel A, Langer K, Fekyl W and Jakesz R (1988) Influence of two glutamine containing dipeptides on growth of mammalian cells In Vitro Cell. Develop. Biol. 24: 696.

Ryan WL and Cardin C (1966) Amino acids and ammonia of fetal calf serum during storage. Proc. Soc. Exp. Biol. Med. 123: 27–30.

Taya M, Mano T and Koybayashi, T (1986) Kinetic expression of human cell growth in a suspension culture system. J. Ferment. Technol. 64: 347–350.

Tritsch GL and Moore GE (1962) Spontaneous decomposition of glutamine in cell culture media. Expl. Cell Res. 28: 360–364.

Wice BM, Reitzer LJ and Kennell D (1981) The continuous growth of vertebrate cells in the absence of sugar. J. Biol. Chem. 256: 7812–7819.

Zielke HR, Zielke CL and Ozand PT (1984) Glutamine: a major energy source for cultured mammalian cells. Fed. Proc. Fed. Am. Soc. Biol. Med. 43: 121–131.

Address for offprints: M. Butler, Department of Microbiology, University of Manitoba, Winnipeg, Manitoba, Canada R3T 2N2.

Cytotechnology **15**: 95–102, 1994.

Effect of endogenous proteins on growth and antibody productivity in hybridoma batch cultures

Patrick J. Farrell, Nicolas Kalogerakis and Leo A. Behie
Pharmaceutical Production Research Facility (PPRF) Faculty of Engineering, University of Calgary, AB, Canada T2N 1N4

Key words: Cell cycle, cytokine, flow cytometry, hybridoma cells, ulfrafiltration

Abstract

It has been shown that some B-cell hybridomas secrete autocrine factors *in vitro* which can influence cell metabolic processes. Rather than screen specifically for suspected cytokines, that may or may not affect our cell line, we have examined the lumped effects of intracellular and secreted factors on cell proliferation and monoclonal productivity in hybridoma batch cultures. Firstly, supplements of total soluble intracellular proteins combined with other intracellular metabolites were found to both decrease the specific growth rate and increase the antibody production rate at higher concentrations in batch culture. This is an important consideration in high cell density cultures, such as perfusion systems, where a reduction of growth by the presence of intracellular factors may be compensated by an increase in MAb production. In addition, flow cytometry data revealed that the average cell cycle G_1 phase fraction was unaffected by the variation in the maximum specific growth rates during the exponential growth phase, caused by the addition of intracellular factors; this suggests that higher MAb productivity at lower growth rates are not a result of cell arrest in the G_1 phase. Secondly, secreted extracellular proteins larger than 10,000 Daltons, which were concentrated from spent culture supernatant, were shown to have no significant effect on growth and specific MAb productivity when supplemented to batch culture at levels twice that encountered late in normal batch culture. This indicates that endogenous secreted cytokines, if at all present, do not play a major autocrine role for this cell line.

Abbreviations: FBS – fetal bovine serum; MAb – monoclonal antibody; MWCO – molecular weight cut off; SDS-PAGE – sodium dodecyl-sulphate-polyacrylamide gel electrophoresis

Symbols

k_d	exponential phase death rate, h^{-1}
q_{MAb}	exponential phase specific monoclonal antibody productivity, pg/(cell·h)
t	time, h
X_d	dead cell density, cells/mL
X_v	viable cell density, cells/mL
μ	specific growth rate, h^{-1}
$\mu_{max\ app}$	apparent maximum specific growth rate, h^{-1}
μ_{max}	maximum specific growth rate, h^{-1} $\mu_{max} = \mu_{max\ app} + k_d$

Introduction

In mammalian cell culture, lactate and ammonia are considered the major secondary metabolites affecting growth and protein product formation. However, it is possible that cells secrete proteins, other than the protein of interest, that may also profoundly affect growth and productivity. In addition to secreted proteins, cells in low viability or high cell density cultures may be influenced by the release of soluble intracellular proteins and metabolites upon cell lysis.

Cells of the lymphoid origin, such as the myeloma and lymphocyte fusion partners in a hybridoma, can potentially secrete many potent cytokines, most larger than 10,000 Daltons, which can influence growth and antibody formation in hybridoma cultures, even in the order of a few ng/mL. It has been shown that some B-cell hybridoma lines produce transforming growth factor β_1, (TGFBβ_1) – an inhibitor of cell growth and antibody production (Kidwell *et al.*, 1989), and interleukin-6 (IL-6) – a stimulator to MAb production (Makishima *et al.*, 1990). Several Epstein-Barr virus transformed B-cell lines were found to secrete IL-l, IL-6, and tumour necrosis factor β (a lymphotoxin) (Jochems *et al.*, 1991). Furthermore, hybridomas have been shown to be sensitive to various exogenous cytokines, such as EGF, FGF, IFNγ, IL-2, IL-6, and TGFβ_1 (Pendse and Bailey, 1990; Makishima *et al.*, 1990).

Some support for the presence of endogenous positive and negative regulatory proteins in hybridoma cultures comes from the use of molecular weight cutoff membranes (MWCO) in bioreactors. The use of low MWCO membranes in normal lymphocyte cultures resulted in improved growth, which was shown be due to the accumulation of endogenous growth stimulating factors (Kidwell *et al.*, 1990). On the other hand, the use of higher MWCO membranes in hybridoma cultures can result in improved growth and specific antibody productivity, attributed to the removal of growth inhibiting proteins such as TGFβ_1 (Hagedorn and Kargi, 1990; Kidwell *et al.*, 1990; Lee *et al.*, 1993). The question raised is whether the latter observation is merely due to the more effective removal of very low molecular weight compounds including lactate and ammonia.

In continuous culture, lower dilution rates correspond to a higher accumulation of secreted substances and intracellular metabolites from a lower viability culture. Moreover, it is well established that the specific antibody productivity is higher at lower dilution rates (i.e. specific growth rate) in continuous culture (Linardos *et al.*, 1991; Martens *et al.*, 1993; Miller *et al.*, 1988). Using a 1,000 MWCO dialysis tubing at a fixed dilution rate in continuous culture, Linardos *et al.* (1992) were able to vary the dialysis flow rate and achieve different steady state accumulations of very low molecular weight compounds in the supernatant, such as lactate and ammonia, however accumulation of higher molecular weight compounds including protein should be fairly constant. They observed that the viability, specific growth rate, and antibody productivity were not affected by the variation in lactate and ammonia levels, and thus we wondered whether enhanced specific productivity at lower dilution rates may be explained by accumulations of intracellular or secreted higher molecular weight antibody stimulating factors.

Our research objective was to establish conclusively whether both endogenous intracellular and extracellular factors play a significant role in the proliferation and MAb secretion from our hybridoma cell line. Rather than screen our cell line for various cytokines we decided to test for any significant *lumped* effect of autocrine factors, both intracellular and secreted, on cell growth and specific MAb productivity.

We first tried to assess the accumulation of endogenous proteins by propagating cells in basal media with no serum, and performing protein electrophoresis on supernatant samples. Secondly, we studied the response of our hybridoma cell line to intracellular factors that would naturally be released from lysed dead cells, particularly in perfusion systems or at low dilution rates in continuous culture, and in the late stages of batch culture. Thirdly, we studied the hybridoma cells' response to their own extracellular factors secreted into the culture supernatant. In all cultures, cell cycle phase distributions were quantified using flow cytometry.

Materials and methods

Cell line and culture maintenance

A mouse – mouse SP2/O.Ag14 derived hybridoma cell line (PF-05) producing anti-Lewis^b monoclonal antibody IgM was used in this study. The culture medium was DMEM (Sigma), supplemented with glucose to 4.5 g/L, 0.03 g/L of gentamicin sulphate (Sigma), and 5% fetal bovine serum (FBS, Gibco). Cells were maintained in Nunc T-flasks in a humidified incubator at 37 °C and 5% CO_2, atmosphere.

Preparation of intracellular protein

Cells were inoculated into several 250 mL spinner flasks and grown to the stationary phase (viability > 95%). Cell counts were taken, prior to harvesting cells by centrifugation at 150 g. The supernatant was discarded, and cells were washed once and resuspended in 40 mL fresh basal medium. This solution was subject to 3 cycles of freeze – thaw, with mild sonication for 30 minutes between cycles to assist in cell lysis. The lysate mixture was then centrifuged at 500 g, and supernatant was collected, free of debris, containing soluble protein among other intracellular metabolites. This was supplemented to the basal DMEM in place of some water, to equalize each cultures medium concentration. FBS was added prior to 0.22 μm filtration.

Preparation of extracellular protein

Cells were inoculated into several 250 mL spinner flasks and grown to the late stationary phase (viability > 95%), to maximize the concentration of cytokines without being contaminated by cell lysate from dead cells. The broth was collected, centrifuged at 500 g, and the cell free supernatant was stored at 4 °C. In parallel, a cell free control of fresh medium containing 5% FBS, also subject to the same incubation temperature and shear effects in spinner flasks, as those containing cells, was collected. Ultrafiltration of 325 mL of each supernatant was achieved in a pressurised (300 kPa) 400 mL stirred cell (Cole-Parmer) using a 10,000 MWCO type C cellulose acetate membrane (Spectra-Por). This chosen pore size will retain most cytokines produced in lymphoid culture, as well as effectively removing low molecular weight molecules that may otherwise increase the media osmolality or inhibit growth. When concentrated to 50 mL, the retentate was then flushed with 300 mL sterile water and reconcentrated to 30 mL (> 5% of original volume to prevent precipitation of serum proteins) to remove excess salts and by products. The ultrafiltration took about 6 h. The retentate was collected and stored at 4 °C, and a sample of permeate was also collected to assess the efficiency of the protein recovery process with a Bradford protein assay. For experiments, the retentate, either from culture supernatant or the cell free control, was added to fresh media at various levels as a salt and nutrient free supplement. The fresh medium, which already contained 5% FBS, was then carefully adjusted to the correct volume and pH, prior to 0.22 μm filtration.

Batch culture

A two stage inoculation procedure was used for batch experiments. Initially cells from a T-flask were centrifuged at 150 g, and inoculated into a 250 mL spinner flasks containing fresh warm media, and incubated. When a sufficient cell density was achieved, the population was verified to be exponentially distributed through the cell cycle using flow cytometry, in order to minimize the lag phase for the impending inoculation. Cells were centrifuged and inoculated at 3×10^5 cells/mL into 100 mL spinner flasks (Bellco).

Sample analysis

A 2 mL sample volume was taken each time. Cell number and viability were determined using a haemocytometer and the tryphan blue exclusion method. Sample supernatant was stored at –20 °C for metabolite and MAb analysis. The monoclonal antibody concentration was determined using an ELISA assay. Medium osmolality was measured using an Osmette S osmometer (Precision Systems Inc.).

Flow cytometry

A Becton-Dickinson FACScan equipped with a 488 nm argon laser was used to determine the DNA, and RNA content of individual cells. Simultaneous fluorescent staining of RNA and DNA in unfixed cells was achieved using an acridine orange staining method and cell cycle analysis was then performed as outlined by Darzynkiewicz *et al.* (1982).

Sodium dodecyl-sulphate-polyacrylamide gel electrophoresis (SDS-PAGE)

For SDS-Page analysis of samples, salts were removed and supernatant proteins were concentrated 100 fold by ethanol precipitation. The proteins were then separated on a 10% polyacrylamide gel using a BioRad minigel system under reduced conditions, and stained with Coomassie Brilliant Blue.

Evaluation of kinetic constants

Kinetic parameters, namely maximum specific growth rate, cell death rate, and specific monoclonal antibody production rate, were calculated according to the following equations:

$$lnX_v = lnX_{v0} + \mu_{max\ app}t$$

$$\Delta X_d = k_d \int_{t_1}^{t_2} X_v dt$$

$$\Delta[MAb] = q_{MAb} \int_{t_1}^{t_2} X_v dt$$

The constants were estimated from experimental data during the exponential growth phase by linear regression, and the uncertainty in the slope was calculated based on 95% confidence limits.

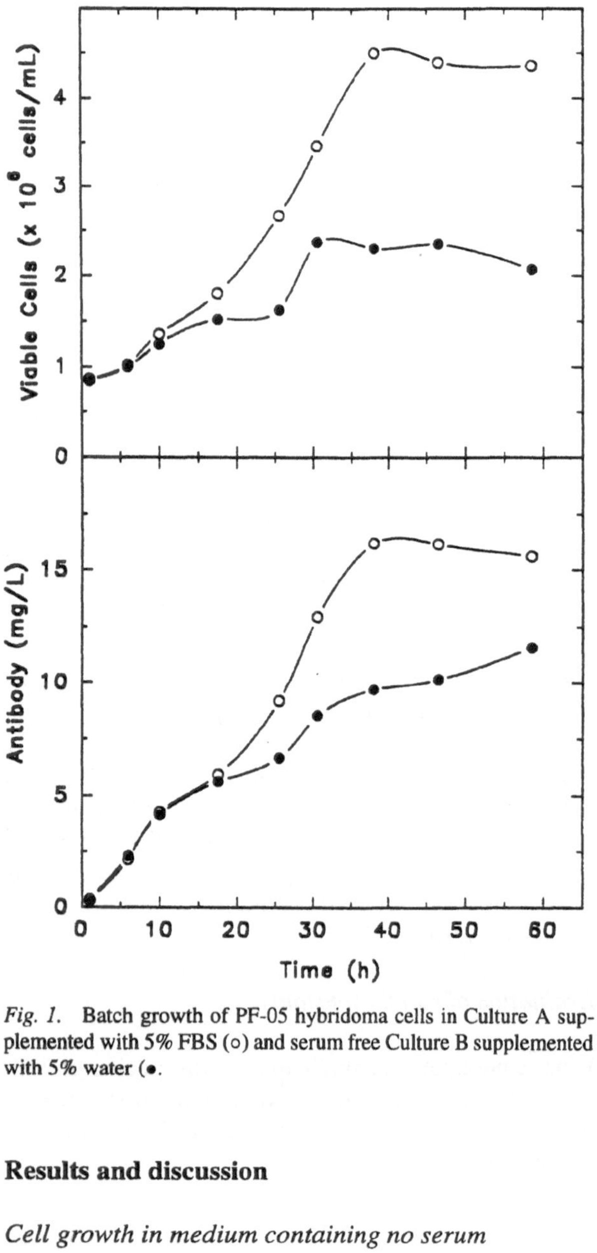

Fig. 1. Batch growth of PF-05 hybridoma cells in Culture A supplemented with 5% FBS (o) and serum free Culture B supplemented with 5% water (•.

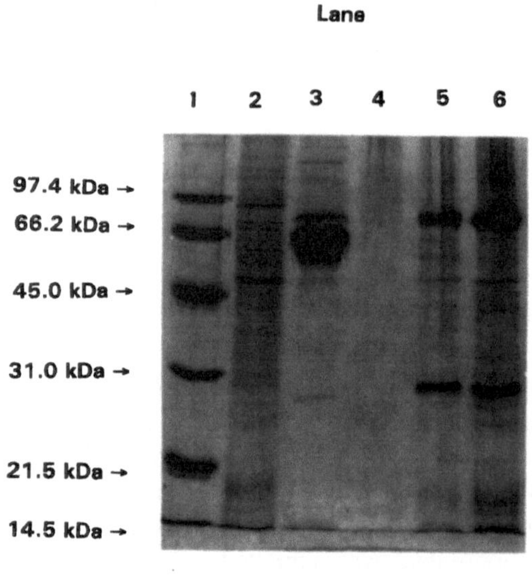

Fig. 2. SDS-PAGE analysis of culture supernatant from PF-05 hybridoma cells supplemented with 5% water to show the accumulation of endogenous proteins. Lane 1 contains molecular weight markers. Lane 2 contains a sample of soluble intracellular proteins. Lane 3 contains a sample of medium containing 5% FBS. Lanes 4, 5, and 6 contain supernatant samples concentrated 100× from 4 h, 24 h, and 48 h respectively.

Results and discussion

Cell growth in medium containing no serum

The presence of endogenous protein secretion was demonstrated by first propagating cells at high densities for several days in the absence of serum, and then performing SDS-Page analysis of the culture supernatant. This allowed us to clearly detect low concentration protein bands in the absence of serum proteins.

The cell density profile presented in Fig. 1 shows that cells can continue to grow for about 1.5 doublings or 30 h even in the complete absence of serum or any serum free supplements (Culture B), if previously exposed to serum in the recent past. However, cell growth was retarded and the maximum cell density was limited to 2.4×10^6 cells/mL compared to 4.5×10^6 cells/mL in the control Culture A containing FBS. Cell cycle arrest in the G_1 phase did not occur in the absence of serum (flow cytometry data not shown), although this occurs in normal untransformed cell cultures (Baserga, 1985). Shear effects in the spinner flasks at 50 RPM were not significant without serum, as the viability remained greater than 90% for about 48 h. In addition, the cells grown without serum (Culture B) continued to produce antibody (Fig. 1) at similar specific productivity [0.17 pg/(cell·h)] compared to control cells [0.18 pg/(cell·h)], and thus should also be capable of producing other proteins, including autocrine factors under these serum free conditions.

The SDS-polyacrylamide gel is shown in Fig. 2 containing samples from Culture B. The sample at 4 h (Lane 4) reveals that no serum proteins (Lane 3) are carried into the new environment with the cell inoculation after washing in basal media. Many protein bands do appear after 24 (Lane 5) and 48 h (Lane 6) of growth under serum free conditions. Many of these correspond to those found in cell lysate (Lane 2) making it difficult to discriminate proteins bands attributed to cytokines, which may be secreted at levels below the detection limit of this staining procedure. However we felt that the appearance of many protein bands in the serum free environment warranted further investigation. Also note that the emergence of thick bands in supernatant samples between 68 and 97.5 kD, and 21.5 and 31.5 kD, that are almost certainly IgM fragments generated by the reducing conditions in the !lectrophoresis.

Effect of intracellular proteins

Soluble intracellular proteins and secondary metabolites were extracted by freeze-thaw from hybridoma cells collected in the stationary phase of batch culture. The soluble lysate from 1, 3, and 5 × 10^8 cells was then added to 100 mL spinner flask-cultures. The objective here was to study the effect of these soluble intracellular proteins and secondary metabolites, released from dead cells in normal cell culture, on hybridoma proliferation and MAb secretion.

Figure 3 shows that the effect of increasing levels of soluble intracellular proteins and secondary metabolites was to retard growth yet increase final antibody titre. As the lysate added was increased to the intracellular equivalent of 5 × 10^8 dead cells, the maximum specific growth rate (μ_{max}) was gradually reduced from 0.049 h^{-1} in control Culture C to 0.023 h^{-1}, while the specific MAb productivity (q_{MAb}) steadily increased from 0.26 pg/(cell·h) to 0.39 pg/(cell·h), as summarised in Table 1. This is an important consideration in high cell density perfusion cultures, where the accumulation of dead cell lysate may significantly retard growth, yet stimulate antibody production. A compromise in the growth and cell viability is almost certainly necessary for optimising the volumetric MAb productivity in perfusion cultures.

We considered two mechanisms to explain the decrease in μ_{max}, and increase in q_{MAb} upon the addition of lysate. Firstly, that specific autocrine factors in the lysate affect q_{MAb}, *independently* of μ. For example, the presence of say 5 × 10^8 dead cells should enhance q_{MAb} 1.5 times over base levels according to Table 1, while the presence of 0.5 × 10^6 cells/mL should enhance q_{MAb} 1.05 times by interpolation. We examined the relative q_{MAb} in published continuous culture data and found that they could not support this mechanism (Linardos *et al.*, 1989; Martens *et al.*, 1993; Miller *et al*, 1988; Ramirez and Mutharason, 1990). The data from Linardos *et al.* (1991) shows that the q_{MAb} was enhanced 1.5 times for a dead cell concentration increase of only 0.5 × 10^6 cells/mL, and thus dead cell levels were too low to support this mechanism. Secondly, we considered that q_{MAb} is *directly coupled* to μ) which is reduced by inhibitory factors

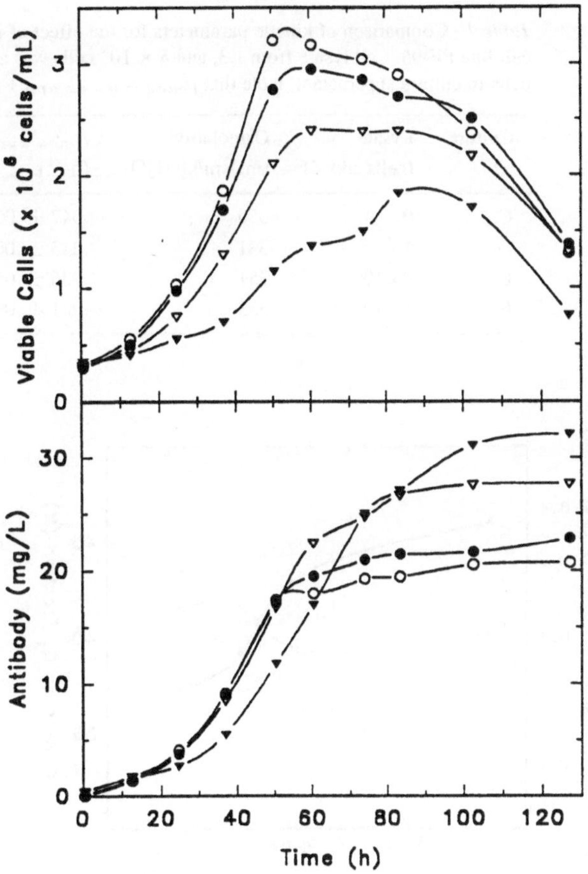

Fig. 3. Batch growth of PF-05 hybridoma cells in 100 mL spinner cultures supplemented with 5% FBS and soluble intracellular proteins and metabolites (lysate) to simulate the presence of dead cells in culture supernatant. Lysate from 10^8 cells (•, Culture D), 3 × 10^8 cells (▽, Culture E), 5 × 10^8 cells (▼, Culture F), was added along with a control (○, Culture C).

in the lysate. A plot of q_{MAb} versus μ supports this mechanism by showing a similar trend to that obtained in continuous cultures (Fig. 4); that is, that the specific antibody productivity is increased at lower specific growth rates, as observed elsewhere (Suzuki and Ollis, 1990).

It has been suggested that the higher fraction of cells arrested at lower growth rates in continuous culture are responsible for higher specific antibody productivity (Suzuki and Ollis, 1989). Flow cytometry data from our experiment are shown in Fig. 5. Some oscillations are apparent due to the induction of some synchrony at higher levels of lysate. However, our batch flow cytometry data reveals that no significant difference occurred in the *average* G_1 phase fraction during the exponential growth period from which q_{MAb} was calculated,

Table 1. Comparison of kinetic parameters for the effect of intracellular proteins and metabolites on the murine hybridoma cell line PF-05. Cell lysate from 1,3, and 5 × 10^8 cells was added to 100 mL spinner flasks to simulate the presence of dead cells in culture supernatant. Note that $\mu_{max} = \mu_{max\ app} + k_d$

Culture	Lysate (cells added)	Osmolality (mOsm/kg H_2O)	$\mu_{max\ app}$ (h^{-1})	k_d (h^{-1})	μ_{max} (h^{-1})	q_{MAb} [pg/(cell·h)]	Relative q_{MAb}
C.	0	350	.047 ± .002	.002	.049 ± .002	.26 ± .01	1.0
D.	10^8	351	.043 ± .003	.001	.044 ± .003	.29 + .01	1.1
E.	3×10^8	351	.036 ± .003	.003	.039 ± .003	.35 ± .01	1.3
F.	5×10^8	353	.021 ± .002	.002	.023 ± .002	.39 ± .02	1.5

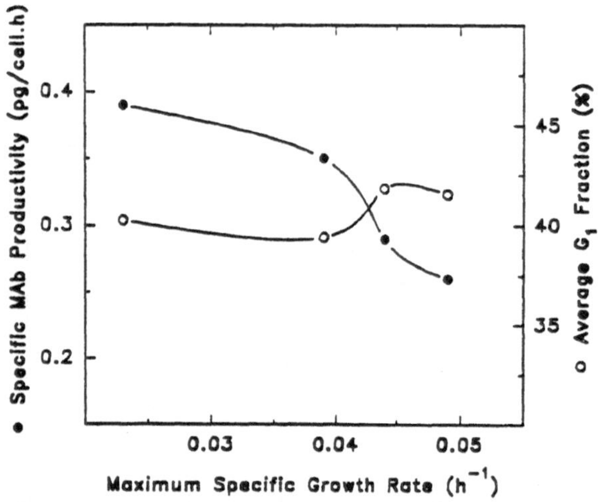

Fig. 4. The variation in specific antibody productivity with maximum specific growth rate (•) resembles profiles in continuous culture suggesting q_{MAb} is dependant on μ. Insignificant variation in the average G, fraction occurred with maximum specific growth rate (o).

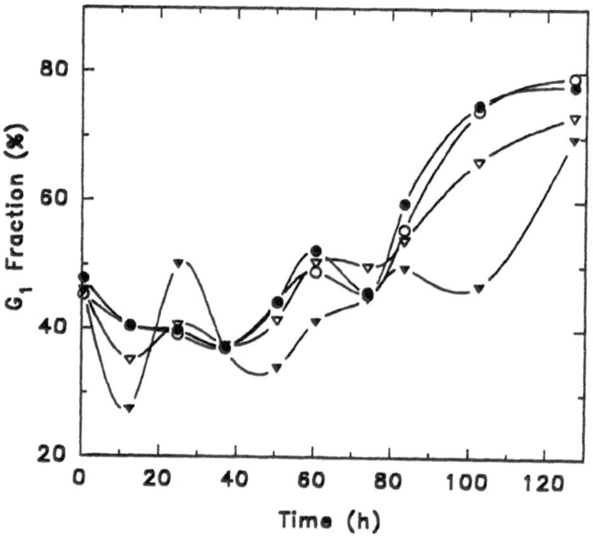

Fig. 5. Cell cycle vatiation in the G_1 phase fraction of PF-05 hybridoma cells in cultures supplemented with 5% FBS and lysate from 10^8 cells (• Culture D), 3 × 10^8 cells (∇, Culture E), 5 × 10^8 cells/mL (▼, Culture F), and a control (o, Culture C).

despite the variation in μ (Fig. 4). Yet in continuous culture an increase in the average G_1 fraction of 15% was observed when μ was reduced by 53% (Martens *et al.*, 1993). However, we have found that effective arrest in the G_1 phase only occurs under nutrient limitation in normal culture conditions (not shown). We have also observed elsewhere that the retardation of growth by the addition of lactate and ammonia in the presence of surplus nutrients, caused no significant increase in the G_1 fraction compared to control cultures during the exponential growth phase. Our only explanation for these observation is that cell progression through all phases of the cell cycle is reduced by the same amount at lower specific growth rates. However, it is generally accepted that cell progression through the S, G_2, M phases are invariant (Smith and Martin, 1973).

Effect of proteins secreted from cells

Supernatant protein, concentrated from the late stages of batch culture and presumably containing endogenous cytokines, was resupplied to new cultures to assess their autocrine effect on growth and antibody production.

Proteins greater than 10,000 MWCO from 325 mL of spent culture supernatant were concentrated to 30 mL in a stirred cell. Bradford protein assays revealed that 97% of the protein was routinely recovered, with 0.1% passing through the filter. The remainder was probably adsorbed on the filter itself. It is assumed that our hybridoma cells produce various cytokines which accumulate to a lumped concentration x by the end of the stationary phase in batch culture. These concentrated proteins were added to the fresh cultures containing 5% FBS at starting levels of x and $2x$. Because our cell line is maintained on 5% FBS and cytokines are usually produced in the order of a few ng/mL, most of the total protein concentrated from spent culture supernatant is serum. Thus, the total serum level in the cultures supplemented with x and $2x$ approached 10% and 15% FBS, and hence it was necessary to run a control culture at the upper limit of 15% FBS, as well as the base case of 5% FBS. An additional control sample of fresh medium, that had been incubated in parallel with cell cultures for supernatant collection, was also ultrafiltered, and total protein equivalent to 10% FBS was added to a fresh culture containing 5% FBS, to account for anomalies in the ultrafiltration process. The osmolalities and culture notation of various cultures are shown in Table 2. A control at high osmolality (375 mOsm/kg H_2O) was also run to account for traces of salt and nutrients present in the ultrafiltration retentate (not shown).

Results of cell growth are shown in Fig. 6, and reveal no significant effect of endogenous secreted proteins on cell growth. It was reported that some hybridoma cell lines produce TGFβ_1, a potent inhibitor to cell growth and antibody production (Kidwell *et al.*, 1989). Although we did not screen specifically for TGFβ_1, or other cytokines, no effect on growth was observed at protein levels over twice ($2x$) that encountered late in normal batch culture for this cell line. An increase in the media osmolality from 350 to 375 mOsm/kg H_2O using NaCl, revealed no significant change in the specific growth rate, specific productivity, or maximum cell density (data not shown). The calculated increase in specific productivity from spent medium (Table 2) could only be attributed to serum proteins. It is possible that the presence of serum protein may mask any effects of low concentration autocrine factors. In addition no significant variation in the cell cycle distribution was observed (data not shown).

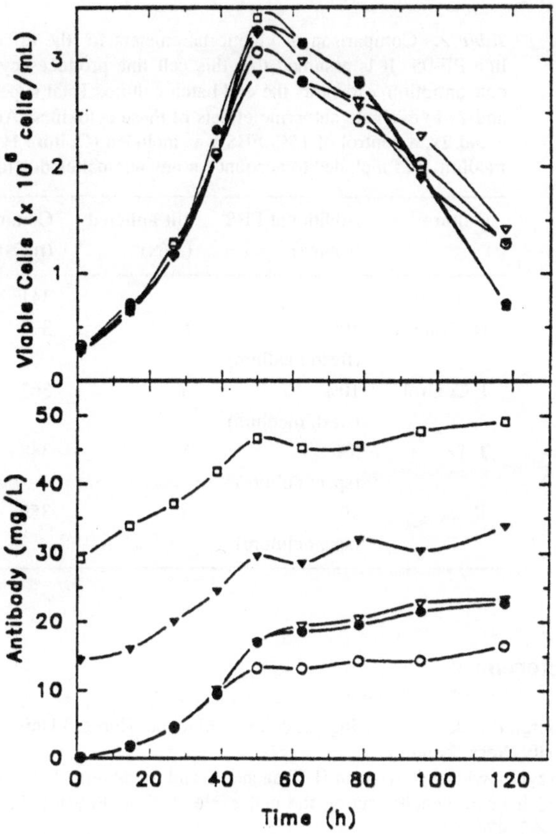

Fig. 6. Batch growth of PF-05 hybridoma cells supplemented with endogenous secreted proteins. It is assumed that this cell line produces cytokines which accumulate to a lumped concentration × towards the end of batch culture. This was concentrated and resupplied a x (▼, Culture J) and $2x$ (□, Culture K) to evaluate the autocrine effects of these cytokines. All cultures contained a base level of 5% FBS (control ○, Culture G). A control of 15% FBS was also included (control ●, Culture H), in addition to a control supplement of ultrafiltered fresh medium to account for anomalies in the ultrafiltration process (▽, Culture I).

Conclusions

A higher specific antibody productivity and reduced growth rate was caused by the presence of soluble intracellular proteins and other intracellular metabolites from cell lysate. It is most likely that these factors act only to reduce μ_{max} which in turn increases q_{MAb}. Despite slower growth in the presence of increasing levels of cell lysate, the average G_1, phase fraction was not significantly affected. This suggests that higher specific antibody productivities in this case are not a result of cell arrest in the G_1 phase, but some other mechanism that seems to slow the progress of cells through all phases of the cell cycle.

In this study we have found no evidence that endogenous secreted cytokines having net autocrine effects on growth or specific antibody productivity at levels normally found in batch culture with our cell line, although the cytokine secretion pattern has been shown to be cell line dependant (Jochems *et al.*, 1991; Kidwell *et al.*, 1989).

Table 2. Comparison of kinetic parameters for the effect of endogenous secreted proteins on the murine hybridoma cell line PF-05. It is assumed that this cell line produces cytokines which accumulate in the culture supernatant to a lumped concentration x towards the end batch culture. Total supernatant protein was concentrated and supplied to new cultures at x and $2x$ to evaluate autocrine effects of these cytokines. All cultures contained a base level of 5% FBS. As FBS is inherent in x and $2x$, a control of 15% FBS was included (Culture H), and an additional control using 10% FBS ultrafiltered from fresh medium was included to account for any anomalies due to ultrafiltration (Culture I).

Culture	Additional FBS (source)	Ultrafiltered (Y/N)	Osmolality (mOsm/kg H_2O)	$\mu_{max\ app}$ (h^{-1})	k_d (h^{-1})	μ_{max} (h^{-1})	q_{MAb} [pg/(cell·h)]
G. Control	–	–	341	.046 ± .03	.001	.047 ± .03	.20 ± .02
H. Control	10% (fresh medium)	N	370	.049 ± .03	.001	.050 ± .03	.24 ± .01
I. Control	10% (fresh medium)	Y	363	.050 ± .03	.001	.051 ± .03	.24 ± .01
J. 1x	5% (spent culture)	Y	348	.049 ± .04	.001	.050 ± .04	.24 ± .04
K. 2x	10% (spent culture)	Y	359	.050 ± .03	.001	.050 ± .03	.23 ± .04

References

Baserga R (1985) The biology of cell reproduction. Harvard University Press, London.

Darzynkiewics Z, Crissman H, Traganos F and Steinkamp J (1982) Cell heterogeneity during the cell cycle. J. Cell. Physiol. 113: 465–474.

Hagedorn J and Kargi F (1990) Production of monoclonal antibodies by hybridoma cells in a flat sheet membrane bioreactor. Biotechnol. Prog. 6: 220–224.

Jochems GJ, Klein MR, Jordens R, Pascual-Salcedo D, van Boxtel-Oosterhof F, van Lier RA and Zeijlemaker WP (1991) Heterogeneity in both cytokine production and responsiveness of a panel of monoclonal human Epstein-Barr virus-transformed B-cell lines. Hum. Antibodies Hybridomas 2(2): 57–64.

Kidwell WR (1989) Filtering out inhibition. Bio Technol. 7: 462–463.

Kidwell W, Knazek R and Wu Y (1990) Effect of fiber pore size on performance of cells in hollow fibre bioreactors. In: Murakami H (ed.) Trends in Animal Cell Culture Technology. Kodansha Ltd., Tokyo, pp. 29–33.

Lee GM, Chuck AS and Palsson BO (1993) Cell culture conditions determine the enhancement of specific monoclonal antibody productivity of calcium alginate-entrapped S3H5/y2bA2 hybridoma cells. Biotechnol. Bioeng. 41: 330–340.

Linardos TI, Kalogerakis N and Behie LA (1992) Monoclonal antibody production in dialyzed continuous suspension culture. Biotechnol. Bioeng. 39: 504–510.

Linardos TI, Kalogerakis N, Behie LA and Lamontagne LR (1991). The effect of specific growth rate and death rate on monoclonal antibody production in hybridoam chemostat cultures. Can. J. Chem. Eng. 69: 429–438.

Makishima F, Mikami T and Terada S (1990) Effect of some cytokines on antibody productivity of a murine B-cell hybridoma. In: Murakami H (ed.) Trends in Animal Cell Culture Technology. Kodansha Ltd., Tokyo, pp. 299–302.

Martens DE, de Gooijer CD, van der Velden-de Groot CAM, Beuvery EC and Tramper J (1993) Effect of dilution rate on growth, productivity, cell cycle and size, and shear sensitivity of a hybridoma cell in continuous culture. Biotechnol. Bioeng. 41: 429–439.

Miller WM, Blanch HW and Wilke CR (1988) A kinetic analysis of hybridoma growth and metabolism in batch and continuous culture: effect of nutrient concentration, dilution rate and pH. Biotechnol. Bioeng. 32: 947–965.

Negri C, Chiesa R, Giulia CB and Ricotti A (1991) Factor(s) required by EBV transformed lymphocytes to grow under limiting dilution conditions. Cytotech. 7: 173–178.

Ozturk OS and Pallson BO (1991) Growth, metabolic, and antibody production kinetics of hybridoma cell culture: 1. Analysis of data from controlled batch reactors. Biotechnol. Prog. 7: 471–480.

Pendse GJ and Bailey JE (1990) Effects of growth factors on cell proliferation and monoclonal antibody production of batch hybridoma cultures. Biotechnol. Lett. 7(12): 487–492.

Ramirez OT and Mutharason R (1990) Cell cycle and growth phase dependant variations in size distribution, antibody productivity, and oxygen demand in hybridoma cultures. Biotechnol. Bioeng. 36:839–848.

Smith JA and Martin L (1973) Do cells cycle? Porc. Natl. Acad. Sci. USA 70: 1263–1267.

Suzuki E and Ollis DF (1989) Cell cycle model for antibody production kinetics. Biotechnol. Bioeng. 34: 1398–1402.

Suzuki E and Ollis D (1990) Enhanced antibody productivity at low growth rates: Experimental demonstration and a simple structured model. Biotechnol. Prog. 6:231–236.

Yamada K, Ikeda I, Sugahara T, Shirahata and Murakami H (1990) Enhancement of immunoglobuiin production of human-human hybridomas by immunoglobulin production stimulating factors. In: Murakami H (ed.) Trends in Animal Cell Culture Technology. Kodansha Ltd., Tokyo, pp. 291–297.

Address for offprints: L.A. Behie, Pharmaceutical Production Research (PPRF), Faculty of Engineering, University of Calgary, AB, Canada T2N 1N4.

Cytotechnology **15:** 103–109, 1994.

Extracellular insulin degrading activity creates instability in a CHO-based batch-refeed continuous process

Denis Drapeau[1], Yen-Tung Luan[1], Joanne A. Popoloski[2], D. Troy Richards[1], David C. Cohen[3], Martin S. Sinacore[1] and S. Robert Adamson[1]
[1] *Genetics Institute, Inc., 1 Burtt Road, Andover MA 01810, USA;* [2] *Biogen, Inc., 14 Cambridge Center, Cambridge MA 02142, USA;* [3] *Miles Biotechnology, Inc., 4th and Parker Streets, Berkeley CA 94701, USA*

Key words: Bioreactor, recombinant CHO cells, insulin, large-scale cell culture, medium

Abstract

In a batch-refeed continuous process involving a recombinant Chinese hamster ovary cell line, a brief upset was occasionally observed during which cell growth halted and cell viability dropped. This was found to be associated with depletion of insulin from the medium early during the affected passage. Insulin depletion was found to be primarily the result of insulin degrading activity released by the cells during the preceding passage.

Abbreviations: CHO – Chinese hamster ovary; rhM-CSF – recombinant human macrophage colony stimulating factor; RIA – radioimmunoassay; RP-HPLC – reverse phase high performance liquid chromatography.

Introduction

The batch-refeed continuous approach for production of cell-generated products is an alternative to the more common terminal batch approach. The terminal batch approach involves cleaning and resterilization of the production bioreactor following each harvest. Prior to each harvest, the culture is generally allowed to progress into a growth-arrested state. In contrast, the batch-refeed approach involves retaining a portion of the culture in the production bioreactor at the end of each harvest to serve as inoculum for the subsequent batch. This enables a series of harvests to be taken from a single production bioreactor without cleaning and resterilizing the bioreactor between batches. Since the culture is maintained in exponential growth, this approach is ideal for the production of labile products that would be damaged by long residence times in the bioreactor or by the intracellular enzymes that are released by growth-arrested cells.

A risk associated with the batch-refeed approach is that an event occurring during one batch can potentially affect performance during a subsequent batch. Such an event was found to be the probable cause of a brief process upset that occurred on several occasions during the manufacture of rhM-CSF at Genetics Institute. Each upset was characterized by a temporary cessation of growth and drop in cell viability that occurred during a single batch cycle within a series. Upon investigation, we found the probable cause to be a temporary shortage of insulin in the medium during the upset. The cause of the insulin shortage appears to have been release of an unusually large amount of insulin degrading activity into the medium during the preceding batch.

Materials and methods

The CHO cell line, coexpressing rhM-CSF and dihydrofolate reductase, was developed at Genetics Institute by transfection of CHO cells with expression plasmids as described earlier (Wong *et al.*, 1987) followed by the adaption of the resulting cell line to serum-free suspension culture. The medium contained bovine insulin at 10 μg/ml and no other proteins. All cultures

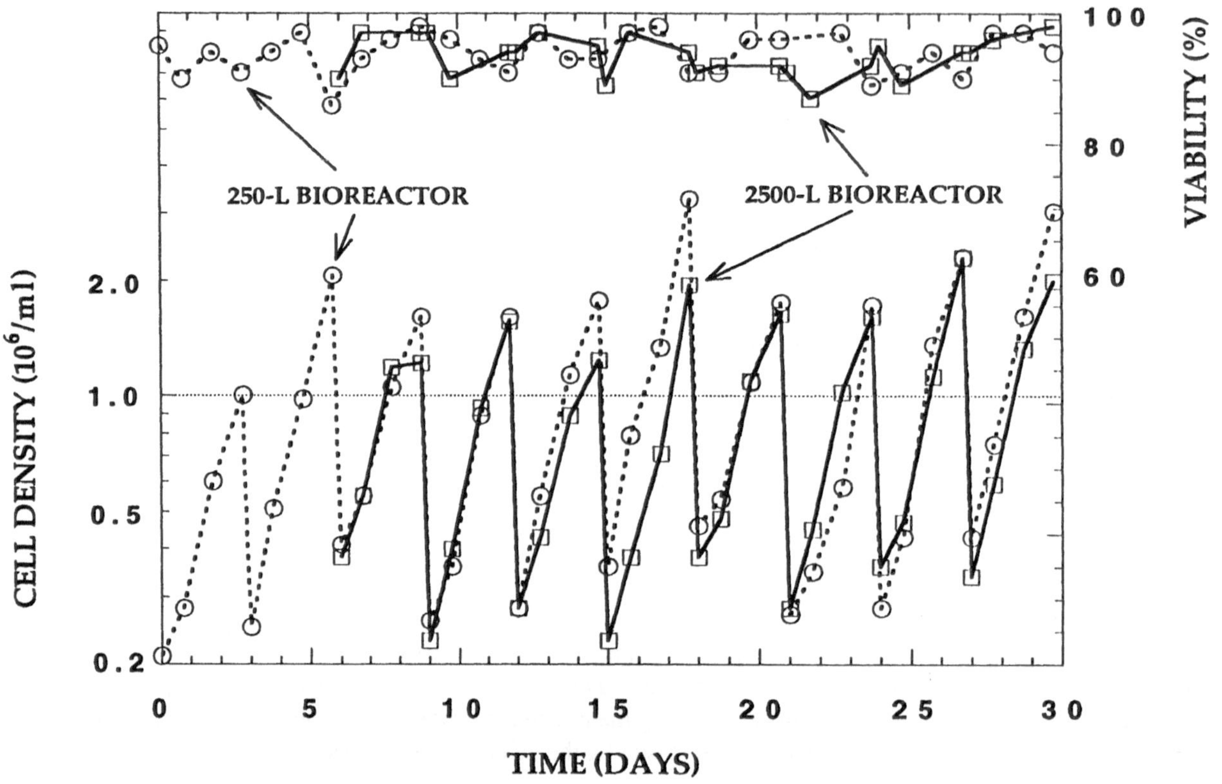

Fig. 1. Cell densities and viabilities during Run A.

were maintained at 37 °C and at pH less than 7.3 by delivery of carbon dioxide to the headspace. In addition, bioreactor cultures were maintained at pH no lower than 7.0 by injection of a solution containing sodium carbonate and sodium bicarbonate, and at dissolved oxygen concentration no lower than 23% by sparging of oxygen. Viable cell densities and viabilities (proportion of total cells viable) were determined by hemacytometer using trypan blue.

A typical production run began with thaw of cells from a working cell bank. The resulting culture was passaged in progressively larger spinner flasks until the number of cells was sufficient for inoculation of a 40 liter culture in a 250-L bioreactor. Three or four days after inoculation of the 250-L bioreactor, the culture was passaged in a manner that brought its volume up to 230 liters. Thereafter it was passaged every third day, initially to serve as a source of inoculum to begin a production run in a 2500-L bioreactor, and later to serve as a back-up if the culture in the 2500-L bioreactor was lost due to contamination or mechanical failure. The 2500-L bioreactor was also passaged every third day. At each passage, a portion of the culture in the 2500-L bioreactor was retained to serve as inoculum for the next batch while the remainder was harvested for isolation of rhM-CSF from the conditioned medium. The protocol for passaging of both bioreactors called for diluting the retained culture by a factor of at least 6, and by a greater factor if necessary to start the next batch at a targeted cell density no higher than 4×10^5 cells/ml. Agitation was provided by Rushton impellers operating at 60 rpm. Conditioned medium samples were stored at –80 °C prior to insulin assays

The response of the cell line to initial insulin concentration was measured using suspension cultures in 25 ml spinner flasks. Cells were suspended at an initial density of 1.6×10^5 cells/ml in fresh medium containing insulin at various concentrations. Each insulin concentration was represented by two spinner flask cultures. Cell densities and viabilities were determined 72 hours later.

For measuring kinetics of insulin degradation, a culture was begun in a 1000-ml spinner flask at an initial cell density of 4.2×10^5 cells/ml. 50-ml portions were drawn from this culture at day 0, day 2, and day 3, clarified by centrifugation, supplemented with insulin

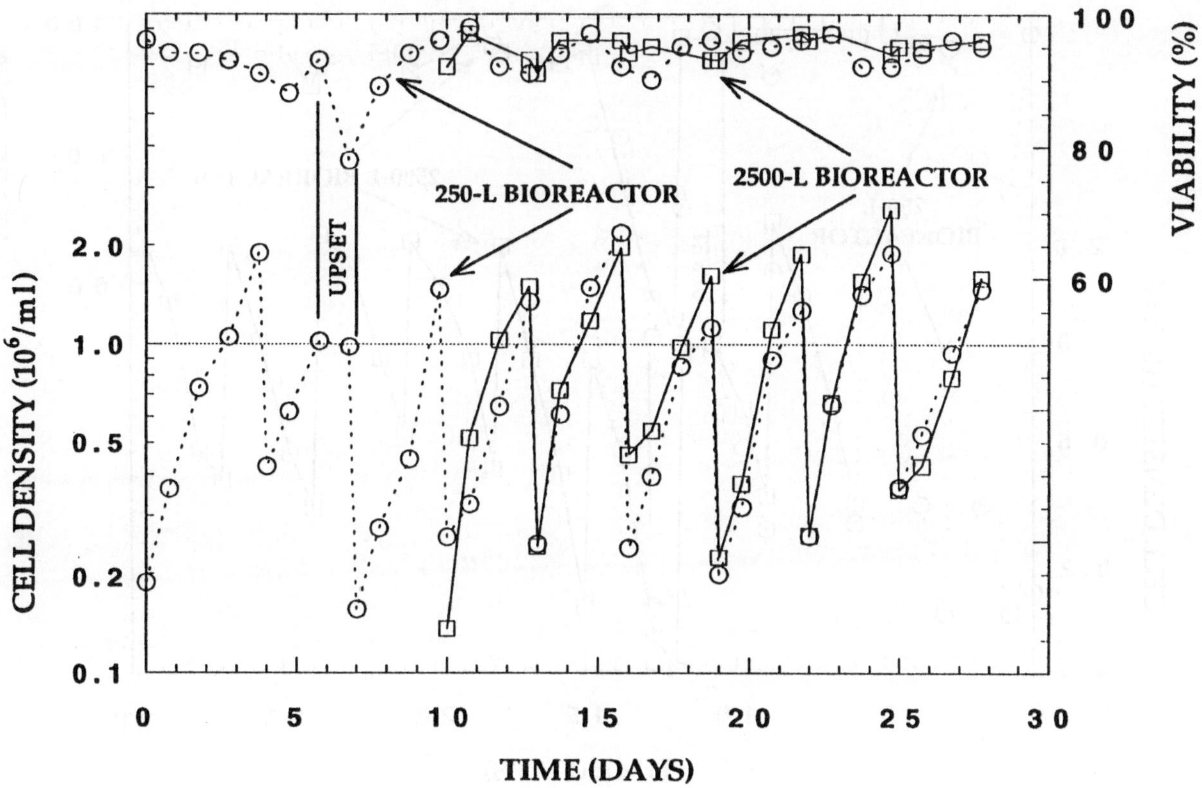

Fig. 2. Cell densities and viabilities during Run B.

as indicated, placed in spinner flasks, and returned to the incubator. Conditioned medium samples were acidified with HCl to pH 3 prior to storage at –80 °C.

Insulin concentrations in medium samples were measured either by RP-HPLC (C4 column, 0.1% trifluoroacetic acid-acetonitrile gradient with UV detection at 205 nm) or by RIA (Diagnostics Products Corp., Los Angeles, CA).

Table 1. Insulin measurements by RP-HPLC during Run B (μg/ml).

Passage	Bioreactor	Day 1	Day 2	Day 3
1	250-L	3.6	2.2	<1.5
2	250-L	<1.5	<1.5	<1.5
3	250-L	2.5	<1.5	<1.5
4	2500-L	2.6	<1.5	<1.5
5	2500-L	1.7	<1.5	<1.5
6	2500-L	2.5	<1.5	<1.5
7	2500-L	3.6	1.9	<1.5
8	2500-L	2.9	<1.5	<1.5

Results

Approximately half of the rhM-CSF production runs at Genetics Institute were characterized by uninterupted cell growth and consistently high cell viability in both the 250-L bioreactor and the 2500-L bioreactor. An example of this was Run A (Fig. 1). In each of the other production runs, however, a brief upset was observed during which growth halted and cell viability dropped. An example of this is the upset that occurred six days after the culture entered the 250-L bioreactor during Run B (Fig. 2). If an upset occurred in the 250-L bioreactor after inoculation of the 2500-L bioreactor, a simultaneous upset occurred in the 2500-L bioreactor. An example of this is the upset that occurred eleven days after the culture entered the 250-L bioreactor (five days after inoculation of the 2500-L bioreactor) during Run C (Fig. 3). On no occasion did an upset occur more than 14 days following inoculation of the 250-L bioreactor.

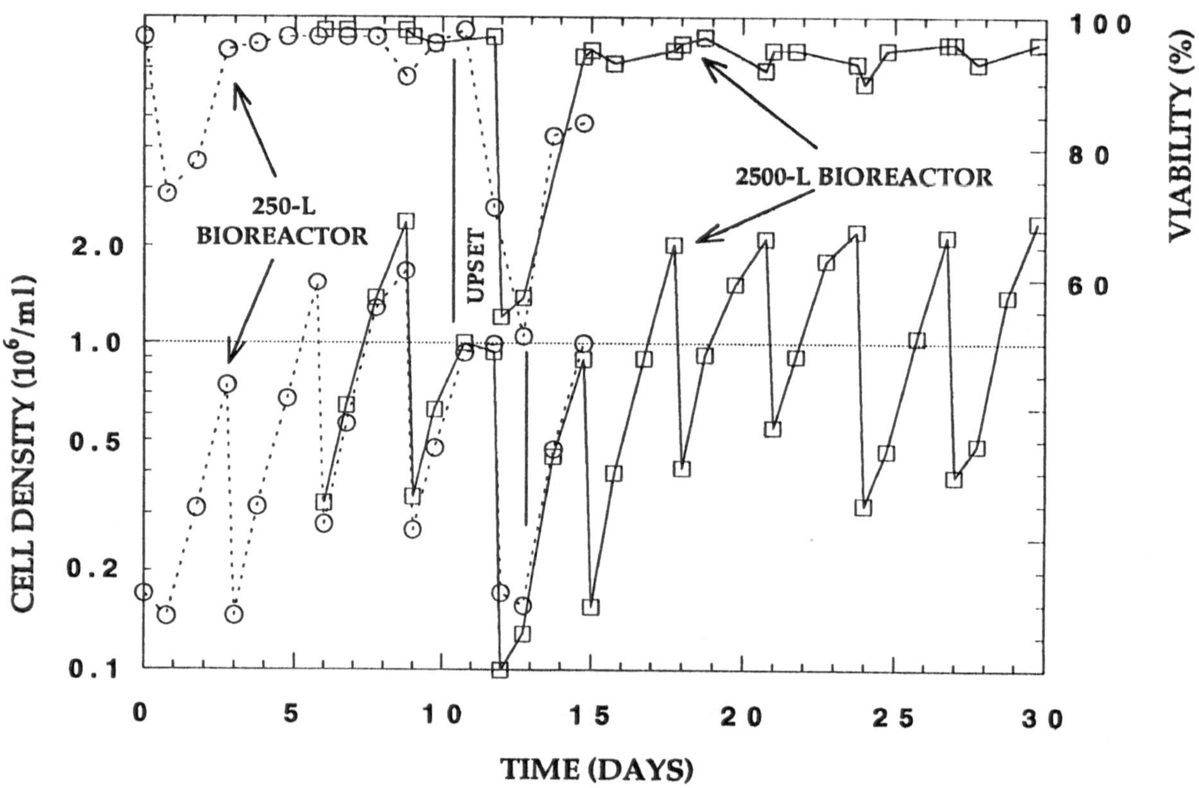

Fig. 3. Cell densities and viabilities during Run C.

Table 2. Insulin measurements by RIA during Run C (μg/ml).

Passage	Bioreactor	Day 1	Day 2	Day 3
3	2500-L	1.78	1.06	0.027
4	2500-L	1.28	0.020	0.00065
5	2500-L	4.07	1..57	0.76
6	2500-L	4.59	2.50	1.21
7	2500-L	2.59	0.88	0.048

Assay of insulin levels in bioreactor medium during Runs B and C indicated that in each case the batch during which the upset occurred was characterized by abnormally low insulin levels. For Run B, medium samples assayed were drawn from the 250-L bioreactor prior to inoculation of the 2500-L bioreactor and from the 2500-L bioreactor thereafter. An RP-HPLC method with a detection limit of 1.5 μg/ml indicated that insulin was already below 1.5 μg/ml by day 1 of the batch during which the upset occurred, whereas it was still above 1.5 μg/ml at day 1 of all other batches (Table 1). For Run C, medium samples assayed were drawn from the 2500-L bioreactor during the first five batches after its inoculation. A much more sensitive RIA method indicated that again insulin was already below 1.5 μg/ml by day 1 of the batch during which the upset occurred but not the other batches (Table 2). Furthermore it indicated that insulin was down to 0.02 μg/ml by day 2 in the batch during which the upset occurred, whereas it was still above this level even at day 3 in all other batches.

Although we have no proof that the shortage of insulin observed during these upsets was actually the cause of growth cessation and dropping viability, we have evidence that an insulin shortage *can* cause such effects. When cultures of the same cell line were grown in 25 ml spinner flasks at various initial insulin concentrations, those starting at concentrations below 0.5 μg/ml showed reduced growth rates and viabilities (Fig. 4). We do not know how low insulin levels had dropped in these cultures at the point when growth and viability effects began to occur, since insulin concentrations after the time of inoculation were not measured.

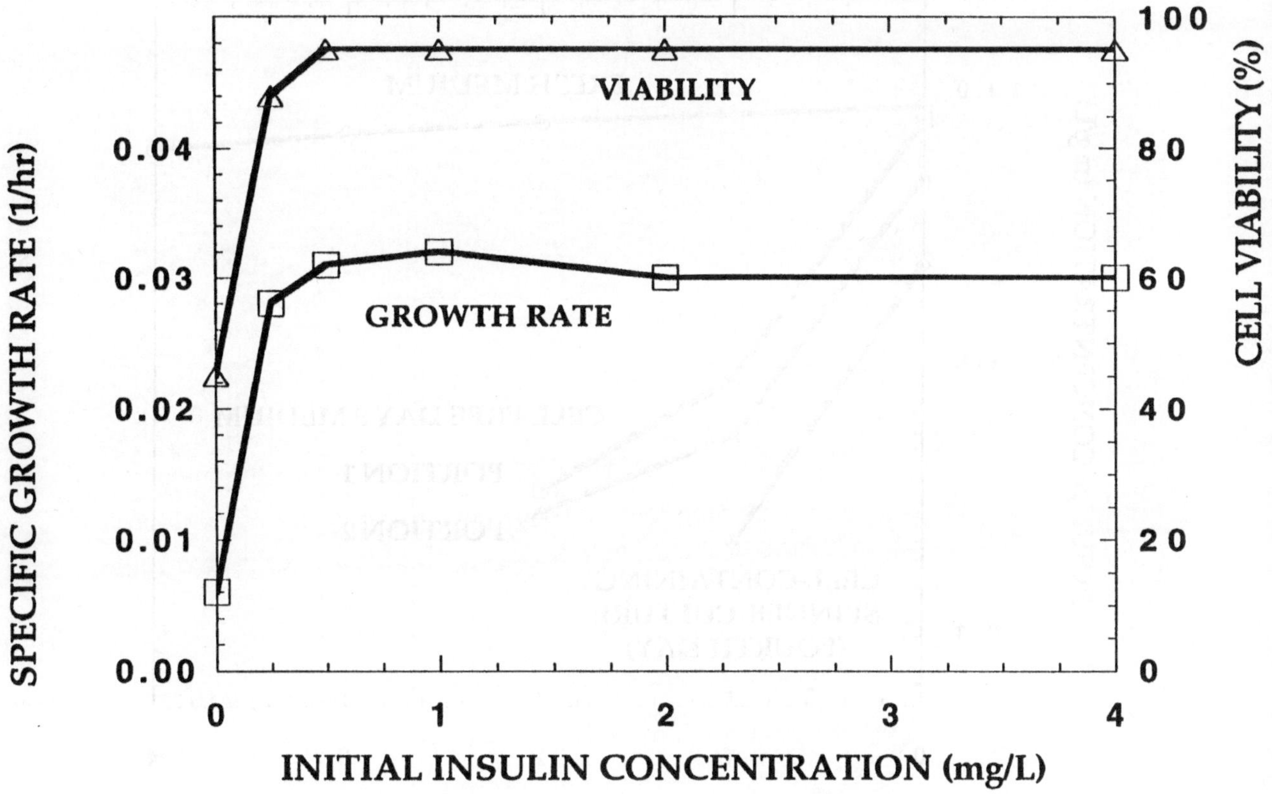

Fig. 4. Insulin dose response in duplicate 25 ml spinner flasks.

In order to identify the cause of the abnormally low insulin levels in batches during which upsets occurred, an experiment was conducted to characterize the degradation of insulin in cell culture medium. This experiment used an RP-HPLC insulin assay that was sensitive enough to quantitate insulin at levels well below 1 μg/ml.

Insulin degradation was found to be slow in fresh medium. When fresh medium containing insulin at 8.1 μg/ml was incubated at 37 °C for 4 days, the insulin concentration only dropped to 5.7 μg/ml (Fig. 5).

Insulin degradation was found to be much faster in cell-free conditioned medium from the third day of a spinner flask culture. Insulin was added back to a level of 7.2 μg/ml in one portion of this conditioned medium and to a level of 4.5 μg/ml in another. After 24 hours incubation at 37 °C, insulin concentrations in these two portions had dropped to 0.80 μg/ml and 0.49 μg/ml, respectively (Fig. 5). In each of these portions the drop in insulin concentration was 89% of the initial amount, suggesting that the rate of insulin degradation is proportional to the insulin concentration and is thus first order. The insulin concentration in cell-containing conditioned medium in the spinner flask culture dropped by roughly the same proportion during the fourth day (Fig. 5), suggesting that most of the insulin degradation that occurs in a culture is extracellular. Thus the cells impart an insulin degrading activity to the medium.

Insulin degrading activity is imparted to the medium much faster, even on a per cell basis, when cell density nears the ceiling imposed by depletion of nutrients and accumulation of inhibitors. To assess the amount of insulin degrading activity present in any given sample of conditioned medium, insulin was added to the sample to bring the insulin level to approximately 8 μg/ml. Then the proportion lost during 24 hours of cell-free incubation was determined. By this measure, the content of insulin degrading activity present in a spinner flask culture remained roughly constant between day 0 and day 2, but rose dramatically between day 2 and day 3 (Fig. 6).

The protocol for the rhM-CSF process was subsequently modified in a way that was intended to prevent the release of large amounts of insulin degrading activity into the medium. The modified protocol called for limiting the cell density at which batches were started

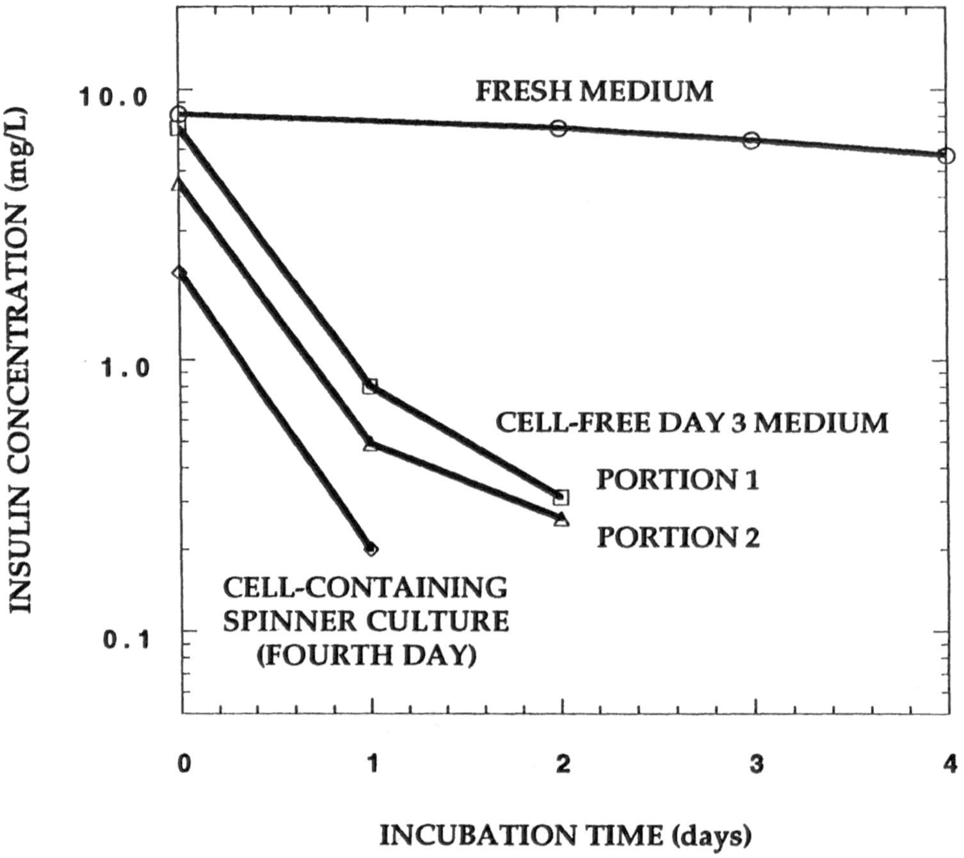

Fig. 5. Degradation of insulin in fesh medium during four days of incubation, in a day 3 spinner flask culture during a fourth day of incubation, or in two insulin-supplemented portions of cell-free day 3 conditioned medium from the spinner flask culture during two days of incubation.

to 2×10^5 cells/ml rather than the normal 4×10^5 cells/ml during the first 14 days following inoculation of the 250-L bioreactor. However insufficient data have been accumulated following this change to confirm its effectiveness.

Discussion

These data support the following hypothesis. During the first 14 days following inoculation of the 250-L bioreactor with the rhM-CSF cell line, the cells are sensitive to conditions present in the 250-L and 2500-L bioreactors. This sensitivity is manifested by a tendency to impart large amounts of insulin degrading activity to the medium at times when cell density is high. Occasionally during this period the amount of insulin degrading activity imparted to the medium by the end of one batch is so great that the amount in the retained culture is sufficient to destroy most of the insulin that enters during the refeed. This leaves the subsequent batch with an insulin shortage, which results in growth cessation and dropping viability.

The beneficial effects of insulin on CHO cell cultures are well known (Gasser *et al.*, 1985; Mendiaz *et al.*, 1986). However, the sensitivity of the rhM-CSF cell line to a shortage of insulin, represented in Figure 4, is greater relative to other recombinant cell lines we have worked with that have been derived from the same CHO DUKX-B11 parent cell line. The vulnerability of the rhM-CSF process to upsets has likewise been greater relative to processes using the other recombinant cell lines.

The degradation of insulin in rhM-CSF cultures occurred primarily extracellularly. This contrasts with *in vivo* insulin degradation, which occurs primarily intracellularly by receptor-mediated endocytosis (Duckworth, 1988; Authier *et al.*, 1994). The extracellular insulin degrading activity in rhM-CSF cultures

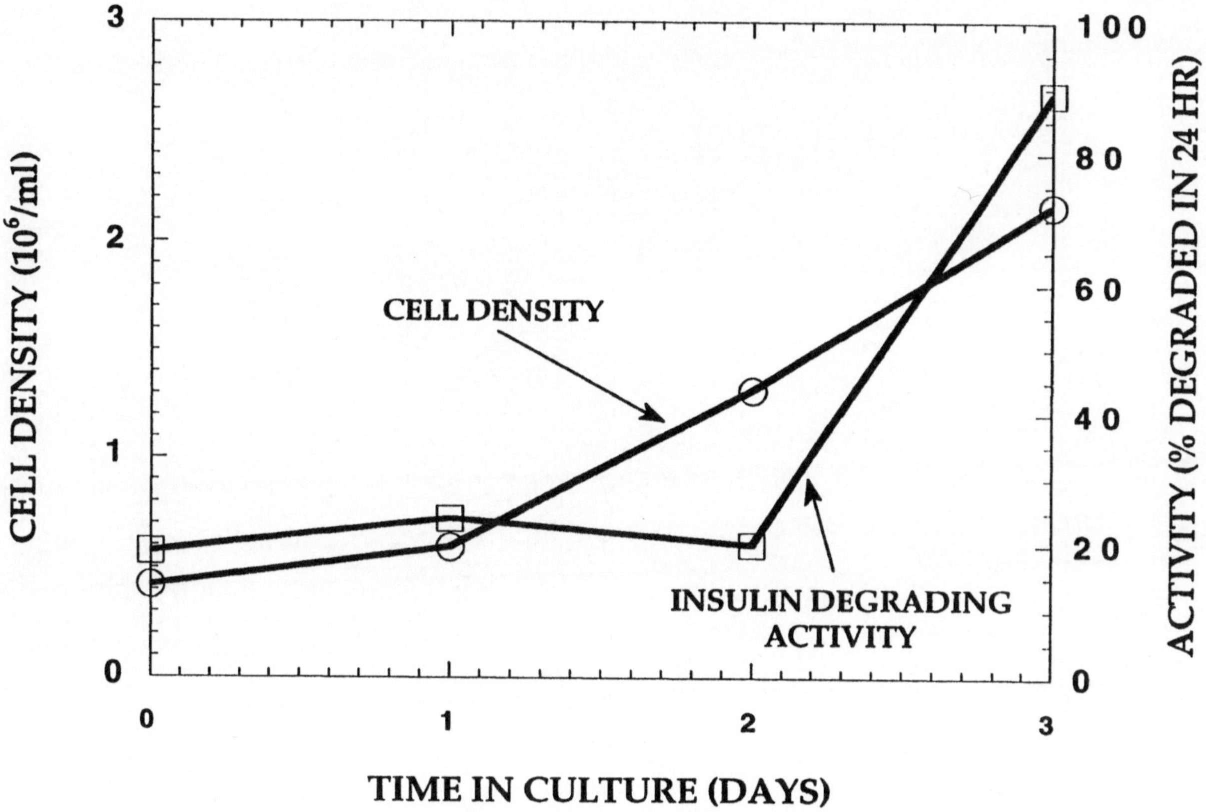

Fig. 6. Cell densities and amounts of insulin degrading activity present in samples drawn from a spinner flask culture.

may, however, have been due to release of intracellular enzymes.

Acknowledgements

Cell density and viability data representing 250-L and 2500-L bioreactor cultures were provided by the Commercial Manufacturing Department at Genetics Institute. RP-HPLC assays were performed by Anne Collins (Run B) or in collaboration with Yangkil Kim (degradation kinetics).

References

Authier F, Rachubinski RA, Posner BI and Bergeron J (1994) Endosomal proteolysis of insulin by an acidic thiol metalloprotease unrelated to insulin degrading enzyme. J. Biol. Chem. 269: 3010–3016.

Duckworth WC (1988) Insulin degradation: mechanisms, products, and significance. Endocrine Rev. 9: 319–345.

Gasser F, Mulsant P and Gillois M (1985) Long-term multiplication of the Chinese hamster ovary (CHO) cell line in a serum-free medium. *In Vitro* Cell Dev. Biol. 21: 588–592.

Mendiaz E, Mamounas M, Moffett J and Englesberg E (1986) A defined medium for and the effect of insulin on the growth, amino acid transport, and morphology of Chinese hamster ovary cells, CHO-K1 (CCL 61) and the isolation of insulin 'independent' mutants. *In Vitro* Cell Dev. Biol. 22: 66–74.

Wong GG, Temple PA, Leary AC, Witek-Giannotti JS, Yang Y, Ciarletta AB, Chung M, Murtha P, Kriz R, Kaufman RJ, Ferenz CR, Sibley BS, Turner KJ, Hewick RM, Clark SC, Yanai N, Yokota H, Yamada M, Saito M, Motoyoshi K and Takaku F (1987) Human CSF-1: molecular cloning and expression of 4-kb cDNA encoding the human urinary protein. Science 235: 1504–1508.

Address for correspondence: S.R. Adamson, Genetics Institute, Inc., 1 Burtt Road, Andover MA 01810, USA.

Cytotechnology **15:** 111–116, 1994.

Low temperature cultivation – A step towards process optimisation

Ralf Weidemann, Andreas Ludwig and Gerlinde Kretzmer
Institut für Technische Chemie, Callinstr. 3, 30167 Hannover FRG

Key words: Adherent animal cells, glucose, lactate, productivity, temperature

Abstract

Adherent recombinant BHK cells were cultivated at temperatures between 30 and 37°C. Batch and repeated-batch-cultivations in a 2-litre bioreactor showed a significant influence on metabolism and cell growth. The low-temperature-cultivations showed a lower growth rate and a lower glucose consumption rate and, therefore, less lactate production. On the other hand, the maximum cell density and productivity seemed not to be affected by the temperature reduction.

Introduction

For cultivating adherent animal cells in large-scale, bioreactors are necessary. Optimal nutrient and oxygen supply is realised by stirring the medium. The movement of the medium creates shear forces on the sensitive phospholipid membranes of animal cells.

It has been shown that temperature and cholestorol influence membrane fluidity by making the membranes less flexible and more rigid. Both parameters intensify the resistance against shear stress (Ludwig *et al.*, 1992; Tomeczkowski *et al.*, 1993).

Of course, it has to be proved that low temperature cultivation does not influence the viability or productivity of the cells. The membrane is also responsible for interchanging certain substances from the cell content to the environment and vice-versa. In the case of hybridoma cells few investigations were reported concerning the influence of reduced temperature on cell growth. Reuveny *et al.* (1986) found no influence on maximum cell density whereas Bloemkolk *et al.* (1992) described higher growth rate for stationary but not for mixed cultures.

A recent paper (Ludwig *et al.*, 1992) described experiments carried out with adherent Baby Hamster Kidney cells at temperatures of 28, 30, 33, 37, and 39 °C. As anticipated, all temperatures under the physiological temperature of 37 °C showed a negative influence on the doubling time of the cells, but the values of the maximal cell density were the same for all temperatures except for 28 °C.

This paper discusses the influence of temperature on process parameters like cell growth, metabolic rates and productivity for recombinant adherent Baby Hamster Kidney cells.

Materials and methods

Experiments were carried out with adherent Baby Hamster Kidney cells (BHK 21 clone 13) producing Antithrombin III (AT III, secreted). They were grown in Dulbecco's Modified Eagle's Medium (DMEM) with 10% Tryptose Phosphate Borth (TPB) and 10% Fetal Calf Serum (FCS). Stock cultures and precultures were maintained in monolayer flasks. Microcarriers (Cytodex III, Pharmacia) were used to carry out the batch and repeated-batch cultures. Batch and repeated batch cultivations were performed in a 2-litre bioreactor (Biostat MC, B. Braun, Melsungen). In repeated batch mode stirring was stopped. After settling of the microcarrier 1200 ml of the supernatant was replaced by fresh medium. The schedule of replacement was set by the lactate concentration which was found to be toxic above 2 gl^{-1}.

Cell count was carried out with the trypan blue dye exclusion method. The activity of lactate dehydrogenase in the medium was measured with an enzyme

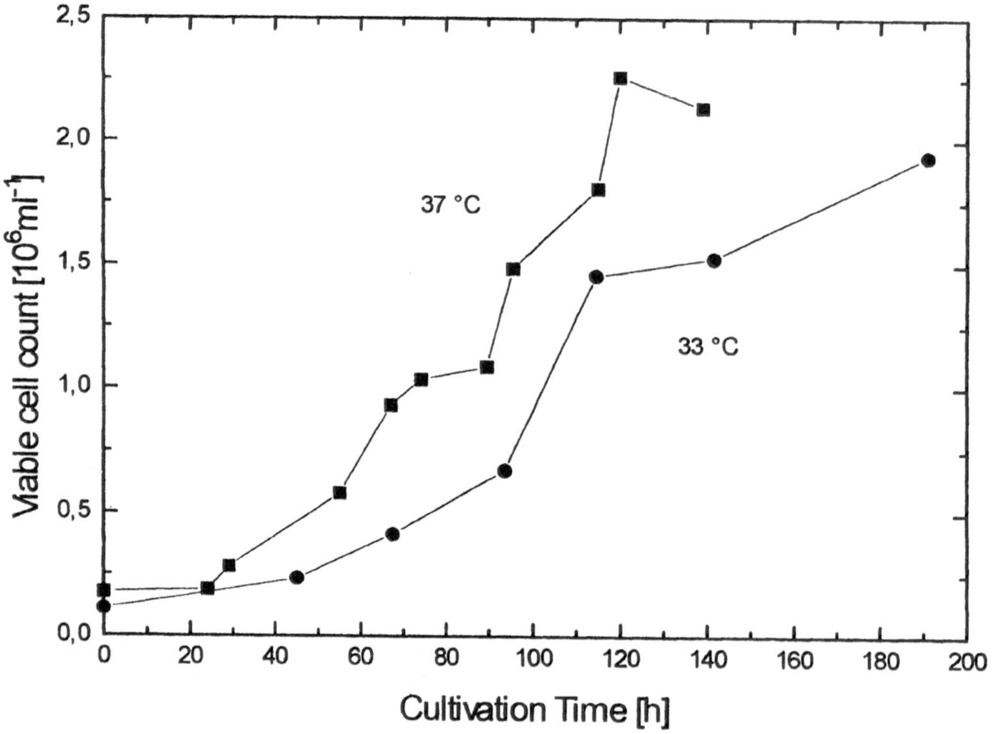

Fig. 1. Batch cultivation at 37 and 33 °C: Viable cell count [10^6 ml^{-1}] versus cultivation time.

test kit (Merckotest 3339). Lactate concentration was determined with a food analysis test kit by Boehringer. The amount of AT III was measured with a non-competitive ELISA-test (Behringwerke AG). Glucose concentration was determined with a Yellow Springs glucose analyser (Model 27).

Results and discussion

Batch cultivations at 37 and 33 °C showed similar behaviour (Fig. 1). After running through a 24-h-lag phase at 37 °C, the cells proliferated with a specific growth rate of 0.62 d^{-1}. Cell viability was about 86%. The maximum cell number (2.26 × 10^6 ml^{-1}) was reached after 120 h and the declining phase started. The specific glucose consumption rate during exponential growth increased to 0.58 mg per 10^6 cell per day. Cultivating the cells at 33 °C resulted in a prolonged lag phase (about 48 h); the specific growth rate (0.50 d^{-1}) during exponential growth was lower than at 37 °C. The maximum cell number reached after 191 h was nearly as high as at the end of the 37 °C cultivation. The cultivation was stopped at this time, therefore the maximum cell count was possibly not reached at this moment. Corresponding to the slower growth of the culture glucose consumption decreased. The calculation of the consumption rate showed a lower specific rate as well, 0.45 mg per 10^6 cells per day. The AT III-productivity was 2.9 μg per 10^6 cells per day for the 37 °C cultivation and 3.2 for the 33 °C cultivation. At both temperatures the productivity of the cells decreased during exponential growth, but this reduction was less for the lower temperature.

At 37 °C the relative accumulated amount of AT III at the end of the batch (139h) was set to be 100%. At 33 °C nearly 110% was achieved after 191 h. The reduction of the temperature prolonged the cultivation time causing a slight increase in the accumulated product (Fig. 2).

The repeated batch cultivations started in a different manner. Cultivation started at 37 °C, but after 90

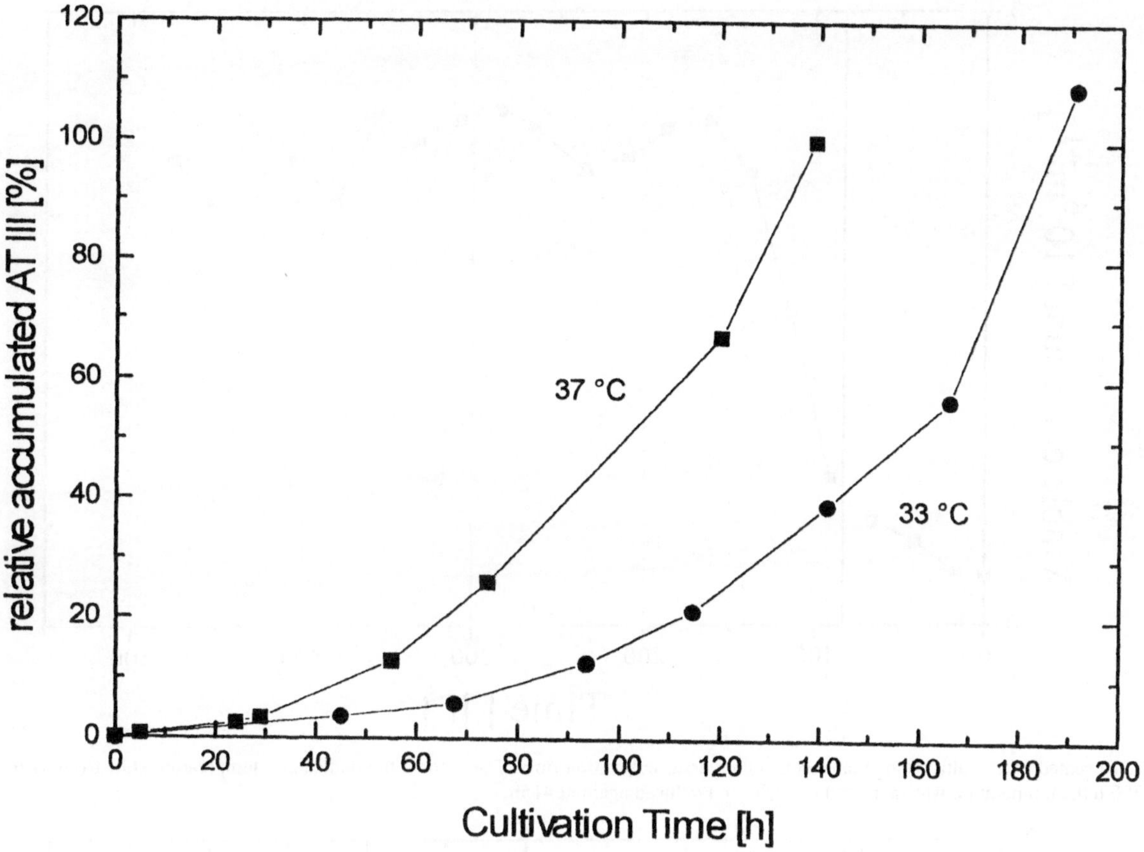

Fig. 2. Batch cultivation at 37 and 33 °C: Relative accumulated product amount (AT III) with cultivation time. The maximum product amount after 120 h at 37 °C was set to 100%.

hours the temperature was reduced to 30 °C. The prolonged lag phase and the reduced growth rate observed in the batch cultivations were avoided by changing the temperature at the mid-exponential phase. Maximum cell density (2.47 × 10^6 cells ml^{-1}) was reached after 160 h (Fig. 3). Unlike the cultivation carried out completely at a low temperature (33 °C), the growth rate of the repeated batch cultivation was not influenced by the temperature reduction. The specific metabolic rates decreased upon reducing the temperature (Fig. 4). During stationary phase the glucose consumption rate was about 0.2 mg per 10^6 cells per day, whereas the lactate productivity was slightly less. Figure 5 depicts the specific AT III-productivity and the specific growth rate of the repeated batch cultivations. The specific productivity seemed not to depend on cultivation temperature but on growth rate. At 30 °C the productivity was half as much as before the temperature reduction. This difference was not due to the temperature reduction because the productivity started to diminish before the temperature shift. This decrease in productivity was also observed in batch cultivations. Independently of the cultivation temperature in the stationary phase the productivity decreased to a constant value of 1–1.5 μg per 10^6 cells per day. Additionally a prolonged phase of high productivity was observed using the repeated batch modus. In the stationary phase the modus of cultivation seems to be less important. After 390 h the cultivation temperature was set to 37 °C again; the glucose consumption rate and the lactate productivity immediately increased. Reducing the temperature again resulted once more in decreased metabolism rates.

Conclusion

Batch cultures showed a negative influence of temperature on the specific growth rate. This could be evaded by starting the culture at 37 °C and reduc-

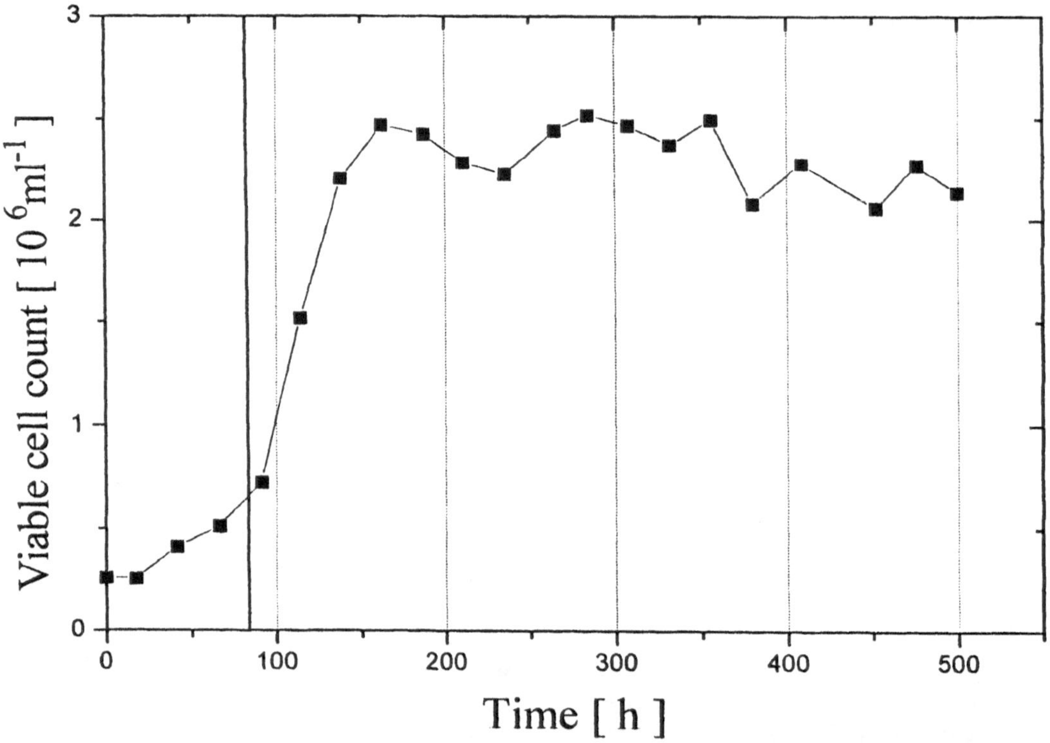

Fig. 3. Repeated batch cultivation: Viable cell count versus cultivation time. The vertical line indicates a temperature shift from 37 to 30 °C. After 390 h the temperature was increased to 37 °C and reduced again at 415h.

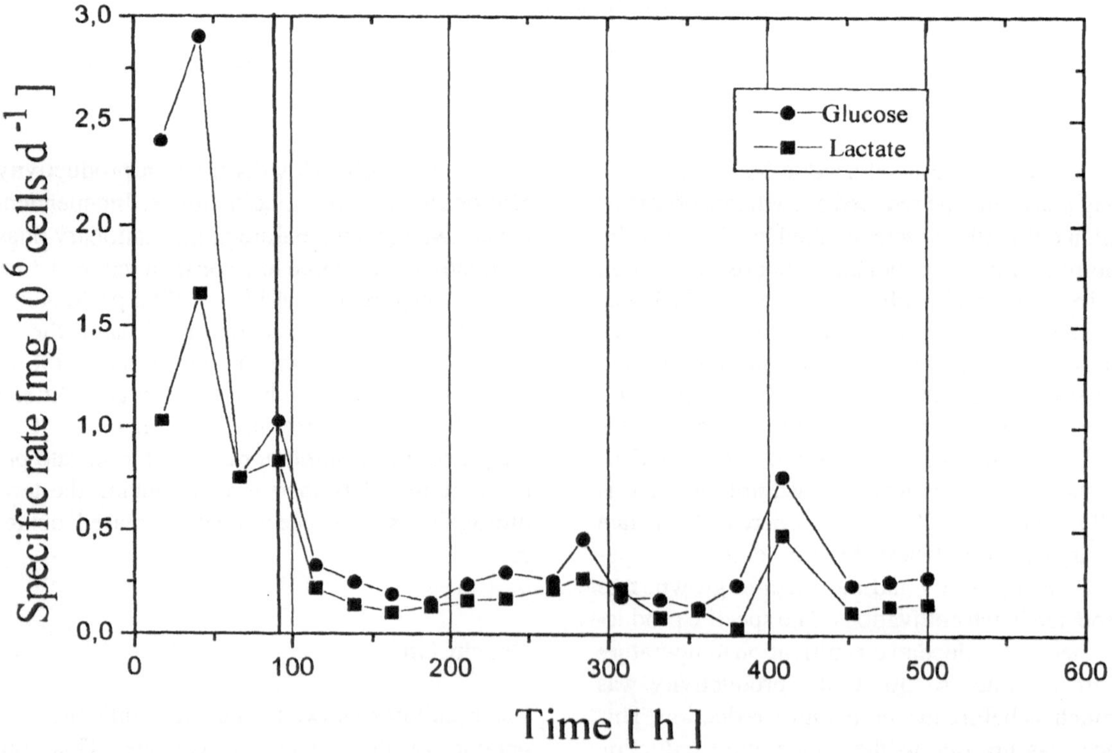

Fig. 4. Repeated batch cultivation. Glucose consumption rate and lactate productivity versus cultivation time. The vertical line indicates a temperature shift from 37 to 30 °C. After 390 h the temperature was increased to 37 °C and reduced again at 415h.

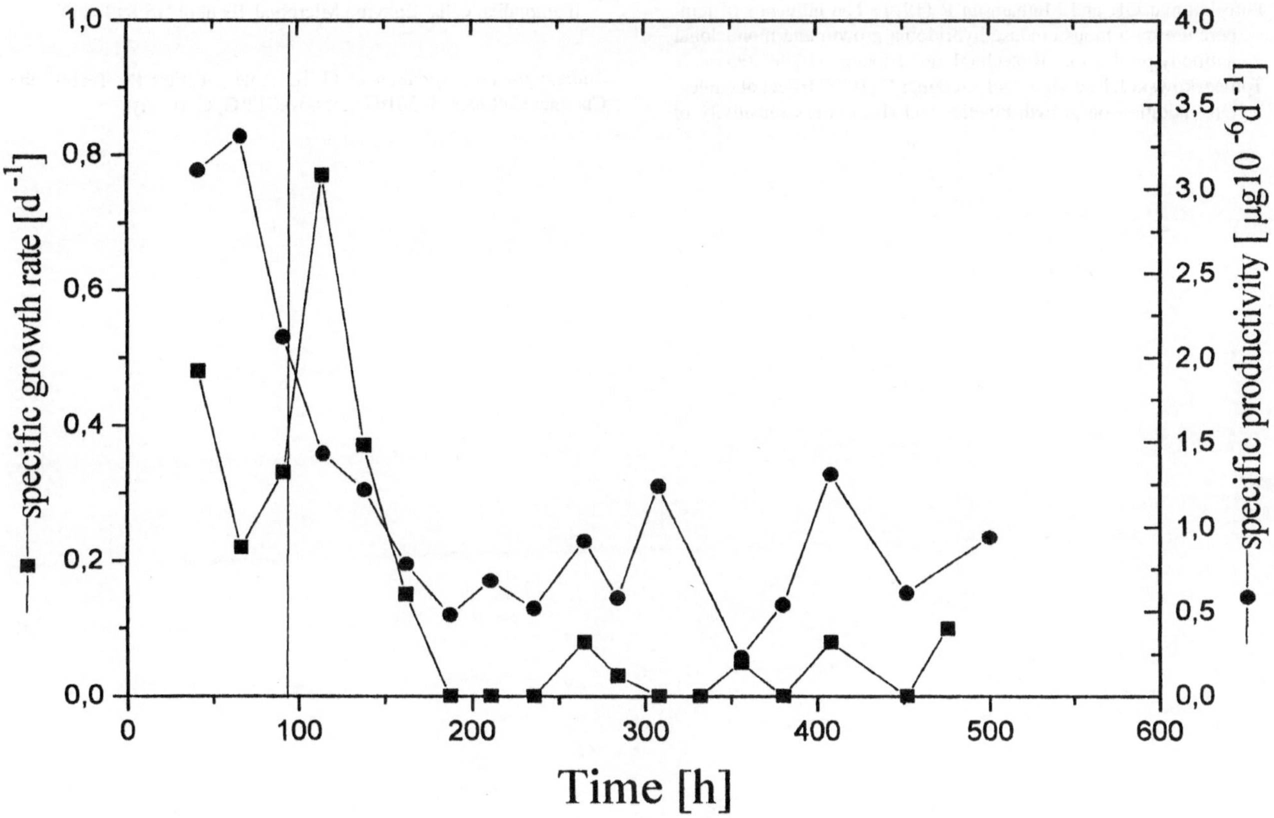

Fig. 5. Repeated batch cultivation. Specific growth rate and specific productivity versus cultivation time. The vertical line indicates a temperature shift from 37 to 30 °C. After 390 h the temperature was increased to 37 °C and reduced again at 415h.

ing the temperature in the middle of the exponential phase as was done in the repeated batch cultures. The temperature shift led to reduced metabolism and therefore, to lower lactate production while monitoring the same specific productivity. The critical lactate level which was determined to be 2 g l^{-1} could be avoided and the medium exchange rate was set at 0.5 per day. The specific productivity of the BHK cells was independent of the cultivation temperature but not of the growth phase. Productivity decreased with increasing cell density. Reuveny *et al.* (1986) found a drastical reduction of the specific productivity with decreasing temperature (12.5 μg (10^6 cells d)$^{-1}$ at 37 °C and 4.9 μg (10^6 cells d)$^{-1}$ at 34 °C). Sureshkumar and Mutharasan (1991) found no influence on the productivity of hybridoma between 35 and 39 °C. These results suggest that the influence of temperature on the process has to be determined in each case.

Concerning the recombinant BHK cells, low temperature cultivation economised the medium demand without loss of productivity. Additionally prolongation of the cultivation time resulted in a higher total product amount.

Acknowledgements

The authors wish to thank the Behringwerke AG, Marburg, Germany for supplying the recombinant cell line and the Bundesministerium für Forschung und Technologie (BMFT) for financial support.

References

Bloemkolk J-W, Gray MR, Merchant F and Mosmann TR (1992) Effect of temperature on hybridoma cell cycle and Mab production. Biotechnol. and Bioeng. 40:427–431.

Ludwig A, Tomeczkowski J and Kretzmer G (1992) Influence of the temperature on the shear stress sensibility of adherent BHK 21 cells. Appl. Microbiol. Biotechnol. 38:323–327.

Reuveny S, Velez D, Macmillan JD and Miller L (1986) Factors affecting cell growth and monoclonal antibody production in stirred reactors. J. Immunol. Methods 86: 53–59.

Sureshkumar GK and Mutharasan R (1991) The influence of temperature on a mouse mouse hybridoma growth and monoclonal antibody production. Biotechnol. and Bioeng. 37:292–295.
Tomeczkowski J, Ludwig A and Kretzmer G (1993) Effect of cholesterol addition on growth kinetics and shear stress sensitivity of mammalian cells. Enzyme Microbial Technol. 15:849–853.

Address for correspondence: G. Kretzmer, Institut für Technische Chemie, Callinstr. 3, 30167 Hannover FRG, Germany.

Cytotechnology **15**: 117–128, 1994.

Induction of apoptosis in oxygen-deprived cultures of hybridoma cells

Sylvain Mercille[1] & Bernard Massie[2]
[1] *Biomira Inc., 6100 Avenue Royalmount, Montréal, Québec, Canada, H4P 2R2;* [2] *Groupe d'Ingénierie des cellules animales, Institut de recherche en Biotechnologie, Conseil National de Recherches du Canada, 6100 Avenue Royalmount, Montréal, Québec, Canada H4P 2R2*

Key words: Animal cell culture, anoxia, apoptosis, cell death, hybridoma, hypoxia

Abstract

It is now well documented that apoptosis represents the prevalent mode of cell death in hybridoma cultures. Apoptotic or programmed cell death occurs spontaneously in late exponential phase of batch cultures. Until lately, no specific triggering factors had been identified. Recently, we observed that glutamine, cystine or glucose deprivation induced apoptosis in both hybridoma and myeloma cell lines whereas accumulation of toxic metabolites induced necrotic cell death in these cells. Other triggering factors such as oxygen deprivation might also be responsible for induction of apoptosis. In the present study, induction of cell death by exposure to anoxia was examined in batch culture of the SP2/0-derived hybridoma D5 clone. The mode of cell death was studied by morphological examination of acridine orange-ethidium bromide stained cells in a 1.5 L bioreactor culture grown under anoxic conditions for 75 hours. Under such conditions, viable cell density levelled off rapidly and remained constant for 25 hours. After 45 hours of anoxia, cell viability had decreased to 30% and the dead cell population was found to be 90% apoptotic. In terms of cellular metabolism, anoxia resulted in an increase in the utilization rates of glucose and arginine, and in a decrease in the utilization rate of glutamine. The lactate production rate and the yield of lactate on glucose increased significantly while the MAb production rate decreased. These results demonstrate that glycolysis becomes the main source of energy under anoxic conditions.

Cells incubated for 10 hours or less under anoxic conditions were able to recuperate almost immediately and displayed normal growth rates when reincubated in oxic conditions whereas cells incubated for 22 hours or more displayed reduced growth rates. Nonetheless, even after 22 h or 29 h of anoxia, cells reincubated in oxic conditions showed no further progression into apoptosis. Therefore, upon removal of the triggering signal, induction of apoptosis ceased.

Abbreviations: VNA – Viable non-apoptotic cells; VA – Viable apoptotic cells; NVNA – Nonviable non-apoptotic or necrotic cells; NVA – Nonviable apoptotic cells; CF – Chromatin-free cells (late nonviable apoptotic cells); AO –Acridine orange; EB – Ethidium Bromide; MAb – Monoclocnal antibody; D.O. – Dissolved oxygen; q_{MAb} – Specific MAb production rate (mg. $(10^9 \text{ cells})^{-1}.\text{day}^{-1}$); μ – Specific growth rate (h^{-1}); X_v – Viable cell number (10^5 cells.mL^{-1}); X_t – Total cell number (10^5 cells.mL^{-1}); $Y_{lac/glc}$ – Yield coefficient of lactate on glucose (mM lactate produced/mM glucose consumed)

Introduction

In the last few years, cells culturists have devoted much effort to study induction of cell death by hydrodynamic shear stress, metabolic intoxication, or trauma associated with large-scale culturing. In all of these studies, cell death was considered as a totally passive process. However, death can also occur through an active intrinsically controlled phenomenon in which the cell actually commits suicide in response to particular modifications in its environment, such as depletion of growth factors (Harrington *et al.*, 1994), serum (Evan *et al.*,

1992), nutrients (Mercille and Massie, 1994) or oxygen.

Under a variety of physiological conditions programmed cell death, or apoptosis, is a finely regulated way to eliminate cells that are old, unnecessary, or harmful (Cotter *et al.*, 1990). A specific sequence of events that requires active participation from the dying cell leads to the phagocytosis of cell constituents without accompanying inflammatory response. During embryonic development, programmed cell death acts as an important homeostatic regulator of tissue mass and architecture through the deletion of a variety of cell types. The apoptotic process, substantially conserved throughout multicellular evolution, is observed under certain pathological conditions where tissue damage occurs and where the removal of injured cells and their replacement with new ones is needed. Apoptosis plays an important role in hormone-dependant atrophy and is regulated by signal-transduction coupled events. It occurs frequently in adult tissues that have a high cell turnover rate, like the liver and assorted lymphoid organs and in the clonal selection of both T- and B-lymphocytes self-antigen recognizing clones (McDonald *et al.*, 1990). Cytotoxic T-lymphocytes and natural killer cells induce apoptosis in target cells such as tumour cells.

In general, cell death may follow two distinct patterns: apoptosis and necrosis. Apoptosis was originally defined and distinguished from necrosis on the basis of its morphology, its ultrastructure suggesting an active intrinsically controlled phenomenon (Wyllie, 1988). A characteristic feature of an apoptotic cells is that it extrudes water in an energy-requiring process in which intact mitochondria are needed. The cell shrinks, resulting in increased buoyant density. The collapse of cytoskeletal structures is associated with cell shrinkage. Nuclear condensation is accompanied by double-strand endonuclease cleavage of cellular DNA at linker regions between nucleosomes, generating fragments of 180–200 bp multiples. These will form the chromatic ladder of DNA typically seen in agarose gel electrophoresis of apoptotic cells (Duke and Cohen, 1992; Collins *et al.*, 1992). High levels of transglutaminase activity can also be detected, which results in the cross-linkage of cytoplasmic and nuclear proteins, converting the cell body into a highly rigid and relatively stable structure. One prevalent hypothesis is that sustained elevation of cytosolic Ca^{2+} concentration could stimulate Ca^{2+}-dependent degenerative enzymes, such as proteases, phospholipases and nucleases, resulting in irreversible cell damage.

Necrosis, in contrast, is a typically passive pathological event that occurs under severe trauma causing a sudden and significant cell damage. It can be caused by membrane attack from complement, toxic metabolites or lytic vital infection. This 'uncontrolled' cell death is characterized by swelling of the cell and mitochondria followed by breakdown of the plasma membrane. Cells ultimately release their cytoplasmic contents to the extracellular fluid, causing damage to neighbouring cells and generating an inflammatory reaction. No evidence for specific signaling pathway exists. In necrotic cell death, DNA is cleaved non-specifically (Koli and Keski-Oja, 1992).

It is now well documented that apoptosis represents the prevalent mode of cell death in T-cell hybridomas (Smith *et al.*, 1989; Shi *et al.*, 1989), B-cell hybridomas, (Al Rubeai *et al.*, 1990; Franek *et al.*, 1992), myelomas (Mercille and Massie, 1994) and other lymphoid cell lines. The occurrence of apoptosis in batch and perfusion cell culture is a highly undesirable phenomenon. Apoptosis accounts for the abrupt decrease in viability observed in late exponential phase of hybridoma or myeloma batch cultures (reviewed in Mosser and Massie, 1994; Vomastek and Franek, 1993). Singh *et al.* (1994) have found a level of apoptosis of 81% in late stages of batch culture of TB/C3 hybridoma cells. Control of dissolved oxygen concentration or pH in these batch cultures does not prevent this decline in viability. The duration of the stationary phase in cultures of hybridomas and myelomas rarely extends beyond 10 to 15 hours.

Apoptosis is especially undesirable in bioprocesses where production is not growth-associated. For example, antibody secretion in many hybridoma cell lines appears to be non-growth-associated. These cells exhibit higher specific productivities at slow growth rates (Shi *et al.*, 1993), especially in batch when transiting from the exponential phase to the stationary and decline phases. Many strategies, such as blockage of cells in the G1 phase of the cell cycle, fed-batch, pulsed-batch or perfusion cultures, have been devised to maintain the cells in that low-proliferative high-producing state. However, agents or strategies targeted at reducing proliferation often trigger apoptosis. The apoptotic nature of hybridomas, myelomas and other cell types with active cell death programmes would therefore contribute to reduce their productivity. Different approaches can be taken in an effort to reduce or eliminate the apoptotic behavior or lymphoid cell lines used in production. Several laboratories (including our own) are now working on cellular engineering targeted at increasing resistance to apoptosis (Mosser and Massie, 1994). However, it is also of critical importance to characterize apoptosis and to clearly identify the inducers of both apoptosis and necrosis in cell cultures. Recently, we observed that glutamine, glucose and cystine deprivation induced apoptosis in both hybridoma and myeloma cell lines whereas accumulation of toxic metabolites such as ammonia and lactic acid induced necrotic cell death in these cells (Mercille and Massie, 1994).

Effects of oxygen limitations on cell growth, production and metabolism have been extensively studied in hybridoma cultures (Ozturk *et al.*, 1990; Miller *et al.*, 1989). Oxygen limitations are however not restricted to the bioprocess environment and can also occur in the human body. Low oxygen tension is a feature of many physiologic and pathologic conditions, including wound healing, fibrosis and neoplasia. Indeed, acute hypoxia and nutrient restriction can occur in normal tissues, particularly those that undergo clinically important ischemic insults (heart, brain (Dragunow *et al.*, 1993) and also liver and kidney during transplantation) or those that become hypoxic under normal conditions (exercising skeletal muscle). A pathologic role of ischemia has been implicated in many cases as a stimulus for the onset of apoptosis. Cultures of MCDK cells treated in an hypoxic atmosphere exhibit significant death by apoptosis (Allen *et al.*, 1992). Embryonic craniofacial malformations involving programmed cell death are induced by hypoxic conditions (Sulik *et al.*, 1988). In tumours, there is also a chronic hypoxia arising from an inadequate blood supply. It was suggested that tumour apoptosis may represent a residual attempt at autoregulation within the expanding tumour population (Fukuda *et al.*, 1993).

In high cell density mammalian cell bioreactors, cellular oxygen demand may exceed oxygen delivery. Oxygen limitations may arise in microporous beads, hollow fibers or other immobilized systems. In addition, temporary anoxia may occur when cells are sent to cell-media separation devices such as centrifuges and sedimentors used in perfusion culture. In centrifuge separators, cells are pelleted and highly concentrated for short periods of time, and nutrient and oxygen deprivations are likely to occur in these situations. The object of the present work is to assess the relative contribution of oxygen deprivation on the induction of apoptosis in hybridoma cultures.

Materials and methods

Cell line and inoculum preparation

The isolation and characterization of the mouse/mouse (SP2/0) D5 hybridoma cell line producing an IgM against blood group A red cell antigen was previously described (Martel *et al.*, 1988). This clone tested negative for mycoplasma contamination using a Hoechst staining and co-culturing detection assay (ATCC quality control methods for cell lines, 1st edition, (1985) pp 12–15).

Media and additives

Cultures were carried out in high glucose Dulbecco's modified Eagle's medium (DMEM) (Gibco, Grand Island, NY, Cat# 11995–032) supplemented with 5% fetal calf serum (FCS) (Hyclone, Logan, Utah, Cat# A–1115–L) and 1:100 antibiotic/antimycotic solution (Gibco, Cat# 15240–013). The antibiotic/antimycotic stock solution consisted of 10,000 units.mL^{-1} penicillin G, 10,000 $\mu g.mL^{-1}$ streptomycin sulphate, and 25 $\mu g.mL^{-1}$ amphotericin B in 0.85% saline.

Culture system

Bioreactor cultures were performed using a 1.5-L CelliGen^{TM} (New Brunswick Sci., Edison, NJ) with an effective volume of 1.2 L. Under oxic conditions, the CelliGen controller maintained the dissolved oxygen at 50 ± 10% of air saturation by adjusting the air/O_2/N_2 ratio in the inlet gas stream. Under anoxic conditions, cells were grown in a nitrogen atmosphere. The standard cell-lift impeller of the Celligen was replaced with two 3-blade marine impellers and a 6-blade Rushton turbine positioned at the surface of the liquid for increased gas transfer capabilities. The pH was also controlled by the CelliGen at 7.0 ± 0.1. The agitation speed was set at 75 rpm and the temperature was maintained at 37°C. Cell samples collected aseptically from the bioreactor at various time intervals following oxygen removal were incubated in oxic conditions in 25 cm^2 stationary flasks and maintained at 37°C in a 5% CO_2/air atmosphere. For each time point, duplicate stationary flasks were seeded.

Analytical methods

Viable and total cell concentrations were evaluated using erythrosin B (Sigma) dye exclusion tests and haemacytometer counts. MAb concentrations in culture supernatants were determined using gel permeation chromatography, as previously described (Chauret *et al.*, 1992). Concentrations of glutamine and other amino acids were determined using orthophtalaldehyde derivatization and reversed-phase HPLC (Cattaneo *et al.*, 1992). Cystine concentration was measured as follows: samples were deproteinized with an equal volume of 10% tricarboxylic acid, diluted with 0.1 M sodium citrate buffer (pH 3), and injected on a Beckman High Performance Analyser, according to Veeraragavan *et al.* (1990). Glucose, lactate and pyroglutamate concentrations were determined using a HPLC method developed by Tedesco (1987).

Computational methods

The following mathematical expressions were used to calculate kinetic variables. Specific growth rate (μ, in h^{-1}) was defined as:

$$\mu = \frac{dX_t}{dt}\frac{1}{X_v}.$$

where X_t ($\times$ 10^5 cells.mL^{-1}) is the total cell population (including both viable and nonviable cells) and X_v ($\times$ 10^5 cells.mL^{-1}) is the viable cell population. The specific growth rate can be evaluated for each data point by taking a 3-point slope of X_t as a function of time (dX_t/dt) and dividing the value of this slope by the corresponding average value of X_v (middle point). Specific MAb production rate (q_{MAb}, in mg.(10^9 cells)$^{-1}$.day^{-1}) was defined as:

$$q_{MAb} = \frac{dMAb}{dt}\frac{1}{X_v}$$

Similarly, the value of q_{MAb} can be obtained by taking a 3-point slope of MAb as a function of time (dMAb/dt) and dividing the value of this slope by the corresponding average value of X_v, (middle point). The specific consumption rates of nutrients (glucose, glutamine, argininine, etc.) and the specific lactate production rate were defined as:

$$q_M = \frac{dM}{dt}\frac{1}{X_v}$$

where M is the concentration of the metabolite (glucose, glutamine, arginine, lactate, etc.). The value of q_M is evaluated similarly as for q_{MAb}.

The specific MAb production rate and all other specific rates have been calculated on the basis of the total viable cell population, including both non-apoptotic and apoptotic viable cells. These rates have been calculated during the first 30 h following oxygen removal.

Quantification of apoptosis and cell viability using fluorescent dyes

In this protocol, a cell suspension was mixed with fluorescent DNA-binding dyes and examined by fluorescence microscopy to visualize and count cells with aberrant chromatin organization (Mishell *et al.*, 1980; Duke and Cohen, 1992). A dye-mix working solution of 100 μg.mL^{-1} acridine orange (AO) (Molecular Probes, Eugene, OR, Cat# A1301) and 100 μg.mL^{-1} ethidium bromide (EB) (Sigma, Cat# E-8751) was prepared in PBS (Gibco, Cat# 310-4190AJ). A volume of 4 μL of this dye mixture was added to 100 μL of cell suspension at 5 $\times$ 10^5 to 5 $\times$ 10^6 cells.mL^{-1}. This mixture was then examined with a 40X objective using epiillumination and a filter combination suitable for observing fluorescein. Estimation of viability following AO/EB fluorescent staining closely matched what was found using standard erythrosine B staining and hemacytometer counts.

A minimum of 200 total cells was counted, recording the number of each of the following four cellular states:

(i) VNA: viable cells with non-apoptotic nuclei;
(ii) VA: viable cells with apoptotic nuclei;
(iii) NVNA: necrotic cells, i.e. nonviable cells with normal nuclei;
(iv) NVA: nonviable cells with apoptotic nuclei.

The percentages of each of these four cellular states in relation to the total of cells are obtained as follows (and similarly for the other three cellular states):

$$\% \text{ VNA } =$$

$$= \text{VNA}/(\text{VNA} + \text{VA} + \text{NVNA} + \text{NVA}) \times 100$$

Absolute cell number values (10^5 cells.mL^{-1}) are then obtained for these four populations by multiplying these percentages values by the total cell number (and similarly for the other three cellular states):

Viable non-apoptotic cell number = % VNA $\times$ Total cell number / 100.

Finally, the % of total nonviable cells that died from either apoptosis or necrosis can be computed as follows at every time point in the culture:

% of total nonviable cells that died from apoptosis = (% NVA/(% NVA + % NVNA)) $\times$ 100

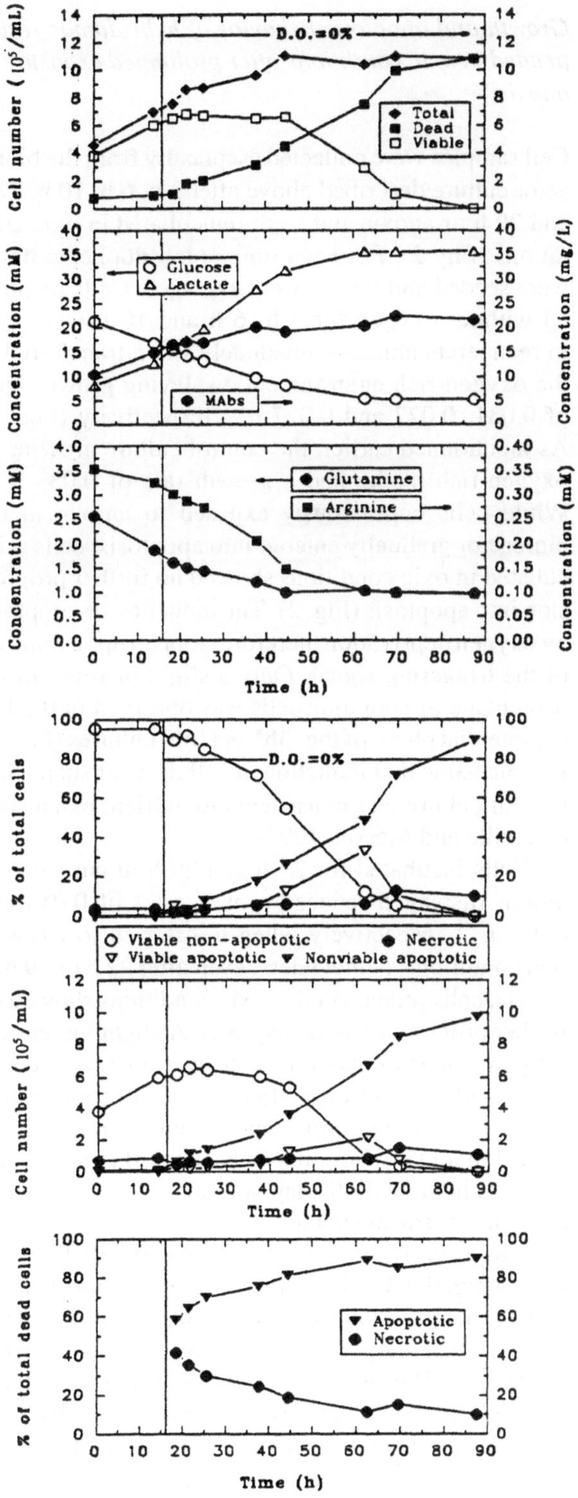

Fig. 1. Induction of apoptosis in D5 hybridoma cells following oxygen deprivation. At t = 15.5 hours, the D.O. level was reduced to 0% by feeding nitrogen and carbon dioxide gases to the reactor. This is indicated by the continuous vertical line.

% of total nonviable cells that died from necrosis = (% NVNA/(% NVA + % NVNA)) × 100

Results

Induction of apoptosis in hybridoma cells following oxygen deprivation

The mechanism and kinetics of cellular death under anoxic conditions were studied in a bioreactor batch culture operated with pH and D.O. control. Cells were inoculated at 3.8×10^5 viable cells.mL^{-1} and left for 15 hours in an oxygen-rich environment with a D.O. of 50 ± 10% (Fig. 1). The viable cell concentration increased steadily with a specific growth rate estimated as 0.033 h^{-1}. A roller bottle control culture run in parallel with a specific growth rate of 0.035 h^{-1} confirmed that cell growth was unaffected under the environment of the bioreactor (data not shown). At t = 15.5 hours, the D.O. level was reduced to 0% by feeding nitrogen and carbon dioxide gases to the reactor. Shortly after oxygen deprivation, the viable cell concentration levelled off at a plateau of 6.6×10^5 cells.mL^{-1} and remained constant for the following 30 hours (Fig. 1). It then declined steadily until t = 90 hours. The total cell number increased slowly throughout the plateau phase from 7×10^5 cells.mL^{-1} at t = 15.5 hours to 11×10^5 cells.mL^{-1} at t = 45 hours. A specific growth rate of 0.015 h^{-1} was estimated throughout this 30-hour plateau phase. During the decline phase, the total cell number remained relatively constant. During the initial 6 hours under anoxia, 90% of the cells were viable and non-apoptotic (Fig. 1). Then, viability started to decline steadily and both dead and viable apoptotic cells started to accumulate in the culture. While viable apoptotic cells never reached beyond 2.5×10^5 cells.mL^{-1}, nonviable apoptotic cells accumulated continuously up to a concentration of 10×10^5 cells.mL^{-1} at t = 90 hours. Within 35 hours of exposure to anoxia 50% of the cells were apoptotic. During the course of the plateau and decline phases, necrotic cells accumulated gradually and reached a density of 1×10^5 cells.mL^{-1}, which was less than 10% of total cells (Fig. 1). Morphological analysis of the nonviable cell population throughout the batch culture revealed that the percentage of cells dying from apoptosis increased gradually and levelled off at 90% at t = 65 hours. On the other hand, the percentage of cells dying from necrosis decreased gradually, levelling off at 10% of total dead cells (Fig. 1). It must be noted that data points taken early in the first 20 hours of culture were not significant since the amount of nonviable cells present during the exponential and early plateau phases was relatively minute. From these results, it can be concluded that D5 hybridoma cells exposed to anoxia die mainly from apoptosis. Comparable experiments performed using NS/0 myeloma cells also revealed that deprivation of oxygen induced apoptosis in these cells (data not shown).

Analysis of D5 hybridoma metabolism under anoxic conditions

Since nutrient deprivation has been shown to induce apoptosis in cultures of both D5 hybridoma cells and NS/0 myeloma cells (Mercille and Massie, 1994), it was very important to establish that the induction of apoptosis observed in the oxygen-deprived culture was not caused by exhaustion of glucose, glutamine or other nutrients. Indeed, it is well known that glucose can also be a limiting substrate in hybridoma culture, especially when rates of glycolysis are abnormally elevated such as in conditions of reduced dissolved oxygen concentration (Shi *et al.*, 1993) or severe hydrodynamic shear stress (Smith and Greenfield, 1992). After analysis of glucose and glutamine in the cell culture media (Fig. 1), it was found that these nutrients were not limiting at any point in the culture. In addition, all other amino acids initially present in the fresh media were only partially consumed (data not shown). Under anoxia, both ammonia and lactate accumulated in the culture media throughout the plateau phase up to concentrations of 1.1 mM and 35 mM respectively (Fig. 1). We have shown previously that ammonia and lactic acid were not responsible for induction of apoptosis in batch culture, but rather induced necrotic cell death in both D5 hybridomas and NS/0 myelomas (Mercille and Massie, 1994).

In terms of cellular metabolism (Fig. 1, Table 1), anoxia resulted in a 66% decrease in the utilization rate of glutamine. While the consumption patterns of most amino acids remained roughly similar in oxic and anoxic conditions, a 2- to 3-fold increase in the utilization rate of arginine was found under anoxia. In addition, the glucose utilization rate, lactate production rate and yield of lactate on glucose ($Y_{lac/glc}$) all increased significantly under anoxia. It was also observed that the specific MAb production rate was decreased by 67% under anoxia.

Growth and apoptotic behavior of hybridomas resuspended in oxic conditions after prolonged exposure to anoxia

Cell samples were collected aseptically from the bioreactor culture described above after 3 h, 6 h, 10 h, 22 h and 29 h of anoxia and were reincubated in oxic conditions (Fig. 2). For each time point, duplicate flasks were seeded and results were averaged. Cells incubated without oxygen for 3 h, 6 h and 10 h were able to recuperate almost immediately when transferred to the oxygen-rich environment displaying growth rates of 0.031, 0.027 and 0.027 h^{-1} respectively (Fig. 3). As mentionned earlier, the control culture growing in oxygen-rich media had a growth rate of 0.033 h^{-1}. While cells continuously exposed to anoxia in the bioreactor gradually entered into apoptosis, cells reincubated in oxic conditions showed no further progression into apoptosis (Fig. 2). The induction of apoptosis by oxygen deprivation therefore stopped upon removal of the triggering signal. Only a slight increase in the percentage of apoptotic cells was observed in the late exponential phase of the '3h' and '6h' cultures (Fig. 2). It is plausible that induction of cell death at such a late time in culture was consequent to nutrient exhaustion (Mercille and Massie, 1994).

Cells incubated for 22 h and 29 h in anoxic conditions displayed reduced growth rates of 0.014 and 0.002 h^{-1} respectively when transferred to oxygen-rich conditions. Nonetheless, even after 22 h or 29 h of anoxia, cells reincubated in oxic conditions showed no further progression into apoptosis. A slight increase in the percentage of apoptotic cells was observed for the '22 h' culture, but only after 50 hours of incubation in oxic conditions. From these results, it can be concluded that D5 hybridoma cells exposed to anoxia for up to 10 hours will display normal growth rates when oxygen is made available.

Interestingly, the cells cultured for 29 hours in anoxia remained viable for more than 72 hours (Fig. 2) when reincubated in oxic conditions, and this, despite the fact that they were not actively dividing (μ = 0.002 h^{-1}). The culture was not followed beyond these 72 hours but it is conceivable that these cells were readapting to the presence of oxygen and that possible regrowth may have been observed.

Table 1. Effect of anoxia on metabolism of D5 hybridoma cells

Metabolic parameter	Oxic environment (D.O. = 30 ± 10%)	Anoxic environment (D.O. = 0%)
glucose utilization rate[a]	8.26	10.37
glutamine utilization rate[a]	2.31	0.79
arginine utilization rate[a]	0.09	0.21
lactate production rate[a]	14.21	22.70
MAb production rate[b]	14.21	4.71
$Y_{lac/glc}$[c]	1.72	2.07

[a] These rates are in millimoles per 10^9 cells per day;
[b] This rate is in milligrams per 10^9 cells per day;
[c] This coefficient is in mmoles of lactate produced per mmole of glucose consumed.

Discussion

From the results presented above, it is clear that deprivation of oxygen in hybridoma cells brings on a series of metabolic changes gradually leading to decreased antibody production and induction of programmed cell death. In terms of energy metabolism, it results in a significant decrease in the utilization rate of glutamine. This has been reported previously (Thömmes *et al.*, 1993) and can be explained by the fact that hybridoma cells degrade most of the glutamine in an oxidative manner (Zielke *et al.*, 1984). Since oxygen is not available as terminal electron acceptor for oxidative phosphorylation, the glutamine utilization rate decreases under anoxic conditions. The redox state of cytochrome c (an electron carrier in the respiratory chain) appears to be very sensitive to oxygen concentration. As oxygen concentration declines, increased reduction of cytochrome *c* will slow down the rate of oxidative phosphorylation (Wilson *et al.*, 1979). However, other researchers have found increased utilization of glutamine in anoxic conditions (Ozturk and Palsson, 1990), possibly reflecting cell line specificities in metabolism.

Anoxia also resulted in a significant increase in the glucose utilization rate, the lactate production rate and the yield of lactate on glucose ($Y_{lac/glc}$). This has also been reported previously (Ozturk and Palsson, 1990; Thömmes *et al.*, 1993; Shi *et al.*, 1993) and demonstrates that glycolysis becomes the main source of energy under anoxic conditions. In fibroblasts, following 6 to 8h of anoxia, DNA synthesis stops and adaptation to glycolytic metabolism occurs, with elevation of hexose transport and induction of lactate dehydrogenase. Additional glycolytic enzymes are induced to a lesser extent (Stoler *et al.*, 1992). Generation of energy from glycolysis is less effective than from glutaminolysis. Indeed, 2 mmoles of ATP are obtained per mmole of glucose degraded through the glycolytic pathway as compared to 27 mmoles of ATP per mmole of glutamine for complete oxidation via the glutaminolytic pathway (McKeehan, 1986). In many animal and human tumours, it has been reported that nutrient deprivation and anoxia arising from inadequate blood supply results in a decreased level of high energy phosphates. Thus cellular ATP levels will be dependant upon oxygen concentration. Under conditions of severe hypoxia, ADP levels would be expected to increase, directly or indirectly activating glycolysis (Tozer and Griffiths, 1992). Reduced availability of high energy phosphates can explain the decrease in specific growth rate of hybridomas observed in the present study.

Although in many cases, total protein synthesis is found to decline under anoxia (Carlos *et al.*, 1991), stress proteins similar to those synthesized following heat shock are produced (Zimmerman *et al.*, 1991). These stress-inducible proteins are abondant in non-stressed cells and play an essential role in protein folding, maturation and translocation (reviewed in Mosser and Massie, 1994). A coupling between heat shock protein (HSP) synthesis and energy metabolism has been proposed by Lanks (1983, 1986) who hypothesizes that HSP synthesis is induced by conditions that restrict the availability of reduced nucleotides, resulting in an increased $NADP^+/NADPH$ ratio. Immediately following the transition from aerobic to anaerobic conditions, increased synthesis of the 70 and 90 kD HSPs is observed in Chinese Hamser Ovary cells. HSP synthesis then diminishes, while glucose-

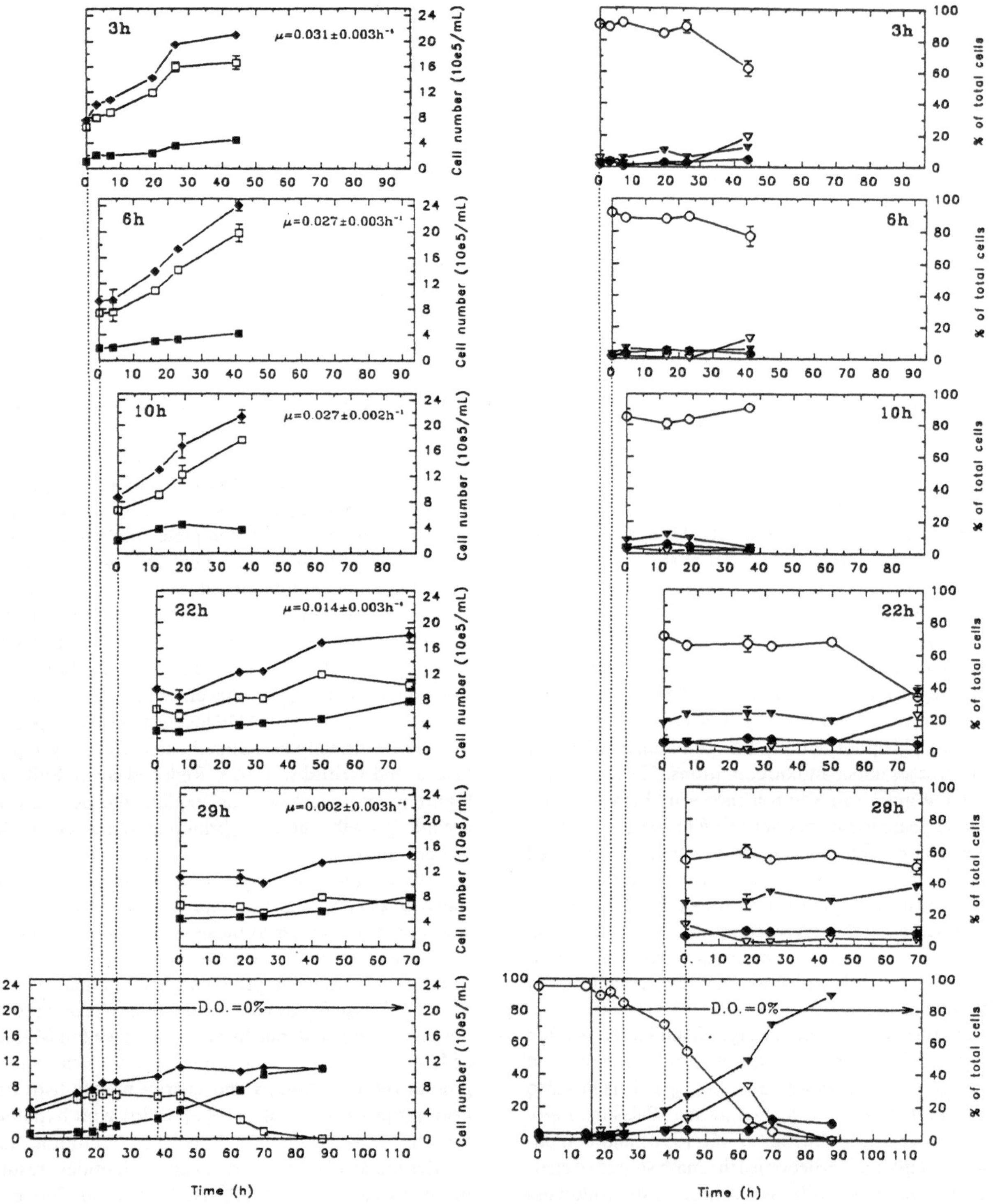

Fig. 2. Reincubation of cells exposed to anoxia for various periods of time. Evolution of the viable (□), dead (■), and total (♦) cell number following erythrosin B staining (left panels) and evolution in the percentage of viable non-apoptotic (○), viable apoptotic (▽), necrotic (●) and nonviable apoptotic (▼) cells following acridine orange/ethidium bromide staining (right panels).

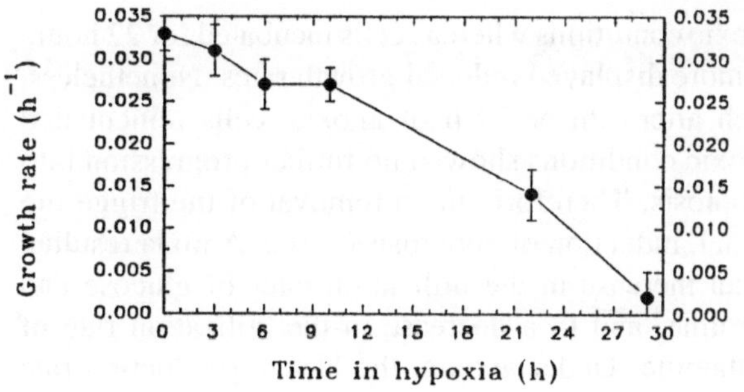

Fig. 3. Effect of prolonged anoxia on growth rate of cells reincubated in oxic conditions.

regulated protein (GRP) synthesis increases (Sciandra *et al.*, 1984). GRPs are located in the endoplasmic reticulum (ER) and are homologous to HSPs (located in the cytoplasm). In chronic hypoxia of tumour cells, the GRPs are also highly expressed (Murphy *et al.*, 1991; Cai *et al.*, 1993). Reports indicate that oxygen depletion inhibits protein maturation within the ER leading to the production of malfolded proteins. This results in the formation of GRP78-protein complexes and increases synthesis of the GRPs (Lee, 1992). The rapid reduction of the specific MAb production rate observed in this study for the D5 hybridoma and elsewhere for other hybridomas subjected to anoxia (Miller *et al.*, 1987; Ozturk and Palsson, 1990; Shi *et al.*, 1993; Thömmes *et al.*, 1993) could be explained, at least in part, by the rapid retention of GRP-antibody complexes in the ER.

The reduced intracellular energetic balance and decreased protein synthesis may also be responsible for induction of apoptosis under anoxic conditions. It has been suggested that ischemia-anoxia induced apoptosis in tumours represents an attempt at autoregulation of tumour size which eventually fails (*et al.*, 1993). Indeed, one possible antineoplastic mechanism involves activation of cell suicide pathways in those cells at risk of neoplastic transformation (Barraf *et al.*, 1986).

There are two types of models for induction of apoptosis. In some cases it appears that apoptosis depends on new gene expression after exposure to the apoptotic stimulus and is blocked if mRNA or protein synthesis is inhibited. These are collectively referred to as 'induction' mechanisms (Tomei, 1991). However, in some cells, for example the human promyelocytic leukemia HL-60 (Martin *et al.*, 1990) and B-cell hybridomas (Perrault and Lemieux, 1994), apoptosis can be triggered by the inhibition of mRNA or protein synthesis. Because these cells behave as though the suicide program is constitutively expressed but inhibited by factors with short half-lives, this is referred to as 'release' mechanism. Indeed, inhibition of protein synthesis by addition of cycloheximide induced apoptosis in the D5 hybridoma cell line used in the present study (Mercille and Massie, 1994). In addition, recent observations (Evan *et al.*, 1992) suggest that cells expressing high levels of c-myc may undergo apoptosis upon removal of signals driving proliferation such as serum or growth factors. In plasmacytomas and myelomas, the c-myc gene is translocated near the immunoglobulin locus and is under the control of the immunoglobulin promotor (Sheng-Ong *et al.*, 1982; Marcu *et al.*, 1992). It is therefore expressed in high levels in these cells. Any perturbation in the energetic balance of the cells such as that triggered by removal of oxygen may be equivalent to the removal of other signals driving proliferation such as serum or growth factors.

Upon removal of oxygen, intracellular ATP levels and protein synthesis rates are likely to decrease gradually in all cells. Since the cell population is heterogeneous and distributed in the various phases of the cell cycle, protein synthesis is prone to decrease at various rates from one cell to the other. In order to enter apoptosis, it is likely that hybridomas must reach a critical state that will result in the inefficient translation of putative short-lived apoptosis inhibitory proteins. This critcal state may also vary from one cell to another. Once this level is reached, the apoptotic machinery would be activated irreversibly and the cell would rapidly undergo the morphological changes associated with the action of degenerative enzymes. Our results show that cultures reincubated in oxic conditions following brief or prolonged exposures to anoxia will arrest progression into apoptosis (Fig. 2). These results suggest that a cell that has reached a critical energetic state will exhibit typical apoptotic morphology within a very short time period. Indeed, when potent protein synthesis inhibitors such as cycloheximide, actinomycin D and heat shock are used, induction of apoptosis in 50% of D5 hybridoma cells is observed within less than 4 hours (Mercille and Massie, 1994; Perrault and Lemieux, 1993, 1994).

Under anoxia, viable apoptotic hybridomas never reached beyond 35% of total cells. Conversely, nonviable apoptotic cells accumulated continuously up to 90% of total cells after 75 hours of anoxia. Such a low level of viable apoptotic cells in the culture suggested a relatively rapid transition period between viable apoptotic and nonviable apoptotic states. This suggests that a viable cell that exhibits DNA condensation will very rapidly reach a state at which it cannot maintain membrane integrity.

In terms of the implication to bioprocess development, the results presented in this study indicate that short periods of anoxia of 10 hours or less, although resulting in temporarily reduced growth and productivity, do not lead to irreversible damage or to significant induction of programmed cell death. Episodic anoxia generated in cell-media separating devices used in perfusion such as centrifuges should therefore not significantly affect the cells. However, prolonged exposure to anoxia in microporous beads or in cellular aggregates could potentially be counterproductive.

As mentioned above, cells incubated for 29 h in anoxic conditions and transferred to oxygen-rich conditions were unable to grow. Surprisingly, they remained viable for more than 72 hours following transfer. Prolonged exposure to anoxia may alter the regulation of oxygen consumption (Lin *et al.*, 1992) and decrease the level of antioxidant molecules (superoxide dismutase, catalase, glutathione peroxidase, glutathione), sensitizing cells to oxidant damage when the oxygen tension is subsequently increased. The culture was not followed beyond these 72 hours but it is conceivable that these cells were readapting to the presence of oxygen and that possible regrowth may have been observed. Further investigation on that phenomenon may also tell us something about the mechanisms of induction of apoptosis, recently linked to free-radicals and lipid peroxidation (Hockenbery *et al.*, 1993).

Conclusions

In the present study, cell death was investigated in cultures of D5 hybridomas exposed to anoxia through morphological examination of cells stained with acridine orange and ethidium bromide. Within 35 hours, viability decreased to 50% and the dead cell population displayed typical apoptotic morphology, with crescents around the periphery of the nucleus, or with the entire nucleus present as one or a group of featureless, bright spherical beads. Hybridoma cells deprived of oxygen therefore undergo apoptotic cell death. Moreover, under anoxia, the specific monoclonal antibody production rate decreases rapidly.

Cells incubated for 10 hours or less under anoxic conditions were able to recuperate almost immediately and displayed normal growth rates when reincubated in oxic conditions whereas cells incubated for 22 hours or more displayed reduced growth rates. Nonetheless, even after 22h or 29 h of anoxia, cells reincubated in oxic conditions showed no further progression into apoptosis. Therefore, upon removal of the triggering signal, induction of apoptosis ceased. Anoxia resulted in an increase in the utilization rates of glucose and arginine, and in a decrease in the utilization rate of glutamine. Under anoxia, the lactate production rate and yield of lactate on glucose increased significantly. These results demonstrate that glycolysis becomes the main source of energy under anoxic conditions. Despite this metabolic shift under anoxia, a close monitoring of nutrient concentration confirmed that induction of apoptosis observed in the oxygen-deprived culture was not the result of nutrient exhaustion.

Acknowledgements

This work was supported by a contribution agreement between Biomira Inc. and the National Council of Canada. We would like to thank Ann Burns-Tardif, Mark Johnson, Stéphane Lanthier and Louise Paquet for their collaboration. We are also grateful to Dick D. Mosser for critical review of this manuscript. Part of this work was presented at the 'Applications of Apoptosis, programmed cell death' CHI meeting held in La Jolla, CA, January 17–19, 1994.

References

Al-Rubeai M, Mills D and Emery AN (1990) Electron microscopy of hybridoma cells with regard to monoclonal antibody production. Cytotechnology 4: 13–28.

Allen J, Winterford C, Axelsen RA and Gobe GC (1992) Effects of hypoxia on morphological and biochemical characteristics of renal epithelial cell and tubule cultures. Ren. Fail. 14: 453–460.

Barraf CE and Bowen ID (1986) Kinetic studies on a murine sarcoma and an analysis of apoptosis. Br. J. Cancer 54: 989–998.

Cai JW, Henderson BW, Shen JW and Subjeck JR (1993) Induction of glucose regulated proteins during growth of a murine tumor. J. Cell. Physiol. 154: 229–237.

Carlos RQ, Seidler FJ and Slotkin TA (1991) Fetal dexamethasone exposure sensitizes neonatal rat brain to hypoxia effects on protein and DNA synthesis. Dev. Brain Res. 64: 161–166.

Cattaneo MV, Luong JHT and Mercille S (1992) Monitoring glutamine in mammalian cell cultures using an amperometric biosensor. Biosensors and Bioelectronics 7: 329–334.

Chauret N, Coté J, Archambault J and André G (1992) High-performance gel-permeation chromatographic analysis of IgM produced by hybridoma culture. J. Chromatography 594: 179–185.

Collins RJ, Harmon BV, Gobé GC and Kerr JFR (1992) Internucleosomal DNA cleavage should not be used as the sole criterion for identifying apoptosis. Int. J. Radiat. Biol. 61: 451–453.

Cotter TG, Lennon SV, Glynn JG and Martin SJ (1990) Cell death via apoptosis and its relationship to growth, development and differentiation of both tumor and normal cells. Anticancer Res. 10: 1153–1160.

Dragunow M, Young D, Hughes P, MacGibbon G, Lawlor P, Singleton K, Sirimanne E, Beilharz E and Gluckman P (1993) Is c-Jun involved in nerve cell death following status epilepticus and hypoxic-ischaemic brain injury? Brain Res. Mol. Brain Res. 4: 347–352.

Duke RC and Cohen JJ (1992) Morphological and biochemical assays of apoptosis. In: Janssen K (ed.) Current protocols in immunology. (pp 3.17.1–3.17.16) Wiley, New York.

Evan GI, Wyllie AH, Gilbert CS, Littlewood TD, Lard H, Brooks M, Waters CM, Penn LK and Hancock DC (1992) Induction of apoptosis in fibroblasts by c-myc protein. Cell 69: 119–1128.

Franek F, Vomastek T and Dolnikova J (1992) Fragmented DNA and apoptotic bodies document the programmed way of cell death in hybridoma cultures. Cytotechnology 9: 117–123.

Fukuda K, Kojiro M and Chiu JF (1993) Demonstration of extensive chromatin cleavage in transplantated Morris hepatoma 7777 tissue: apoptosis or necrosis? Am. J. Pathol. 142: 935–946.

Harrington EA, Fanidi A and Evan GI (1994) Oncogenes and cell death. Current opinion in Genetics and development 4: 120–129.

Hockenbery DM, Oltvai ZN, Yin XM, Milliman CL and Korsmeyer SJ (1993) BCl-2 functions in an antioxidant pathway to prevent apoptosis. Cell 75: 241–251.

Koli K and Keski-Oja J (1992) Cellular senescence. Annals of Medicine. 24: 313–318.

Lanks KW (1983) Metabolic regulation of heat shock protein levels. Proc. Natl. Acad. Sci. 80: 5325.

Lanks KW (1986) Modulators of the eukariotic heat shock response. Exp. Cell. Res. 165: 1.

Lee AS (1992) Mammalian stress response: induction of the glucose-regulated protein family. Current opinion in cell biology 4: 267–273.

Lin AA and Miller WM (1987) CHO cell responses to low oxygen: Regulation of oxygen consumption and sensitization to oxidative stress. Biotechnol. Bioeng. 40: 505–516.

Marcu KB, Bossone SA and Patel AJ (1992) Myc function and regulation. Annual Rev. Biochem. 61: 809–860.

Martel M, Bazin R, Verrette S and Lemieux R (1988) Characterization of higher avidity monoclonal antibodies produced by murine B-Cell hybridoma variants selected for increased antigen binding of membrane Ig. J. Immunol. 141: 1624–1626.

Martin SJ, Lennon SV, Bonham AM and Cotter TG (1990) Induction of apoptosis (programmed cell death) in leukemic HL-60 cells by inhibition of RNA or protein synthesis. J. Immunol. 145: 1859–1862.

McDonald HR and Lees RK (1990) Programmed death of autoreactive thymocytes. Nature 343: 623-644.

McKeehan WL (2986) Glutaminolysis in animal cells. In: Morgan MJ (ed.) Carbohydrate metabolism in cultured cells. Chapt. 4. Plenum Press, New York.

Mercille S, Johnson M, Lemieux R and Massie B (1994) Filtration-based perfusion of hybridoma cultures in protein-free medium: Reduction of membrane fouling by medium supplementation with DNase I. Biotechnol. Bioeng. 43: 833–846.

Mercille S and Massie B (1994) Induction of apoptosis in nutrient-deprived cultures of hybridoma cells. Biotechnol. and Bioeng. 44: 1140–1154.

Miller WM, Wilke CR and Blanch HW (1987) Effects of dissolved oxygen concentration on hybridoma growth and metabolism in continuous culture. J. Cell. Physiol. 132: 524–530.

Mishell BB, Shiigi SM, Henry C, Chan EL, North J, Gallily R, Slomich M, Miller K, Marbrook J, Parks D and Good AH (1980) Preparation of mouse cell suspensions. In: Mishell BB and Shiigi SM (eds.), Selected Methods in cellular immunology, pp 21–22 WH Freeman, New York.

Mosser DD and Massie B (1994) Genetically engineering mammalian cell lines for increased viability and productivity. Biotech. Adv. 12: 253–277.

Murphy BJ, Laderoute KR, Short SM and Sutherland RM (1991) The identification of heme oxygenase as a major hypoxic stress protein in Chinese hamster ovary cells. Br. J. Cancer 64: 69–73.

Ozturk SS and Palsson BO (1990) Effects of dissolved oxygen on hybridoma growth, metabolism and antibody production in continuous culture. Biotechnol. Prog. 6: 437–446.

Perreault J and Lemieux R (1994) Essential role of optimal protein synthesis in preventing the apoptotic death of cultured B cell hybridomas. Cytotechnology 13: 99–105.

Perreault J and Lemieux R (1993) Rapid apoptotic cell death of B-cell hybridomas in absence of gene expression. J. Cell Physiology 156: 286–293.

Sciandra JJ, Subjeck JR and Hughes CS (1984) Induction of glucose-regulated proteins during anaerobic exposure and of heat-shock proteins after reoxygenetation. Proc. Nat. Acad. Sci. USA 81: 4843.

Sheng-Ong GLC, Keath EJ, Piccoli SP and Cole MD (1982) Novel myc oncogene RNA from abortive immunoglobulin-gene recombination in mouse plasmacytomas. Cell 31: 443–452.

Shi Y, Ryu DDY and Park SH (1993) Monoclonal antibody productivity and the metabolic pattern of perfusion cultures under varying oxygen tensions. Biotech. Bioeng. 42: 430–439.

Shi Y, Sahai BM and Green DR (1989) Cyclosporin A inhibits activation-induced cell death in T-cell hybridomas and thymocytes. Nature 339: 625–626.

Singh RP, Al-Rubeai M, Gregory CD and Emery AN (1994) Cell death by necrosis and apoptosis during the culture of commercially important cell lines. In: Spier RE (ed.), Animal Cell Technology: Products of Today, Prospects for Tomorrow. Butterworth-Heinemann, Oxford, England.

Smith CA, Williams GT, Kingston R, Jenkinson EJ and Owen JJ (1989) Antibodies to CD3/T-cell receptor complex induce death by apoptosis in immature T cells in thymic cultures. Nature 337: 181–184.

Smith CG and Greenfield PF (1992) Mechanical agitation of hybridoma suspension cultures: Metabolic effects of serum Pluronic F68, and albumin supplements. Biotechnol. Bioeng. 40: 1045–1055.

Stoler DL, Anderson GR, Russo CA, Spina AM and Beerman TA (1992) Anoxia-inducible endonuclease activity as a potential basis of the genomic instability of cancer cells. Cancer Research 52: 4372–4378.

Sulik KK, Cook CS and Webster WS (1988) Teratogens and cranial malformations: relationships to cell death. Development 103: 213–231.

Tedesco JL (1987) Analysis of glucose and lactic acid in cell culture media by ion moderated partitioning high performance liquid chromatography. Biotechniques 5: 46–51.

Thömmes J, Gatgens J, Biselli M, Runstadler PW and Wandrey C (1993) The influence of dissolved oxygen tension on the metabolic activity of anim immobilized hybridoma population. Cytotechnology 13: 29–39.

Tomei LD (1991) Apoptosis: a program for death or survival? In: Apoptosis: The molecular basis of cell death, edited by LD Tomei and FO Cope (Cold Spring Harbor Laboratory Press, New York) pp 279–316.

Tozer GM and Griffith JR (1992) The contribution made by cell death and oxygenation to ^{31}P MRS observations of tumour energy metabolism. NMR in Biomedicine 5: 279–289.

Veeraragavan K, Colpitts T and Gibbs BF (1990) Purification and characterization of two distinct lipases from Geotrichum Candidum. Biochem. Biophys. Acta 1044: 26–33.

Vomastek T and Franek F (1993) Kinetics of development of spontaneous apoptosis in B cell hybridoma cultures. Immunol. Lett. 35: 19–24.

Wilson DF, Erecinska M, Drown C and Silver IA (1979) Arch Biochem Biophys 195: 485–493.

Wyllie AH (1980) Glucocorticoid-induced thymocyte apoptosis is associated with endogenous endonuclease activation. Nature 284: 555–556.

Zielke HR, Zielke CL and Ozand PT (1984) Glutamine: a major energy source for cultured mammalian cells. Fed. Proc. 43: 121–125.

Zimmerman LH, Levine RA and Farber HW (1991) Hypoxia induces a specific set of stress proteins in cultured endothelial cells. J. Clin. Invest. 87: 908–914.

Address for offprints: B. Massie, Groupe d'Ingénierie des cellules animales, Institut de recherche en Biotechnologie, Conseil National de Recherches du Canada, 6100 Avenue Royalmount, Montréal, Québec, Canada H4P 2R2.

Cytotechnology **15**: 129–138, 1994.

Maximization of recombinant protein yield in the insect cell/baculovirus system by one-time addition of nutrients to high-density batch cultures

C. Bédard, A. Kamen, R. Tom and B. Massie
Animal Cell Engineering Group, Biotechnology Research Institute, National Research Council Canada, 6100 Royalmount Avenue, Montréal, Québec, Canada, H4P 2R2

Key words: Baculovirus expression vector system, insect cell culture, multiplicity of infection, nutrients, metabolic byproducts, culture medium

Abstract

Suspension cultures of Sf-9 cells at different stages of growth were infected with a recombinant baculovirus expressing β-galactosidase, using a range of multiplicities of infection (MOI) of 0.05 to 50. Following infection, the cells were resuspended either in the medium in which they had been grown or in fresh medium. Specific β-galactosidase yields were not markedly affected by either MOI or medium change in cultures infected in early exponential phase ($\leq 3 \times 10^6$ cells mL^{-1}). In cultures infected at later growth stages, β-galactosidase yields could only be maintained by medium replacement. The possibility that this requirement for medium replacement is due either to the accumulation of an inhibitory byproduct or nutrient limitation was examined. Alanine, a major byproduct of cultured insect cell metabolism, did not significantly reduce recombinant protein yield when added to infected cultures in concentrations of up to 40 mM. Following a factorial design, various nutrient concentrates were added alone or in combination to cultures infected in late exponential phase. Additions that included both yeastolate ultrafiltrate and an amino acid mixture restored specific β-galactosidase yields to levels observed at earlier growth stages or in late stages with medium replacement; the addition of these concentrates, by permitting production at higher cell density, led to increases in the volumetric yield of recombinant protein. Together or separately, the concentrates when added to uninfected late exponential phase cultures, lead to a doubling of the maximum total cell protein level normally supported by unamended medium.

Introduction

The insect cell/baculovirus system is one of the most popular systems presently available for the production of eukaryotic proteins. Cell culture and infection conditions have been shown to influence recombinant protein yields in the system [recently reviewed in O'Reilly *et al.* (1992) and van Lier *et al.* (1992)]. Two of these conditions, the multiplicity of infection and the growth stage of the culture, are the focus of the present study.

Recombinant protein yield tends to decline in cultures infected in the later stages of growth in batch culture (Caron *et al.*, 1990; Lazarte *et al.*, 1992; Lindsay and Betenbaugh, 1992). These drops in protein yield can be partially or completely reversed by replacing spent medium with fresh (Caron *et al.*, 1990; Lindsay and Betenbaugh, 1992), by fresh medium addition (Lazarte *et al.*, 1992), or by simultaneous medium replacement and supplementation with nutrients (Reuveny *et al.*, 1993). The continuous replacement of spent medium with fresh in culture, a process known as perfusion, has successfully been used to produce recombinant proteins in high density insect cell culture (Caron *et al.*, 1994; Jäger *et al.*, 1994).

The positive effect of medium replacement on protein production may be attributed to the overcoming of nutrient limitation or, alternatively, the removal of accumulating. inhibitory byproducts. The number of studies on the nutrition and excretion of insect cells infected with baculovirus is limited [recently reviewed in (Taticek *et al.*, 1994)]. It is known, however, that infected cells rapidly consume glucose and glutamine (Kamen *et al.* 1991; Wang *et al.*, 1993a, b) and excrete elevated amounts of alanine (Kamen *et al.*, 1991; Wang *et al.*, 1993b).

The multiplicity of infection (MOI), i.e. the number of plaque forming units (PFU) added to a culture per cell, has also been found to influence recombinant protein yield, at least in some situations (Maiorella *et al.*, 1988; Murhammer and Goochee, 1988; Caron *et al.*, 1990; Schopf *et al.*, 1990; Licari and Bailey, 1991; Nguyen *et al.*, 1993). An important finding with implications for the present work is that recombinant protein yield is relatively insensitive to MOI in cultures infected in early exponential phase whereas, in late exponential phase, a strong positive correlation exists between the two (Licari and Bailey, 1991).

The initial objective of the study was to maximize the production of a recombinant protein using the Sf-9 cell/baculovirus system by choosing the appropriate MOI and growth stage, and, if necessary, replacing the medium. In particular, we wished to clarify how these different factors interact to determine recombinant protein yield, something not explicitly examined in most of the above mentioned studies on MOI. Unlike the present study, none of the previously cited studies on MOI included all the following conditions: wide ranges of MOI and cell densities, cell densities greater than 5×10^6 mL^{-1}, medium change, and culture in suspension. Easily quantifiable β-galactosidase was employed as a model recombinant protein. Sf-900 II medium (Gibco, Grand Island, NY), was chosen because it supports some of the highest maximum cell densities in batch culture ($\approx 8 \times 10^6$ cells ml^{-1}) reported for any insect cell medium.

Because we found that medium replacement significantly improved protein yields of high cell density late exponential phase cultures, we determined that a further objective of the study should be to investigate whether the limitation imposed by the medium was due to the depletion of one or more nutrients or to the accumulation of an inhibitory byproduct. Nutrient concentrates were added to cultures in late stages of growth and the effects on growth and recombinant protein production measured using a factorial design (Box *et al.*, 1978). Fed-batch approaches to improving recombinant protein yields in the insect cell/baculovirus system have recently been explored (Nguyen *et al.*, 1993; Wang *et al.*, 1993). In these studies, however, a somewhat restricted number of nutrients as chosen and their use in differing combinations was not examined with a systematic statistical approach. A broad range of additives as employed here in order to maximize the probability that the one or more additives responsible for the limitation of growth or protein production would be included.

We report that specific protein yields in normally low-producing late exponential phase cultures can be restored to high levels by the addition of an ultrafiltrate of yeastolate and a concentrated mixture of amino acids, thus leading to high volumetric yields. This finding demonstrates that it is the absence of one or more nutrients and not the accumulation of inhibitory byproducts that limits recombinant protein production in batch cultures of Sf-9 cells. The finding has also simplified large-scale recombinant protein productions in our laboratory by allowing production at high cell density without the need for the cell separation step required when replacing medium.

Materials and methods

Cell line, virus, and medium

Spodoptera frugiperda Sf-9 cells were maintained in Sf-900 II medium (Gibco), a serum-free formulation. Sf-9 cells adapted to growth in Sf-900 II medium were kindly provided by G. Godwin (Gibco). Maintenance cultures were grown at 27 °C in shakeflasks rotating at 120 rpm. Cells were passaged in exponential phase by ten-fold or greater dilution.

A recombinant *Autographa californica* multinuclear polyhedrosis virus (AcMNPV) expressing *Escherichia coli* β-galactosidase was constructed in the laboratory of Dr. C. Richardson of our institute. Large, low passage (< 5) viral stocks were prepared in shake-flask Sf-9 cultures. The stocks were conserved at 4 °C. Viral titers were determined by plaque assay (Summers and Smith, 1987).

Effects of growth stage, MOI, and medium change on β-galactosidase yields

Four 1-L shake-flasks, each containing 400 mL of medium, were seeded with Sf-9 cells in exponential phase at a density of 1×10^6 mL^{-1}. Starting on the day of inoculation, one flask was processed daily as follows. Cells from the flask were equally distributed between four 100-mL cultures in 250 mL shake-flasks

Table 1. Composition of concentrate A

Amino acid	Concentration (mM)
L-Arginine	40.2
L-Asparginine	26.5
L-Aspartic acid	26.3
L-Glutamic acid	40.8
L-Glutamine	68.4
L-Glycine	86.6
L-Histidine	12.9
L-Isoleucine	57.2
L-Leucine	19.1
L-Lysine	34.2
L-Methionine	3.4
L-Phenylalanine	9.1
L-Proline	30.4
DL-Serine	104.7
L-Threonine	14.7
L-Tryptophan	4.9
L-Valine	8.5
L-Cysteine · HCl · H_2O	4.3

infected at different MOI (0.05, 0.5, 5, and 50) with the recombinant baculovirus expressing β-galactosidase. Each culture was prepared by centrifuging 100 mL of the mother culture at 500 g for 15 minutes and resuspending the cells in the appropriate mixture of viral stock and fresh medium. Following a 1 hour infection period, the cells in each culture were pelleted. One half of the cells were resuspended in fresh medium and the other half resuspended in used medium kept from the mother flask for this purpose. Cells were resuspended at the same density as in the mother flask. From these suspensions, triplicate 10 mL cultures in 50 mL shake-flasks were prepared and sampled daily for β-galactosidase activity and cell protein determinations.

Effect of alanine on β-galactosidase yields

Sf-9 cells were centrifuged and resuspended at 2×10^6 mL^{-1} in fresh medium. The cultures were aliquoted into shake-flasks to which varying amounts of a 200 mM alanine solution were added. Triplicate flasks were prepared for each added alanine concentration chosen. These cultures were infected at an MOI of 5 with the recombinant baculovirus expressing β-galactosidase. The activity of this protein was measured at 72 hours post-infecion.

Table 2. Composition of the solutions used to prepare concentrate D[a]

Components	Concentration (mM)
Solution 1	
Thiamine · HCl · 1/2 H_2O	0.237
Riboflavin	0.213
D-Calcium pantothenate	0.181
Pyridoxine HCl	1.945
Para-aminobenzoic acid	2.334
Nicotinic acid	1.300
i-Inositol	2.220
Biotin	0.655
Choline chloride	143.000
Vitamin B_{12}	0.177
Solution 2	
Folic acid	0.181
Solution 3	
Molybdic acid, ammonium salt	0.035
Cobalt chloride hexahydrate	0.210
Cupric chloride · $2H_2O$	0.117
Manganese chloride · $4H_2O$	0.104
Zinc chloride	0.294
Solution 4	
Ferrous sulfate · $7H_2O$	1.980
Aspartate	2.675

[a] Concentrate D was prepared immediately before addition to cultures by mixing together equal amounts of solutions 1 through 4.

Effect of nutrient additions on growth and β-galactosidase yields

A 2-Level factorial design was employed to examine the effects of 4 nutrient concentrates of different composition on growth and recombinant β-galactosidase production in late exponential phase Sf-9 cultures. The 2 levels correspond to either addition or non-addition of the concentrate. Factorial designs are extensively discussed in Box *et al.* (1978). The composition of concentrate A, a 10 × mixture of amino acids, is given in Table 1. Concentrate B is a 50 × yeastolate ultrafiltrate (Gibco cat. nb. 6708200AG) prepared following the method of Maiorella *et al.* (1988). Concentrate C is a 100 × lipid emulsion (Maiorella *et al.* 1988). Solution D, a 250 × concentrate of vitamins, trace metals, and iron, was prepared by combining 4 different solutions immediately before addition to the cultures (Table 2). Once prepared, concentrates were immediately frozen at –20 °C to insure the stability of their components. Concentrates were thawed immediately prior to addition to the cultures. The formulation of the various concentrates could not be based on the composition of Sf-900 II medium since this is proprietary. The composition of concentrates B, C, and D was derived from the composition of a serum-free medium published by Maiorella *et al.* (1988). Concentrate A was our own design. Major salts, organic acids of the TCA cycle, and glucose were not included in any of the concentrates. Major salts, assumed to be present in ample supply even in spent medium, were not added in order not to increase medium osmolarity excessively. The Sf-9 cells used in this study have been successfully grown in the absence of organic acids of the TCA cycle (Schlaeger *et al.*, 1993) often included in insect cell culture media (Hink and Bezanson, 1985). Glucose, was measured in spent Sf-900 II and found to be present in ample amount for the cells (data not shown). 10 mL Sf-9 cultures in 50 mL flasks were prepared from a mother shake-flask culture in late exponential phase. The starting cell density was 7×10^6 cells mL^{-1}. 2 separate sets of 23 10-mL cultures were prepared as follows. Each one of the 16 possible combinations of the 4 concentrates including the absence of all concentrates was tested. Only the combination including all 4 concentrates was replicated; in this case, 7 replicate flasks were prepared. These replicates were used for estimating the standard error of the effects of the concentrates and the effects of the interactions between concentrates. In a separate single control culture, cells were pelleted and resuspended in fresh medium. Concentrates A, B, C, and D were added to the cultures in amounts of 1 mL, 200 μL, 100 μL, and 40 μL, respectively. Following nutrient addition or medium replacement, one set of 23 cultures was infected at an MOI of 5 with the β-galactosidase expressing baculovirus, the other set was left uninfected. In the former set, β-galactosidase was quantitated in each flask at 72 hours post-infection. In the latter set, total cell protein was measured after 2 days. In the flasks receiving all the concentrates the osmolarity increased by approximately 40 mOsm.

A separate experiment to confirm that concentrates A and B added together restore β-galactosidase yields in Sf-9 cell cultures in late exponential phase to the levels observed when the medium is replaced was carried out as follows. 3 treatments were employed: no addition (cultures left untreated), medium replacement, and addition of concentrates A and B. Each treatment was done in triplicate in 50 mL flasks, each containing 10 mL of culture. Cultures were infected at an MOI of 5.

Analytical

Methods for culture sampling, sample preparation, viable cell counting and cell protein determinations have been described previously (Kamen *et al.*, 1991; Bédard *et al.*, 1993). β-Galactosidase was quantitated by a modification of the method of Miller (1972). 10 μL of culture (cells and medium) or culture appropriately diluted in Z-buffer were added to a mixture of 2 mL Z-buffer, 100 μL of chloroform, and 50 μL of 0.15% (w/v) SDS. The reaction was started by addition of 400 μL of 4 mg mL^{-1} *o*-nitrophenyl β-D-galactopyranoside (ONPG). Following color development at 28 °C, the reaction was arrested by addition of 1 mL of molar sodium carbonate solution and the absorbance at 420 nm was read. Activity in units per mL (U mL^{-1}) was calculated as follows:

$$U\,mL^{-1} = (OD_{420}/\text{minutes incubation}) \times \times \text{dilution factor} \times 57 \qquad (1)$$

Results and discussion

Effects of growth stage, MOI, and medium change on β-galactosidase yields

Recombinant β-galactosidase accumulation in infected Sf-9 cell cultures, when it occurred, was first detectable at 48 hours post-infection and reached a maximum at 72 hours or beyond (Fig. 1). Levels of the protein remained approximately constant following attainment of the maximum. In cultures infected in early exponential phase ($\leq 3 \times 10^6$ cells mL^{-1}), final (maximum) yields were inversely correlated to MOI although β-galactosidase accumulation was initially slower at lower MOI than at higher values (Fig. 1). The maximum volumetric yield at the highest MOI was never less than 40% of the yield at the lowest MOI (Fig. 1). A possible explanation for the inverse correlation between MOI and volumetric yield is offered by Licari and Bailey (1992). The greater yields in cultures infected at low MOI are the result of greater cell densities that occur because of the continued division of uninfected cells following virus addition: in cultures infected at high MOI such continued multiplication is much more limited, resulting in lower cell densities.

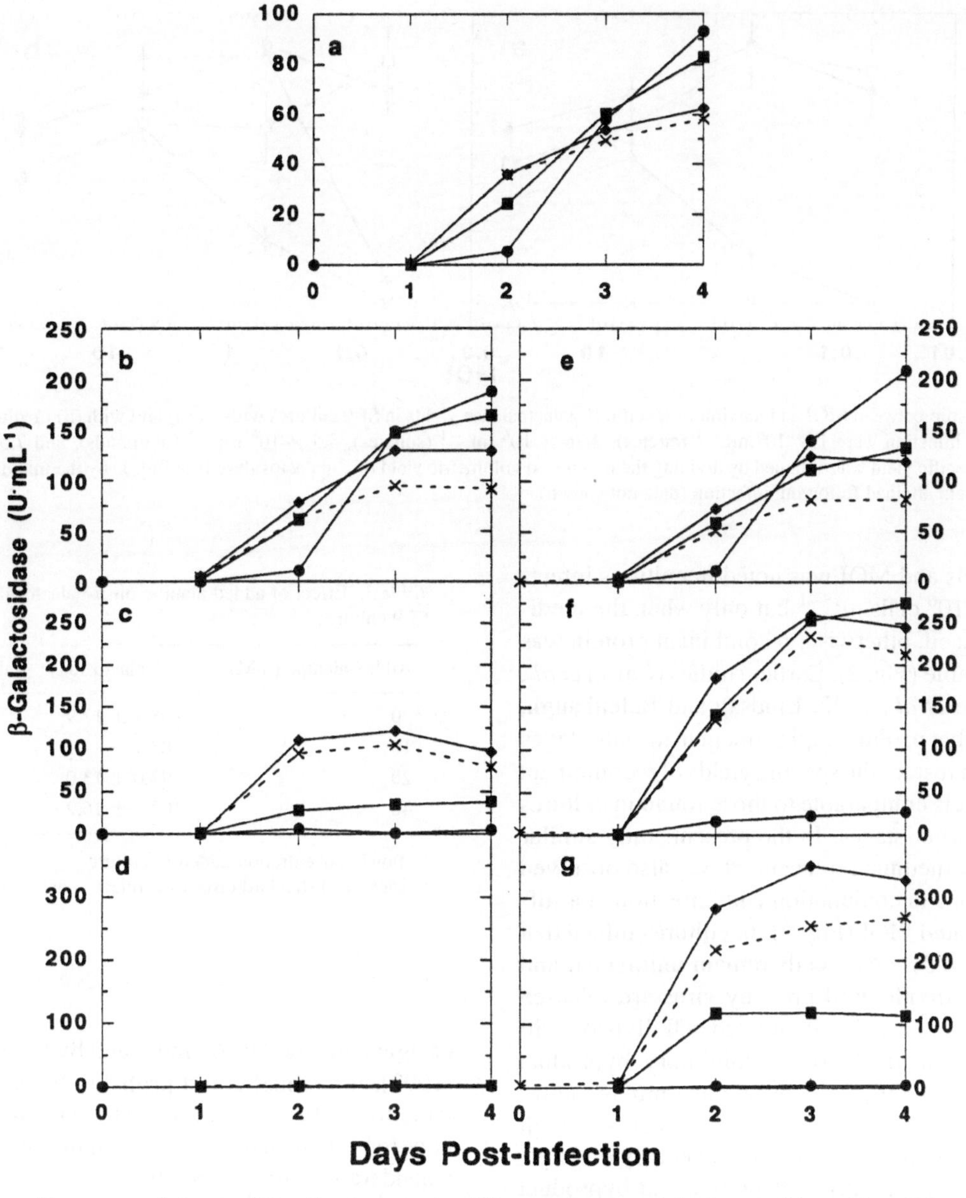

Fig. 1. β-galactosidase accumulation in Sf-9 cell cultures under different culture and infection conditions. The MOIs chosen were 0.05 (circles), 0.5 (squares), 5 (diamonds), and 50 (crosses). Cell densities at infection were 1×10^6 mL^{-1} (a), 2.76×10^6 mL^{-1} (b and e), 5.5×10^6 mL^{-1} (c and f), and 7.07×10^6 mL^{-1} (d and g). Cells were resuspended either in used medium (b, c, and d) or fresh (a, e, f, and g). Values provided are the means of triplicate flasks. Standard errors (always less than 20% of the mean for β-galactosidase values > 10 U mL^{-1}) have been omitted for the sake of clarity. See Methods for further details. Note that some accumulation curves are indistinguishable from their corresponding X-axis.

Consistent with this explanation, the maximum specific yields of β-galactosidase in cultures infected in early exponential phase ($\leq 3 \times 10^6$ cells mL^{-1}) were all maintained at approximately comparable levels irrespective of infection and culture conditions (Fig. 2). Systematic but relatively minor decreases (< 30%) in specific yield occurred between MOI values of 5 and 50 in these cultures.

In cultures infected at a cell density of 5.5×10^6 mL^{-1}, β-galactosidase yields increased dramatically with increasing MOI within the lower range of MOI values employed, whether or not the medium was replaced (Fig. 2). A similar positive correlation

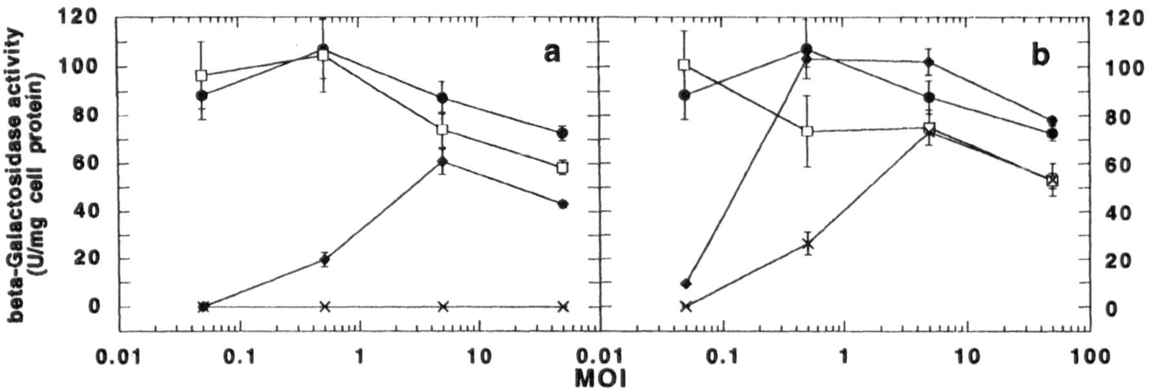

Fig. 2. Relationship between MOI and maximum specific β-galactosidase yields in Sf-9 cultures without (a) and with (b) medium replacement. Cell densities at infection were 1×10^6 mL^{-1} (circles), 2.76×10^6 mL^{-1} (squares), 5.5×10^6 mL^{-1} (diamonds), and 7.07×10^6 mL^{-1} (crosses). The specific yield was obtained by dividing the maximum volumetric yield of β-galactosidase (see Fig. 1) by the maximum volumetric yield of cell protein attained following infection (data not shown).

between yields and MOI was noted in cultures infected at 7.07×10^6 cells mL^{-1} but only when the medium was replaced, otherwise, recombinant protein was barely detectable (Fig. 2). Earlier studies (Caron *et al.*, 1990; Lazarte *et al.*, 1992; Lindsay and Betenbaugh, 1992) found that medium replacement in the late stages of growth can restore the specific yields of recombinant protein to levels comparable to those found in cultures infected in earlier stages. In the present study similar restoration by medium replacement was also observed but only if done in conjunction with infection at a sufficiently elevated MOI (Fig. 2). In cultures infected at low MOI ($\leq$ 0.5), many cells remain uninfected and can therefore divide until progeny virus are released by the infected cells. This continued cell division will result in nutrient depletion or inhibitory byproduct accumulation, either of which could limit recombinant protein production. In a culture infected at high MOI, because all the cells are simultaneously infected, cell division-related nutrient depletion and byproduct accumulation will be considerably diminished, allowing recombinant protein production by infected cells to proceed. A similar type of explanation, developed into a mathematical model of the population dynamics of baculovirus-infected insect cells, was originally proposed by Licari and Bailey (1992). However, their model explicitly assumed that continued cell division following infection at low MOI was limited not by nutrient availibility or the accumulation of inhibitory byproducts as in the suspension cultures used here, but by the surface area remaining available to the cells; this surface area was thought to be the likeliest limiting factor in the stationary cultures for which the authors designed the model. As indicated by Licari and Bailey (1992), the model could probably be easily reformulated for suspension culture. The reformulation would have to be done in relation to the nutrient and byproduct considerations described above.

Table 3. Effect of added alanine on β-galactosidase yields in Sf-9 cultures

Added alanine (mM)[a]	β-Galactosidase (units mL^{-1})[b]
0	105.1 ± 12.9
10	83.7 ± 3.7
20	93.0 ± 13.0
40	102.5 ± 16.2

[a] Final concentration added to culture.
[b] Mean and standard error of triplicate flasks.

Effect of alanine additon on β-galactosidase yields

Alanine accumulations of up to 12 mM have been noted in both uninfected and infected Sf-9 cell cultures (Bédard *et al.*, 1993; Wang *et al.*, 1993b). β-Galactosidase production in Sf-9 cells was not affected by levels of added alanine of up to 40 mM (Table 3), a level unlikely ever to be encountered in batch culture. Other potentially limiting metabolic byproducts, ammonia and lactate, generally do not accumulate significantly in either uninfected (Bédard *et al.*, 1993) or infected Sf-9 cultures (Kamen *et al.*, 1991; Wang *et al.*, 1993b).

Table 4. Calculated main and interraction effects of nutrient concentrates A, B, C, and D on total cell protein levels in the Sf-9 cultures[a]

Effect	Estimate ± Standard error
Main effects	
A	1.02 ± 0.25*
B	1.06 ± 0.25*
C	0.31 ± 0.25
D	0.34 ± 0.25
Two-factor interractions	
A × B	0.34 ± 0.25
A × C	–0.25 ± 0.25
A × D	0.11 ± 0.25
B × C	–0.12 ± 0.25
B × D	–0.11 ± 0.25
C × D	–0.13 ± 0.25
Three-factor interractions	
A × B × C	0.06 ± 0.25
A × B × D	0.39 ± 0.25
A × C × D	0.02 ± 0.25
B × C × D	–0.11 ± 0.25
Four-factor interraction	
A × B × C × D	–0.25 ± 0.25

[a] See text for the composition of the various concentrates and details concerning the method of calculation of the effects and their standard error. Effects were calculated from the values in Figure 3.

* Significant at the 95% level based on a reference t-distribution with 6 degrees of freedom and a scale factor of 0.25 (the standard error of the effects).

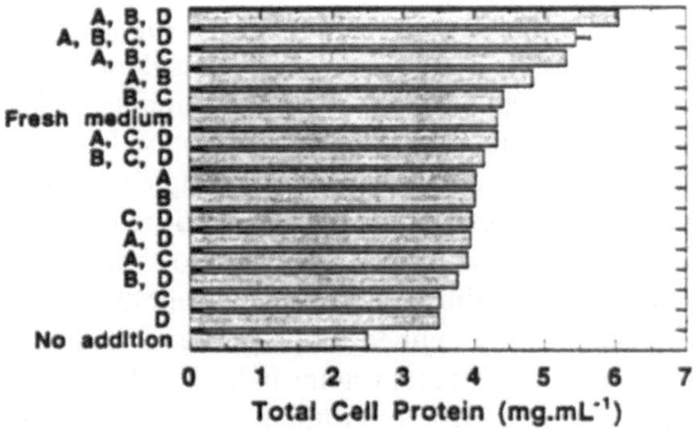

Fig. 3. Total cell protein in Sf-9 cultures following nutrient concentrate additions made in late exponential growth phase. Concentrates are labeled A through D. A standard error bar is provided for the one combination (A. B, C, D) that was replicated. See the text for details.

Effect of nutrient additions on growth and β-galactosidase yields

(Fig. 3) depicts measurements of total protein levels in late exponential phase cultures that received various additions of nutrient concentrates. The figure clearly shows that cell growth in such cultures was markedly enhanced by certain nutrient additions. Furthermore, a number of additions led to protein levels comparable to or better than those obtained with medium replacement. The main and interaction effects of the concentrates (Table 4) were computed and analyzed following the methods of Box *et al.* (1978). Analysis of the effects indicates that the amino acid concentrate (A) and the yeastolate ultrafiltrate (concentrate B), whether added alone or with other concentrates, significantly enhanced culture growth at the 95% probability level (Table 4). At this level, none of the other effects were judged significant.

The mean total cell protein level in the cultures that received both concentrates A and B, 5.45 mg mL^{-1}, was more than double the level in the culture without additions (Fig. 3). Assuming a per cell total protein content of 0.23 ng if the cultures were in exponential phase (Bédard *et al.*, 1993) or a protein content of 0.33 if the cultures were in stationary phase (Bédard, unpublished estimate), the average total cell density in cultures receiving A and B would have fallen in the range of 16×10^6 mL^{-1} to 23×10^6 mL^{-1}. Maximum cell densities previously reported from Sf-9 batch cultures were in the order of 10×10^6 mL^{-1} to 13×10^6 mL^{-1} King *et al.*, 1992; Schlaeger *et al.*, 1993).

Fig. 4 depicts measurements of β-galactosidase in cultures infected in late exponential phase following addition of nutrient concentrates. A number of the various additions resulted in marked enhancement of β-galactosidase yield. Certain combinations of concentrates can restore β-galactosidase yields to levels comparable to those obtained when spent medium is replaced with fresh. The data in Fig. 4 were subjected to the same type of computations and analysis as the total cell protein data. The analysis (Table 5) indicates that A and B when added together significantly enhanced β-galactosidase yields at the 95% probability level. When a statistically significant interaction effect is observed between two factors, it is not possible to determine whether the main effect of each of the interacting factors is itself significant (Box *et al.*, 1978). The nature of the interaction between concentrates A and B appears to be synergistic since the combinations of concentrates that included A and B together yielded

Table 5. Calculated main and interraction effects of nutrient concentrates A, B, C, and D on β-galactosidase yields in the Sf-9 cultures[a]

Effect	Estimate ± Standard error
Main effects	
A	96.0 ± 17.8
B	150.8 ± 17.8
C	4.2 ± 17.8
D	24.0 ± 17.8
Two-factor interractions	
A × B	47.9 ± 17.8*
A × C	16.7 ± 17.8
A × D	-5.9 ± 17.8
B × C	-15.3 ± 17.8
B × D	-11.9 ± 17.8
C × D	15.2 ± 17.8
Three-factor interractions	
A × B × C	-5.9 ± 17.8
A × B × D	-30.8 ± 17.8
A × C × D	15.1 ± 17.8
B × C × D	1.3 ± 17.8
Four-factor interraction	
A × B × C × D	-6.0 ± 17.8

[a] See text for the composition of the various concentrates and details concerning the method of calculation of the effects. Effects were calculated from the values in Figure 3.

* Significant at the 95% level based on a reference t distribution with 6 degrees of freedom and a scale factor of 17.8 (the standard error of the effects).

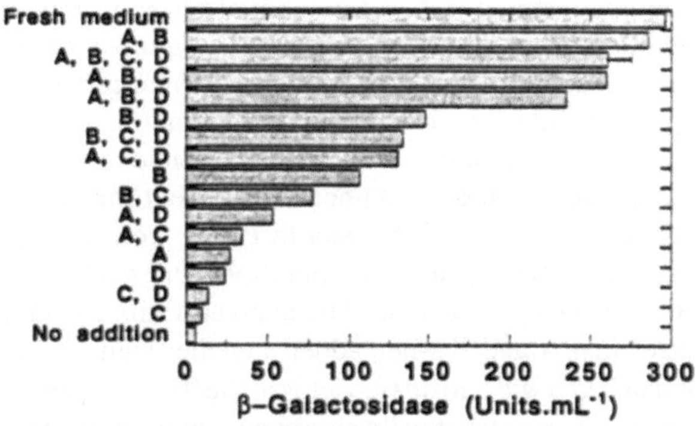

Fig. 4. β-Galactosidase yields in Sf-9 cultures following nutrient concentrate additions made in late exponential growth phase. Concentrates are labeled A through D. Cell density at infection was 7×10^6 mL^{-1}. β-Galactosidase was measured 72 hours after infection. A standard error bar is provided for the one combination (A, B, C, D) that was replicated. See the text for details.

Table 6. Effects of the addition of concentrates A and B and medium replacement on β-galactosidase yields (U mL^{-1}) in Sf-9 cultures infected in late exponential phase[1]

Control[2]	Nutrient addition[3]	Medium replacement[4]
6.5 ± 0.3	240.7 ± 10.3	272.3 ± 20.3

[1] Means and standard errors of triplicate flasks are provided. The cell density at infection was 6.6×10^6 cells mL^{-1}.
[2] No addition, no medium replacement.
[3] The nutrient added to 10 mL of culture consisted of 1 mL of concentrate A (see Table 1) and 0.2 mL of concentrate B.
[4] Used medium was removed by centrifugation and the cells were resuspended in fresh medium.

on average twice the sum of the β-galactosidase levels obtained with A alone and B alone. Other effects were judged not to be significant (Table 5).

Because the culture in which only A and B were added and the culture in which medium was replaced were not replicatd in the above described experiment, it could not be ascertained whether there was any significant difference between these treatments. A separate, replicated experiment allowing statistical comparison was therefore carried out. Table 6 summarizes the results. A t-test showed no significant difference between the means of the two treatments.

Recent studies also report improved cell densities and recombinant protein production in Sf-9 cultures following addition of various nutrients (Nguyen *et al.*, 1993; Wang *et al.*, 1993a). As mentioned previously, the present work differs from these investigations in that a broader range of nutrients was examined and a systematic approach based on a factorial design was employed. Also, the media employed by Nguyen *et al.* (1993) and Wang *et al.* (1993) supported maximum Sf-9 cell densities that are considerably lower ($\leq 3 \times 10^6$ cells mL^{-1}) than those attainable in the Sf-900 II medium ($\approx 8 \times 10^6$ mL^{-1}) used here. Furthermore, both studies relied on multiple additions of nutrients. The one-time addition used here is simple and easily scaleable as shown by work done in our laboratory; one-time addition of concentrates A and B was successfully applied to the production of several different recombinant proteins in high cell density late growth stage bioreactor cultures (B. Jardin, S. Perret, and R. Tom, unpublished results). Nutrient addition to these multiliter cultures eliminated the need for the medium replacement otherwise required to maximize recombinant protein production; the replacement of such large volumes of medium, can only be done with a cell separation device that can be complicated to set up and operate, and can increase the chance of culture contamination.

The limits of the improvements in recombinant protein yields attainable in the insect cell/baculovirus system with either one-time nutrient addition or more conventional fed-batch are not known. These questions are presently being investigated in our laboratory.

Conclusion

In this study, we have clarified how MOI growth stage, and the nutrient requirements of cells interact to determine recombinant protein yields in insect cell cultures using the baculovirus expression vector system. We have shown that these yields can be maximized if, upon attainment of maximum density in a culture, the cells are infected at a sufficiently high MOI (≥ 5), and the spent medium is either replaced with fresh medium or amended with yeastolate ultrafiltrate and certain amino acids. If these procedures are followed, the specific protein yield will be maintained at levels compatable to those obtained at earlier stages of culture growth. Producing at high cell density allows considerable improvement of volumetric yields. The use of nutrient addition rather than medium replacement is clearly an advantage in bioreactor culture since there is no need to separate the cells from the medium.

Acknowledgments

We thank G. Godwin of Gibco for providing the recombinant AcNPV expressing β-galactosidase and B. Jardin and S. Perret for their assistance.

References

Bédard C, Tom R and Kamen A (1993) Growth, nutrient consumption, and end-product accumulation in Sf-9 and BTI-EAA insect cell cultures: insights into growth limitation and metabolism. Biotechnol. Prog. 9: 615–624.

Box GEP, Hunter WG and Hunter JS (1978) Statistics for Experimenters. An Introduction to Design, Data Analysis, and Model Building. John Wiley and Sons, New York.

Caron AW, Archambault J and Massie B (1990) High-level recombinant protein production in bioreactors using the baculovirus-insect cell expression system. Biotechnol. Bioeng. 36: 1133–1140.

Caron AW, Tom RL, Kamen AA and Massie B (1994) Baculovirus expression system scaleup by perfusion of high density Sf-9 cultures. Biotechnol. Bioeng. 43: 881–891.

Hink WF and Bezanson DR (1985) Invertebrate cell culture media and cell lines. Techniques in the Life Sciences, C1. Setting Up and Maintenance of Tissue and Cell Cultures C111: 1–30.

Jäger V, Kobold A, Köhne C, Deutschmann SM, Grabenhorst E, Karger C and Conradt HS (1994) High density insect cell culture for the production of recombinant proteins with the baculovirus expression system. In: Spier, R. E., Griffiths, J. B. and Berthold, W. (eds) Animal Cell Technology: Products of Today, Prospects of Tomorrow. (pp. 207–211) Butterworth-Heinemann, Oxford.

Kamen AA, Tom RL, Caron AW, Chavarie C, Massie B and Archambault 1(1991) Culture of insect cells in a helical ribbon impeller bioreactor. Biotech. Bioeng. 38: 619–628.

King GA, Daugulis AJ, Faulkner P and Goosen MFA (1992) Recombinant β-galactosidase production in serum-free medium by insect cells in a 14-L airlift bioreactor. Biotechnol. Prog. 8: 567–571.

Lazarte JE, Tosi P-F and Nicolau C (1992) Optimization of the production of full-length rCD4 in baculovirus-infected Sf9 cells. Biotechnol. Bioeng. 40: 214-217.

Licari P and Bailey JE (1991) Factors influencing recombinant protein yields in an insect cell-baculovirus system: multiplicity of infection and intracellular protein degradation. Biotechnol. Bioeng. 37: 238–246.

Licari P and Bailey JE (1992) Modeling the population dynamics of baculovirus-infectd insect cells: optimizing infection strategies for enhanced recombinant protein yields. Biotechnol. Bioeng. 39: 432–441.

Lindsay DA and Betenbaugh MJ (1992) Quantification of cell culture factors affecting recombinant protein yields in baculovirus-infected insect cells. Biotechnol. Bioeng. 39: 614–618.

Maiorella B, Inlow D, Shauger A and Harano D (1988) Large-scale insect cell-culture for recombinant protein production. Bio/Technology 6: 1406–1410.

Miller JH (1972) Assay of β-galactosidase. In: Experiments in molecular genetics. (pp. 352–355) Cold Spring Harbor Laboratory Press, Cold Spring Harbor, NY.

Murhammer DW and Goochee CF (1988) Scaleup of insect cell cultures: protective effects of pluronic F68. Bio/Technology 6: 1411–1418.

Nguyen B, Jarnagin K, Williams S, Chan H and Barnett J (1993) Fed-batch culure of insect cells: a method to increase the yield of recombinant human nerve growth factor (rhNGF) in the baculovirus expression system. J. Biotechnol. 31: 205–217.

O'Reilly DR, Miller LK and Luckow VA (1992) Baculovirus Expression Vectors: a laboratory manual. W. H. Freeman and Company, New-York.

Reuveny S, Kim YJ, Kemp CW and Shiloach J (1993) Production of recombinant proteins in high-density insect cell cultures. Biotechnol Bioeng 42: 235–239.

Schlaeger E-J, Foggetta M, Vonach JM and Christensen K (1993) SF-1, a low cost culture medium for the production of recombinant proteins in baculovirus infected insect cells. Biotechnology Techniques 7: 183–188.

Schopf B, Howaldt MW and Bailey JE (1990) DNA distribution and respiratory activity of *Spodoptera frugiperda* populations infected with wild-type and recombinant *Autographa californica* nuclear polyhedrosis virus. J. Biotechnol. 15: 169–186.

Summers MD and Smith GE (1987) A manual of methods for baculovirus vectors and insect cell culture procedures. Texas Agricultural Experiment Station Bulletin No. 1555.

Taticek RA, Lee CWT and Shuler ML (1994) Large-scale insect and plant cell culture. Current Opinion in Biotechnology 5: 165–174.

Van Lier FLJ, Vlak JM and Tramper J (1992) Production of baculovirus-expressed proteins from suspension cultures of insect cells. In: Spier, R. E. and Griffiths, J. B. (eds) Animal Cell Biotechnology. Vol. 5 (pp. 169–188) Academic Press, London.

Wang M-Y, Kwong S and Bentley WE (1993) Effects of oxygen/glucose/glutamine feeding on insect cell baculovirus protein expression: a study on epoxide hydrolase production. Biotechnol. Prog. 9: 355–361.

Wang MY, Vakharia V and Bentley WE (1993) Expression of epoxide hydrolase in insect cells. A focus on the infected cell. Biotechnol Bioeng 42: 240–246.

Address for offprints: A. Kamen, Animal Cell Engineering Group, Biotechnology Research Institute, National Research Council Canada, 6100 Royalmount Avenue, Montréal, Québec, Canada, H4P 2R2.

Cytotechnology **15:** 139–144, 1994.

Recombinant protein expression in a *Drosophila* cell line: comparison with the baculovirus system

A.R. Bernard[1], T.A. Kost[2], L. Overton[2], C. Cavegn[1], J. Young[1], M. Bertrand[1], Z. Yahia-Cherif[1], C. Chabert[1] and A. Mills[1,3]
[1] *Glaxo Institute for Molecular Biology, Chemin des Aulx 14, CH1228, Geneva, Switzerland;* [2] *Glaxo Research Institute, 5 Moore Drive, Research Triangle Park, NC 27709, USA;* [3] *Present Address: Wellcome Research Laboratories, Langley Court, Beckenham, Kent, BR33BF, UK*

Key words: Baculovirus, cell culture, *Drosophila*, gene expression, insect cell, metallothionein promoter, recombinant protein

Abstract

In this report, we compare two different expression systems: baculovirus/*Sf9* and stable recombinant *Drosophila Schneider 2* (S2) cell lines. The construction of a recombinant S2 cell line is simple and quick, and in batch fermentations the cells have a doubling time of 20 hours until reaching a plateau density of 20 million cells/ml. Protein expression is driven by the *Drosophila* Metaliothionein promoter which is tightly regulated. When expressed in S2 cells, the extracellular domain of human VCAM, an adhesion molecule, is indistinguishable from the same protein produced by baculovirus-infected Sf9 cells. Additionally, we present data on the expression of a seven trans-membrane protein, the dopamine D4 receptor, which has been successfully expressed in both systems. The receptor integrates correctly in the S2 membrane, binds [^{3}H]spiperone with high affinity and exhibits pharmacological characteristics identical to that of the receptor expressed in Sf9 and mammalian cells. The general implications for large scale production of recombinant proteins are discussed.

Introduction

The two most widely used systems for the large scale expression of recombinant proteins in higher eucaryotes, namely recombinant mammalian cells and baculovirus-infected insect cells, present distinct disadvantages. Recombinant mammalian cell lines are obtained only after a long process of transfection, selection, amplification, clonal selection and optimisation. The inducible promoters often exhibit basal constitutive activity (especially when present at high copy number), and in their induced state, often drive only moderate levels of protein expression (Pavalakis *et al.*, 1983; Johansen *et al.*, 1989). Once constructed, the recombinant line frequently needs to be maintained under selective pressure and, even under these conditions, may be unstable over prolonged culture.

The baculovirus/insect cell system, which is increasingly used for expressing many different recombinant proteins offers a simple and quick approach to making mg quantities of protein. However, its main drawback is that expression of the desired protein, usually driven by a very late viral promoter, peaks at a point where cells are dying from the viral infection. The amplification, titration and storage of virus stocks add more steps that slow down the whole process.

We therefore decided to evaluate an alternative system which potentially alleviates most of the problems discussed above. *Drosophila melanogaster* cell lines (Schneider 1972; Sang 1981) have been known for many years to be suitable hosts for heterologous expression. In addition, one of the *Drosophila* metallothionein promoter (Mtn) has been thoroughly studied, shown to be tightly regulated, and capable of driving high level heterologous transcription upon induction by metals such as Cu or Cd (Maroni *et al.*, 1986; Otto *et al.*, 1987).

Several investigators (Johansen *et al.*, 1989; Culp *et al.*, 1991; Olsen *et al.*, 1992) have also demonstrated that expression vectors utilising this promoter integrate as multiple copies into the *Drosophila* cell chromosome. Here we show that this system compares well with the baculovirus/Sf9 system in terms of the time required to construct the cell line, as well as the quantity and biological activity of the heterologous protein produced. Two very different proteins have been investigated: the extracellular domain of the human vascular cell adhesion protein, (VCAM), (Walsh *et al.*, 1991), which is secreted, and a membrane spanning, seven trans-membrane protein, the dopamine D4 receptor (Van Tol *et al.*, 1991). We have previously reported the expression of each of these proteins using the baculovirus system (Cavegn *et al.*, 1994; Mills *et al.*, 1993) and here we present the results of a comparison with recombinant S2 cell lines.

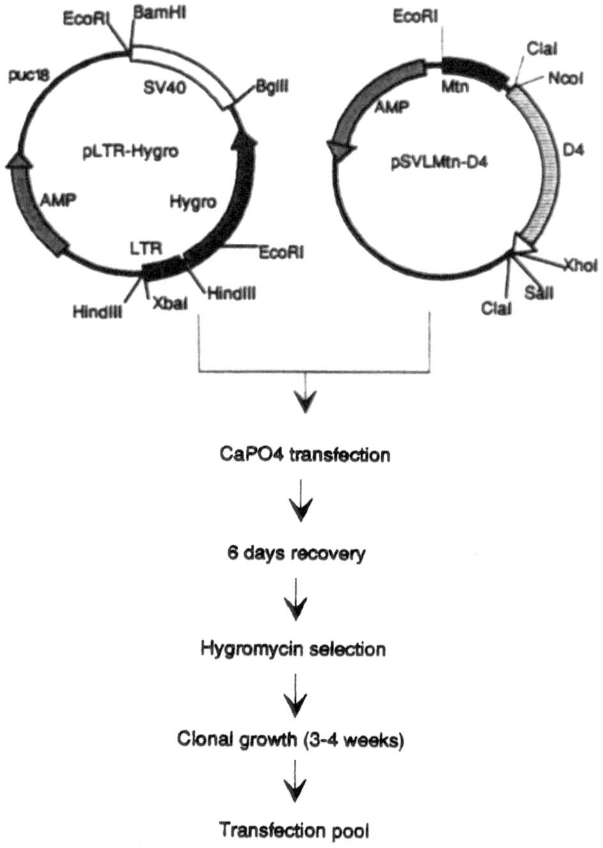

Fig. 1. Construction of recombinant S2 lines. pLTR-Hygro carries the selectable marker, and pSVL-Mtn carries the gene of interest (here D4) which can be cloned at a unique ClaI site. AMP: Ampicillin resistance, Hygro: Hygromycin resistance gene. LTR: Long Terminal Repeat of the *Drosophila* Copia sequence. SV40: PolyA and splice signal sequences from SV40.

Materials and methods

Cells and media

Schneider 2 cells (ATCC CRL 1963) were grown in commercial Schneider medium (Gibco or Sigma) supplemented with 5–10% FCS (Seromed, Switzerland). Sf9 cells (ATCC CRL 1711) were maintained in serum free medium Ex-cell 401 (JRH Biosciences, USA).

Plasmids and transfections

The extracellular portion of the human VCAM gene was cloned into a baculovirus transfer vector (pAC373-BA2-VCAM) as previously described (Cavegn *et al.*, 1994). In this construction, the gene is fused at its 3' end with an IgG binding sequence (zz) derived from protein A, which facilitates detection and purification. The recombinant VCAM baculovirus was obtained after co-transfection of Sf9 cells with wild type virus DNA and pAC373-BA2-VCAM transfer vector according to standard procedures (Summers *et al.*, 1988).

The hygromycin selection vector (in which the hygromycin resistance gene is under control of a strong *Drosophila* Copia LTR promoter) was constructed by synthesising the *Drosophila* Copia LTR promoter (Mount *et al.*, 1985), excising the hygromycin resistance gene from plasmid pSV2-hph (ATCC #37647), and inserting these two fragments into pUC18. The Mtn promoter, kindly supplied by Dr G. Maroni (University of North Carolina, Chapel Hill) was cloned into the pSVL-Mtn expression vector (in which the *Drosophila* metallothionein promoter drives expression of cloned genes). The VCAM coding sequence was excised from pAC373-BA2-VCAM and inserted into the pSVL-Mtn vector. S2 cells were co-transfected with the hygromycin selection vector and the pSVL-MtnVCAM vector by the calcium phosphate method (Fig. 1). After 3 weeks of selection (with hygromycin at a concentration of 300 μg/ml), clonal growth was observed and several clones (S2-VCAM) were pooled, amplified and stored in liquid nitrogen. The D4 expressing cell line (S2-D4) was obtained by an identical protocol in which the D4 coding sequence

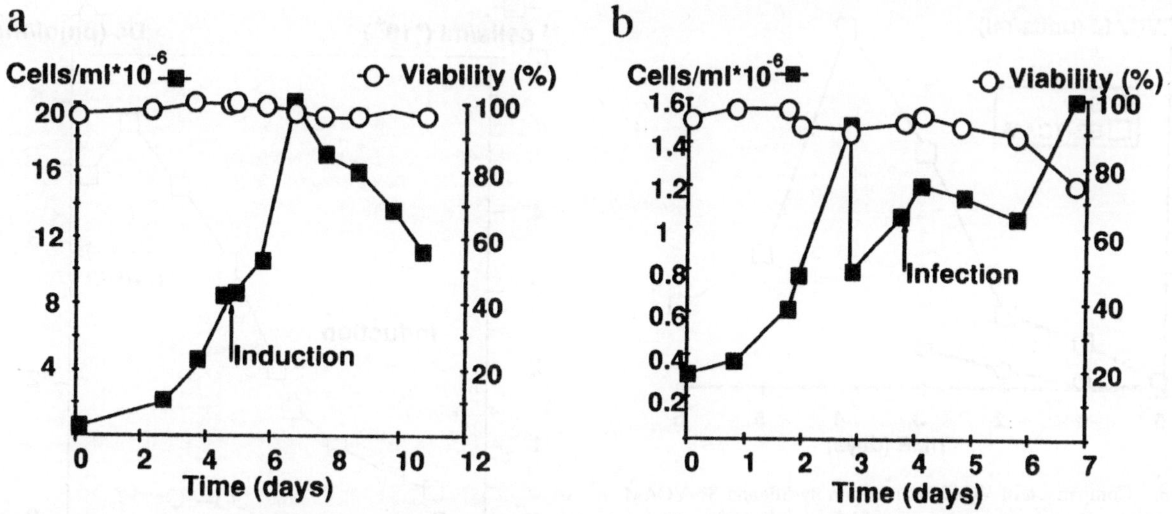

Fig. 2. Growth and expression of VCAM by S2-VCAM and Sf9. a) S2-VCAM. Induction was started at day 4,5 by addition of a concentrated $CuSO_4$ solution so as to reach a final concentration of 500 μM final; b) Sf9 cells were initially grown to 1.5 10^6/ml, split two-fold and then infected at day 4 with the recombinant VCAM baculovirus at a MOI of 5.

had been extracted from a baculovirus expression vector (Mills *et al.*, 1993).

Fermentations

Fermentations were carried out in 15 litre working volume stirred tank bioreactors operated in batch mode. Dissolved oxygen tension was maintained at 50% air saturation by direct oxygen gas sparging upon demand. Sf9 cells were infected by addition of a virus stock suspension so as to reach a multiplicity of infection of 5-10. S2-VCAM and S2-D4 cells were induced by the addition of $CuSO_4$ to a final concentration of 500 μM. Total cell density and viability were measured in a hemocytometer by the Trypan Blue exclusion method.

Product characterisation

The analysis of VCAM was carried out by western blots and biological activity estimated by a cell binding assay (Cavegn *et al.*, 1994). The quantification of D4 receptors was performed using a [^{3}H]spiperone binding assay on membrane preparations as described previously (Mills *et al.*, 1993).

Results and discussion

Construction of recombinant S2 lines

From a single transfection, 2–10 clones were recovered after only 3 weeks of hygromycin selection. In contrast to other selection methods (neomycin for example), the selection by hygromycin resistance is very efficient, and no spontaneous resistance could be observed with non transfected cells (frequency below 1 in 10^6). The short time (approximately one week) required to kill all the background of untransfected cells is also a major advantage. As there is no need for amplification of the foreign integrated DNA, this is a much more rapid method than those involving mammalian cell lines. Construction of a recombinant S2 cell line takes approximately the same time as generating a recombinant baculovirus, making a virus stock suitable for large scale expression in Sf9, and evaluating its titre. However, one level of variability (the virus stock) is eliminated since the recombinant S2 cell line contains all the necessary genetic information for expressing the desired protein.

Expression of VCAM

S2-VCAM and StP cells were grown in spinners, inoculated in stirred tank reactors and growth was monitored (Fig. 2). In both instances product expression is initiated when cells reach about 80% of their plateau

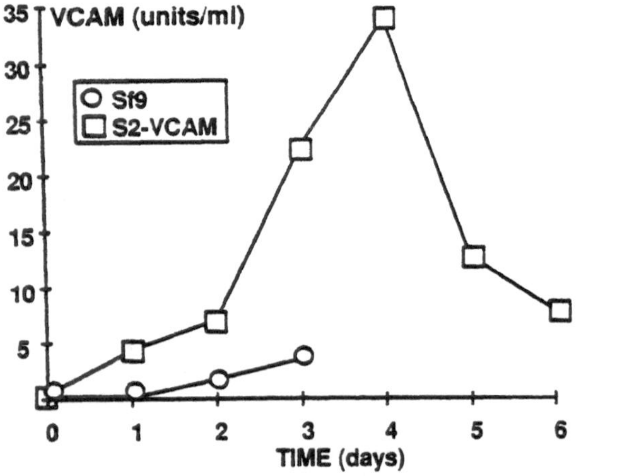

Fig. 3. Comparison of VCAM expression by Sf9 and S2-VCAM. Time origin (X axis) represents onset of infection (for Sf9) or induction (for S2-VCAM). The Y axis values represent arbitrary units per millilitre in the biological activity assay.

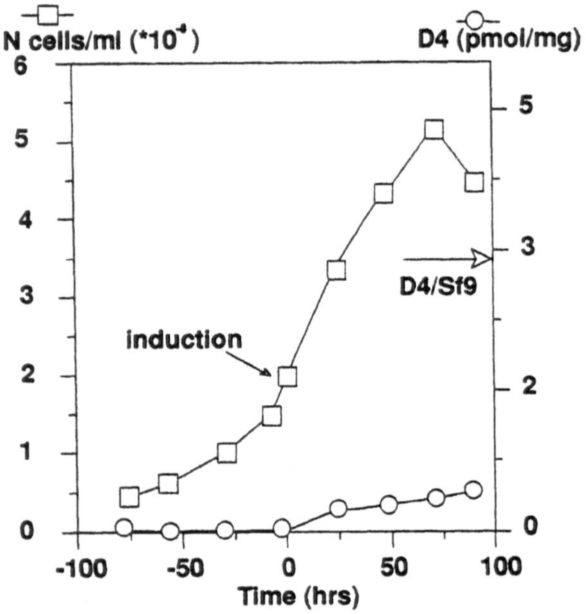

Fig. 4. Kinetics of growth and expression by S2-D4. The time axis is referenced to the induction time point. The Y axis indicates the number of viable cells per millilitre (in millions). Arrow indicates expression level in Sf9.

density (1 10^6 cell/ml for Sf9, 8 10^6 for S2-VCAM). S2-VCAM cells continue to grow with another doubling to 20 10^6 cell/ml, after which point the cell number decreases rapidly. However, cell viability remains above 90% at all times. In contrast, Sf9 cell viability decreases progressively after virus infection. The potential of S2-VCAM to express VCAM was compared to Sf9 by measuring the kinetics of product accumulation in these batch fermentations. In Western Blot analysis (data not shown) we observed a single band at the expected size of 95-100 kD in the conditioned medium of induced S2-VCAM cultures. This immunoreactive band was absent in non transfected S2 cells, in conditioned medium from induced cells transfected with a vector where the gene was in the wrong orientation, and in conditioned medium from non induced S2-VCAM cells. These results demonstrate that the expression of VCAM is due to the integration of the VCAM expression vector, that it is tightly regulated by the Mtn promoter and that the protein is secreted into the extracellular medium.

This observation was confirmed by activity measurements (Fig. 3). No active VCAM protein was detectable before S2-VCAM induction or Sf9 infection (day 0 on Fig. 3). The peak activity of VCAM occurs 4 days after induction of S2-VCAM, and at three days after infection of Sf9 cells. At these respective time points, S2-VCAM cells secrete about 10–20 fold more active VCAM than do virus infected Sf9 cells. However, upon prolonged induction with $CuSO_4$, (for 3–6 days after induction), we observe a decrease of cell number with no corresponding decrease in cell viability which is due to the toxicity of the inducer and has also been reported by others (Culp *et al.*, 1991; Olsen *et al.*, 1992). VCAM activity is also reduced during this late phase of induction, probably as a result of proteolytic degradation.

To confirm that the VCAM was biologically active, we used a monoclonal antibody (Ig11) which we had previously shown (Cavegn *et al.*, 1994) could block VCAM activity. This antibody inhibited the protein produced either by S2-VCAM or Sf9 cells with the same efficiency (0.25 μg/ml, 40% inhibition). This indicates that the epitope for IG11 is conserved on both sources of protein and implies that the protein is correctly folded.

Expression of D4

In another test case, we evaluated expression of the D4 receptor in recombinant S2 cells (S2-D4). The recombinant S2-D4 cell line was obtained by the same protocol as the one described for S2-VCAM. Figure 4 shows the time course of growth and product formation as a function of time. Here again, S2-D4 cell growth is not inhibited by the induction process since

Table 1. **pKi of dopaminergic compounds for recombinant D4. Values are reported as means of 3–6 separate experiments. The Sf9 column refers to cells infected with the recombinant D4 baculovirus**

COMPOUND	S2-D4	Sf9
Butaclamol	6.67	6.59
Clozapine	7.37	7.11
Eticlopride	7.05	6.94
Haloperidol	8.42	8.08
Thioridazine	7.58	7.40
YM-09151-2	9.52	9.36
Quinpirole	7.14	7.32
Apomorphine	8.09	8.52
dpADTN	7.93	8.28

viable cell number continues to increase to 5 10^6/ml. Prior to induction of the Mtn promoter, no ligand binding could be detected in cell membrane preparations. After 24 hours of induction, D4 can be detected in cell membranes by ligand binding assay. In this case, the maximal level of protein produced (about 1 pmol/mg of membrane protein) is much lower than that produced by baculovirus infected Sf9 cells (3 pmol/mg of membrane protein). Also in these experiments, we found an important difference in the cell viability between the two systems. After 72–96 hours of induction, when D4 production is at its peak in S2-D4 cells, the cell viability is still very high (above 85%), whereas the infected Sf9 cells, after 48–72 hours of infection are only 60–70% viable.

In order to verify the properties of the D4 receptor, the Kd for [^{3}H]spiperone was measured and found to be 0.22 nM. This compares well with the values obtained in Sf9: 0.3 nM (Mills *et al.*, 1993) and COS cells: 0.08 nM (Van Tol *et al.*, 1991). Additionally, the pK_i values for a range of agonists and antagonists were determined in S2-D4 and infected Sf9 cells (Table 1). The rank order of affinity for these compounds was the same in both cell lines and identical to that observed in COS cells (Van Tol *et al.*, 1991; Van Tol *et al.*, 1992), indicating that the protein displays the correct pharmacological characteristics in both expression systems.

Conclusions

We have shown that *Drosophila* S2 cells are amenable to recombinant protein production with two test proteins: a secreted protein and an integral membrane receptor. The cells grow to high density in simple batch fermentations, are robust and tolerate high levels of heavy metal ions such as Cu, thus allowing the use of the Mtn promoter to drive expression of heterologous genes. The activity of the promoter is tightly regulated with undetectable basal level of transcription and yet high inducibility. In S2 cells, the recombinant protein is expressed from a culture at very high viability, which is a distinct advantage over the baculovirus insect cell system in which viability of the culture is below 80% at the peak of expression. The VCAM protein expressed in this system is correctly processed and secreted in the extracellular medium and is identical in all biological functions tested so far to the same protein expressed by the baculovirus/insect cell system. Our results also show that the recombinant D4 receptor produced in S2 cells exhibits the expected pharmacological properties. This suggests that the recombinant D4 receptor is properly folded and integrated into S2 membranes, as has been shown for baculovirus expressed D4. The reproducibility of the process is enhanced in the case of S2 cell lines, which contain all the information needed, in contrast to the baculovirus/insect cell system in which virus stocks can differ one from another. However, in terms of productivity per unit volume, it is difficult to generalise our conclusions on the relative benefits of each expression system. In the case of VCAM, the secreted protein, S2 is clearly more advantageous than baculovirus/Sf9, whereas the opposite is true for the membrane spanning dopamine D4 receptor.

Acknowledgements

We wish to thank Bernard Allet for engineering all expression vectors, Mark Payton and Kim Hardy for their constructive comments on the manuscript.

References

Cavegn C, Young J, Bertrand M and Bernard AR (1994) *Drosophila* cell lines as hosts for recombinant protein expression. in: Berthold W (ed.) Animal Cell Technology, products for today, prospects for tomorrow. In press. Butterworth-Heineman, Oxford.

Culp JS, Johansen H, Hellmig B, Beck J, Matthews TJ, Delers A and Rosenberg M (1991) Regulated expression allows high level production and secretion of HIV-1 gp120 envelope glycoprotein in *Drosophila* Schneider cells. Bio/Technology 9: 173–177,

Johansen H, Van der Straten A, Sweet RI, Otto E, Maroni G and Rosenberg M (1989) Regulated expression at high copy number allows production of a growth inhibitory oncogene product in *Drosophila* Schneider cells. Genes and Development 3: 882–889.

Maroni G, Otto E and Latowski-Perry D (1986) Molecular and cytogenetic characterization of a metallothionein gene of *Drosophila*. Genetics 112: 493–504.

Mills A, Allet B, Bernard A, Chabert C, Brandt E, Cavegn C, Chollet A and Kawashima E (1993) Expression and characterization of human D4 dopamine receptors in baculovirus infected insect cells. FEBS Lett. 320: 130–134.

Mount SM and Rubin GM (1985) Complete nucleotide sequence of the *Drosophila* transposable element copia: homology between copia and retroviral proteins. Mol. Cell. Biol. 5:1630–1638,

Olsen MK, Rockenbach SK, Fischer HD, Hoogerheide G and Tomich CSC (1992) Stable production of an analog of human tissue plasminogen activator from cultured *Drosophila* cells. Cytotechnology 10: 157–167.

Otto E, Allen JM, Young JE, Palmiter RD and Maroni G (1987) A DNA segment controlling metal-regulated expression of *Drosophila melanogaster* Metallothionein gene *Mtn*. Mol. Cell. Biol. 7: 1710–1715.

Pavalakis G and Mamer DH (1983) Regulation of a metallthionein-growth hormone hybrid gene in bovine papilloma virus. Proc. Natl. Acad. Sci. 80: 397–401.

Sang JH (1981) *Drosophila* cells and cell lines. Adv. Cell Culture 1: 125-177.

Schneider I (1972) Cell lines derived from late embryonic stages of *Drosophila melanogaster*. J. Embryol. Exp. Morph. 27 (2): 353–365.

Summers MD and Smith GE (1988) Texas Agricultural Experiment Station Bulletin 1555, Texas A&M University College Station TX USA.

Van Tol HH, Bunzow JR, Guan HC, Sunahara RK, Seeman P, Niznik HB and Civelli 0 (1991) Cloning of the gene for a human dopamine D4 receptor with high affinity for the antipsychotic clozapine. Nature 350: 610–614.

Van Tol HH, Wu CM, Guan HC, Ohara K, Bunzow JR, Civelli O, Kennedy J, Seeman P, Niznik HB and Jovanovic V (1992) Multiple dopamine D4 receptor variants in the human population. Nature 358: 149–152.

Walsh GM, Mermod JJ, Hartnell A, Kay AB and Wardlaw AJ (1991) Human eosinophil, but not neutrophil, adherence to IL-l-stimulated human umbilical vascular endothelial cells is alpha 4 beta 1 (very late antigen-4) dependent. J. Immunol. 146: 3419–3423.

Address for offprints: A.R. Bernard, Glaxo Institute for Molecular Biology, Chemin des Aulx 14, CH1228, Geneva, Switzerland.

Cytotechnology **15:** 145–155, 1994.

Scale-up of the adenovirus expression system for the production of recombinant protein in human 293S cells

Alain Garnier, Johanne Côté, Isabelle Nadeau, Amine Kamen and Bernard Massie
Institut de recherche en biotechnologie, CNRC, 6100 Royalmount, Montréal, Québec, Canada, H4P 2R2

Key words: Adenovirus, human 293S cells, recombinant protein, scale-up, metabolism

Abstract

Human 293S cells, a cell line adapted to suspension culture, were grown to 5×10^6 cells/mL in batch with calcium-free DMEM. These cells, infected with new constructions of adenovirus vectors, yielded as much as 10 to 20% recombinant protein with respect to the total cellular protein content. Until recently, high specific productivity of recombinant protein was limited to low cell density infected cultures of no more than 5×10^5 cells/mL. In this paper, we show with a model protein, Protein Tyrosine Phosphatase 1C, how high product yield can be maintained at high cell densities of 2×10^6 cells/mL by a medium replacement strategy. This allows the production of as much as 90 mg/L of active recombinant protein per culture volume. Analysis of key limiting/inhibiting medium components showed that glucose addition along with pH control can yield the same productivity as a medium replacement strategy at high cell density in calcium-free DMEM. Finally, the above results were reproduced in 3L bioreactor suspension culture thereby establishing the scalability of this expression system. The process we developed is used routinely with the same success for the production of various recombinant proteins and viruses.

Abbreviations: CFDMEM – calcium-free DMEM; CS – bovine calf serum; hpi – hours post-infection; J+ – enriched Joklik medium; MLP – major late promoter; MOI – multiplicity of infection (# of infectious viral particle/cell); q – specific consumption rate (mole/cell.h); pfu – plaque forming unit (# of infectious viral particle); Y – yield (μg/E6 cells or mole/cell)

Introduction

The technological potential of adenovirus vectors (AV) in various applications such as 1) recombinant protein production, 2) live viral sub-unit vaccines production and 3) gene transfer for establishing stable cell lines or for gene therapy (reviewed in Berkner, 1988, 1992; Gerard and Meidell, 1993; Graham and Prevec, 1992) currently gives rise to growing interest from biotechnologists. All of these applications will require the production of large quantities of either recombinant proteins or AV stocks. However, so far, no significant research efforts have been directed towards the scale-up of the AV expression system.

Helper-independent AVs were developed in the early 80's for high-level expression of recombinant proteins in human cells. By deleting the E1 and E3 regions of the adenovirus genome, transcription cassettes up to 7.0Kbp could be inserted in AV (Gluzman *et al.*, 1982). While the deletion of the E3 region only affects the ability of the AV to efficiently propagate in whole animals (Berkner, 1988), the deletion of the E1 region prevents it's replication in all mammalian cells (either in vivo or in vitro). However, AV lacking the E1 and E3 regions can be propagated in the human 293 cell line which constitutively expresses the adenovirus E1 polypeptides (Graham *et al.*, 1977, Berkner, 1988). Thus, the AV/293 expression system has a double lock security feature built-in that restricts the propagation of replication defective recombinant viruses to the complementing 293 cell line. Typically, the construction of replication defective AV for recombinant protein production is accomplished by inserting Major Late promoter-based (MLP) expression cassettes in place of the deleted adenovirus El region. The adenovirus MLP is one of the strongest mammalian promoters and its transcriptional activity is responsible for the accumulation of the abundant adenovirus late proteins which represent collectively as much as 30–40% of total cellular proteins in adenovirus-infected cells (Ginsberg, 1984). However, due to the complexity in the regulation of gene expression in adenovirus, the recombinant protein production, using the first generation of expression vectors, has never exceeded 4% of the total proteins (estimated from data in Berkner, 1992). Consequently, the development of the full potential of AV as a high-level expression system has lagged behind other similar expression vectors such as the baculovirus/insect cells system (reviewed in O'Reilly *et al.*, 1992).

Recently, we have reported the construction of a new adenovirus expression vector (pAdBM5) that allows for the production of unprecedented levels of recombinant protein in AV-infected 293 cells (Massie *et al.*, 1994). In 293 cells infected with AV derived from the pAdBM5 transfer vector (AdBM5), the recombinant protein can accumulate at levels up to 10-20% of total cellular proteins (equivalent to 30–60 μg/10^6 cells), which makes it the most abundant protein in the infected cells. This yield compares advantageously to established expression systems, such as baculovirus/insect cell (Bac), for the production of non-secreted protein. Massie *et al.* (1994) compared the production of herpes simplex virus ribonucleotide reductase R1 and R2 subunits in both optimized culture of AV/293 and Bac systems. While the R2 subunit was about 5 fold more abundant and active in AV/293 than in Bac infected Sf9 cells, the R1 subunit was produced at roughly similar level in both systems. However, the amount of active soluble R1 obtained from AV/293 was at least 5 times higher than in Bac/Sf9 presumably due to better folding of the R1 protein in 293 cells. In terms of scale-up, the fact that, contrarily to Bac, AV virions remain concentrated within the cell long after yields have reached maximum levels, facilitating virus collection and concentration (Graham and Prevec, 1992) is also an advantage over Bac.

Another scale-up issue was the difficulty to design a large scale unit for the culture of adherent cells. The adaptation of the original 293A (anchorage-dependant) cells to suspension culture was a pre-requisite for the scale-up of the AV/293 system. The 293N3S subline developed by Graham (1987) by passage of the 293A cells through nude mice, was the first subclone of 293 cells successfully adapted to suspension culture. In our hands however, the 293N3S cells had a relatively long initial lag phase in suspension, a low growth rate, and a strong tendancy to clump, even in calcium-free medium. We then tested another subline, the 293S cells (Cold Spring Harbor Laboratories), obtained by gradual adaptation to suspension growth. The 293S cells grew more readily in suspension with no initial lag phase, a doubling time of 24 h and minimal clumping in calcium-free medium. Furthermore, 293S cells produced equivalent level of recombinant proteins compared to 293A (Massie *et al.*, 1994). The 293S cell line was therefore chosen for further process development.

Medium limitation and/or by-product inhibition is an important scale-up problem in animal cell culture in general and the AV/293 system is not an exception. Although 293S cells could reach plateau density of 2–5 $\times$ 10^6 cell/mL depending on the culture medium, productive infection with AV was restricted to cell density lower than 5×10^5 cells/mL in batch culture without medium replacement. In this paper, we present a two-step approach, undertaken to improve the volumetric yield of the AdBM5/293S recombinant protein production system with the model protein Protein Tyrosine Phosphatase 1C (PTP1C). This 68kDa enzyme is a highly phosphorylated intracellular phosphatase that plays a crucial role in signal transduction and is a potential target for cancer therapy (Shen *et al.*, 1991). In the first step a medium replacement strategy has been applied, in order to rapidly overcome any medium-related limitation/inhibition problems. We will show the success and the limitations of this strategy. In a second step, an extensive metabolic analysis has been undertaken in order to identify more precisely what was limiting or inhibitory in the medium. This knowledge was then applied to a specific addition and control strategy of the infected culture. This longer procedure has lead to higher volumetric productivity with lower medium expenses.

Materials and methods

Cells, medium and virus

The 293A cells were used for plaque assay. The cells are derived from human kidney fibroblast transformed with Ad5 DNA and express the E1A and E1B proteins constitutively (Graham *et al.*, 1977). 293A were obtained from ATCC and sub-cultured twice weekly in DMEM with 10% fetal bovine serum in 25 cm^2 T-flasks. The 293S were obtained from Dr. Michael

Matthew (Cold Spring Harbor Laboratories). 293S were kept frozen in liquid nitrogen until used. A fresh cell aliquot was thawed every two months and maintained in 100 mL spinner flask at 37 °C, 5% CO_2 by diluting twice a week to cell densities 1–3×10^5 cells/mL with complete Joklik + medium (J +) described below.

J+ medium was inspired from Chillakuru *et al.* (1991) who used enriched DMEM for cultivation of vaccinia virus in HeLa cells. It was made of Joklik medium (calcium-free modification of MEM, Sigma) supplemented with 2.5 g/L glucose (total 4.5 g/L, 25mM, Sigma), 1X MEM essential amino acids (Gibco), IX MEM non-essential amino acids (Gibco), 1X MEM vitamin solution (Gibco), 0.11 g/L Na.pyruvate(Gibco), 5.7 g/L $NaHCO_3$ (Sigma) and 2.5 g/L HEPES buffer (Sigma). The mixture was then adjusted to pH=6.75 and filter-sterilized. Calcium-free DMEM (CFDMEM, custom made, Gibco) was also tested. This medium was supplemented to yield a final concentration of 4.5 g/L glucose and 0.11 Na.pyruvate, equivalent to J+. However, CFDMEM was different from J+ as it did not contain any other non-essential amino acids except serine (0.1mM in J+ vs 0.4mM in CFDMEM) and it's buffering capacity consisted in 3.7 g/L of $NaHCO_3$. Both media were always completed with 5% iron supplemented bovine calf serum (CS)(Hyclone) and 0.1% (w/v) pluronic F-68 (Gibco) unless otherwise stated.

The replication defective AV, AdBM5-PT, have been constructed in Dr Shen's laboratories, to produce protein tyrosine phosphatase (PTP1C)(Zhao *et al.*, 1993). A stock of the virus has been constituted and used throughout all of the experiments: 6×10^9 cells were infected at a multiplicity of infection (number of viruses/number of cells or MOI) of 1 and harvested 72 hours post-infection (hpi). The cell pellet was then diluted to 10^7 cells/mL with J+ and then freezed-thawed three times to liberate the virus. The stock has then been titrated to 1.2×10^9 pfu/mL by standard plaque assay method.

Culture and infection in spinner flasks

Unless otherwise stated, culture and production runs were done in 100 mL siliconized spinner flasks (Bellco) with 50 mL of cell suspension in a 37 °C, 5% CO_2, humidified incubator. Samples were taken on a daily basis for viable and total cell count and were kept at –80 °C for further analyses. Aliquots of 1×10^6 cells were centrifuged (13,000 g), the cell pellet was washed twice in PBS and then frozen at –80 °C.

Growth of 293S cells in batch cultures were initiated by inoculating fresh J + medium with 1–2×10^5 cell/mL in exponential growth phase.

Production runs were prepared by first centrifuging (600g, 15 min) aliquots of a cell culture in the exponential phase or in the very beginning of the plateau phase. To insure that the viral adsorption phase was identical for each assay, uniform conditions were imposed for the initial incubation of the cell/virus mixture. The cell pellets were then resuspended with the AdBM5-PT virus in either spent or fresh medium at a cell density of 10^7 cells/mL and a MOI of 10 to insure synchronous infection. These concentrated cell/virus suspensions were incubated 2 hours and then diluted at various cell densities with spent or fresh medium. The infected cultures were incubated 3–5 days while samples were taken once or twice daily. For medium replacement experiments, infected cultures were centrifuged at 600g for 15 min and the spent medium discarded and replaced with the same volume of fresh medium. In a few cases, pH was periodically adjusted (2–3 times/day) in spinner flasks by addition of 7.5% $NaHCO_3$ until the color of the culture returned to 7.1. In spinner flasks, pH was estimated by the medium color compared to standard flasks (red-orange at pH $\approx$ 7.1, yellow at pH $\leq$ 6.5).

Bioreactor description and operation

A 3.5 L bioreactor (Chemap CF-3000 with a CBC-10 control unit) was used with 2.7 L of culture volume in order to scale-up the spinner productions. The tank was equipped with 3 surface baffles to break the liquid surface and increase mass transfer. Mixing was performed with a marine impeller rotating at 100 RPM. The temperature was maintained at 37 °C with a water jacket. D.O. and pH probes (Ingold) were mounted for monitoring and control purposes. The pH was controlled at 7.0 by intermittent addition of 7.5% $NaHCO_3$ solution. Feed gas composition was regulated by the sequential opening of electro-valves: CO_2 was kept at 7% and O_2 set-point was under the control of dissolved oxygen (DO) in order to maintain DO above 20%. Operation parameters were sent to a Compaq Deskpro PC for data acquisition.

The bioreactor infection protocol was identical to the one used for spinner cultures except that transfer of fluid to and from the bioreactor was achieved through sterile connections instead of under a biological hood.

Analytical methods

Viable and total cells were counted on a haemacytometer. Viability was assessed by dye exclusion using erythrosine B. The 293S cells having the tendancy to agglomerate, special care was taken to separate the clumps without affecting viability.

Medium composition was analyzed via HPLC. The various amino acid concentrations were measured by a reversed phase method as described previously by Kamen *et al.* (1991). The glucose and organic acid concentrations were obtained using an Interaction Ion-300 organic acid column (Chemicals Inc) with 0.0033N sulphuric acid as a mobile phase and two detectors: a refractive index detector (model 410, Millipore) and a spectrophotometer detector (model 490, Millipore) at 210 nm.

The oxygen uptake rate was measured using a YSI model 53 biological oxygen monitor, following the protocol provided with the system.

SDS-PAGE electrophoresis of cellular proteins was performed as follows. Frozen cell samples were thawed and diluted to 10^7 cell/mL in extraction buffer (80 mM Tris-HCI pH 6.8, 2% (w/v) SDS and 10% (v/v) glycerol) and then sonicated (Heat Systems-Ultrasonics Inc, model W-375) 5s, 90W. Cell extracts (10 μL) were diluted with 10μL of NOVEX (San Diego, CA) sample buffer, containing 0,5% (v/v) β-mercaptoethanol. The diluted samples were heated at 85 °C for 5 min. and centrifuged 15s in an Eppendorf centrifuge before being loaded on a 8% acrylamide NOVEX precasted gel (10^5 cells per lane). The SDS-PAGE was run for 90 min. at 125 V following the NOVEX procedures.

Protein Tyrosine Phosphatase activity

PTP1C activity was measured according to the method described by Pot *et al.* (1991) with the following modifications. Batch analysis was performed by doing multiple dilutions in 96-well plates where samples were quantified against a PTP1C standard (kindly provided by Dr S.Shen). However, it was found that when diluted in the extraction buffer alone, the activity of the purified enzyme was drastically reduced. In order to stabilize the enzymatic activity, the purified PTP1C was diluted in a cell lysate obtained by adding 400 μL of the extraction buffer (described below) per 1×10^6 non-infected cell pellet. Samples, stored at –80 °C, were thawed on ice and resuspended at 2.5×10^6 cells/mL in extraction buffer: 25mM Tris-HC1, pH = 7.5, 10 mM β-mercaptoethanol, 2mM EDTA and 0.5% (v/v) Triton X-100. Aliquot volumes of 1 to 6 μL were transferred in the 96 well plate followed by 95μL of pNPP reagent: 25mM pNPP, 1.6mM DTT, 40 mM MES, pH=5. The plate was incubated at room temperature for 10 min and then 100 μL/well of 0.2N NaOH solution was added to stop the reaction. The plate was read at 405 nm using a Titertek Multiskan MCC microplate reader. A calibration curve was obtained from dilutions of the standard and the PTP1C content of the samples was calculated from that curve.

The specific PTP1C activity has been verified to be equivalent for cell samples and purified standard. For dilutions of identical activities, the PTP1C band obtained on PAGE for the purified standard was always equal or less than the PTP1C band for a cell sample. The stability of the frozen PTP1C standard was also assessed by series to series reproducibility of the activity calibration curve.

Yield (Y) and specific consumption rate (q) calculations

During growth experiments, the limits of the exponential growth phase were identified by first determining the zone of linear relationship on the plot of the natural log of cell concentration ln(X) vs time (t). The specific growth rate (μ) was then estimated as the slope of that ln(X) vs t plot and the doubling time (t_d) was computed: $t_d = \ln(2)/\mu$. The cellular yield per mole of consumed substrate (Y_s) was calculated by dividing the quantity of cells produced by the quantity of substrate consumed during the exponential growth phase, while the product per cell yield was the mole produced divided by the quantity of cell produced during the exponential growth phase. The specific substrate consumption (q_s) rate was obtained by using: $q_s = \mu/Y_s$.

During infection, since cells do not grow significantly, Y and q were calculated differently. Specific substrate consumption rates were estimated during the period of initial linear consumption by dividing the quantity consumed by the time interval and the mean cell concentration during that period. Product yield was obtained by dividing the maximum quantity produced by the total cell concentration at that time.

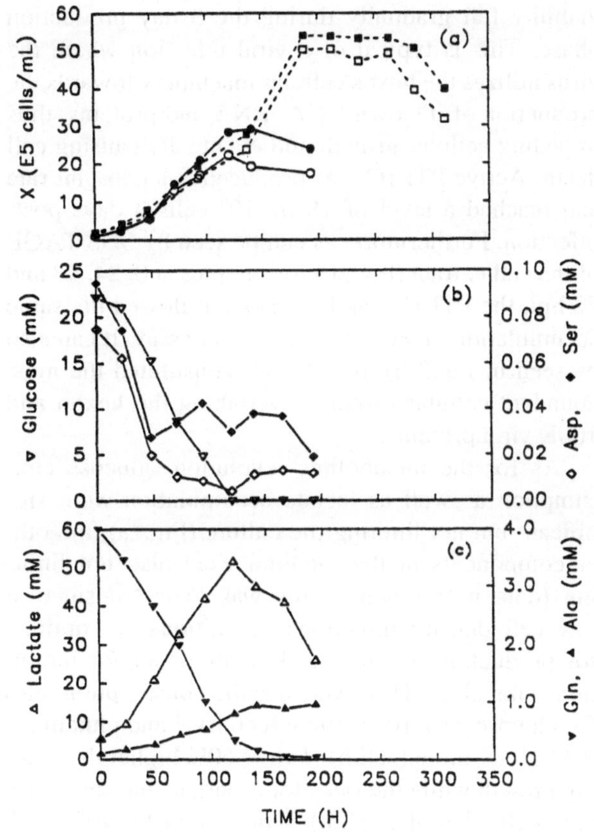

Fig. 1. Typical growth curves of 293S cells in Joklik+ and CFD-MEM. (a) ○, ● viable and total cells in J+, □, ■ viable and total cells in CFDMEM. Key metabolites profiles for cultures grown in Joklik+ medium: (b) ▽ glucose, ◇ aspartate, ◆ serine, and (c) △ lactate, ▼ glutamine, ▲ alanine.

Results and discussion

Growth kinetic of 293S cells in suspension culture

Typical growth curves for the 293S cells in J+ and CFDMEM, together with the variation of key medium components in J+ is shown in Fig. 1(a) through (c). Viable and total cell densities are shown in Fig.l(a). 293S cells were inoculated at 0.15x10^6 cells/ml in complete J+ medium. Exponential growth started soon after cell inoculation and was maintained for a period of 3 days with a doubling time of 24 h. The growth was then linear for the next two days followed, at day 5, by a plateau of 2–3$\times$10^6 cells/mL. At that point, the viability of the culture did not decrease sharply but rather stayed at the plateau for a few days before declining.

As can be seen in Fig.l(b) and (c), the growth of 293S cells in J+ was characterized by a substantial consumption of: glucose ($Y_{glucose}$ 10^{11} cell/mole or $q_{glucose} = 29\times10^{-14}$ mole.(cell.h)$^{-1}$ in the exponential phase), aspartate ($Y_{asp} = 7.1\times10^{12}$ cell/mole or $q_{asp} = 0.4\times10^{-14}$ mole.(cell.h)$^{-1}$), serine ($Y_{ser} = 6.7\times10^{12}$ cell/mole or $q_{ser} = 0.43\times10^{-14}$ mole.(cell.h)$^{-1}$) and glutamine ($Y_{gln} = 5\times10^{11}$ cell/mole or $q_{gln} = 5.8\times10^{-14}$ mole.(cell.h)$^{-1}$). Although glucose and glutamine were consumed at a higher specific rate, aspartate and serine were depleted first since they are at a much lower concentration in the medium (0.1 mM compared to 25mM for glucose and 4mM for glutamine). Ammonia never exceeded 2mM at which concentration it was tested to be non-inhibiting for 293S growth (data not shown). On the other hand, lactate was the main by-product of the culture ($Y_{lact} = 2\times10^{-11}$ mole produced/cell) while a marginal amount of alanine was produced (1mM at day 6). Indeed, 293S cells did not oxidize significant amounts of glucose; most of it was used through glycolysis. Lactate has been found to impede 293S cell growth at concentrations of 20mM or more (data not shown). Consequently, lactate accumulation might have caused the shift from exponential to linear growth around day 3, although depletion of aspartate and serine may also be involved. Lactate accumulated to 50 mM by day 5, at which point glucose was depleted and the cells stopped dividing. Glucose and/or glutamine depletion around day 5 was most probably the cause of the culture entering the plateau phase, after which the cells started consuming lactate.

The 293S cells were also grown in complete CFD-MEM (Fig.l(a)). As can be seen, while the growth kinetics were equivalent in both media up to day 5, beyond that point, cells in CFDMEM kept growing for another 3 days reaching a cell density of 5$\times$10^6 cells/mL. In contrast, cells in J+ entered the plateau phase at this time. A 3 day plateau then occurred before the cell density started decreasing. As descrihed in the previous section, the only difference between CFD-MEM and J+ is the lower buffering capacity and the absence of non-essential amino acids with the exception of serine which is four times more concentrated. The need for higher levels of serine and/or the effect of the buffers (especially HEPES) on cell/substrate yields (glucose or glutamine) could explain the better performance of CFDMEM for 293S cell growth.

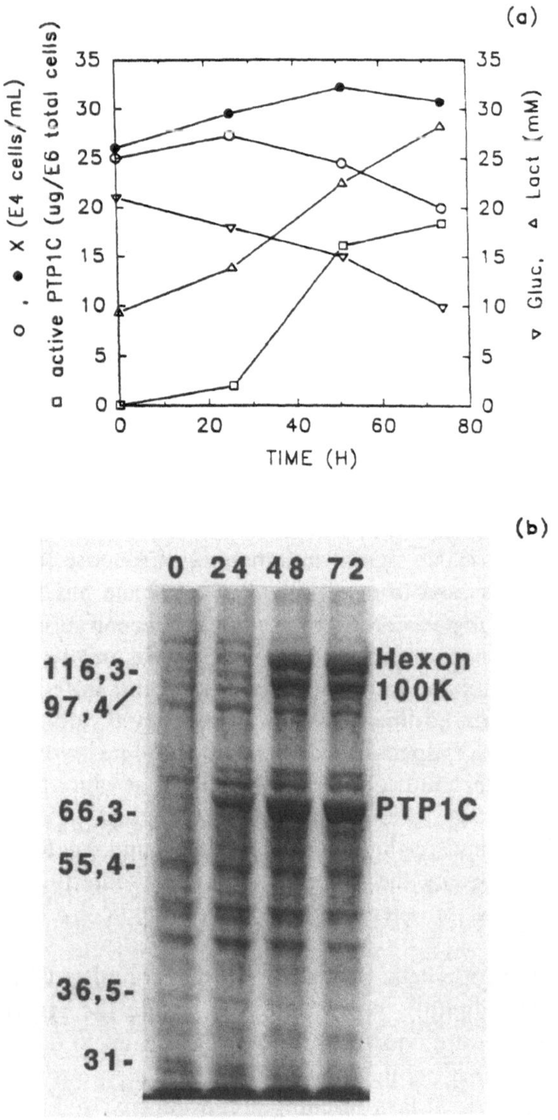

Fig. 2. Typical batch production of PTP1C at low cell density (0.25×10^6 cells/mL) in Joklik+ ○, ● viable and total cells, □ PTP1C, ▽ glucose, △ lactate; (b) SDS-PAGE of cell extracts sampled at different time the infection, 10^5 cells per lane: 0, 24, 48 and 72 hpi.

Production of PTP1C in AdBM5-PT infected 293S at low cell density

Fig. 2(a) shows the increase in cell density with time for infecfed 293S cells seeded at 0.25×10^6 cells/mL. The active PTP1C cell content, glucose consumption as well as the lactate accumulation are also shown. The growth rate for infected cells was close to zero and cell viability fell gradually during the 3 day production phase. This is typical of a viral infection where the virus utilizes the host's cellular machinery towards the production of it's own DNA, RNA and proteins, thus impeding cellular growth and eventually causing cell death. Active PTP1C was produced at a constant rate and reached a level of 18 μg/10^6 cells 3 days post-infection. Furthermore, as can be seen by SDS-PAGE of the cell extract (Fig.2b) for samples at 0, 24, 48 and 72 hpi, the PTP1C band at 68kDa followed the same accumulation kinetics as the activity assay. It can also be seen in Fig.2(b) that PTP1C constituted the most abundant cellular protein, overtaking the hexon and 100K viral proteins.

As for the metabolites' evolution, glucose consumption as well as lactate accumulation were significant but not limiting the culture (Fig.2a); the other components in the medium were also not limiting (data not shown). This was expected since at low cell density infections, conditions are optimal for production because no limitations and/or inhibitions take place. However, specific consumption rates for glucose, 46×10^{-14} mole.(cell.h)$^{-1}$ and glutamine, 9.4×10^{-14} mole.(cell.h)$^{-1}$ were 60% higher than during growth while the rates for aspartate and serine, q= 0.25×10^{-14} mole.(cell.h)$^{-1}$ decreased by 40% with respect to growth. The significant increase in glucose and glutamine consumption rate, both primary sources of energy, indicate a general acceleration of the cell metabolism during infection. Once again, lactate accumulated at a specific rate twice that of glucose consumption (1×10^{-12} mole.(cell.h)$^{-1}$) implying complete glycolysis.

Production of PTP1C at higher cell densities: effect of medium replacement

A first series of experiments was undertaken in order to precisely establish the potential of the production medium in terms of the maximum cell density at infection that would not impair product yield. 293S cells were infected at different days in a culture and therefore at different densities (day 2, 0.6×10^6, day 3, 1.3×10^6 and day 4, 1.7×10^6 cells/mL) have been infected and resuspended at their initial cell densities in either (a) their spent medium, (b) fresh medium or (c) fresh medium followed with a medium replacement at 24 hpi. The resulting PTP1C yields at various time points are presented in Fig. 3 (a) through (c) and compared to a culture control infected a low cell density (0.25×10^6 cells/mL). It can be seen in Fig. 3(a) that at 0.6×10^6

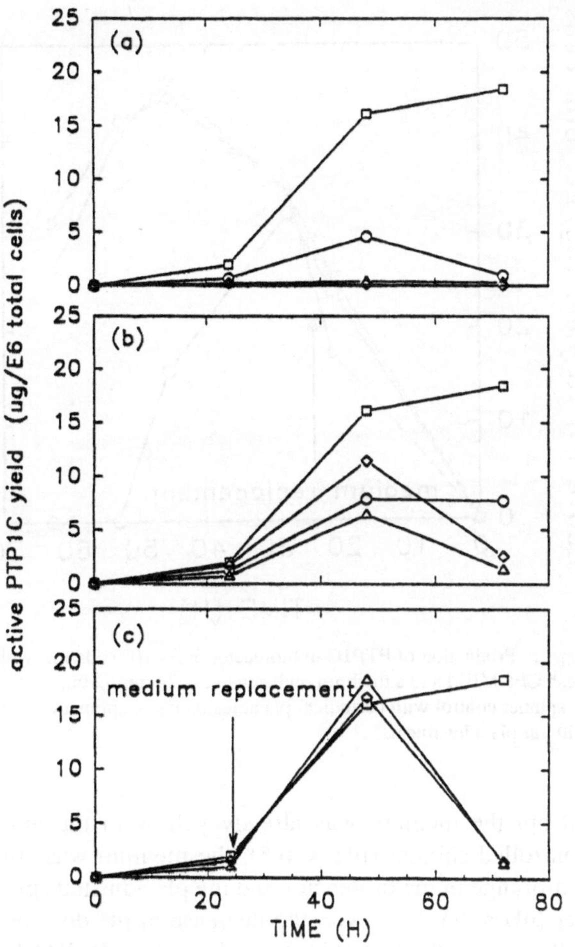

Fig. 3. PTP1C production in J+ at different cell densities. Infected cells resuspended in: (a) spent medium, (b) fresh medium, and (c) fresh medium with medium replacement at 24 hpi. Cell densities at infection: ○ 0.6×10^6, ◇ 1.3×10^6 and △ 1.7×10^6 cells/mL. □ duplicate controls at 0.25×10^6 cells/mL resuspended in spent or fresh medium.

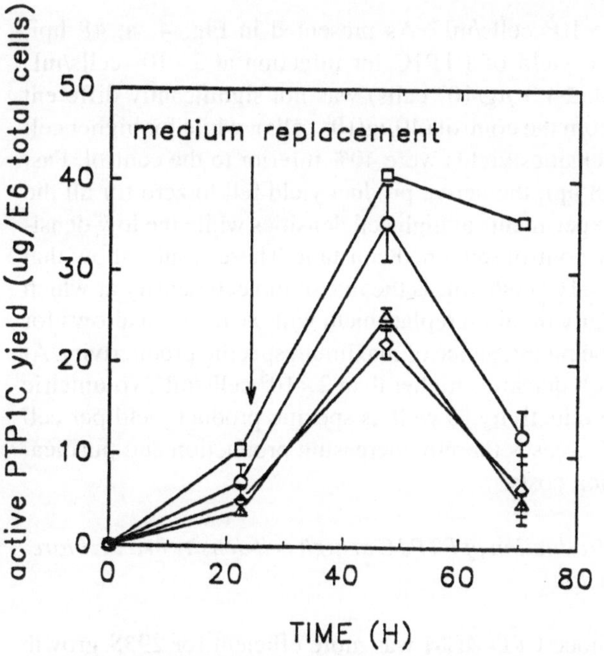

Fig. 4. PTP1C production in fresh J+ at different cell densities with one medium replacement at 24 hpi. ○ 2×10^6, ◇ 3×10^6, △ 4×10^6 cells/mL. □ control at 3×10^5 (without medium replacement). Means of duplicates are represented (± S.D.).

cells/mL or above, no significant amount of PTP1C was produced when the infected cells were resuspended in their spent medium. However, when resuspended in fresh medium (Fig.3b) production regained 50% of the maximum productivity with respect to the control at all cell densities. Furthermore, a second medium replacement at 24 hpi (Fig.3c) allowed for a sustained maximum specific productivity, even at the highest cell density (initial 1.7×10^6 led to a final 2.2×10^6 cells/mL).

These results clearly establish the existence of a substrate limitation and/or a by-product inhibition at high cell densities, a problem which can be partially remediated by an initial cell resuspension in fresh medium (Fig.3b) and completely restored with a medium change at 24 hpi (Fig.3c), resulting in a maximum productivity comparable to optimal low cell density infection. It also establishes that the cell culture growth stage (from early exponential at day 1 to beginning of plateau phase at day 4) do not influence protein production since cells infected in their 4th day of culture produced as much PTP1C as one day old infected culture, as long as the medium is not limiting and/or inhibiting. However, this medium replacement strategy was apparently only effective for a period of 24 to 48 h since the PTP1C activity decreased abruptly at 72 hpi. Analysis of total cell protein by SDS-PAGE showed that the loss in activity at 72 hpi was not concurrent with an equivalent loss in total PTP1C content (data not shown). The decrease in PTP1C activity ohserved at 72 hpi was therefore not due to protein degradation, but rather to an unknown mechanism such as protein aggregation, as previously observed for the HSV R1 subunit expressed with a similar AV (Massie *et al.*, 1994) or a major change in the phosphorylation state of the protein. Resolving this issue will require further investigation.

In order to evaluate the limit of this initial and 24 hpi medium replacement strategy, PTP1C yield was tested following infection of 293S cells at 2, 3 and

4×10^6 cells/mL. As presented in Fig. 4, at 48 hpi, the yield of PTP1C for infection at 2×10^6 cells/mL (35 ± 4.3 μg/10^6 cells) was not significantly different from the control (40μg/10^6 cells), while for higher cell densities yields were 40% inferior to the control. Past 48 hpi, the active product yield fell to zero for all the experiments at high cell densities while the low density control remained constant. These results show that 2×10^6 cells/mL is the maximum cell density at which daily medium replacement with J+ medium allows for the maintenance of maximum specific productivity. At cell densities higher than 2×10^6 cells/mL, volumetric productivity as well as specific product yield per cell decreases thereby increasing production and purification costs.

Production of PTP1C at high cell density in a 3L bioreactor

Since CFDMEM was more efficient for 293S growth than J+, its performance during an infection was tested. However, a preliminary experiment comparing both media for an infection at high cell density showed that CFDMEM acidified more rapidly than J+, which was not the case during cellular growth. At 72 hpi, in cell cultures infected at 1.3×10^6 cells/mL in fresh medium with a medium replacement at 24 hpi, the pH dropped below 6.5 (yellow medium) in CFDMEM while pH was roughly equal to 6.8 (orange medium) in J+ (data not shown). As a result, the PTP1C activity was also lower in CFDMEM (20 μg/10^6 cells) than in J+ (32 μg/10^6 cells) at 48 hpi.

In order to assess the correlation between active PTP1C production and pH as well as the scalability of the process, an infection experiment was performed in a pH-controlled 3L Chemap bioreactor. The results are shown in Figure 5 for a culture infected at 2×10^6 cells/mL (MOI=10) in CFDMEM. The bioreactor was compared to infected cell controls: two 50mL spinner flasks initially taken from the bioreactor, one with and the other without periodical pH adjustment.

The PTP1C specific productivity was equivalent in the bioreactor and in the pH controlled spinner with a peak of 45 μg active PTP1C/10^6 cells at 48 hpi, followed by a slight decrease in the active PTP1C concentration. By contrast, in the spinner flask without pH control, the accumulation of active PTP1C stopped at 30 hpi with a peak of 30 μg active PTP1C/10^6 cells, followed by a rapid reduction in activity, falling close to zero by 52 hpi. A correlation between the PTP1C activity loss and pH decrease was observed. While at

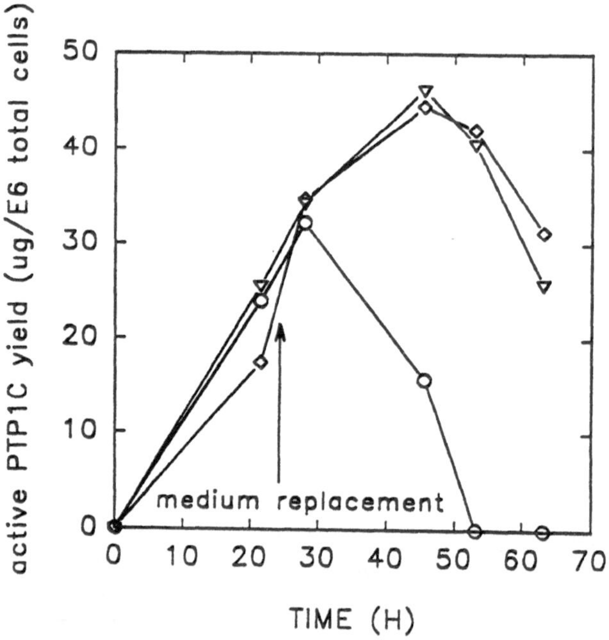

Fig. 5. Production of PTP1C in bioreactor at 2×10^6 cells/mL with fresh CFDMEM and a medium replacement at 24 hpi. ◇ bioreactor, ▽ spinner control with periodical pH adjustments, o spinner control without pH adjustments.

48 hpi the medium was already yellow in the non-controlled spinner (pH $\leq$ 6.5), the medium was still red-orange in the bioreactor and the pH adjusted spinner (pH $\approx$ 7). However, the decrease in pH does not fully explain the PTP1C activity loss since the PTP1C activity also decreased slowly in a pH controlled environment past 48 hpi. The pH control only delays the lytic process that inevitably takes place during adenoviral infection, while permitting the maximum product yield to be attained.

These results show that it is possible to maintain maximum specific production rate at high cell densities, by medium replacement at 0 and at 24 hpi, in a pH controlled culture at the 50 mL spinner flask scale as well as 3L bioreactor scale. Although peak product yield varied among the experimental runs, one can expect to obtain 40 to 45 μg PTP1C/10^6 cells, equivalent to 15% of the total cellular protein content (based on 300μg total protein/10^6 cells) or 90 mg PTP1C per Litre of culture (at 2×10^6 cells/mL) compared to 13.5 mg/l for 0.3×10^6 cells/mL. These figures are comparable to productions of Herpes Simplex Virus ribonucleotide reductase subunits R1 and R2 obtained previously in our lab with other AdBM5 AV's (Massie *et al*, 1994). Although this process is very effective with respect to cell yield and protein purification for an intracellular product, it is not optimal in terms of medium expenses. In fact, with two medium changes, the yield of product per spent medium is equivalent to 30 mg PTP1C per L of medium (at 2×10^6 cells/mL) which is not much higher than 15 mg/L of medium for productions at 3×10^5 cells/mL without medium replacement. We then turned our attention to the analysis of key metabolites in order to improve yield based on spent medium.

Analysis of key metabolites during the infection phase

Samples from the infection of 1.3×10^6 293S cells/mL in fresh J+ medium without medium replacement at 24 hpi (for which PTP1C yield has been presented in Fig.3b) have been analyzed for their content in glucose, organic acids, and amino acids. In this experiment, PTP1C was only 60% the level of it's maximum specific activity due to limitation of nutrients and/or inhibition of by-products. The results are presented in Fig. 6.

The total cell density increased from 1.3×10^6 to a mean value of 1.9×10^6 cells/mL during the first 24 h and remained constant thereafter. This slight initial increase of about 20% in cell count is routinely observed for infected culture. However, the maintenance of viability up to 96 hpi is peculiar to infection in limiting and/or inhibiting environment. This could be explained by the fact that under sub-optimal conditions of infection, the overall cycle of virus reproduction would occur at a lower rate, thereby reducing the infection stress on the cell which in turn would result in a prolonged viability.

Glucose was completely depleted before 48 hpi. Based on the first 24 hour period, the specific glucose consumption rate, was 38×10^{-14} mole glucose.(cell.h)$^{-1}$ which is a 30% increase compared to $q_{glucose}$ during cell growth. However, glucose consumption rate was reduced by 17% compared to infection at low cell density. Furthermore, while glucose was totally transformed into lactate during growth and infection at low cell density, only 45% of it was metabolized through glycolysis during the infection phase at high cell density; 22mM of glucose yielding only 20mM of lactate instead of 44mM for a complete glycolysis. It appears that while the general metabolic activity was higher compared to growth phase, in this case there was a glucose limitation, thereby reducing glucose consumption rate as well as lactate production rate. This in turn should increase the cellular oxygen requirement. The specific oxygen consumption rate

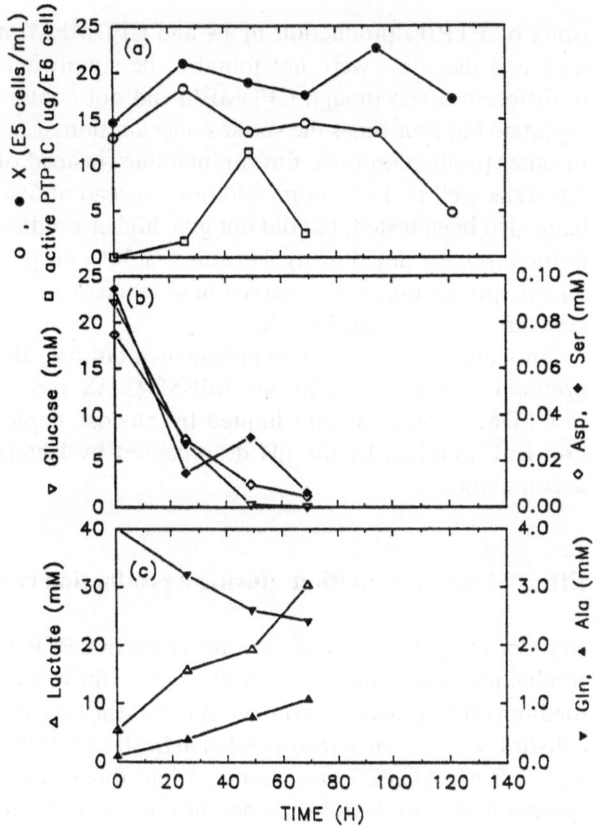

Fig. 6. Key metabolites evolution in a batch production of PTP1C in fresh J+ at initial cell density of 1.3×10^6 cells/mL: a)○, ● viable and total cells in J+, □ PTP1C yield; (b) ▽ glucose, ◇ aspartate, ◆ serine, and (c) △ lactate, ▼ glutamine, ▲ alanine.

(q_{O2}) has been measured in growth phase as well as during infection. Indeed, the average value obtained during infection, $q_{O2} = 16\times10^{-14}$ mole O_2.(cell.h)$^{-1}$ was twice as high as for exponential growth, $q_{O2} = 8\times10^{-14}$ mole O_2.(cell.h)$^{-1}$. This is consistent with a significant increase in metabolic rate during the infection phase compared to the growth phase and a reduced glycolysis rate with respect to nonlimiting medium conditions.

During infection, aspartate and serine specific consumption rates were low ($\approx 0.2\times10^{-14}$ mole.(cell.h)$^{-1}$ for both), but, given the high cell density, were rapidly depleted by 48 hpi. Glutamine was consumed (2.2×10^{-14} mole.(cell.h)$^{-1}$) but not depleted and alanine was produced (1×10^{-14} mole.(cell.h)$^{-1}$). These specific consumption rates were smaller than those obtained for infection at low cell density and exponential growth. This might be due to glucose limitation. However, aspartate and serine, although depleted, were not limiting. Indeed, comparisons of PTP1C production in J+ and CFDMEM at high cell densities were not found to be significantly different, even though CFDMEM did not contain aspartate but four times the serine concentration of J+. In other production runs, 6mM glutamine (instead of 4mM) as well as 10% serum addition (instead of 5%) have also been tested, but did not give higher productivities (results not shown). However, a 50% drop in PTP1C production was observed in absence of serum compared to the usual 5% CS.

In summary, in serum supplemented culture, the production of PTP1C with the AdBM5/293S system in CFDMEM was mainly limited by glucose depletion and inhibited by the pH drop caused by lactate accumulation.

Effect of specific additions during a production run

In order to apply and verify the above results, PTP1C production was compared in a glucose addition vs a medium replacement experiment. A culture at 1.5×10^6 cells/mL was infected, resuspended in fresh CFDMEM with 2.5 g/L HEPES and aliquoted into three 50mL spinner flasks. In the first spinner the culture was centrifuged and the medium replaced with fresh CFDMEM + HEPES at 24 hpi. In the other two, 0.5 mL of a 200 g/L glucose solution was added to the culture at 24 hpi (+2 g/L, or 11 mM glucose addition). In one of these pH was periodically adjusted.

Figure 7 shows that, PTP1C production followed a similar profile in both the medium replacement and the glucose addition experiment where pH was controlled. The two feeding strategies yielded a maximum active PTP1C content of 25 μg/10^6 cells at 36–48 hpi. Since it has been shown that at high cell density without medium replacement or glucose addition, active PTP1C yield declined after 24 hpi, it is clear that glucose addition is responsible for sustained PTP1C production, equivalent to production with medium replacement. It therefore confirms that glucose is most probably the major limiting substrate of PTP1C production.

In the third spinner (glucose addition without pH control) the PTP1C activity decreased linearly after 24 hpi and was absent by 48 hpi. This severe drop in activity is again correlated with a decrease in pH; indeed, the medium was already yellow at 36 hpi or earlier (pH $\leq$ 6.5) in the glucose addition spinner without pH control while it was maintained around red-orange (pH $\approx$ 7) in the other spinner with glucose addition. The exclusive relation between pH and PTP1C activity

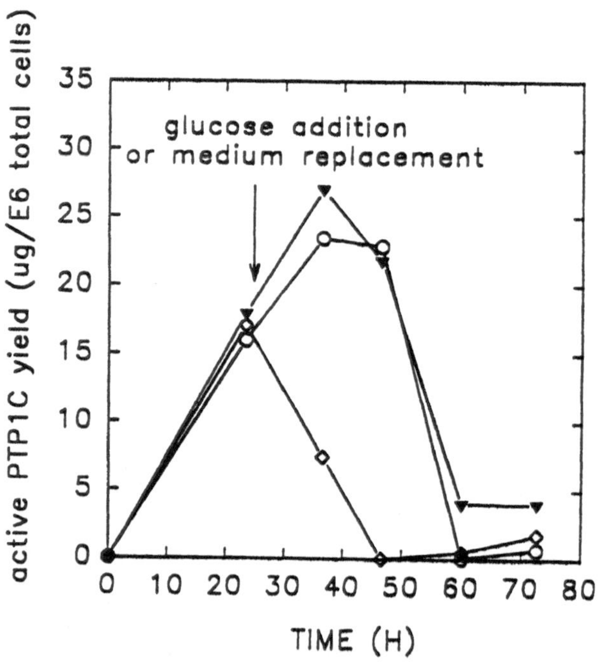

Fig. 7. Production of PTP1C in fresh CFDMEM at 1.6×10^6 cells/mL. o medium replacement at 24 hpi, ▼ glucose addition with periodical pH adjustments, ◊ glucose addition without pH control.

was confirmed by lactate analysis since lactate concentrations attained equivalent level (40–50 mM) in both glucose-supplemented spinners (data not shown). It is therefore the pH decrease as such and not lactate production, that has a negative effect on active PTP1C yield. This is encouraging since pH is easier to control than lactate production.

Conclusion

In this paper we have presented the first results concerning the scale-up of a high-level recombinant protein production AV/293 system. The 293S cells have been shown to be able to grow to plateau cell densities of 5×10^6 cells/mL in calcium-free DMEM. With an initial and a 24 hpi medium replacement, the specific PTP1C yield could be maintained at it's maximum level up to infected cell densities of 2×10^6 cells/mL. Under these conditions, volumetric productivities of 90 mg/L could be attained. At an infected cell density of 1.6×10^6 cells/mL, the replacement of the 24 hpi medium change by a 2 g/L glucose addition, together with periodical pH adjustments, allowed the same specific productivities, but at lower medium expenses.

It is expected that glucose fed-batch in pH-controlled bioreactor will further improve these performances.

Acknowledgments

The authors would like to thank Dr. M. Matthew from Cold Spring Harbor for providing 293S cells, Dr S-H. Shen and P. Bouchard for providing AdBM5-PT recombinant adenovirus, B. Ouellette and D. Mosser for critical reading of the manuscript.

References

Berkner KL (1988) Development of adenovirus vectors for the expression of heterologous genes. Biotechniques 6:616–628.

Berkner KL (1992) Expression of heterologous sequences in adenoviral vectors. Curr. Top. Microbiol. Immunol. 158:39–66.

Chillakurur RA, Ryu DDY, Yilma T (1991) Propagation of recombinant vaccinia virus in HeLa cells: Adsorption kinetics and replication in batch cultures. Biotechnol. Prog. 7: 85–92.

Gerard RD and Meidell RS (1993) Adenovirus-mediated gene transfer. Trends Cardiovasc. Med. 5:171–177.

Ginsberg HS (ed.) (1984) The adenoviruses. Plenum Publishing Corp., New-York.

Gluzman Y, Reichl H and Solnick D (1982) Helper-free adenovirus type 5 vectors. In Eucaryotic viral Vectors, pp. 187–192. Edited by Y. Gluzman, New York: Cold Spring Harbor Laboratory.

Graham FL, Smiley JR, Russell WC and Nairn R (1977) Characteristics of a human cell line transformed by DNA from human adenovirus type 5. J. Gen. Virol. 36:59–72.

Graham, FL (1987) Growth of 293 Cells in Suspension Culture. J. gen. Virol., 68:937–940.

Graham FL and Prevec I (1992) Adenovirus-based expression vectors and recombinant vaccines, p.363–390. In RW Ellis (ed.), Vaccines: new approaches to immunological problems. Butterworth-Heinemann, Boston.

Kamen A, Tom R, Caron A, Chavarie C, Massie B, Archambault J (1991) Culture of insect cells in a helical ribbon impeller bioreactor. Biotechnol. Bioeng., 38:619–628.

Massie B, Dionne J, Lamarche N, Fleurent J, Langelier Y (1994) Improved adenovirus vector produces herpes simplex virus ribonucleotide reductase R1 and R2 subunits more efficiently than baculovirus vector. Submitted to Bio/Technology.

O'Reilly KR, Miller LK and Luckow VA (1992) Baculovirus expression vectors, a laboratory manual. W.H. Freeman and Company, New-York.

Shen S-H, Bastien L, Posner BI, Chrétien P (1991) A protein-tyrosine phosphatase with sequence similarity to the SH2 domain of the protein-tyrosine kinases. Nature, 352: 736–739.

Zhao Z, Bouchard P, Diltz CD, Shen S-H and Fischer EH (1993) Purification and characterization of a protein tyrosine phosphatase containing SH2 domains. J. Biol. Chem. 268:2816–2820.

Address for offprints: A. Garnier, Institut de recherche en biotechnologie, CNRC, 6100 Royalmount, Montréal, Québec, Canada, H4P 2R2.

Cytotechnology **15**: 157–167, 1994.

Relationship between oxygen uptake rate and time of infection of Sf9 insect cells infected with a recombinant baculovirus

T.K. Kathy Wong, Lars K. Nielsen, Paul F. Greenfield and Steven Reid
Department of Chemical Engineering, The University of Queensland, Queensland 4072, Australia.

Key words: Insect cells, oxygen uptake rate, baculovirus, time of infection, nutrient consumption.

Abstract

Oxygen uptake rates (OUR) of Sf9 insect cells propagated in a serum-free medium (SF900II, Gibco) and of cells infected with a recombinant AcNPV were investigated before and after infection in a laboratory-scale bioreactor. The volumetric OURs of uninfected and exponentially growing cells were found to be proportional to the cell density. For infected cultures, the specific OUR of cells increased immediately after addition of virus and a maximum of 1.3 times the value of uninfected cells was noted for all the cultures between 8 to 30 hours post infection, which coincides with the period at which most viral replication and the majority of DNA synthesis takes place. It was observed that the rate of rise in the specific OUR decreased as the cell density at the time of infection increased, which meant that the later the infection, the later the maximum sOUR was observed. We therefore suggest that OUR measurement can be used to reflect the efficiency of a batch infection. Carbohydrate and amino acid consumption rates from an infected run were analysed in an effort to identify substrate(s) that may be used at increased rates to fuel the rise in oxygen demand observed early in the infection cycle. No observable rise in the consumption rates of glucose or glutamine, which are the major energy sources for animal cells, were seen after infection but an increase in the consumption rates of some amino acids suggests that infected Sf9 cells may utilise amino acids at an enhanced rate for energy post infection.

Introduction

The interest in the insect cel/baculovirus expression vector system (BEVS) has become increasingly widespread during the past decade. The potential industrial applications range from production of bio-insecticides to expression of a wide spectrum of high valued recombinant proteins (Luckow *et al.*, 1988; Goosen *et al.*, 1991). A variety of engineering and biological considerations for optimising the overall process have been studied extensively and several attempts at large scale production using the BEVS have already been successfully implemented (Maiorella *et al.*, 1988; Weiss *et al.*, 1990; King *et al.*, 1992) To fully optimise the process for large-scale production, much information on the characteristics of these cultures are still required. Oxygen demand by insect cells was reported to have a great influence on the final product titer and subsequently, much effort has been focused on the sensitivity of insect cells to shear stress in agitated and sparged environments, which is inevitable when considering oxygenation in the large scale cultivation of insect cells (Murhammer *et al.*, 1988). Recent investigations on the relationship between the oxygen uptake rate and cellular activity of mammalian cells, such as variation of cell density, cell cycle, and the available energy source levels, is beginning to reveal that an oxygen uptake rate measurement may be a promising candidate for providing an insight into the metabolic state of the cells (Ramirez *et al.*, 1990; Yamada *et al.*, 1990; Worhlpart *et al.*, 1990). Similar behaviour in insect cells is therefore expected. Jain *et al.*, (1991) reported that the specific oxygen uptake rates of Sf9 cells varied with the level of dissolved oxygen. Although several researchers reported observations of an increase in the specific oxygen uptake rate after cells were infected with a baculovirus (Weiss *et al.*, 1982; Maiorella *et al.*, 1988; Schopf *et al.*, 1990), detailed studies and comparisons on the oxygen consumption rate of uninfected and infected insect cells and its relationship with cellular activity before and after infection is limited. Also, no study has been attempted to investigate the effect that different times of infection may have on the oxygen uptake rate of infected cells. Nielsen *et al.* (1993a, b) recently identified the potential of an on-line indirect oxygen uptake rate measurement for monitoring viable cell density levels and as a diagnostic tool for both insect cell/BEVS and hybridoma cultures. The studies suggested that the volumetric oxygen uptake rate could be used to identify the time of infection (TOI), which was required to give an optimal yield and is known to occur around the mid-exponential phase (Caron *et al.*, 1990; Radford *et al.*, 1992). The present article addresses the change in patterns of oxygen uptake rates of Sf9 cultures before and after infections carried out at various TOIs. Further potential applications of monitoring the oxygen uptake rate is also suggested. On the basis of the information obtained in this study, an initial attempt has been made to identify the relationship between the oxygen uptake rate of an infected culture and its substrate utilisation rates.

Materials and methods

Cell line and maintenance

Spodoptera frugiperda Sf9 insect cells were obtained from the American Type Culture Collection (ATCC no. CRL-1711), and were maintained in SF900II serum-free medium (Gibco, Grand Island, NY) at 27 °C in 250 ml glass Erlenmeyer flasks (Schott) shaking at 120 rpm on an orbital shaker. The cells were passaged to a density of 4 $\times$ 10^5 cells ml^{-1} when the cell density reached 4-5 $\times$ 10^6 cells ml^{-1}. Cell density was determined using a haemocytometer and the viability was assessed by trypan blue (0.2%) exclusion.

Virus stock, assay and infection procedures

A recombinant baculovirus (β-gal – AcNPV) was obtained from Amrad (Melbourne, Australia). The β-gal – AcNPV was constructed by homologous recombinantion using the E2 strain wild type virus and the pAc360 plasmid containing the β-galactosidase gene (Summers *et al.*, 1987). A stock of this virus at passage number seven was replaque purified and subsequently amplified via 3 passages only, to yield the virus stock used for all experiments in this report. Virus infectivity titer was determined on sample supernatants using a modified endpoint dilution method (Nielsen *et al.*, 1992).

Infection of bioreactor cultures was performed at a multiplicity of infection (MOI) of 5–8 plaque forming units per cell. Such high MOIs ensured synchronous infection in all experiments. The volume of virus added at the time of infection (TOI) did not exceed 5% of the total culture volume on any occasion and so any effect on nutrient concentrations from the added virus was minimal.

β-galactosidase (β-gal) assay

β-gal concentrations were determined using the o-nitrophenyl-galactopyranoside (ONPG) assay (Miller, 1972). To 950μl of Z buffer (0.06 M Na_2HPO_4, 0.04 M NaH_2PO_4, 0.01 M KCl, 0.001 M $MgSO_4$ and 0.05 M β-Mercaptoethanol) were added 50 μl of diluted sample and 200 μl of ONPG (4 mg ml^{-1}). After incubation for 30 minutes at 37 °, 500 μl of 1 M Na_2CO_3 was added to stop the reaction. The absorbance at 420 nm was measured on a double beam spectrophotometer. Concentration of β-gal in Units ml^{-1} was determined by using an extinction coefficient of 4500 l $mole^{-1}$ cm^{-1}. Total samples were prepared by placing samples, still containing cells, at 4 °C. These samples represented the total activity and were measured after the cells were lysed in deionised water as described by Radford *et al.* (1992).

Bioreactor

A Setric Genie 2C bioreactor (Setric Genie Industriel, Toulouse, France) operated with a 900–1100 ml working volume was used in all experiments and the temperature was controlled at 27 °C. The dissolved oxygen tension (DOT) was monitored by a polarographic electrode (Ingold) and maintained at 50% saturation by an oxygen controller (Realtime Engineering, Sydney, Australia). The impeller used for agitation was 7 cm in diameter with 4 blades pitched at 570 to the horizontal, each blade being 7 by 1.8 cm in dimension. The agitation speed was maintained at 140 RPM. This speed provided sufficient oxygen transfer to maintain the desired dissolved oxygen tension (DOT) via headspace aeration, which used a controlled stream of an air-oxygen mixture. The oxygen controller was interfaced to a computer fitted with a prototype data acquisition card and an analog-to-digital converter. Data was acquired, displayed and saved by a driver program. pH was not controlled but was recorded externally. It was observed that pH variation during the course of each experiment was less than 5% of the original value.

Measurement of mass transfer coefficient (Kla′)

The Kla′ of the surface of the culture medium, SF900II, was determined by the static gassing-out method at 140 RPM, using the mass balance equation:

$$\frac{dy}{dt} = Kla'(y_m^* - y) \tag{1}$$

where y is the DOT (% air saturation), t is time (h) and $y_m^* = 1$, which is the DOT of the medium which is in equilibrium with the bulk gas phase (air). Oxygen concentration in the medium was lowered by gassing it with nitrogen, and the increase in DOT was monitored following the start of aeration with air at 0.5 – 0.8 l min^{-1}. Kla′ was evaluated from the slope of a plot of ln(1 – y) *vs* t (yielded from integration of Equation 1) using linear regression. Due to the variation of the culture volume resulting from the withdrawal of samples, Kla′ was not constant throughout the experiment. An empirical equation was therefore developed for Kla′ by taking into account the effect of the changing culture volume, V, dm^3 (Nielsen, 1993a)

$$Kla' = \frac{Kla^*}{V}\exp(-\alpha(V - 1)) \tag{2}$$

where Kla* and α were fitted by non-linear regression and calculated to be 2.50 h^{-1} dm^3 and 1.57 dm $^{-3}$ respectively.

Measurement of volumetric oxygen uptake rate (vOUR)

The DOT dynamics in the culture medium in the presence of cells is given by:

$$\frac{dy}{dt} = Kla'(y_m^* - y) - \frac{vOUR}{C_0} \tag{3}$$

where C_0 is the solubility of oxygen in SF900II, which was assumed to equal the solubility of oxygen in water at 27 °C (0.256 mM, Schumpe *et al.*, 1982). vOUR is the volumetric oxygen uptake rate in mmol h^{-1}. The specific oxygen uptake rate of cells (sOUR), mmol (10^6 cells)$^{-1}$ h^{-1}, was calculated by dividing the vOUR at any given time by the viable cell density at that time.

One of the most widely documented methods for determining the vOUR in an actively respiring culture is the dynamic gassing out method, which assumes that Kla′ is negligible when aeration is terminated. vOUR can then be calculated directly from Equation (3). In the present system, Kla′ cannot be assumed negligible because of the relatively low culture volume (0.9–1.1 L), low viable cell density achieved by insect cells and the high stirring rate used. Therefore, a method that utilises the whole form of Equation (3) was adopted and described as follows.

Integrating Equation (3) from t = 0 to t yields the following equation:

$$y = \left(1 - \frac{vOUR}{C_0}\right)\left(1 - e^{-Kla't}\right) + y_0 e^{-Kla't} \tag{4}$$

where y_0 is the DOT at t = 0. This equation applies only when the headspace of the bioreactor is fully calibrated with air, which makes y_m^* known and equal to 1 at t_0. By using non-linear regression, the vOUR was estimated from a fit of the DOT and time data collected during a measurement. This was achieved by disconnecting the controlled headspace aeration, and in its place air at 0.5 l min^{-1} or 0.8 l min^{-1} was sparged into the headspace. Data logging began when the headspace was equilibrated with air, which was assumed to occur after three headspace volumes (3 L) had been delivered. All measurements were terminated at a DOT not lower that 15% or more than 80%. Within this range, no apparent change in cell growth and productivity has been observed in our laboratory (data not shown). OUR measurements were taken in 6–12 hour intervals during each experiment.

Carbohydrate and amino acid determinations

Glucose, sucrose, maltose and lactate concentrations were analysed by HPLC. Supernatant samples were deproteinated using ultrafiltration membranes (10,000 MW cut-off, Millipore, Bedford, MA) and glucose concentration was obtained by separtation on a Shodex cation-exchange column in Na^+ form. Deproteinated samples were also incubated with 6 N HCl at 60 °C for 15 minutes, which hydrolyses sucrose completely to glucose and fructose. The samples were then separated on a HPX-87 H^+ column (Bio-Rad, Richmond,CA), which gives peaks of glucose, fructose, maltose and lactate. The difference between the glucose concentration from this column and that of the Na^+ column represents the amount of glucose converted from sucrose. Sucrose concentration could then be calculated stoichiometrically from the concentrations of sucrose-converted glucose and fructose. Amino acid concentrations were analysed by HPLC using the Pico-tag system (Waters, Millipore) as described by Reid *et al.* (1987). The overall specific rates of substrate utilisation, q_s (mmol $(10^9$ cells$)^{-1}$ h^{-1}), were determined using

$$\frac{dS}{dt} = q_s X_v \qquad (5)$$

where S is the substrate concentration (mM) and X_v,is the viable cell density (10^6 cells ml^{-1} During the exponential growth phase, a plot of the substrate concentration versus viable cell density, yields a slope of q_s/μ_{app}. The apparent specific growth rate, μ_{app} (h - 1) was determined from linear regression of the logarithmic increase in viable cells with respect to time. As cell growth was largely but not completely halted after infection, the post infection q_s was expressed as the mean q_s value calculated from four sampling time points during the first 48 hours post infection. For each sample,

$$\text{Post infection } q_s(t) = \frac{S_t - S_{t'}}{X_{v'}(t - t')} \qquad (6)$$

where t is the time when the sample was collected (hours post infection), t′ is the sampling time previous to t, S_t and $S_{t'}$ are substrate concentrations (mM) at t and t′ respectively and $X_{v'}$ is the average cell number (10^9 cells) between t and t′.

Results and discussion

Validity of oxygen uptake rate (OUR) measurements as an indicator of cell density and cellular activity

In Fig. 1, the results of two typical uninfected batch suspension cultures of Sf9 cells are shown. The average specific growth rate was 0.031 h^{-1} during exponential growth. The volumetric oxygen uptake rate (vOUR) steadily increased from inoculation until the onset of the stationary phase. The vOUR began to decrease shortly after the maximum cell density had been reached. The consistency of the results during the exponential growth phase for the two cultures demonstrates that when the culture environments are similar, both cell growth and their corresponding oxygen uptake rates are reproducible between runs. From the results of run 2 (■), it can be seen that the vOUR began to decline rapidly after reaching the maximum value, which was prior to the onset of the cell death phase, or drop in viable cell numbers. Therefore, the specific oxygen uptake rate (sOUR) of the cells began to decline at the start of the stationary phase while the viable cell density remained largely unchanged, during the initial stages of the vOUR decline. Hybridomas were found to show similar behaviour, (Ramirez *et al.*, 1990; Yamada *et al.*, 1990; Worhlpart *et al.*, 1990). The profile of vOUR during the growth phase for run 1 (●) was comparable but it decreased somewhat at a slower rate than run 2 after the maximum value had been reached. For this run the vOUR decline coincided with the time at which viability of the cells began to fall. This resulted in a rise of the calculated sOUR during the death phase which contradicted the pattern seen in run 2. The higher values of the sOUR were likely caused by inaccuracies of cell viability assessments during the death phase, rather than a higher cell energy demand being met during the death phase. As frequently discussed by others (Jain *et al.*, 1991) and clearly observed in our laboratory, cells in late stationary phase and infected cells appear to be stained by trypan blue more readily than uninfected cells due to the loss of integrity of the plasma membrane. This often results in under-estimating the 'true' cell density, contradicting other measurements such as oxygen

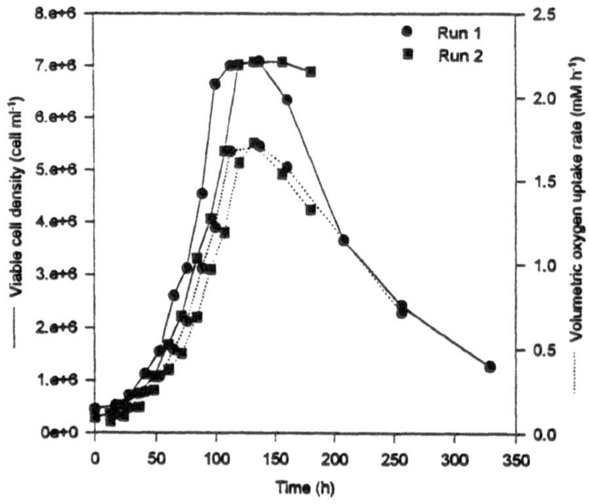

Fig. 1. Relationship between the volumetric oxygen uptake rate (vOUR) and the viable cell density in two uninfected batch cultures of Sf9 cells.

uptake rates and intracellular product release rates via cell lysis. Therefore, the proportionality between cell density and vOURs can only be relied on until the onset of the stationary phase or before the cell culture viability begins to decline, beyond which the inaccuracy of cell viabilities determined via trypan blue staining can result in erroneous conclusions on cellular activity in terms of sOURs.

The average sOUR of uninfected Sf9 cells, calculated from the slope of a linear fit of the vOUR on cell density during the exponential growth phase for both runs, was 0.22 mmol $(10^9\ \text{cell})^{-1}\ h^{-1}$. This was comparable to the sOUR of 0.19 mmol $(10^9\ \text{cell})^{-1}\ h^{-1}$ reported by King *et al.* (1992) for exponentially growing and uninfected Sf9 cells in SF900 serum-free medium. Maiorella *et al.* (1988) reported a value of 0.16 mmol $(10^9\ \text{cell})^{-1}\ h^{-1}$ for exponentially growing and uninfected Sf9 cells in IPL-41 medium supplemented with 10% fetal bovine serum. A lower sOUR for cells cultured in serum containing medium was also observed by Reuveny *et al.* (1992). Their study shows that for cells growing in a modified version of IPL-41 supplemented with 10% fetal calf serum, the sOUR was 0.22 mmol $(10^9\ \text{cell})^{-1}\ h^{-1}$ and cells growing in a serum free medium, ICSF-WB, had a value of 0.39 mmol $(10^9\ \text{cell})^{-1}\ h^{-1}$.

The reproducibility of the vOUR and sOUR measurements indicate that these can be readily used for monitoring insect cell cultures and their subsequent infection with a baculovirus. The vOUR is a good indicator for cell density and culture conditions whereas the sOUR can be used to reflect the metabolic activity of infected or uninfected cells, as long as viabilities are sufficiently high (> 95%) to prevent misleading results obtained from partial staining of viable cells.

Effects of time of infection (TOI) on OUR

To investigate the effects of infection on OURs at different TOIs, four separate batch infections at TOIs of 1.6×10^6, 3.0×10^6, 4.2×10^6 and 5.1×10^6 cells ml^{-1} were performed in a 2 L bioreactor (0.9–1.1 L working volume). Figs. 2a and 2b show the production time courses for extracellular virus and total β-gal of the four infections. Results are expressed on a per cell basis, which were calculated by dividing the volumetric concentration at all time points by the maximum viable cell count recorded in each of the infections. It can be seen that the cells infected at 1.6×10^6, 3.0×10^6 and 4.2×10^6 cells ml^{-1} yielded high levels of extracellular virus and β-gal whereas production

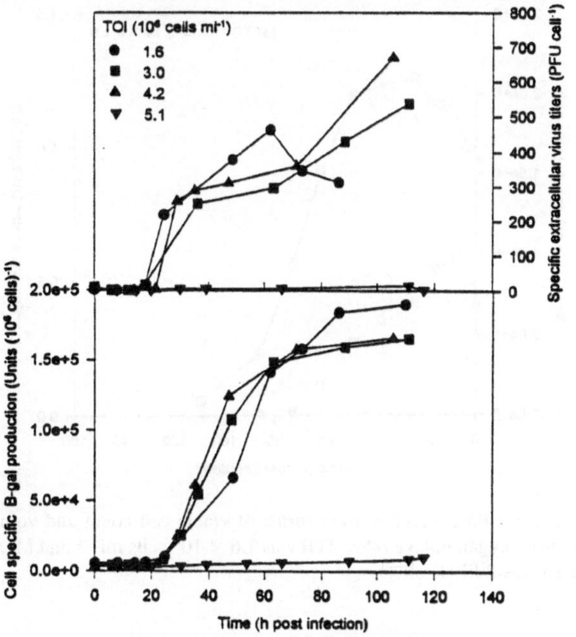

Fig. 2. Post infection trajectories of the specific productions of extracellular pAc-360 β-gal infections performed at 1.6×10^6, 3.0×10^6 4.2×10^6 and 5.1×10^6 cells ml^{-1}. MOI used was 5–8 PFU $cell^{-1}$. β-gal concentrations represent total (intracellular and extracellular) activities as outlined in the Materials and Methods.

from the cells infected at 5.1×10^6 cells ml^{-1} were negligible. This is in agreement with results observed by others (Wood *et al.*, 1982; Maiorella *et al.*, 1988; Caron *et al.*, 1990; Radford *et al.*, 1992), whereby significant reductions in recombinant protein and virus production levels were found in cultures infected during the late exponential and stationary phases. The specific total β-gal productivity of the three high yielding infections was on average 170,000 units $(10^6\ \text{cells})^{-1}$, which is consistent with most published data on Sf9 batch cultures infected with a recombinant baculovirus expressing β-gal (Weiss *et al.*, 1990; King *et al.*, 1992; Power, 1993; Reuveny *et al.*, 1993a). The volumetric productivity of β-gal (Units ml^{-1}) for the infections at TOIs of 3.0×10^6 and 4.2×10^6 cells ml^{-1} are relatively high when compared with most published data to date which is mainly due to the capability of SF900II to support high cell density infections in suspension cultures. As the purpose of this study was to investigate the OUR characteristics of infected Sf9 cells under operating conditions normally employed in batch cultures, no attempt was made to restore the maximum cell specific yield of high cell density infec-

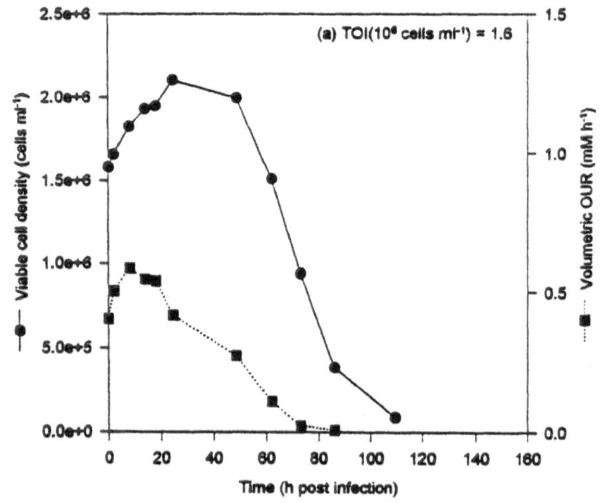

Fig. 3a. Post infection trajectories of viable cell count and volumetric oxygen uptake rates. TOI was 1.6×10^6 cells ml^{-1} and MOI used was 8 PFU $cell^{-1}$.

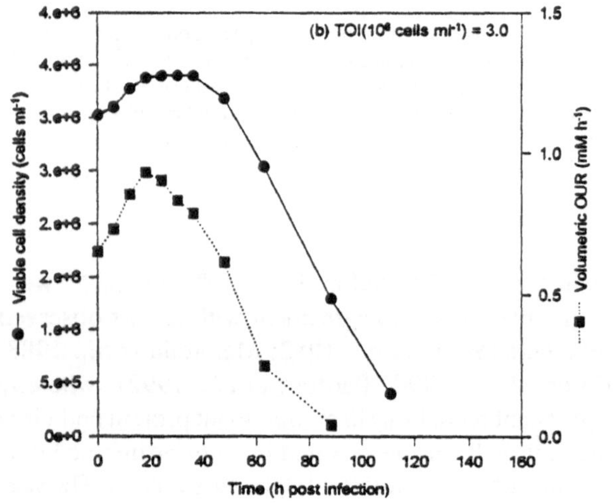

Fig. 3b. Post infection trajectories of viable cell count and volumetric oxygen uptake rates. TOI was 3.0×10^6 cells ml^{-1} and MOI used was 5 PFU $cell^{-1}$.

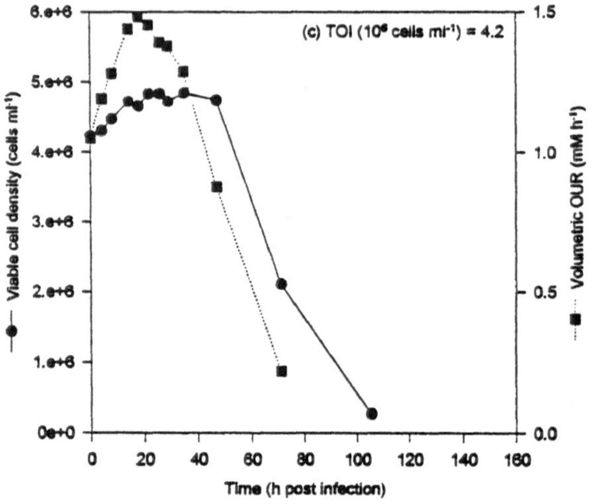

Fig. 3c. Post infection trajectories of viable cell count and volumetric oxygen uptake rates. TOI was 4.2×10^6 cells ml^{-1} and MOI used was 8 PFU $cell^{-1}$.

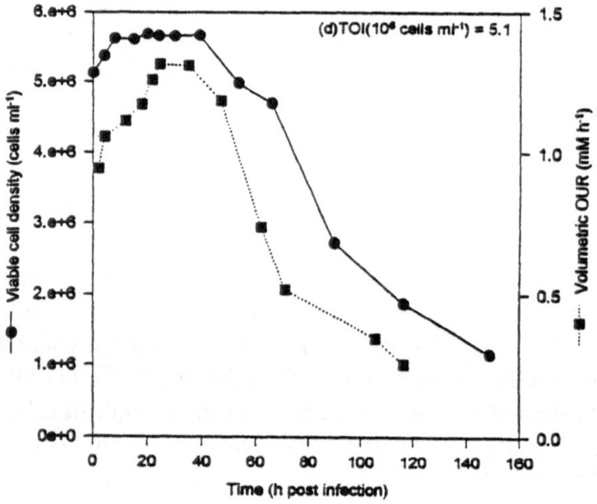

Fig. 3d. Post infection trajectories of viable cell count and volumetric oxygen uptake rates. TOI was 5.1×10^6 cells ml^{-1} and MOI used was 8 PFU $cell^{-1}$.

tions by replenishment of spent media, following cell centrifugation or perfusion processes.

Figs. 3a–3d show the trajectories of viable cell count and the corresponding vOUR profiles for the four infections. Infected cells continued to grow slightly until 20–24 hours post infection, cell density then remained stationary for a further 24 hours. The figures show that vOURs in all of the four infections increased immediately after addition of virus. The initial elevated respiration rate could be due to the total energy demands placed on the cells which continued to grow for the first 20 hours post infection coupled with the penetration events of the virus particles. Virus particles enter the cell via an endocytotic pathway into the cytoplasm, which requires ATP to lower the pH in the endosomes during fusion between the viral envelope and the inner membrane of the endosome (Marsh *et al.*, 1989; Charlton *et al.*, 1993). For the three high yielding infections, the vOUR passed through a maximum at approximately 8–18 hours post infection, and the vOUR peak values observed were 0.58, 0.93 and 1.47 mM h^{-1} respectively for the infections at 1.6, 3.0

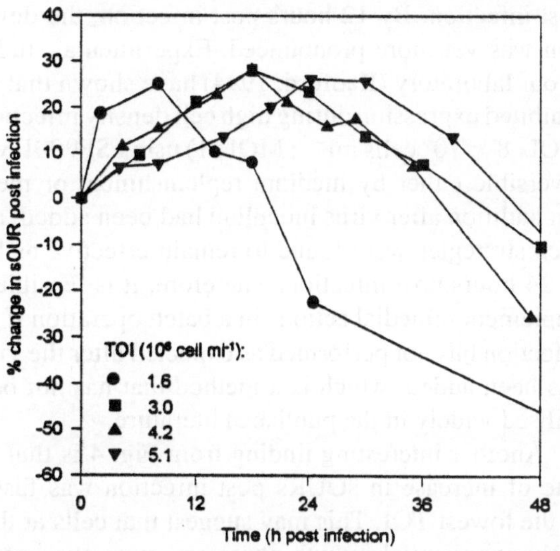

Fig. 4. Percentage change in specific oxygen uptake rate (sOUR) form 0 to 48 hours post infection, relative to the sOUR determined from the same culture, just prior to addition of virus. Cultures were infected at 1.6×10^6, 3.0×10^6, 4.2×10^6 and 5.1×10^6 cells ml^{-1}. MOI used was 5–8 PFU $cell^{-1}$.

and 4.2×10^6 cells ml^{-1} In all cases, the maxima was followed by a rapid decline to levels which indicated that most of the cells were lysed by 80–90 hours post infection. This reflects the success of the lethal events that took place in the cells as a result of viral replication. In contrast, the vOUR for the infection at 5.1×10^6 cells ml^{-1} increased only gradually and it did not reach its maximum (1.31 mM h^{-1}) until 25–30 hours post infection. Although the vOUR declined immediately after reaching its maximum, the rate of decrease was slower than for the other high yielding infections and it began to level off at 70 hours post infection. Both viable cell count data and vOURs decreased from this point but still remained at significant levels at 120–140 hours post infection. This suggests that a portion of this infected cell population did not reach the rapid lytic state that is normally inflicted by a baculovirus infection.

For all of the runs, viability began to decline from approximately 48 hours post infection. As mentioned above, sOUR calculations become unreliable when cell viabilities drop below a certain level as a result of reduced membrane integrity. Therefore, only the first 48 hours post infection allowed detailed analysis of the sOUR characteristics of infected cells. Fig. 4 illustrates the changes in sOUR in the four cultures immediately after infection as a percentage of the reference value determined just before the addition of virus in each culture. As described in the previous section (Fig. 1), the average sOUR of uninfected Sf9 cells during the exponential growth phase was found to be constant at 0.22 mmol $(10^9 \text{ cell})^{-1} h^{-1}$, which was consistant with the sOUR value recorded for each of the four infections just prior to addition of virus.

The profiles of sOUR for all of the infections were similar to those plotted for the vOUR in Fig. 3. A maximum of 30% rise in sOUR in all of the infections were observed, which demonstrates that the extent of alteration in cellular metabolism initiated by baculovirus infection was the same in all of the experiments, regardless of their cell densities at which infection was initiated and their subsequent recombinant protein and extracellular virus productivities. For the high yielding infections, the sOUR peaked at 8 hours post infection (TOI: 1.6×10^6 cell ml^{-1}) and 18 hours post infection (TOI: 3.0 and 4.2×10^6 cells ml^{-1}) which coincides with the period at which most viral replication and the majority of DNA synthesis takes place (Ooi *et al.*, 1988). This period was also characterised by the commencement of progeny virus budding through the plasma membrane of the infected cells (Fig. 2) and the termination of host protein synthesis (Ooi *et al.*, 1988).

The observation of an elevated sOUR shortly after infection is generally in agreement with the published data (Street *et al.*, 1978; Schopf *et al.*, 1990; Reuveny *et al.*, 1993a, b), which were obtained from cells infected during their early to mid-exponential growth phase at MOIs of 1–10. The sOUR for the low yielding infection reached its maximum at 25–30 hours post infection. This delay was a result of the relatively slow increase in sOUR during the early period of infection, indicating that the rate of some early infection processes were somewhat reduced when the TOI was high. The observed decline in the efficiency of early metabolic alteration through viral infection indicates that the processes of virus internalisation and/or replication were possibly inhibited at high cell densities.

A wide range of explanations have been given to this phenomenon of inhibition of infection/production at high cell densities by serval researchers. Lynn *et al.* (1976) suggested that the efficiency of virus attachment and/or penetration of cell membranes might be higher during the S phase and Wood *et al.* (1982) suggested that infected cells must pass through a certain cell cycle stage before viral DNA replication can be initiated. Stockdale *et al.* (1977) also hypothesised that high cell density infections induced the synthesis of an intracellular metabolite which inhibits virus replication. Some early processes might also be inhibited because of depletion of certain nutrients as it was reported that, the presence of cholesterol in the target membrane of vertebrate cells appears to be a requirement for the attachment of enveloped viruses to these cells (Catrasco, 1987). In our laboratory, exposure of Sf9 cells to a cholesterol deficient media also resulted in a significant reduction of baculovirus infectivity in exponentially growing cultures (data not shown). Although the reason for the inhibition of infection at high cell densities remains unclear, recent reports on the reversibility of product expression using media replenishment (Wood *et al.*, 1982; Caron *et al.*, 1990; Ruvenry *et al.*, 1993b; Radford *et al.*, 1994), and the modelling of the baculovirus expression system (Power, 1993) seem to suggest strongly that nutrient depletion plays a role in the observed inhibition.

The potential of using OURs as an indicator for baculovirus infection kinetics in a batch culture is clearly demonstrated here. Batch operations are still commonly used for commercial products such as viral vaccines due to their flexibility in both operational and economical benefits. Inspection of an infected culture by sampling viable cell counts alone, which is routinely performed in many laboratories, would not distinguish a high yielding and desirable infection from a low yielding infection until detection of extracellular virus and recombinant protein is due to begin at approximately 24 hours post infection (Fig. 2). This could be evidenced by the relatively similar profiles of viable cell counts post infection in Figures 3a–d. This is especially true when one is to optimise the production in a batch suspension culture by infecting cells late in the exponential growth phase, where production can be dramatically reduced if cells are infected at a cell density that is slightly over the threshold level. If OUR monitoring is performed during the early stages of the infection, it is possible to identify whether the infection will proceed as predicted by comparing the culture OUR with the expected rate of increase in sOUR. Certain procedures, such as nutrient feeding and partial or complete medium replenishment, which is not normally conducted in a batch operation, can then be undertaken in an attempt to restore the productivity if the rate of increase in sOUR is found to reflect a potentially low yielding infection. Such precautions should be undertaken as early as possible so that the ability of the infected cells to express viral products remains reversible. Fig. 4 indicates that the rise of sOUR of the run at 5.1×10^6 cells ml^{-1} began to differ from that of the run at 4.2×10^6 cells ml^{-1} as early as 6 hours post infection. By 12 hours post infection, the deviation was yet more pronounced. Experimental studies in our laboratory (Radford, 1994) have shown that the inhibited expression during high cell density infections (TOI : 8×10^6 cells ml^{-1}; MOI : 1) using SF900II was reversible either by medium replenishment or medium addition after virus inoculum had been added, and such strategies were found to remain effective as late as 24 hours post infection. Therefore, it is feasible to implement remedial actions in a batch operation if the infection has not performed as expected after the virus has been added, which is a method that has not been utilised widely in the published literature.

Another interesting finding from Fig. 4 is that the rate of increase in sOURs post infection was fastest at the lowest TOI. This may suggest that cells at their early exponential growth phase are more susceptible to virus infection and/or viral replication is more efficient, compared to cells at the mid-exponential phase. Most researchers have recorded a maximum sOUR at 14–18 hours post infection for infections carried out during the mid-exponential phase using MOIs ranging from 1–10 (Street *et al.*, 1977; Schopf *et al.*, 1990; King *et al.*, 1992; Reuveny *et al.*, 1992), which is in accordance with the results observed in this study. However, no measure of OURs for infections during the early exponential phase were found in the literature and thereby a possible relationship between the efficiency of viral infection and cell density has previously been overlooked, in relation to early exponential versus mid-exponential infections.

Relationship between carbohydrate/amino acid consumption and the observed increase in sOUR post infection

The rise in specific oxygen uptake rate immediately after infection and during the early stages of replication may imply that more energy derived from certain nutrients is needed to increase the production of ATP from the TCA cycle/electron transport chain. To determine the relationship between the rise in oxygen uptake rate and consumption of some major energy sources, carbohydrate and amino acid analysis were conducted on supernatant samples from the infection carried out at the TOI of 4.2×10^6 cells ml^{-1}. The samples from this high TOI culture were chosen because the magnitude of changes in the consumption of the nutrients would be more apparent relative to the other infections conducted at lower cell densities. Lactate production was not observed in any of the samples analysed and there-

Table 1. Percentage of nutrients remaining and their specific consumption rates before and after infection at 4.22×10^6 cells ml^{-1}

Substrate	%* remaining at 105 hpi	Specific consumption rates ($\times 10^{-12}$ mmol $cell^{-1}$ h^{-1}) before infection	after infection
Carbohydrate			
Glucose	72	49.9	33.7
Sucrose	100	0.0	0.0
Maltose	0	7.2	9.5
Amino Acid			
Alanine	335	–23.5	–6.2
Arginine	86	2.8	4.5
Asparagine	96	1.6	12.6
Aspartate	87	4.1	1.0
Cystine	64	0.3	2.2
Glycine	92	0.2	3.1
Glutamate	100	2.6	1.9
Glutamine	39	23.9	17.9
Histidine	84	0.7	1.9
Hydroxyproline	100	0.0	0.0
Isoleucine	88	2.8	10.0
Leucine	73	8.3	8.5
Lysine	87	2.2	11.8
Methionine	97	1.7	4.2
Phenylalanine	97	0.7	7.9
Proline	86	1.9	3.5
Serine	48	9.5	9.0
Threonine	70	2.5	3.7
Trytophan	78	1.4	2.9
Tyrsine	67	1.9	1.5
Valine	81	3.4	3.9

* % of levels present in the medium at the start of the culture, not at the start of infection.

fore the contribution of glycolysis to the total energy production was presumably minimal.

Table 1 lists the percentage of the measured nutrients remaining in the media by the end of the culture and their consumption rates before and after infection. It was observed that 70% and 40% of the initial glucose and glutamine levels respectively remained in the media by the end of the culture. Sucrose was not utilised and maltose was exhausted by the end of the culture. There was no apparent exhaustion of any of the measured amino acids. Alanine accumulated in large amounts (330%) during the culture which was also observed by others (Radford *et al.*, 1990; Weiss *et al.*, 1990; Ferrance *et al.*, 1993; Bédard *et al.*, 1993). Ferrance *et al.* (1993) reported that alanine accumulation might be serving as a sink for the ammonia produced from amino acid catabolism. The specific consumption rates of glucose and glutamine, surprisingly did not increase, and indeed they both decreased slightly after infection, but consumption rates of 13 out of the 20 amino acids increased. Maltose consumption also accelerated after infection. In contrast to our data, a rise in the glucose consumption rate post infection was observed by Reuvely *et al.* (1992). The serum-free media employed in this study contained glucose as the only carbohydrate source, however. As there was no indication that glucose and glutamine were responsible for providing the extra energy the infected cells needed as reflected by the 30% increase in the sOUR following infections, it was likely that some amino acids and maltose were meeting this demand, in addition to the amino acids being utilised for viral macromolecules and recombinant protein production. This agrees well with the suggestion reported by Ferrance *et al.* (1993) that uninfected Sf9 cells utilise both glucose and amino acids for energy, and the amino acids that could provide energy via the TCA cycle directly or indirectly are: aspartate, asparagine, glutamate, glutamine, glycine, arginine, serine and threonine. Of these amino acids, the specific consumption rates of asparagine, arginine, glycine and threonine increased significantly after infection (Table 1).

Although the observed increase in amino acid and maltose consumption rates coincided with the 30% increase in the specific oxygen uptake rate after infection, it is unclear why infected Sf9 cells would prefer to generate extra energy via such less efficient routes, when sufficient glucose and glutamine are still available. It is also possible that lipid catabolism is contributing to the energy supply post infection. Lipids are supplied in the SF900II medium but were not analysed in this work.

Conclusions

The periodical measurements of OURs in a laboratory-scale bioreactor in this study have demonstrated the potential of using OURs for monitoring large-scale insect cell cultures to provide on-line information on cell density and cellular activity in baculovirus-infected cells. The OUR results obtained from this study show that baculovirus infection of Sf9 cells initiates an early energy demanding event as indicated by the 30% rise of sOURs in all of the infections. This also occurred in the nonproductive culture at a TOI of 5.1×10^6 cells ml^{-1} which could suggest that the majority of the cells underwent normal, albeit delayed early infection events. It was observed that the pattern of the sOUR rise follows a general trend for infections carried out from early to mid to late exponential phases: ie. the rate of increase in sOUR of the infected cells decreases as the cell density at the TOI increases. Such a trend therefore can be utilised as a guideline for determining the efficiency of the initial kinetics of an infected batch culture, and remedial steps such as medium replenishment and medium addition can then be considered early if an infection appears to be failing. Glucose and glutamine consumption rates did not correlate with the rise in the sOUR after infection but some amino acids were utilised to a larger extent. Therefore, infected and uninfected Sf9 cells may use both carbohydrate and amino acids for energy. As the main concern, of this report has focused on the OUR characteristics of Sf9 cells in batch cultures infected at high MOIs, further investigations will be performed to elucidate the effect of low multiplicities of infection on OURs. Incorporation of a perfusion system will also allow us to further confirm the finding that high cell density infection is achievable by replenishing the spent medium sometime after the virus has been added in a bioreactor system.

Acknowledgments

The authors wish to thank W.P. Abeydeera for excellent technical assistance. The contribution of the Australian Research Council to this work is gratefully acknowledged.

References

Bédard C, Tom R and Kamen A (1993) Growth nutrient consumption, and endproduct accumulation in Sf9 and BTI-EAA insect cell culture: insights into growth limitation and metabolism. Biotechnol. Progr. 9:615–624.

Caron AW, Archambault J and Massie B (1990) High-level recombinant protein production in bioreactors using the baculovirus-insect cell expression system. Biotechnol. Bioeng. 36:1133–1140.

Carrasco L (1987) Mechanisms of virus-induced cell toxicity: a general overview. In: Mechanisms of viral toxicity in animal cells. CRC Press, Inc., Boca Raton, Florida.

Charlton CA and Volkman LE (1993) Penetration of *Autographa californica* nuclear polyhedrosis virus nucleocapsids into IPLB Sf21 cells induces actin cable formation. Virol. 197:245–254.

Ferrance JP, Goel A and Ataai MM (1993) Utilisation of glucose and amino acids in insect cell cultures: quantifying the metabolic flows within the primary pathways and medium development. Biotechnol. Bioeng. 42:697–707.

Goosen MFA (1991) Insect cell cultivation techniques for the production of high-valued products. Can. J. Chem. Eng. 69:450–456.

Jain D, Ramasubramanyan K, Gould S, Seamans S, Wang S, Lenny A and Silberklang M (1991) Production of Antistasin using the Baculovirus Expression System. In: Expression systems and processes for rDNA products, ACS Symposium Series No. 477, ed. Hatch RT, Goochee C, Moreira A and Alroy Y, ACS, Chs, pp. 97–110.

King GA, Daugulis kT, Faulkner P and Goosen MFA (1992) Recombinant β-galactosidase production in serum-free medium by insect cells in a 14-L airlift bioreactor. Biotechnol. Progr. 8:567–571.

Luckow VA and Summers MD (1988) Trends in the development of baculovirus expression vectors. Bio/Technology 6:47–55.

Lynn DE and Hink WF (1976) Replication of alfalfa looper (*Autographa californica*) nuclear polyhedrosis virus in synchronous insect cell cultures. In: Proceedings of the First International Colloquium on Invertebrate Pathology, pp. 395–396. Queen's University, Kingston.

Maiorella B, Inlow D, Shauger A and Harano D (1988) Large-scale insect cell-culture for recombinant protein production. Bio/Technology 6:1406–1410.

Marsh M and Helenius A (1989) Virus entry into animal cells. Advan. Virus Res. 36:107–151.

Miller JH (1972) Assay of β-galactosidase. In: Experiments in Molecular Genetics. (pp. 352–355) Cold Spring Harbor Laboratory, Cold Spring Harbor, N.Y.

Murhammer DW and Goochee CF (1988) Scale-up of insect cell cultures : protective effect of Pluoronic F-68. Bio/Technology 6:1411–1418.

Nielsen LK, Smyth GK and Greenfield PF (1992) Accuracy of the endpoint assay for virus titration. Cytotechnology 8:231–236

Nielsen LK, Wong TKK, Power J, Reid S and Greenfield PF (1993a) Using oxygen uptake rates to time infections in the baculovirus expression vector system. In: Okumura (ed.) Animal Cell Technology : Basic and Applied Aspects. Vol. 6 Kluwer Academic Publishers, Dordrecht. In Press.

Nielsen LK and Greenfield PF (1993b) On-line indirect measurement of oxygen uptake rate in bench scale hybridoma cultures. In: Animal Cell Technology : Products for Today, Prospects for Tomorrow. Butterworth-Heinemann. In Press.

Ooi BG and Miller LK (1988) Regulation of host RNA levels during baculovirus infection. Virol. 166(2):515–523.

Power J (1993) Modelling and Optimisation of the Baculovirus Expression Vector System in Suspension Culture. Ph.D thesis, The University of Queensland, Brisbane, Australia.

Ramirez OT and Mutharasan R (1990) Cell cycle-and growth phase-dependent variations in size distribution, antibody productivity, and oxygen demand in hybridoma cultures. Biotechnol. Bioeng. 36:839–848.

Radford KR, Reid S and Greenfield PF (1992) Improved production of recombinant proteins by the baculovirus expression system using nutrient enriched serum free media, pp. 297–303. In: JM Vlak, E-J Schlaeger, AR Bernard (ed), Baculovins and Recombinant Protein Production Processes. Proceedings of the Baculovirus and Recombinant Protein Production Workshop, March 29-April 1, Interlaken, Switzerland. F Hoffmann-La Roche Ltd, Basel, Switzerland.

Radford KR, (1994) PhD Thesis, The University of Queensland, Brisbane, Australia. Reid S, Randerson DH and Greenfield PF

(1987) Amino acid determination in mammalian cell culture supernatants. Austral. J. Biotechnol. 1:9–72.

Reuveny S, Kemp CW, Eppstein L and Shiloach J (1992) Carbohydrate metabolism in insect cell cultures during cell growth and recombinant protein production, pp. 230–237 In: H Pederson, R Mutharasan and D Dibiaso, (eds), Biochemical engineering VII, vol. 665. N.Y. Academy of Sciences, New York.

Reuveny S, Kim YJ, Kemp CW and Shiloach J (1993a) Effect of temperature and oxygen on cell growth and recombinant protein production in insect cell cultures. Appl. Microbiol. Biotechnol. 38:619–623.

Reuveny S, Kim YJ, Kemp CW and Shiloach J (1993b) Production of recombinant proteins in high-density insect cell cultures. Biotechnol. Bioeng. 42:235–239.

Schopf B, Howaldt MW and Bailey JE (1990) DNA distribution and respiratory activity of *Spodertera frugiperda* populations infected with wild-type and recombinant *Autographa californica* nuclear polyhedrosis virus. J. Biotechnol 15:169–186.

Schumpe A and Quicker G (1982) Gas solubilities in microbial culture media. Advances in Biochemical Engineering 24:1–38.

Stockdale H and Gardiner GR (1977) The influence of the condition of cells and medium on production of polyhedra of *Autographa californica* nuclear polyhedrosis virus *in vitro* Invertebr. Pathol. 30:330–336

Street DA and Hink WF (1977) Oxygen comsumption of *Trichoplusia ni* (TN-386) insect cell line infected with *Autographa californica* Nuclear Polyhedrosis Virus. J. of Invertebr. Pathol. 32:112–113.

Summers MD and Smith GE (1987) A Maunal of Methods for Baculovirus Vectors and Insect Cell Culture Procedure. Texas Agriculture Experimental. Station., Bullutin No. 1555.

Weiss SA, Orr T, Simth GC, Kalter SS, Vaughn JL and Dougherty EM (1982) Quantitative measurement of oxygen consumption in insect cell culture infected with polyhedrosis virus. Biotechnol. Bioeng. 24:1145–1154.

Weiss SA, Gorfien S, Fike R, DiSorbo D, Jayme D (1990) Large scale production of protein using serum free insect cell culblre. Proceedings of the Ninth Australian Biotechnology Conference, Queensland, Australia, pp, 220–231.

Wohlpart D, Kirwan D and Gainer J (1990) Effects of cell density and glucose and glutamine levels on the respiration rates of hybridoma cells. Biotechnol. Bioeng. 36:630–635.

Wood HA, Johnston LB and Burand JP (1982) Inhibition of *Autographa californica* nuclear polyhedrosis virus replication in high-density *Trichoplusia ni* cell cultures. Virol. 119:245–254.

Yamada K, Furushou S, Sugahara T, Shirahata S and Murakami H (1990) Relationship between oxygen consumption rate and cellular activity of mammalian cells cultured in serum-free media. Biotech. Bioeng. 36:759–762.

Address for offprints: Kathy Wong, Department of Chemical Engineering, The University of Queensland, Queensland 4072, Australia.

Cytotechnology **15:** 169–176, 1994.

Optimization of vaccine production for animal health

W. Noe, R. Bux, W. Berthold and W. Werz
Dr. Karl Thomae G.M.B.H., 88397 Biberach, Germany

Key words: Flaviviridae, BDV, fermentation, metabolism, vaccines

Abstract

Vaccines on the basis of mammalian cell cultures are of major importance for human and animal health. Therefore efforts are undertaken for the improved production of more effective vaccines. Of course, the main purpose of all these approaches is to save lives and improve the quality of life for human beings. However, there is also some remarkable effort in the food industry and the associated animal production, especially in the case of some Flaviviridal viruses (BVD), where >80% of all cattle herds are found to be infected. These viruses can cause tremendous economic losses of calfs and embryos (Ames, 1990). Because of these facts, there is a continuous endeavour for improving the manufacturing of therapeutics or preventing agents such as vaccines for the treatment of cattle. The competitive economic situation and the specific market demands still require effective and high yield production methods, especially in the case of one of the most widespread viral diseases in cattle like BVD (Ames, 1990).

We have succeeded in establishing an improved method for the production of BVD on the basis of a continuous fermentation mode, that consist of modifications of the corresponding process and media improvements.

Introduction

The Flaviviridae virus family, and among them especially the BVD virus, is capable of producing a wide array of disease syndromes in cattle. The disease is also suspected as the cause of a wide range of other multisystemic syndromes in herds by its immunosuppressive properties (Schlesinger and Schlesinger, 1990). Thus, while acute infection is primarily expressed as a mild disease in posnatal calfs, the effects can be very severe in the developing fetus. Also, acute BVD infections occasionally develops into a severe disease in postnatal cattle (enteric and respiratory diseases). Therefore the treatment of acute virus infection has become a major issue for animal health world-wide for years (Ames and Baker, 1990).

During the past years the market material of a variety of vaccines has been mainly produced in roller- and Pfizer bottles, respectively (Fig. 1). One of the most important, well established systems for the production of vaccines is the replication of (attenuated) viruses in MDBK (Madin Darby Bovine KLidney) cell cultures (Franke, 1994). On that basis we have developed a production systemn, which can be adapted to a 'multi-harvest' process with enhanced yield and cost savings (Fig. 2).

Material and methods

MDBK cells have been received from ATCC (Nr.: MDBK, NBL-1, CCL 22) and were characterized for production purposes according to 'points to consider' release procedures (Zoon, 1993).

The attenuated Flaviviridal variants have also been checked for points, relevant for manufacturing items. BDV was used as a model for virus production.

The $TCID_{50}$ (Tissue Culture Infectious Dose) measurement for the determination of active virus has been performed according to (Koerber, 1931).

Amino acid determinations have been performed by means of 'OPA'-HPLC methods according to (Pharmacia/LKB, 1987).

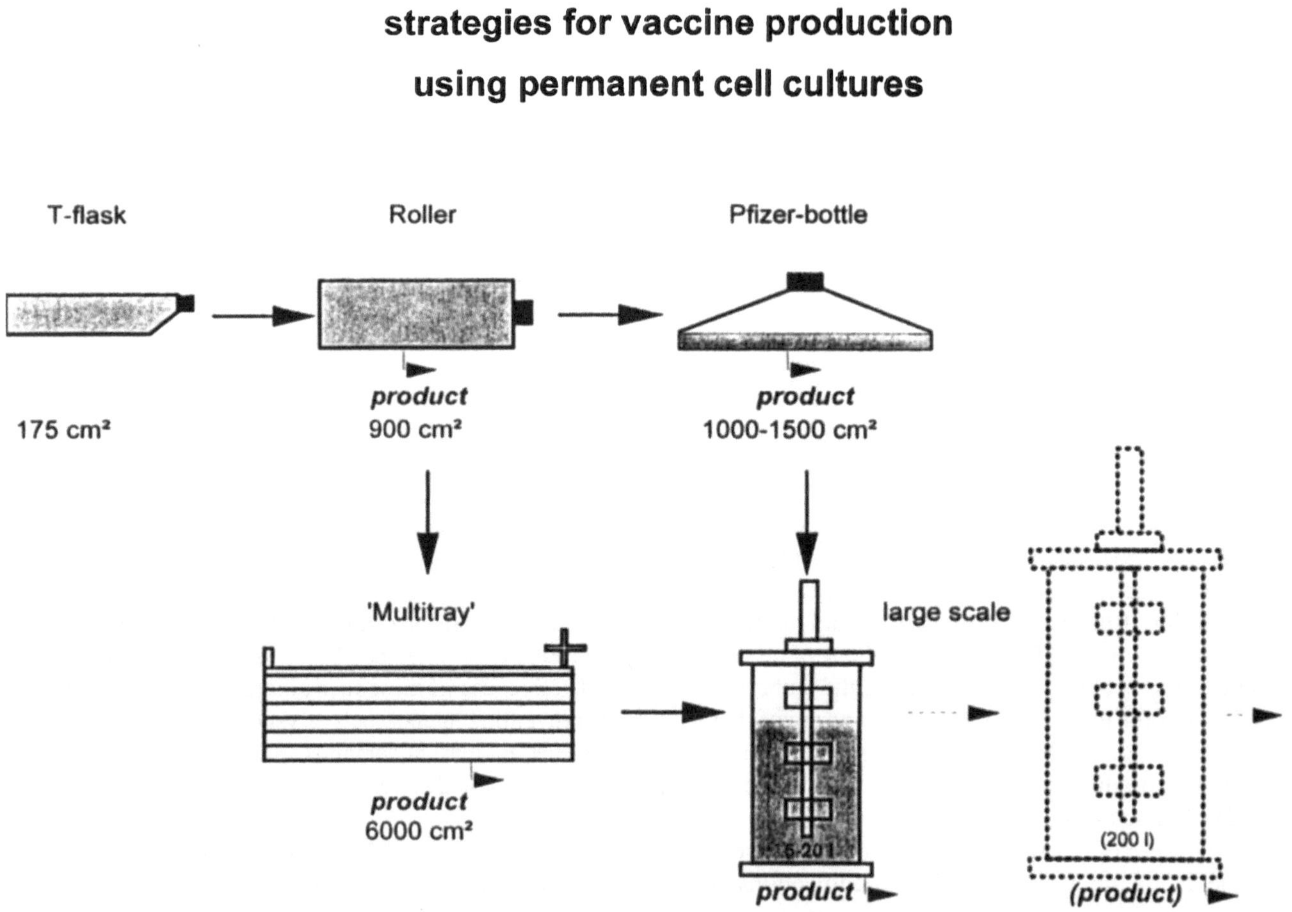

Fig. 1. Principal production strategies for cell culture derived vaccines.

Cell count determinations were performed, after trypsin treatment, according to (Morgan and Darling, 1993), using a Neubauer chamber.

All described experiments have been performed in the air-lift-driven fermentation system on the basis of 'coiled steel wire'-solid bed technology (Fig. 3).

coils:	stainless steel, V_4A, $6 \times 6 \times 0{,}6$ mm
surface area:	1,025 m^2/l settled bed
void volume:	75%, (v/v) calculated on the packed bed basis

Results and discussion

The BVD/MDBK virus replication system is reported to be mostly lytic (Schlesinger and Schlesinger, 1990). From that point of view most virus/vaccine production systems follow the approach of simple batch systems for the harvest of the (virus) product: After the multiplication phase of target cells by means of various methods, the cell culture is infected with the virus. After a variable period of replication the virus product can be harvested (Griffith, 1985). During several optimization studies with the vaccine system we have noticed, that the amino acid metabolism of the MDBK cells shows remarkable differencies when compared to the virus-infected cells (Fig. 4). We therefore modified the basal media composition for MDBK (see Fig. 5). By modulating the frequency of the medium exchange rate (Fig. 6) we could achieve a highly reproducible harvesting schedule. Applying these modifications to the single batch process, a large increase in amount of active virus can be generated from the same seed culture (Fig. 7). Moreover, the 'lytic' process appears to affect much less than 100% of the cells and leave sufficient 'hosts' for further infections. This can be exploit-

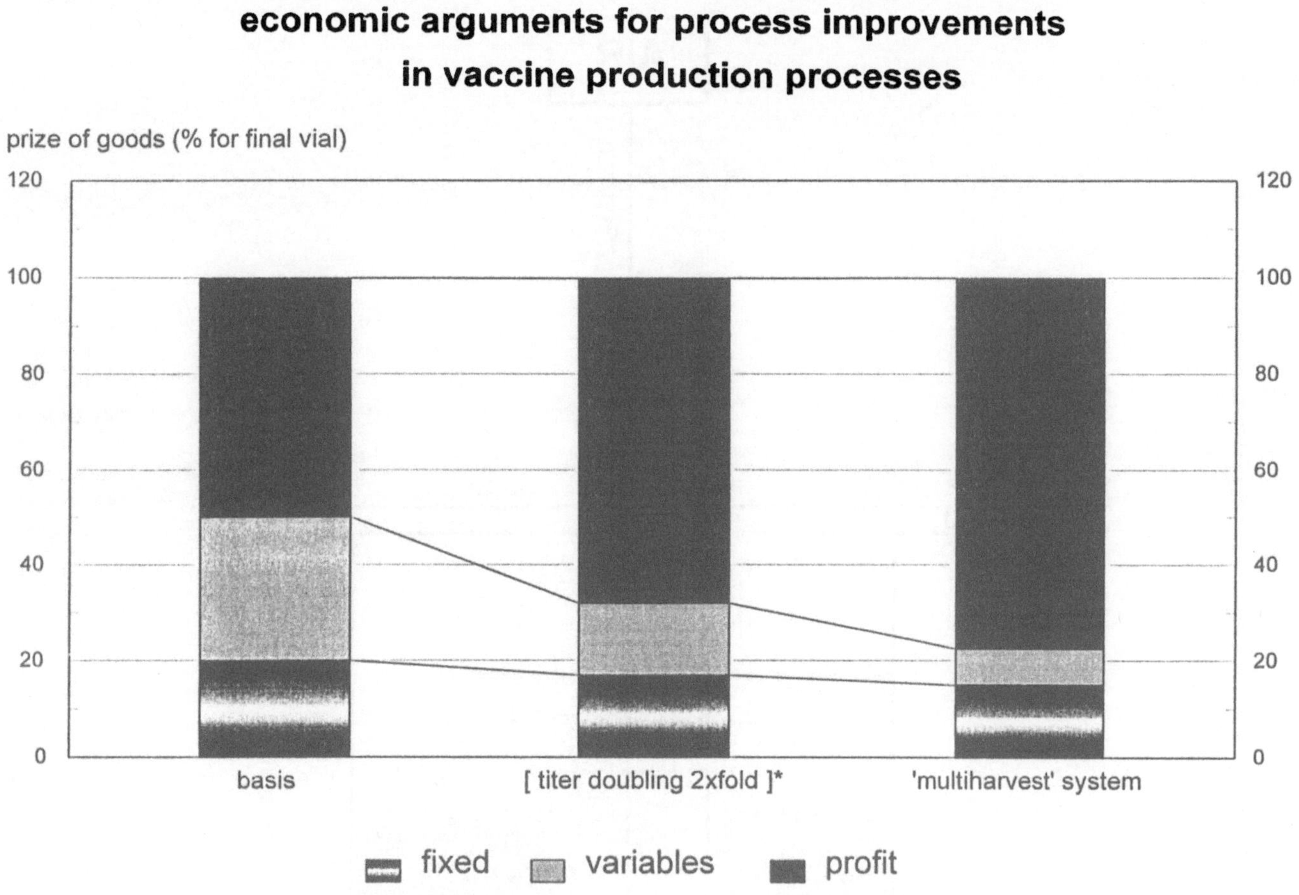

Fig. 2. Effects of various process improvements on the overall product economics.

ed by a repeated harvesting (Fig. 6)! This unexpected behaviour of the virus replicating system suggests that the target cell system can be strengthened by means of supporting optimal 'cellular maintenance' via a modified amino acid composition. Most apparently, the cells are not killed by exhaustion due to virus replication, because improved amino acid supplements compensate the demands of virus metabolism. To resolve these findings, process alternatives will be the objective for further investigations and studies for virus expression with regard to the 'switch' in the target cell metabolism after virus infection.

Conclusion

An air-lift 'solid bed'-system has been scaleds up to 20 l industrial scale and is now well established. It has found approval by the USDA for manufacturing purposes of various vaccines. The product consistency as well as the economic improvements of that system, when compared to traditional manufacturing methods, are clear advantages (see also Figs. 1–3, 6–7).

Starting from a conventional single harvest production method for, e.g. BVD viruses, we have developed a continuous system with improved performance concerning capacity, virus titer, and of raw material consumption. The system will be investigated for the application with other vaccine production methods. The major improvements consist in an adaptation of an (obligatory) lytic virus infection to a continuous virus shedding production system for BVD. We will elucidate these metabolic correlations in terms of common relevance for other 'virus-host'-systems.

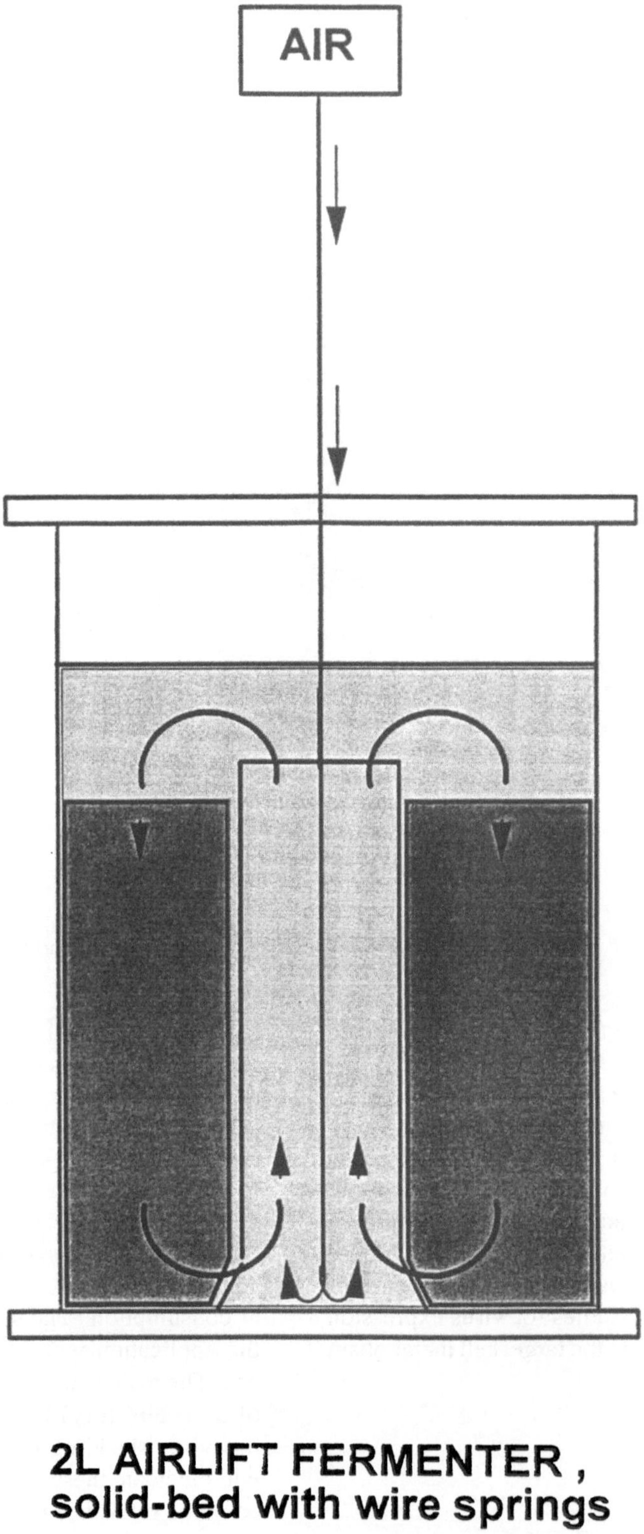

Fig. 3. Schematic overview of a solid-bed fermenter.

SPEC. AMINO ACID - UPTAKE (%/day)
during MDBK GROWTH PHASE / BVD PRODUCTION PHASE

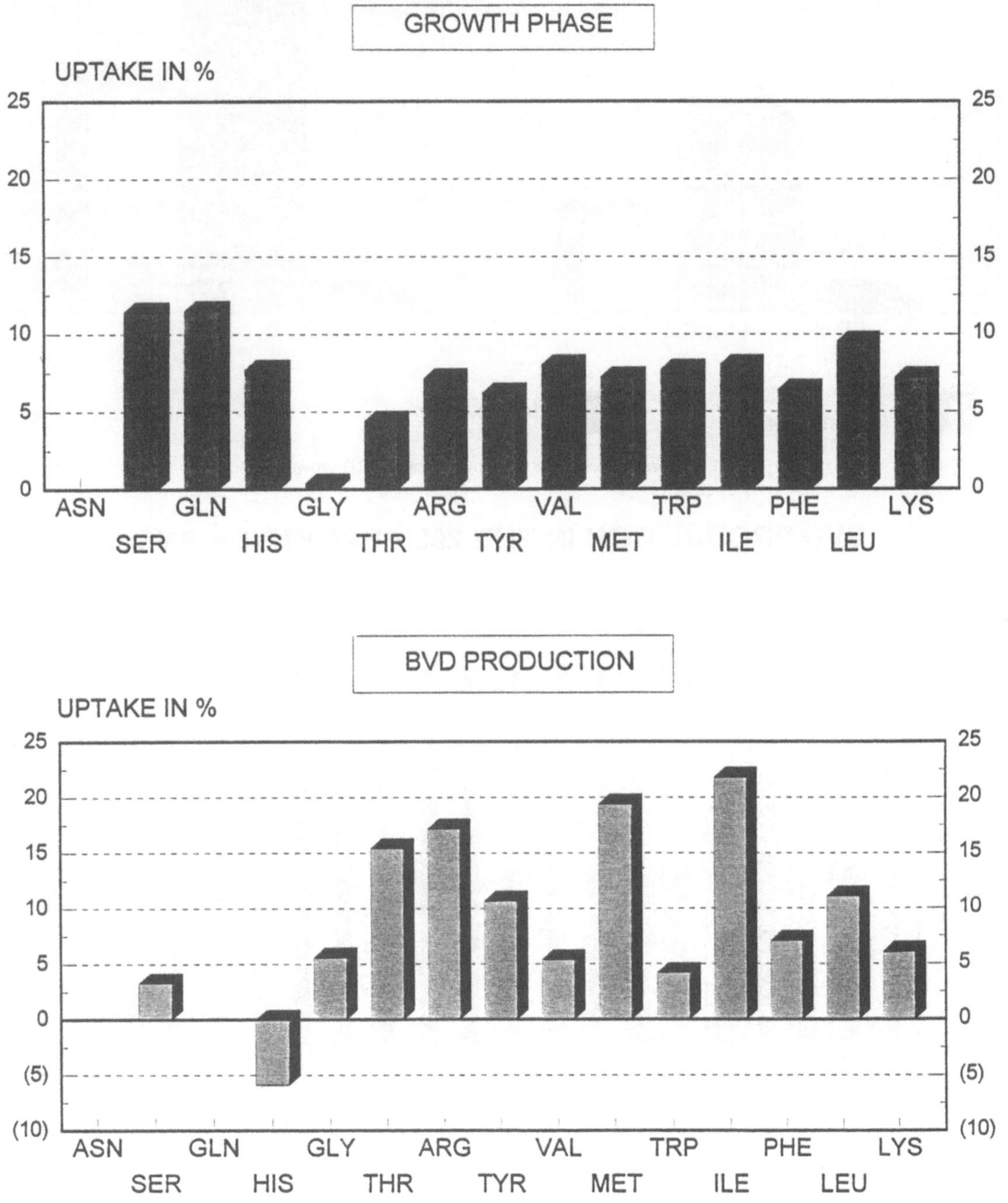

Fig. 4. Comparison of amino acid metabolism in BVD infected MDBK cells and uninfected MDBK.

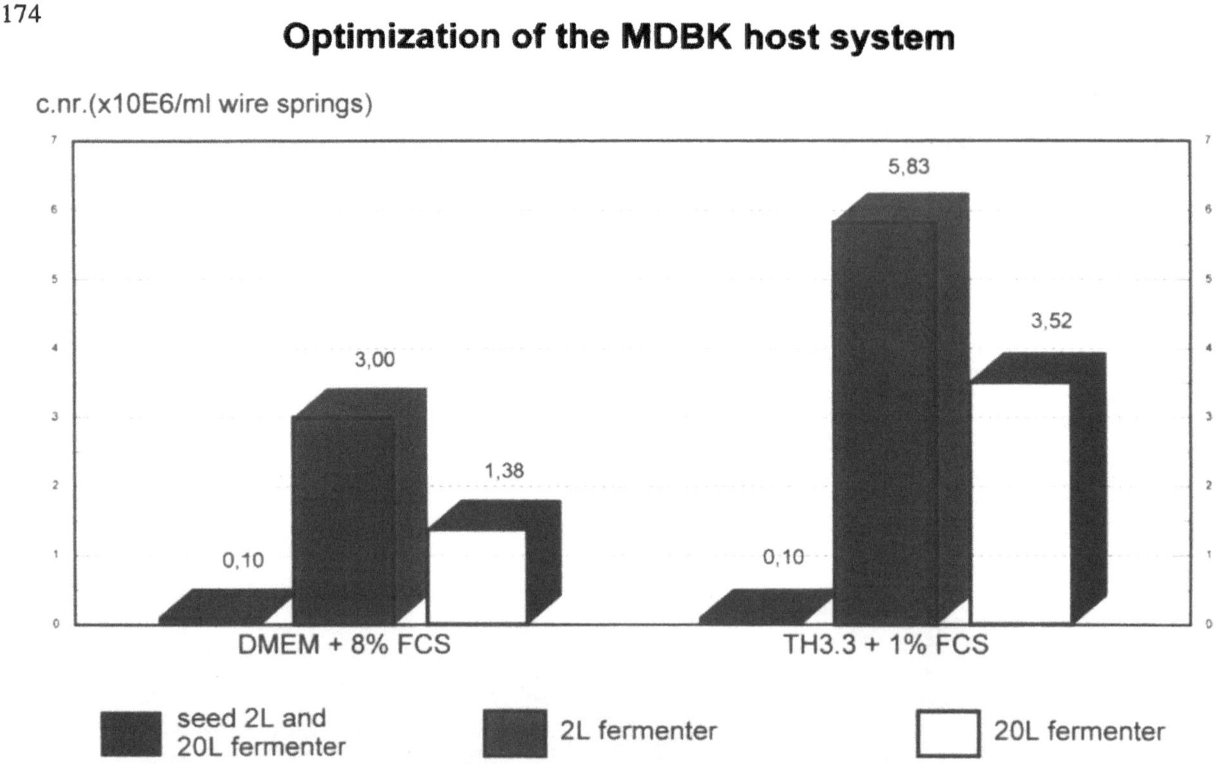

Fig. 5. Optimization of the basal media for MDBK growth phase.

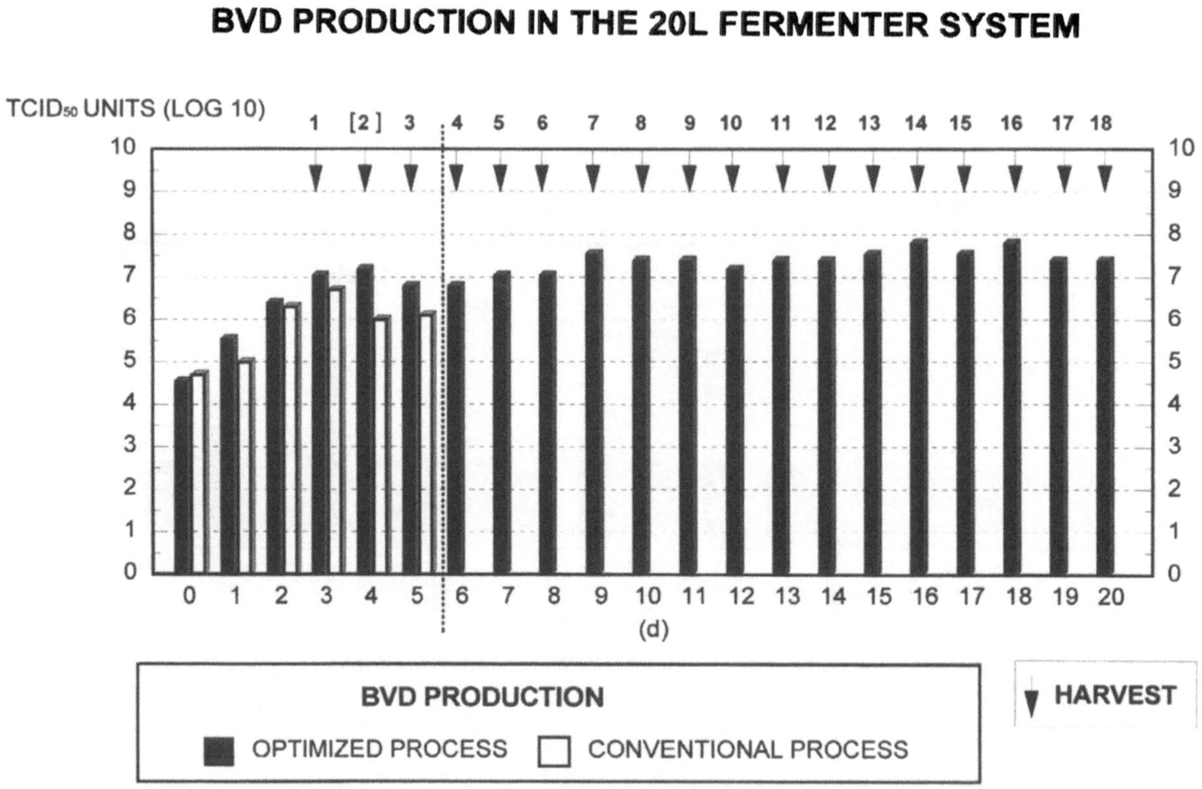

Fig. 6. Comparison of the optimized BVD production scheme with previously applied process mode.

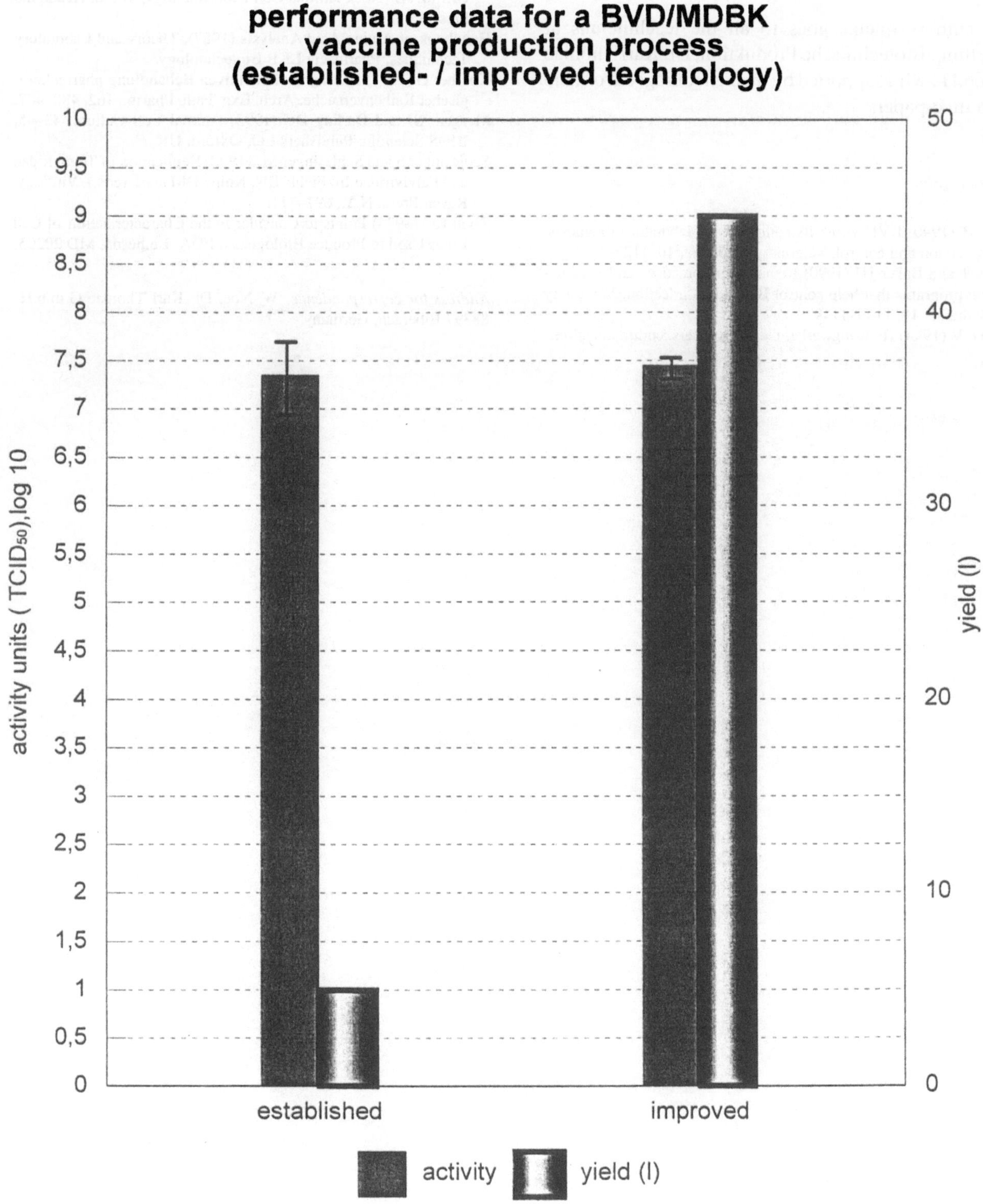

Fig. 7. Volumetric performance of the improved BVD expression system vs. the conventional approach.

Acknowledgements

The authors thanks goes to all the teechnicians at Abteilung Biotechnische Produktion, Dr. Karl Thomae G.m.b.H., who supported by their encouraged work the submitted paper.

References

Ames T (1990) BVD virus; Its pathogenesis, laboratory diagnosis, prevention and control. Veterinary Medicine, 10, 1123.

Ames T and Baker HJ (1990) Management practices and vaccination programs that help control BVD virus infection. Veterinary Medicine, 10, 1140–1149.

Franke V (1994) (Behringwerke, Germany), pers. communication.

Griffith JB (1985) Cell Products, An Overview. In: Spier RE and Griffith JB (eds.), Animal Cell Biotechnology, vol. 2, Academic Press, London.

Handbook of Amino Acid Analysis (1987), Theory and Laboratory Techniques, Pharmacia LKB Biotechnology.

Koerber G (1931) Beitrag zur kollektiven Behandlung pharmakologischer Reihenversuche. Arch. Exp. Path. Pharm., 162, 480–487.

Morgan, SJ and Darling DC (1993) Animal Cell Culture. 43–45, BIOS Scientific Publishers Ltd, Oxford, UK.

Schlesinger S and Schlesinger MJ (1990) Replication of Togaviridae and Falviviridae In: Fields BN, Knipe DM *et al.* (eds.), Virology, Raven Press, N.Y., 697–711.

Zoon KC (1993) Points to Consider in the Characterization of Cell Lines Used to Produce Biologicals. FDA, Bethesda, MD 20205.

Address for correspondence: W. Noe, Dr. Karl Thomae G.m.b.H., 88397 Biberach, Germany.

Cytotechnology **15:** 177–186, 1994.

Evaluation of monitoring approaches and effects of culture conditions on recombinant protein production in baculovirus-infected insect cells

William T. Hensler[1] & Spiros N. Agathos[2]
[1] *Schering Plough Research Institute, 1011 Morris Avenue, Union, NJ 07083, present address: Raytheon Engineers and Constructors, 30 South 17th Street, Philadelphia, PA 19101, U.S.A.;* [2] *Unit of Bioengineering, Catholic University of Louvain, Place Croix du Sud, 2 Bte 19, B-1348 Louvain-la-Neuve, Belgium*

Key words: Baculovirus infection, β-galactosidase, insect cell aggregation, fluorescent microscopy, EXCELL 401 serum-free medium, *Spodoptera frugiperda* (Sf9) insect cells

Abstract

The baculovirus infection process of *Spodoptera frugiperda* (Sf9) insect cells in oxygen-controlled bioreactors in serum-free medium was investigated using a recombinant *Autographa californica* (AcNPV) virus expressing β-galactosidase enzyme as a model system. A variety of monitoring techniques including trypan blue exclusion, fluorescent dye staining, oxygen uptake rate (OUR) measurements, and glucose consumption were applied to infected cells to determine the best way of evaluating cell integrity and assessing the course of baculovirus infection. The metabolism of newly-infected cells increased 90% during the first 24 hours, but as infection proceeded, and cells gradually succumbed to the baculovirus infection, the cytopathic effect of the baculovirus on the cells became evident. Oxygen and glucose uptake rate measurements appeared to more accurately assess the condition of infected cells than conventional trypan blue staining, which tended to overestimate cell viability in the mid stages of infection. The optimal harvest time varied, depending on which technique – SDS-PAGE, chromogenic (ONPG) or fluorometric (C_{12}FDG) – was used to monitor β-galactosidase production. Specific β-galactosidase production was found to be insensitive to a wide range of culture dissolved oxygen tensions, whereas resuspending cells in fresh medium prior to infection increased volumetric productivity approximately two-fold (800,000 units β-galactosidase/ml) compared to cultures infected in batch mode and allowed successful infections to occur at higher cell densities.

Abbreviations: ONPG – ortho-phenyl 2-β-D-galactopyranoside, OUR – oxygen uptake rate (μ-mol O_2/liter/hour), $q_{glucose}$ – specific glucose uptake rate (mg glucose/10^6cell/hour), $q_{glutamine}$ – specific glutamine uptake rate (mg glutamine/10^6cell/hour), qO_2 – specific oxygen uptake rate (μ-mol O_2/10^6cell/hour), MOI – virus multiplicity of infection (viral plaque forming units/cell).

Introduction

Insect cell culture is becoming an important technological tool for the production of biologicals, including recombinant proteins and biopesticides through the application of the baculovirus expression vector system (BEV). One of the more difficult operational aspects of this technology has been the ability to monitor baculovirus-infected cells, where little information in the form of published data is available. For instance, only recently has the importance of oxygen supply during the infection phase been established (Betenbaugh *et al.*, 1991; Lindsay and Betenbaugh, 1992; Scott *et al.*, 1992; Wang *et al.*, 1993a), or the potential need for nutrient feed strategies to replenish those nutrients consumed during cell growth (Caron *et al.*, 1990, Reuveny *et al.*, 1993; Wang *et al.*, 1993a; Wang *et al.*, 1993b). It is now well understood that although infected cells no longer divide, they continue to consume vital nutrients, such as glucose, glutamine and oxygen until the gradual cytopathic effect of the baculovirus infection causes complere cessation of cell-specific functionality, followed by cell death and lysis. Monitoring infected cells up until the time of cell death is crucial, since expression of heterologous genes under the polyhedrin promoter occurs very late in the cell-virus life cycle, typically when cells are in the throes of death. The rationale for this study was to evaluate different methods for monitoring infected cells in ways that might be applicable for bioprocess control.

In this study we compared existing techniques for monitoring baculovirus-infected *Spodoptera frugiperda* Sf9 insect cells in EXCELL 401 serum-free medium, and discuss the relevance of each as they relate to expression of foreign genes under the polyhedrin promoter. We have included in this study the conventional methods of trypan blue dye exclusion, oxygen and glucose uptake rate measurements, and also a novel fluorescent staining technique using recently introduced dyes. Also, we explored the effects of widely varying culture dissolved oxygen conditions and media replenishment on baculovirus-infected cell survival and foreign gene expression in 250 ml oxygen-controlled bioreactors.

Materials and methods

Cell culture

Frozen vials of low passage *Spodoptera frugiperda* Sf9 insect cells were obtained from ATCC (CRL 1711) and grown as monolayers in T-flasks, adapted to suspension, then gradually adapted to EXCELL 401 serum-free medium (JRH Biosciences, Lenexa, KS), a derivative of the basal medium IPL-41 serum-free formulation. Adaptation of cells to EXCELL 401 was found to take approximately six weeks. No antibiotics of any type were added in these experiments. A genetically engineered *Autographa california* AcNPV baculovirus expressing *E. coli* β-galactosidase (Invitrogen, San Diego, Ca, Cat. No. B822-04) was used in the course of this study. The virus was scaled up by two rounds of infection, and was titered at 3×10^8 plaque forming units (pfu) by dilution plaque assays with *X-gal* in the agarose overlay method (Summers and Smith, 1987).

Equipment

All experiments were conducted in oxygen-controlled 250 ml Bellco spinner flasks (Vineland, NJ) in a non-humidified Infors ISF-4 incubator maintained at 28 degrees C. The tops were modified to accommodate dissolved oxygen and pH probes from Ingold (Wilmington, MA). The pH probe was calibrated initially at 4.0 and 7.0 prior to sterilization, then once again after inoculation by removing a grab sample of the medium. The dissolved oxygen probe was calibrated at 100% and 0% using air and nitrogen overlay, respectively. Oxygen uptake rates (OUR) were determined using the dynamic technique of Singh and Gunnarson (1988), by measuring the time it took to deplete the culture DO by 10%. A saturation constant of 0.26 m-mol O_2/liter was assumed. Specific OUR was obtained by dividing the OUR by the viable cell count (10^6 cells/ml). A Biolafitte (Princeton, NJ) polarographic DO module was used in conjunction with a Data Interface Unit (DIU) supplied by Computer Products (Maynard, MA) and control software (written by Dave Willard at Schering) to control DO from the DIU. Surface aeration with saturated air and oxygen pulsing from a solenoid valve was use to maintain DO. Both DO and pH measurements were transmitted to a VAX 3100 Server, where values were stored and could be accessed over a PC Network using Microsoft Excel 4.0^{TM} and Microcal Origin 2.8^{TM} under Windows for data analysis.

Analytical methods

Trypan blue dye exclusion was performed by diluting infected cell samples with 0.4% trypan blue dye so that 50–100 cells per field were present, to minimize counting errors. O-nitrophenyl-β-D galactopyranoside (ONPG – Sigma Chemical Cat. No. N1127) substrate and the method of Miller (1972) were used to determine total β-galactosidase activity in the culture broth. Briefly, samples were treated with sodium dodecyl sulfate and chloroform to extract β-galactosidase into the culture supernatants from cell pellets (results are reported as total units). SDS PAGE analysis was performed on culture samples using standard electrophoretic techniques. A 7.5% polyacrylamide gel was used to resolve the β-galactosidase band under reducing conditions. Glucose was determined using a YSI Model 2700 combination glucose/lactate analyzer. Standards were run with all assays to determine linearity and accuracy of each assay, and correction factors were applied where neceassary.

A combination live/dead fluorescent stain kit (Cat No. L-3224) from Molecular Probes, Inc., (Eugene, OR), was used in conjunction with epifluorescence microscopy as an indicator of metabolic activity before and after infection. The calcein AM green stain portion of the kit was used to assess the activity of cells. Under the action of internal cellular esterases, a nonfluorescing substrate (calcein AM) is cleaved into a highly fluorescent calcein substrate. A second fluorescent stain, ethidium homodimer, was used to assess infectivity in cells. A third fluorescent dye, ImaGene Green Kit Cat No. I-3098 (Molecular Probes. Eugene, OR) was used to qualitatively assess specific (per cell) β-galactosidase expression.

Experimental design

Two sets of runs were conducted during the course of this study. The first set consisted of a series of experiments of batch infections conducted at different times in the cell batch cycle, corresponding to various cell densities. In this study, we evaluated different techniques to monitor the infection cycle, compared the metabolism of infected cells to that of uninfected cells, and found the optimum timing of infection in batch cultures. The second set of experiments was conducted by resuspending cells in fresh media just prior to infection. Since glucose and glutamine (and possibly other) nutrients were found to be limiting from batch studies, renewal of the medium allowed the metabolism of infected cells to be more closely examined under nonlimiting conditions, including the consumption patterns of the key nutrients glucose and glutamine. In all experiments, sufficient virus (MOI) was added so as to cause synchronous infection of the entire population of cells. An MOI of 10 was found to result in a synchronous infection. Finally, for each set of experiments, a single inoculum source containing cells of high viability (>90%) at mid-exponential growth was used in order to minimize the potential for biological variation.

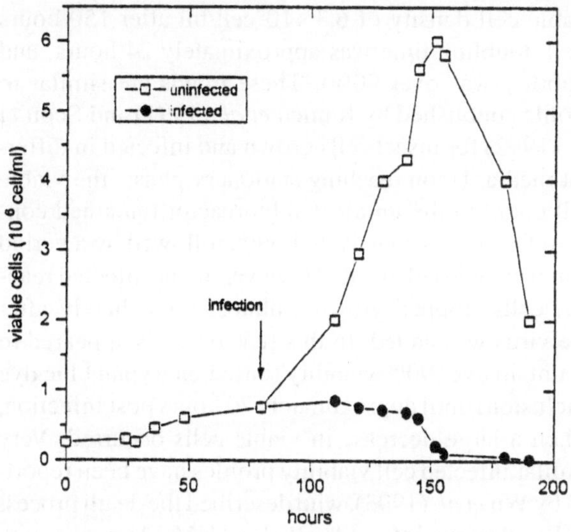

Fig. 1. Viable Sf9 cell profiles of infected and uninfected parallel bioreactors.

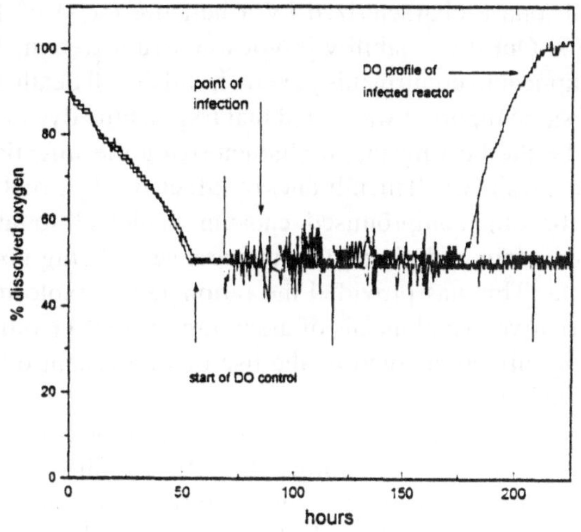

Fig. 2. Dissolved oxygen profiles of infected and uninfected Sf9 cells from parallel bioreactors.

Results and discussion

Monitoring the infection process: infected versus uninfected cells

In the first experiment, two oxygen-controlled bioreactors were inoculated with cells at 2.5×10^5cell/ml and run in parallel. After a fairly substantial lag period, cells in both bioreactors grew more vigorously (Fig. 1) and consumed more oxygen (Fig. 2). At 50 hours, the dissolved oxygen controllers of both reactors were set to 50%, and thereafter bioreactors were sampled for viability (based on trypan blue), glucose, glutamine, and pH. When the cell densities of both reached 0.8×10^6cell/ml, approximately 80 hours into the run, the first reactor was infected with recombinant virus at an MOI of 10, whereas the second was uninfected. Thereafter, growth profiles of the two reactors differed significantly. From Fig. 1, cells in the uninfected bioreactor continued to grow, attaining a final viable cell density of 6.3×10^6cell/ml after 150 hours (cell doubling time was approximately 24 hours, and viability was over 90%). These trends are similar to profiles published by Kamen *et al.* (1991) and Scott *et al.* (1992) for insect cells grown and infected in different media. Upon reaching stationary phase, the viable cell count in the uninfected bioreactor remained constant for approximately 24 hours, followed by a period of fairly rapid cell death. However, in the infected reactor, cells stopped growing almost immediately after the virus was added. In this reactor, cells appeared to maintain over 90% viability (based on trypan blue dye exclusion) until approximately 70 hours post infection, when a large decrease in viable cells occurred. Very similar infected cell viability profiles have been reported by Wu *et al.* (1993), who described the death process of baculovirus-infected cells (in TNM-FH media with 10% fetal bovine serum) in two phases: a constant viability (or delay) phase with a delay time (t_d) lasting approximately 72 hours, followed by a first-order death phase characterized by a half-life ($t_{1/2}$) of 17 hours. Our own viability profiles in serum-free medium appear to confirm this general trend of cell death. In this same report, it was noted that trypan blue dye may not be the best means of characterizing the infection process, since cell membranes of infected cells eventually become compromised, causing viable cell to take up the stain and giving the appearance of being non-viable. This has provided the rationale for exploring alternative visual means of assessing the cell viability during infection, such as the use of fluorescent dyes (see below).

The dissolved oxygen (DO) profiles in the two parallel bioreactors were similar (Fig. 2), with the exception of the phases of the virally infected reactor. In this reactor, the culture DO began to slowly rise at 175 hours (corresponding to 100 hours post infection), as infected cells began to die (there was no provision in this reactor to control DO on the high side). Cell-specific oxygen uptake rate measurements were also taken in both bioreactors (Fig. 3). The cell-specific oxygen uptake rate, qO_2, of the uninfected reactor was measured at 0.33 μ-mol $O_2/10^6$ cell/hour, and did not vary significantly during the run. We have previously found that qO_2 measurements of uninfected Sf9 cells fully adapted to serum-free medium had consistently higher values than the qO_2 of the same cells grown in TNM-FH medium with 5–10% fetal bovine serum (Hensler *et al.*, 1994). However, in the infected reactor, there was an abrupt rise in qO_2 just after infection, indicating that virally infected cells are metabolically

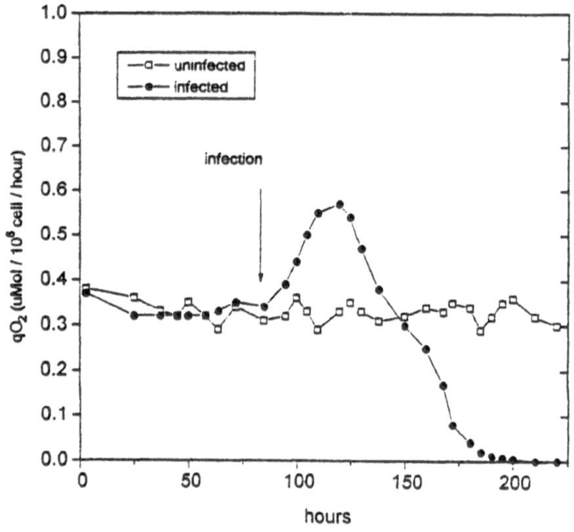

Fig. 3. qO_2 measurements of infected and uninfected parallel bioreactors.

more active. In this reactor, qO_2 rose from 0.33 to 0.60 within 24 hours. The magnitude of this increase is consistent with the 1.8 fold increase in the DNA content of Sf9 cells after infection with wild-type AcNPV as reported by Schopf *et al.* (1990), and similar to the qO_2 profile reported for *Trichoplusia ni* (TN-368) cultured cells by Streett and Hink (1978), as well as the peak in culture OUR of Sf9 cells in IPL-41 serum-free medium as shown by Kamen *et al* (1991). This peak is relatively short-lived, for a gradual decline of qO_2 follows, most likely due to the progressively cytopathic effect of the virus on the cells. During this period, cell specific functions decreased as a successive cascade of virally synthesized genes became active and redirected cellular activity towards producing virus-driven products. By hour 175 (100 hours post infection), qO_2 in the infected reactor had fallen significantly, corresponding to the slow rise in dissolved oxygen.

It is clear from these data that the initial period of higher cell respiratory activity and active viral synthesis early on in the infection, followed by the slow, gradual shift towards cell death can be obtained by oxygen uptake rate measurements, which appear to provide a realistic picture of the infection process in cells, and can easily be converted to on-line data (Singh and Gunnarson, 1988). On the other hand, trypan blue staining does not predict the period of increased activity just after infection, nor does it suggest a gradual progression towards cell death (Fig. 1). Instead, this profile suggests almost constant retention of cell activity followed by a rapid onset of cell death. Since virus-encoded genes are highly temporally regulated, the timing of cell death by either method could be correct, however later in this work, glucose patterns of infected cells were found to more closely mimic the pattern suggested by the oxygen uptake rate method (see below). Recent attempts to model the baculovirus-infected cell death process using data from trypan blue exclusion (Power *et al.*, 1992; Wu *et al.*, 1993) might therefore be improved by using qO_2 data as the means for determining viability in infected cells.

A third monitoring technique, fluorescent dye staining with combination stains, was also applied to infected and uninfected cells. This live/dead fluorescent stain kit (used in conjunction with epifluorescence microscopy) proved to be a very useful indicator of metabolic activity before and after infection. Approximately 5 μmoles of each dye was added to the samples. The calcein AM stain, which is cleaved to a highly fluorescent bright green calcein under the action of cellular esterases, was used to assess the activity of both infected and uninfected cells. Cells early in the infection stage (24-36 hours post infection) stained a brighter green than the uninfected cells, suggesting an increase in activity as indicated by the OUR method. As the infection proceeded, fewer cells were brightly lit, suggesting a gradual diminishing of cellular activity (Fig. 4). Conversely, uninfected cells tended to exhibit the same staining characteristics throughout their life cycle, indicative of constant activity. A second fluorescent stain, ethidium homodimer, proved to be a useful means of assessing infectivity in cells. This dye, which can only penetrate cells with damaged or compromised membranes, indicated that membranes of our infected cells were compromised within 48 hours of infection. This dye intercalated within the cellular DNA to produce a bright orange fluorescent glow, and appeared to support the increase in DNA activity upon baculovirus infection that was reported by Schopf *et al.* (1990). The results of the combination fluorescent dye staining, using both calcein AM and ethidium homodimer simultaneously, were more in line with the qO_2 measurements, and suggest a transient two-fold increase in cellular activity after infection, followed by a long, gradual cytopathic progression. A recent comparison of combination fluorescent staining to trypan blue dye staining for murine hybridoma cells also favored combination fluorescent staining for its greater accuracy in determining cell viabilioty (Altman *et al.*, 1993).

Although the infected cell viability profiles were different depending on which methods are used to characterize the infection process, all methods predicted that *Spodoptera frugiperda* Sf9 insect cells are killed within 100 hours of baculovirus infection. These data are consistent with infected Sf9 cell viability profiles published by a number of authors using different media and expression vectors (King *et al.*, 1992; Licari and Bailey, 1992; Lindsay and Betenbaugh, 1992; Wang *et al.*, 1993b). We next explored the foreign gene production as it related to the condition of the infected cells. This is a very important area of research, for one would expect that infected cells need to be maintained in the best possible physiological state for maximum expression of foreign genes, since protein production ultimately relies on the transcriptional, translational and post-translational capabilities of the host cell. Recently, it has been suggested that the kinetics of lethality of infected insect cells is an important factor in heterologous gene expression, with the highest expression in those cell lines best able to survive the baculovirus infection for the longest period (Betenbaugh *et al.*, 1991). Conversely, Wu *et al.* (1993) found no correlation between foreign protein production and cell viability after infection.

Monitoring recombinant β-galactosidase protein expression

Three techniques were used to assess total β-galactosidase production in the infected bioreactor: the chromogenic ONPG method of Miller (1972), a fluorometric technique using $C_{12}FDG$ (Zhang *et al.*, 1991), and gel electrophoresis under reducing conditions (SDS-PAGE). The kinetics of β-galactosidase expression according to the ONPG assay suggested a 48 hour delay (post infection), followed by a pattern of increasingly intense expression, which continued for 96-120 hours post infection (Fig. 5). These expression kinetics are similar to reports by Betenbaugh *et al.* (1991), King *et al.* (1992), Licari and Bailey (1992), Neutra *et al.* (1992), and Ogonah *et al.* (1991), all of which show β-galactosidase expression in Sf9 cells using the ONPG assay. The fluorometric $C_{12}FDG$ stain gave similar results (data not shown), but was slightly more sensitive in that minute levels of expression could be determined after 36 hours. On the other hand, both chromogenic and fluorometric assays suggested β-galactosidase expression continues past 100 hours post infection. The ONPG kinetic profiles did not correlate with the viability of infected cells, since most of the protein expression appeared to occur when the Sf9 cells were virtually dead (Fig. 5). The ONPG and the fluorometric assay may therefore slightly overes-

Fig. 4. Infected Sf9 cells stained with fluorescent calcein AM dye show a diminished activity at 60 hours post infection

timate the production of β-galactosidase during the latter stages of infection. SDS PAGE results (data not shown), suggested an earlier expression (36 hours post infection) which levelled off by 48–72 hours post infection. Our SDS PAGE profiles were similar to those of Caron *et al.* (1990) for VP6 expression and Scott *et al.* (1992), for recombinant BHC11 virus coding for core regions of the hepatitis C virus, and tended to correlate better with cell viability. Further work would be required to reconcile the differences between these assays and determine the optimum harvest time.

Optimization of infection in batch cultures

Nutrient depletion (especially glucose and glutamine) has previously been identified as a key limiting factor in recombinant protein production in insect cell cultures (Caron *et al.*, 1990; Reuveny *et al.*, 1993, Wang *et al.*, 1993b). Infections in batch cultures must therefore be done relatively early in the growth cycle to ensure an adequate supply of nutrients for expression. To determine the optimum point of infection in batch cultures, aliquots were removed from a control flask at different times in batch growth (corresponding to different cell densities) and infected with virus at an MOI of 10, and assayed for β-galactosidase using the ONPG method after 100 hours post infection. Specific (per cell) productivity was relatively constant at 250,000 (total) units of β-galactosidase/10^6 cell when cells were infected early in the batch cycle (0.5 and 1.5×10^6 cell/ml), but dropped off significantly when cells were infected at 2×10^6 cell/ml (Fig. 6). Thus, the volumetric enzyme production was maximum at 1.5×10^6 cell/ml. Glucose was depleted in each of the flasks at the end of the infection period (data not shown).

Effect of medium replenishment on β-galactosidase production

In the experiments that follow, concentrated cells were diluted with completely fresh medium to initial cell densities of 1,2,3 and 4×10^6 cell/ml in four oxygen-controlled bioreactors. The bioreactors were then infected with virus at an MOI of 10. Glucose profiles from each reactor show that glucose was initially consumed at a 70% higher rate by infected cells, consistent with the observed increase in qO_2, and became limiting above an infected cell density of 3×10^6

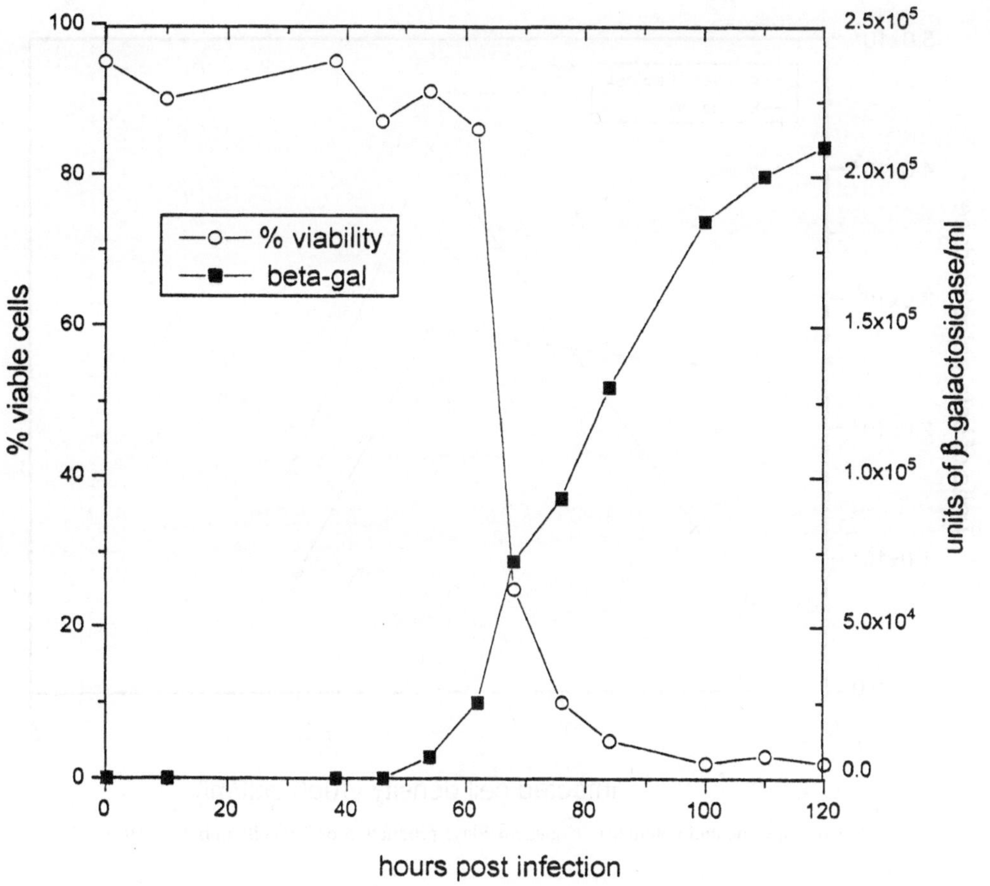

Fig. 5. Total β-galactosidase production as a function of apparent cell viability.

Table 1. Specific glucose, glutamine and oxygen uptake rates of infected and uninfected cells

	qO_2 µmol/10^6 cell/hour	$q_{glucose}$ — mg/10^6 cell/hour	$q_{glutamine}$ —
infected	0.60	13.5	9.0
uninfected	0.33	8.0	5.5

* Time of measurement for infected cells: 24 hours post infection.

cell/ml (Fig. 7). Glutamine was also initially consumed at a higher rate by infected cells, but was not limiting at any of these infected cell densities (data not shown). Table I compares the maximum specific consumption rates of glucose, glutamine and oxygen of uninfected cells and cells 24 hours after infection.

β-galactosidase production for each bioreactor is shown in Figure 8. By renewing the culture medium prior to infection, specific productivity was maintained at infected cell densities even beyond 3×10^6 cell/ml (compared to only 1.5×10^6 cell/ml in the batch culture), thereby extending the maximum volumetric productivity to 8×10^5 units β-galactosidase/ml, and over a wider range of infected cell densities. This experiment emphasized the importance of replenishing the culture medium prior to infecting in order to maximize foreign gene expression, and implicates nutritional deficiencies rather than contact inhibition as the root of previously observed upper limits in appropriate cell density for infection.

Effect of dissolved oxygen on growth and β-galactosidase production

Using oxygen as a control parameter, four oxygen-controlled 250 ml bioreactors agitating at 70 RPM were set to dissolved oxygen levels of 100%, 50%, 10% and 5%. Each bioreactor was inoculated at 4×10^5 cell/ml and grown to a high cell density of 5×10^6 cells/ml. No differences in cell growth, maximum cell density, cell viability, nutrient consumption rates or OUR were observed, indicating a relative lack of sensitivity of cell

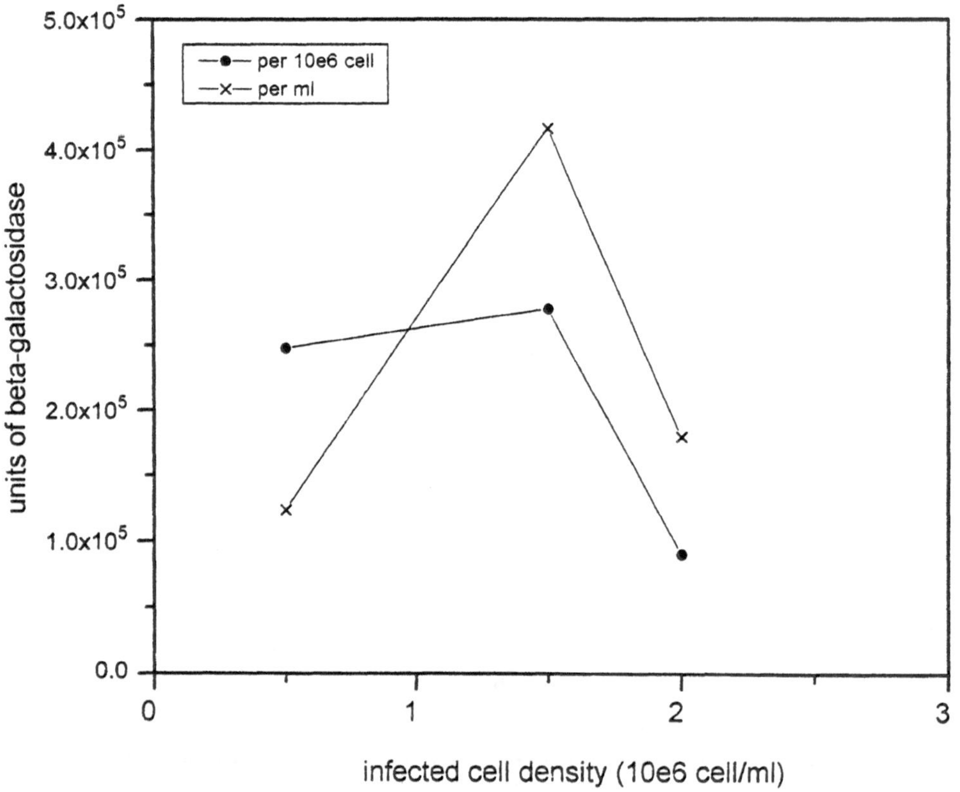

Fig. 6. Specific and volumetric β-galactosidase production of Sf9 cells in batch culture.

growth to oxygen levels in the media. These cells were then resuspended in fresh media to 1×10^6 cells/ml and infected with a recombinant β-galactosidase expressing baculovirus (MOI of 10), still under DO control at the level indicated. All four spinners were visibly infected within 24 hours and expressed β-galactosidase at identical levels. Our conclusion is that the Sf9 cells can tolerate a wide range of dissolved oxygen concentrations during growth and after infection, and are not sensitive to low oxygen tensions. This is in agreement with earlier work by Lin *et al.* (1993), for CHO cells, where metabolism, growth and recombinant tPA production were found to be relatively unaffected by DO tensions as low as 2%, however it does contradict the earlier work of Jain *et al.* (1991), for Sf9 insect cells. In their work, cell growth and antistasin production were adversely affected by low (below 20%) and high (above 100%) dissolved oxygen levels. Our results are difficult to compare directly with those of Lindsay and Betenbaugh (1992), and of Scott *et al.* (1992), who suggested adverse effects of oxygen limitation upon recombinant protein productivity in Sf9 cells, since in neither of these two studies was the dissolved oxygen actually measured, despite some evidence that the prevailing conditions may have been oxygen-limiting.

Conclusions

In this study, we examined several different means of monitoring the infection process in Sf9 insect cells in serum-free medium and also of evaluating β-galactosidase expression. Both specific oxygen uptake measurements and combination staining with the fluorescent dyes calcein AM and ethidium homodimer appeared to more accurately describe the infection process than the trypan blue exclusion method. Specific oxygen uptake rates of infected cells increased almost two-fold within 24 hours of infection, followed by a gradual decrease until, by 100 hours post infection, cell respiratory activity had virtually ceased. This type of behavior was also suggested by the fluorescent stains, whereas trypan blue exclusion predicted a constant cell viability until 72 hours post infection, followed by massive cell death. The optimum harvest time of infected cells for β-galactosidase varied, depending on whether

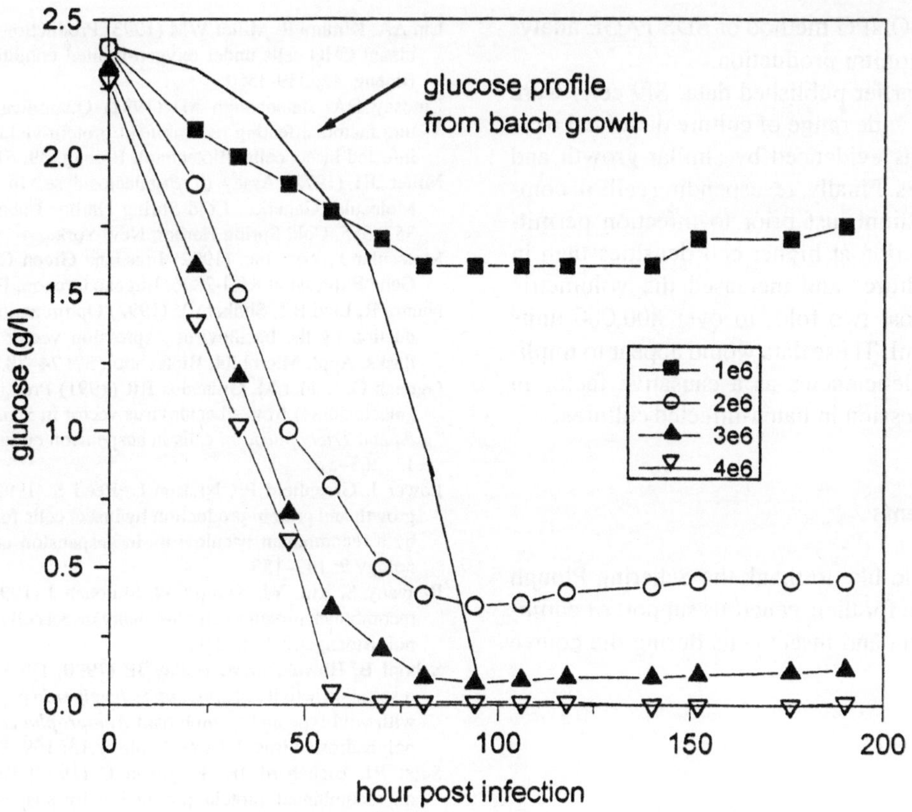

Fig. 7. Glucose profiles of infected and uninfected Sf9 cells.

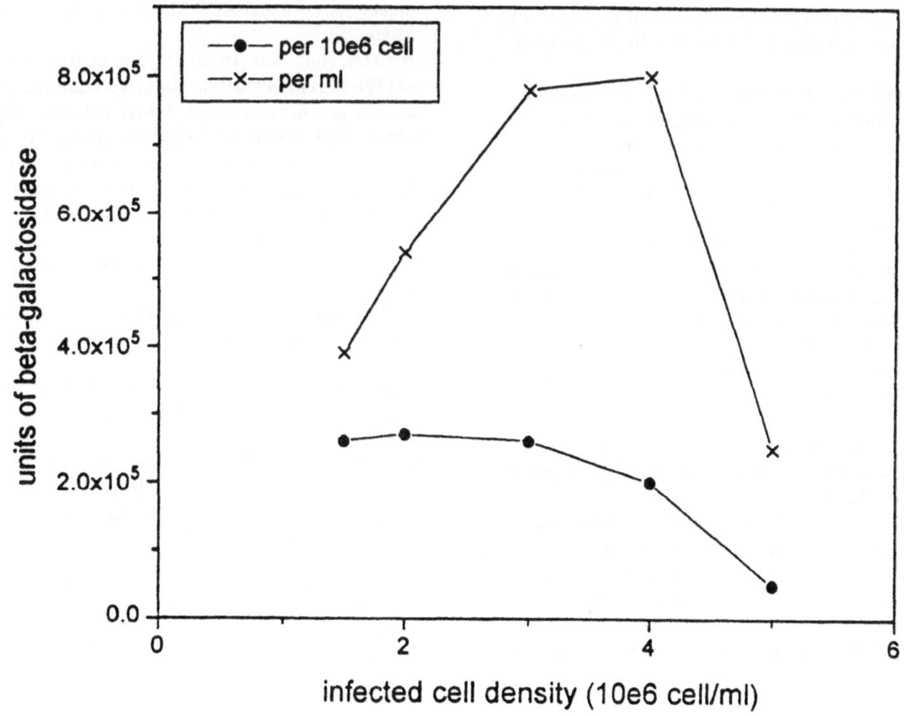

Fig. 8. Specific and volumetric β-galactosidase production of Sf9 cells after resuspension in fresh medium.

the chromogenic ONPG method or SDS PAGE analysis was used to monitor production.

Contrary to earlier published data, Sf9 cells were not sensitive to a wide range of culture dissolved oxygen conditions, as evidenced by similar growth and expression profiles. Finally, resuspending cells in completely fresh medium just prior to infection permitted efficient infection at higher cell densities than in simple batch culture, and increased the volumetric productivity almost two-fold, to over 800,000 units β-galactosidase/ml. These data would appear to implicate nutritional deficiencies as a causative factor in sub-optimal expression in batch-infected cultures.

Acknowledgements

The authors would like to thank the Schering Plough Research Institute for their generous support of equipment, virus, media and insect cells during the course of this work.

References

Altman SA, Randers L, Rao G (1993) Comparison of trypan blue dye exclusion and fluorometric assays for mammalian cell viability determinations. Biotechnol. Prog. 9: 671–674.

Betenbaugh M, Balog L, Lee PS (1991) Production of recombinant proteins by baculovirus-infected gypsy moth cells. Biotechnol. Prog. 7: 462–467.

Caron AW, Archambault J, Massie B (1990) High-level recombinant protein production in bioreactors using the baculovirus-insect cell expression system. Biotechnol. Bioeng. 36: 1133–1140.

Hensler W, Singh V, Agathos SN (1994) Production of β-galactosidase in Sf9 insect cells in serum-containing and serum-free medium. Annals of N.Y.A.S. (in press).

Jain D, Ramasumbramanyan K, Gould S, Lenny A, Candelore M, Tota M, Strader C, Alves K, Cuca G, Tung JS, Hunt G, Junker B, Buckland BC, Silberklang M (1991) Large-scale recombinant protein production using the insect cell-baculovirus expression vector system: antistasin and β-adrenergic receptor. In: Spier RE, Griffiths JB and Meignier B (eds.) Production of Biologicals from Animal Cells in Culture (pp 345–350). Butterworth-Heinemann, Oxford.

Kamen AA, Tom RL, Caron AW, Chavarie C, Massie B, Archambault J (1991) Culture of insect cells in a helical ribbon impeller bioreactor. Biotechnol. Bioeng. 38: 619–628.

King GA, Daugulis AJ, Faulkner P, Goosen MFA (1992) Recombinant β-galactosidase production in serum-free medium by insect cells in a 14-L airlift bioreactor. Biotechnol. Prog. 8: 567–571.

Licari P, Bailey JE (1992) Modelling the population dynamics of baculovirus-infected cells: optimizing infection strategies for enhanced recombinant protein yields. Biotechnol. Bioeng. 39: 432–441.

Lin AA, Kimura R, Miller WM (1993) Production of tPA in recombinant CHO cells under oxygen-limited conditions. Biotechnol. Bioeng. 42: 339-3350.

Lindsay DA, Betenbaugh MJ (1992) Quantification of cell culture factors affecting recombinant protein yields in baculovirus-infected insect cells. Biotechnol. Bioeng. 39: 614–618.

Miller JH (1972) Assay of β-galactosidase. In: Experiments in Molecular Genetics. Cold Spring Harbor Laboratory Press. pp. 352–355. Cold Spring Harbor, New York.

Molecular Probes, Inc. (1992) ImaGene Green C_2-C_{16} FDG *lacZ* Gene Expression Kit I-2896. Eugene Oregon, Bulletin 2896.

Neutra R, Levi BZ, Shoham Y (1992) Optimization of protein production by the baculovirus expression vector system in shake flasks. Appl. Microbiol. Biotechnol. 37: 74–78.

Ogonah O, Shuler M, Granados RR (1991) Protein production (β-galactosidase) from a baculovirus vector in *Spodoptera frugiperda* and *Trichoplusia ni* cells in suspension culture. Biotech. Lett. 13: 265–270.

Power J, Greenfield PF, Nielsen L, Reed S (1992) Modelling the growth and protein production by insect cells following infection by a recombinant baculovirus in suspension culture. Cytotechnology 9: 149–155.

Reuveny S, Kim YJ, Kemp CW, Shiloach J (1993) Production of recombinant proteins in high-density insect cell cultures. Biotechnol. Bioeng. 42: 235–239.

Schopf B, Howaldt MW, Bailey JE (1990) DNA distribution and respiration activity of *Spodoptera frugiperda* populations infected with wild-type and recombinant *Autographa californica* nuclear polyhedrosis virus. J. Biotechnology 15: 169–186.

Scott RI, Blanchard JH, Ferguson C (1992) Effects of oxygen on recombinant protein production by suspension cultures of *Spodoptera frugiperda* (Sf9) insect cells. Enzyme Microb. Technol. 14: 798–804.

Singh V, Gunnarson A (1988) On-line method for measuring oxygen uptake rate. Annual ACS Meeting. Presentation 110. November Miami, FL.

Streett DA, Hink WF (1978) Oxygen consumption of *Trichoplusia ni* (TN-368) insect cell line infected with *Autographa californica* nuclear polyhedrosis virus. J. Invert. Pathol. 32: 112–113.

Summers MD, Smith GE (1987) A manual of methods for baculovirus vectors and insect cell culture procedures. Bulletin No. 1555. Texas Agricultural Experimental Station.

Wang MY, Kwong S, Bentley WE (1993a) Effects of oxygen/glucose/glutamine feeding on insect cell baculovirus protein expression: a study on epoxide hydrolase production. Biotechnol. Prog. 9: 355–361.

Wang MY, Vakharia V, Bentley WE (1993b) Expression of epoxide hydrolase in insect cells; a focus on the infected cell. Biotechnol. Bioeng. 42: 240–246.

Wu SC, Dale BE, Liao JC (1993) Kinetic characterization of baculovirus-induced cell death in insect cell cultures. Biotechnol. Bioeng. 41: 104–110.

Zhang ZY, Naleway JJ, Larison KD, Huang Z, Haughland R (1991) Detecting *lacZ* gene expression in living cells with new lipophilic, fluorogenic β-galactosidase substrates. FASEB 5: 3108–3112.

Address for offprints: W.T. Hensler, Raytheon Engineers and Constructors, 30 South 17th Street, Philadelphia, PA 19101, USA *or* S.N. Agathos, Catholic University of Louvain, Place Croix du Sud, 2 Bte 19, B-1348, Belgium.

Cytotechnology **15:** 187–194, 1994.

The analysis of protein pharmaceuticals: Near future advances

C. Russell Middaugh
Department of Pharmaceutical Research, Merck Research Laboratories, WP78–302, West Point, PA 19486

Key words: Proteins, analysis, x-ray crystallography, nuclear magnetic resonance

Abstract

The analysis of protein pharmaceuticals currently involves a complex series of chromatographic, electrophoretic, spectroscopic, immunological and biological measurements to unequivocally establish their identity, purity and integrity. In this review, I briefly consider the possibility that at least the functional identity and integrity of a protein drug might be established by either a single analysis involving X-ray diffraction, NMR or mass spectrometry, or by a chromatographically based multi-detector system in which a number of critical parameters are essentially simultaneously determined. The use of a protein standard to obtain comparative measurements and new advances in the technology of each of these methods is emphasized. A current major obstacle to the implementation of these approaches is the frequent microheterogeneity of protein preparations. The evolution of biological assays into measurements examining more defined intracellular signal transduction events or based on novel biosensors as well as the analysis of vaccines is also briefly discussed.

Introduction

Although the use of proteins as drugs is by no means new (e.g., insulin, gamma-globulin and protein-containing vaccines have been routinely employed for decades), the advent of recombinant DNA technology has resulted in an explosion of interest in their pharmaceutical applications. We now find ourselves in the situation that we can make most proteins in sufficient quantities for therapeutic use (albeit at times with some difficulty!). It is significantly more difficult, however, to take such a molecular entity and prepare it in the form of a practical drug substance. One problem in this regard concerns difficulties in the analysis of protein pharmaceuticals because of their inherent structural complexity. This is dramatically illustrated by the fact that when a new process is devised to produce a protein pharmaceutical with an already well established safety and efficacy profile, it may be necessary to perform bridging safety and clinical efficacy trials to ensure the identity and quality of the protein derived from the new process. This arises from the fact that it is difficult to unambiguously demonstrate that a protein derived from a new process is, in fact, functionally equivalent to the original molecule. This is clearly an undesirable situation which can result in a significant impediment to the production of proteins of higher quality and decreased cost. Thus, advances in our ability to analyze the structure of proteins in a pharmaceutical context are of potentially great importance.

The analysis of protein pharmaceuticals

The ability to accurately define the chemical nature of any protein that is being contemplated for use as a drug substance is absolutely essential to the pharmaceutical development process. Currently, such a characterization is performed by a series of chromatographic, electrophoretic, spectroscopic, immunological and biological measurements (Jones, 1993). Despite the intrinsically higher accuracy and precision of the former four types of assays, the use of a biological (e.g. cell or organism based) method often remains critical, since there is as of yet little confidence in the ability of chemical and physical methods to detect potentially subtle structural alterations which might be manifested as decreases in either the safety or efficacy of the product. The major exception to the need for a biological assay are proteins in which a well defined enzymatic activity comprises the pharmaceutical activity of the macromolecule. Although such macromolecules still require the same complex, multifaceted characterization as other proteins, a biological assay can usually be replaced by an in vitro enzymatic assay of superior accuracy and precision. The problem I will address is the possibility of providing structural analysis of a protein with sufficient resolution to establish its functional identity.

The critical question one needs to first ask is exactly what kind and how much information is absolutely necessary to unequivocally establish the identity of any particular protein. A formal but simplistic answer is easily provided: a data array (i.e. a matrix) specifying the three spatial coordinates and elemental identity (i.e. C, H, O, N, S, etc.) of each atom in a protein uniquely defines any particular protein molecule. We will ignore for the moment the dynamic, fluctuating nature of proteins and the inherent microhetereogeneity of most protein preparations. Only two methods, X-ray crystallography and NMR, are currently available which provide the type of three dimensional structural information desired. Both are usually considered primarily research tools and for a variety of reasons not applicable to the routine pharmaceutical analysis of proteins. I argue that at least in principle this need not necessarily be the case and discuss each approach below.

The existence of a hierarchy of structure, however, suggests it may not be necessary to directly determine the location of each atom to ensure the identity of a protein. The combination of various types of information acquired at different levels of structure can provide a very detailed picture of a macromolecule. But, can this approach provide a sufficiently high resolution description that it can be used to establish pharmaceutical identity? The answer to this question is uncertain, but it seems to me that it may ultimately be possible if certain carefully delineated requirements are fulfilled. Unfortunately, the theoretical basis for such an attempt has yet to be developed. One can view this at least partially as a complex problem in information theory. To reiterate, what kind and how much information is necessary to define a protein's structure? It is now clear that the chemical (or covalent) identity of a protein can be well-defined by modern methods of mass spectrometry and/or peptide mapping and sequencing as discussed below. Secondary structure can be measured to an accuracy of several percent by CD, FTIR and, in some cases, 1-D NMR. Aspects of the integrity (correctness) of tertiary structure can be probed only at lower resolution by techniques such as ultraviolet absorption, intrinsic fluorescence as well as chemical and immunological reactivity. Unfortunately, knowledge that a protein exists as the proper chemical entity (correct unmodified sequence) with apparently native secondary structure does not guarantee the existence of correct tertiary structure. Nevertheless, it may be the case that less than complete direct knowledge of tertiary structure is actually necessary to ensure identity if the propriety of lower levels of structure are established. Proper quaternary structure is probably the least difficult problem since a combination of permeation chromatography, static and dynamic light scattering as well as equilibrium and velocity analytical ultracentrifugation can rigorously establish the number of subunits (and aggregation state) under most solution conditions. I will not attempt here to answer the theoretical question of how much information is ultimately necessary, but rather below outline how one might experimentally measure a number of different parameters simultaneously that reflect a variety of aspects of protein structure.

X-ray crystallography

X-ray crystallography currently offers the accepted definitive method for determining the three dimensional structure of molecules including proteins. The only major concern about this approach is that the structure elucidated is that of the molecule in the crystalline rather than solution state. A minor caveat is that highly flexible regions of a protein (e.g., the termini) may not be well resolved. The general consensus, however, based on a plethora of evidence, is that the crystalline state is predictive of, if not entirely identical to, the solution state. Nevertheless, for a variety of reasons, this method is assumed not to be applicable to the pharmaceutical analysis of proteins. First, the necessary crystallization of proteins is usually considered to be more art than science. Many proteins are thought to be difficult if not impossible to crystallize with sufficient order and crystal size to permit high resolution data to be obtained. Furthermore, the quantities of protein necessary for successful crystallization (tens of mgs.) are often not readily available. Secondly, to solve a structure using this method, isomorphous heavy metal derivatives of the proteins must be obtained to permit the phase of the diffracted X-rays to be determined. Some knowledge of a protein's polypeptide backbone structure, however, (perhaps by homology to a related molecule of known structure) may obviate this need. Third, the entire process is both exceedingly labor- and time-intensive. Complete determination of a structure of a protein by this method can potentially take many months or even years. Such a situation is clearly unacceptable to scientists working on pharmaceutical macromolecules where data usually needs to be obtained in days to weeks.

Nevertheless, it seems possible that in the near future crystallography has the potential to play a significant role in the pharmaceutical analysis of proteins. Several factors encourage this view. Most importantly, it should not be necessary to perform actual structural determinations (i.e., generation of electron density maps) of molecules of interest. Rather, one need merely compare a series of X-ray diffraction patterns of a test protein with a standard preparation of known integrity. Thus, the problem becomes one of comparing a multi-dimensional data matrix with the elements consisting of crystallographic reflections defined by spatial coordinates and intensities. The similarity of such matrices could then constitute evidence of structural identity at the level of the resolution of the diffraction pattern. Methods such as discriminant analysis could be used for pattern recognition and appropriate statistical analysis. In principle, information about dynamic aspects of the protein's structure is also available through the temperature dependence of the pattern and 'patches' present in the diffraction pattern arising from diffuse X-ray scattering due to correlated intramolecular motions (Faure *et al.*, 1994). The question therefore becomes whether such data can routinely be obtained. In many if not most cases, this may be possible. Protein concentrations of several to tens of mg/ml are necessary for crystallization. Ultrafiltration procedures offer rapid (i.e., hours) methods to obtain such concentrations. This task is also facilitated by the quantities of protein that should typically be available at the pharmaceutical stages of development. The major concern is then our ability to crystallize a particular protein. Recent advances suggest, in fact, that many if not most proteins can be crystallized under one or more sets of conditions. While they may be small, it appears that crystals of sufficient quality can often be reproducibly obtained with persistent effort. This is reflected in the eventual appearance (usually within a few years) of crystallographically determined structures of many soluble proteins of intense, immediate interest, especially during drug discovery processes. Thus, although obtaining diffractable crystals is not automatic, it may rarely prove to be an insoluble problem. Rather the limiting requirement to obtaining the critical data may frequently be the availability of a protein solution of sufficient concentration and purity. Processes such as glycosylation and deamidation often introduce significant microheterogeneity into many protein preparations. One approach could be to minimize this phenomenon, perhaps through chemical or enzymatic conversion to a more homogeneous state prior to crystallization although this is clearly a less than satisfactory solution from a pharmaceutical perspective. Rather, the use of more homogeneous proteins as a consequence of improved production and isolation procedures may eventually provide proteins of sufficient uniformity to minimize this problem. Nevertheless, this microheterogeneity remains a major stumbling-block to the use of this approach.

Recent advances in both radiation sources and detectors have also greatly reduced the time necessary to obtain high resolution diffraction data. Approximately a dozen synchrotron beam lines are now available worldwide with another dozen or so projected to be available in the near future. In fact, over 20% of the protein crystallography papers published during 1992–1993 employed synchrotron radiation (Ealick and Walter, 1993). The high photon flux available from this radiation leads to very rapid data collection as well as better spot to spot resolution and signal to noise ratios. Simultaneously, film detection methods have been extensively replaced by image plates/storage phosphor systems (Pflugrath, 1992). In addition, charged-coupled device (CCD)-based X-ray detectors are beginning to appear which produce a rapid, directly computer processable readout of data. Both of these integrating X-ray position sensitive detectors feature high spatial resolution and dynamic ranges as well as low noise, ability to count at high rates and large active areas. The combination of synchrotron sources and these new detectors means that complete data sets can now be obtained in many cases in a few hours.

In summary, recent advances in X-ray crystallography combined with the need for only comparative data suggest that this method may soon be ready to play a tentative role in the pharmaceutical analysis of proteins. This potentially provides the ability to argue that the three dimensional structure of a test protein is unchanged at least within the (relatively high) resolution of this approach.

Nuclear Magnetic Resonance (NMR)

Like X-ray crystallography, NMR has generally been considered to be of little help in analyzing proteins under pharmaceutical conditions, although its general power as a structure tool for proteins is well recognized (Bax, 1991). Many of the problems such as the need for high protein concentrations and the complexity of the data analysis are similar to those of crystallography. As discussed above, this need not totally discourage the pharmaceutical use of NMR. In this case, a two dimensional NOE (nuclear Overhauser effect) proton spectrum of a test protein can be compared to a standard spectrum and the comparative data analysis reduces to that described above for crystallography with cross-peaks playing an analogous role to X-ray reflections. An advantage of NMR is that the solution state of the protein is examined. Unfortunately, this approach offers a major disadvantage that significantly reduces its potential utility. As the size of a protein increases, the line width of the resonances increases. Although this problem can be overcome to a limited extent by the combined use of isotopically labeled proteins and three and four dimensional techniques, an upper limit of approximately 35–40 kDa is still apparent (Clore and Gronenborn, 1994). Unfortunately, isotopic labeling is generally unacceptable for the pharmaceutical applications of interest. It follows, however, that in the case of smaller proteins of pharmaceutical utility (i.e., <20 kDa), proton NMR does have the potential to provide the type of high resolution structural information desired. For larger proteins, however, some type of breakthrough seems necessary for NMR to become of direct applicability to the pharmaceutical analysis of protein structure. The nature of such an improvement is not obvious. The use of higher magnetic fields seems unlikely to have much of an impact in this case. The significant problem of protein microheterogeneity remains with this technique as well.

Mass spectrometry

Recent advances in mass spectrometry (MS) have led to the application of this high resolution technique to proteins and other large molecules (Carr *et al.*, 1991; Chait and Kent, 1992). In particular, the introduction of electrospray (ES) and matrix-assisted laser desorption (MALD) methods permit the measurement of protein molecular weights with better than single dalton accuracy with molecules of moderate size or less. These techniques therefore permit the *covalent* structure of a protein to be rigorously defined. In general, they provide little conformational information although recent work suggests that limited three dimensional structural information might be obtainable by the electrospray method (Feng and Konishi, 1993; Mirza *et al.*, 1993). The two methods are complementary with each possessing distinctive strengths and weaknesses. In ES-MS, molecules at atmospheric pressure are ionized by injection through an electrostatically charged needle from a flowing liquid and consequently passed directly into the mass spectrometer. A characteristic of this process is that a series of multiply charged molecules are formed due to the many potential sites of cation association on proteins. While each peak permits an independent measurement of molecular weight, the spectra are complicated and thus less amenable to contaminant or degradate detection. The sample requirements for ES-MS are also rather stringent since salts significantly interfere with analysis. This problem is somewhat offset, however, by the ease of coupling of electrospray instruments to HPLCs thereby permitting the analysis of complex mixtures.

In contrast, MALD-MS usually produces only a few peaks and can be performed on samples under a wide variety of solution conditions including in the presence of many solutes such as salts. In this procedure, a high concentration of laser-absorbing matrix material is dissolved with the sample and this mixture is dried. Illumination of this solid by a laser pulse causes volatilization of the protein/matrix mixture and the subsequently ionized protein is accelerated down a time-of-flight analyzer. Most of the protein molecules are singly ionized thus simplifying analysis and often permitting contaminants and degradates to be identified. Overall, MALD-MS is somewhat more sensitive than electrospray methods (<1 pmol of protein can often be detected) but of somewhat lower accuracy and resolution.

Both methods permit mass alterations arising from protein modifications such as glycosylation, phosphorylation, acetylation, proteolysis and oxidation of Met and Trp residues to be easily detected. More subtle changes such as those produced by disulfide formation or breakage (a change [M] of 2 Da), deamidation of Asn or Gln residues (M~1 Da), or the presence of a C-terminal amide (M~1 Da) are more difficult to detect but can, in fact, be reliably determined with the use of proper standards, especially with proteins of molecular weight <20,000. Thus, in many cases the chemical identity of a protein can be reliably established by these methods. Furthermore, the high resolving power and molecular weight accuracy of MS can be employed in peptide mapping analysis (Amatt *et al.*, 1993). ES-MS can be combined with HPLC to identify individual peaks while MALD-MS on its own can be used to directly generate an interpretable peptide fragmentation pattern (Billeci and Stults, 1993). It is possible that MS alone or in combination with time-dependent proteolysis or other chemical modifications can be employed to yield conformational information, but the resolution of such approaches is currently uncertain. We emphasize again, however, that MS procedures are generally insensitive to protein conformational features and therefore are unlikely to be used to demonstrate the unique, three dimensional structure of a macromolecule. Nevertheless, the high information content of MS measurements could potentially be utilized in combination with lower resolution methods to provide a more comprehensive structural characterization, as we discuss next.

Multiple parametric approaches

If the information necessary to define the structure of a protein cannot be obtained by a single, information rich method, then the use of some combination of procedures can be employed. This is, of course, the currently used approach in which chromatographic, electrophoretic, spectroscopic, immunological and biological methods are sequentially implemented (Jones, 1993). As described above, major problems with this approach include the time, effort and complexity involved as well as ultimate reliance upon some type of biological assay to finally ensure structural propriety. One possible step forward would be a single device which simultaneously performs a carefully selected set of the analyses desired (Middaugh, 1990). A number of different types of such machines can be envisioned. Here we describe only one such system based on a selection of some of the more information-rich structural parameters for a typical pharmaceutical protein.

In general, such a device is envisioned to consist of a chromatographic system with multiple detectors. Several options exist for the front-end chromatographic component based on the property of the protein to which we wish the separation process to be sensitive. It would seem simplest to also avoid nonaqueous (i.e., organic) solvent systems because of the structural lability of proteins in the presence of such agents. Therefore, either an ion-exchange or molecular sieve column (separating on the basis of charge and size, respectively) are reasonable choices by these criteria. Alternatively, if the problem of small sample volume can be overcome, capillary electrophoresis could be used. The choice of detectors is currently primarily limited by availability and overall system compatibility.

A logical first choice for a detector is a simple UV/visible absorption device. This should be either of the diode array or fast-scan grating type to permit entire spectra to be rapidly obtained. Analysis of protein UV absorbance spectra in the form of second derivatives can potentially provide simultaneous evaluation of the average local environments of Trp, Tyr and Phe residues in a protein and thus serve as a sensitive monitor of tertiary structure (Mach and Middaugh, 1994). In principle, the spectral properties of the peptide-bond absorption peak near 190 nm could provide information about the protein's secondary structure content. Because of high extraneous absorptive interference in this region, however, this would almost certainly be better accomplished by a second detection component based on far-UV circular dichroism (CD) (Johnson, 1990). Such a measurement also presents several difficulties in obtaining the type of data required (e.g. low sensitivity, interference by many solutes, etc.), but the construction of an appropriate specialized CD detector is technically feasible. The other attractive secondary-structure sensitive method which involves measurement of infrared absorption by the protein's amide bonds is overly complicated by even lower sensitivity in the solution state and the intense infrared absorption of water (Surewicz *et al.*, 1993).

A third attractive detection step would involve measurement of the intrinsic UV fluorescence emission of the protein. Far UV excitation (260–300 nm) produces primarily emission from a protein's Trp residues (Burstein *et al.*, 1973). Although such measurements provide only a probe of the immediate environment of the few indole sidechains in most proteins, these residues are frequently very sensitive to the three dimensional structure of many proteins. Another advantage of this technique is that Raleigh scattered light is also produced by the protein and this can be simply detected (usually at right angles) during the same measurement (Dollinger et al, 1992). The intensity of this scattered light (see below) is proportional to the size of the scatterer (protein) and could thus serve as a convenient monitor of protein size. A major requirement for obtaining useable information from fluorescence emission data is that the entire spectral region of interest needs to be examined. Again, this can now be simply accomplished by either fast scanning or diode array technology.

The size or state of oligomerization/aggregation of a protein is another crucial molecular parameter of interest. Although elution position in a front-end molecular sieve separation provides information in this regard, this is probably best accomplished by measurement of the angular dependence of the steady-state intensity of scattered light and/or a dynamic light-scattering determination in which the autocorrelation function of the scattered light is analyzed. Both techniques have the required sensitivity and are available commercially in a chromatographic detector format (Claes, 1990; Wyatt and Papazian, 1993). In theory, a combination of these two measurements can provide the radius of gyration, hydrodynamic radius and molecular weight of a protein. This level of analysis is probably unnecessary, however, since as argued above, comparative measurements employing a protein standard of the same identity as the test molecule should be all that is necesssary.

Finally, it would then seem optimal to pass the analyzed material to a mass spectrometer for definitive chemical identification. At this point, ES-MS would seem the better approach, but MALD-MS methods involving deposition of samples with auxiliary matrix addition onto some type of receptive device for subsequent analysis may also be possible.

A plethora of variations of the type of instrument outlined can be imagined in which rigorous chemical identification (MS) can be combined with analysis of secondary (e.g., CD), tertiary (UV absorption, fluorescence), and quaternary (e.g., light scattering) structure in a chromatographic format (e.g., electrostatic information) to provide a comprehensive description of a macromolecule. Then by comparison to a standard protein of the same type, the identity of the molecule can arguably be established. One can even imagine the optical portion housed in a single package perhaps employing CCD technology. Such a 'protein machine' should require only submilligram quantities of protein and could also be used to analyze stressed samples (e.g. employing elevated temperature or extremes of pH) providing both real time and accelerated stability information. I conclude this section on a tentative note, however, in the form of a question: How could one unequivocally demonstrate the ability of such a system to detect any and all potential changes in a target protein? As indicated above, the intrinsic microheterogeneity of most proteins presents a significant challenge to such a system although this problem could be addressed by the chromatographic and MS elements. At this point, however, the less than satisfactory answer must be that detection of all potential structural alterations would not be possible. Rather, the same procedure currently utilized requiring a series of different measurements would need to be employed and all possible relevant forms of molecular change generated and then demonstrated to be resolvable by the combined approach.

Miscellaneous comments

An area not directly addressed by the methods described previously is the analysis of contaminants and degradates. It seems at least in the immediate future, highly specific individual assays will still need to be developed for each specific protein pharmaceutical. It is possible, however, that mass spectroscopy or some type of multiple-parameter protein analysis system could be an advance in this regard as well. Neither crystallography or NMR seems likely to prove helpful despite their potential power in structural verification.

Another area in which advances might be anticipated, however, concerns more biologically based assays. A strong trend already exists toward the replacement of animal studies with cell culture based measurements. For example, the biological activity of growth factors is now usually measured by their stimulation of mitogenesis of cultured cells. Similarly, many proteins which initially exert their effects by binding to cell surface receptors can be analyzed in terms of the binding process itself. A simple extrapolation of the success of these approaches suggests that analysis of critical intracellular responses induced by a particular pharmaceutical protein might be of significant utility in monitoring the protein's biological activity. Advances in this area could take several forms. For example, the development of pH sensitive microsensors has permitted the biological responses of several proteins to be quantitated in terms of changes in pH induced in the medium of cultured cells by be proteins (McConnell *et al.*, 1992). Commercial instrumentation has recently become available based on this work which should permit data of adequate precision to be obtained. Similarly, the development of commercial plasmon resonance technology which detects changes in the concentration of agents near an optical surface as an immediate consequence of binding phenomena could also prove of pharmaceutical utility (Granzow and Reed, 1992). A variety of other physically and biologically based sensors such as quartz crystal microbalances

(Lasky and Buttry, 1990) should become available in the next several years which have the potential to be applied to new bioassays. One focus could be on the events which take place during the complex molecular signaling pathways which underlie many intracellular information transduction mechanisms. Already well described are studies of changes in intracellular calcium levels and alterations in membrane potential which are easily detectable by fluorescence methods. Once again, it will be incumbent upon the pharmaceutical community to demonstrate that such approaches both serve as accurate and biologically realistic monitors of protein bioactivity and possess the ruggedness and precision required of pharmaceutical assays, but this seems well within the realm of near-future plausibility.

Finally, it seems probable that the analysis of vaccines is on the verge of significant progress. Vaccines usually consist of a relatively complex mixture of components. Major forms of vaccines include attenuated and inactivated viral particles, high molecular weight polysaccharides, polysaccharides conjugated to various proteins and protein complexes, protein aggregates (natural and recombinant), peptides and peptide conjugates, naked DNA, etc. These materials are usually analyzed by either their induction of antibody titers in test animals or some type of cell based plaque assay. Unfortunately, such techniques are of low throughput and high variability often resulting in major analytical limitations. Structural characterization, if any, is usually limited to simple electrophoretic or chromatographic methods. As simpler peptide and recombinant protein based vaccines become available, the more comprehensive type of analysis currently employed for protein phamraceuticals can be expected to be employed (Volkin *et al.*, 1994). Such vaccines are usually of reduced immunogenicity, however, and higher resolution methods are needed to analyze the more complex types of vaccines. In fact, it would seem that the first steps can immediately be taken. It is clear, for example, that detailed information about the size, shape and aggregation state of even complex entities (e.g., viral or protein particles, polysaccharide-protein conjugates) can be obtained by a combination of molecular sieve chromatography, static and dynamic light scattering, and analytical equilibrium and velocity sedimentation. Similarly, CD and FTIR can be employed to extract secondary structure information, intrinsic fluorescence for tertiary structure characterization and mass spectrometry for detailed chemical data. A variety of extrinsic spectroscopic (e.g., fluorescent) probes sensitive to the polarity of their binding sites can also be used to explore both local and global aspects of structural features of vaccine molecular complexes. These methods could also be applicable to polysaccharides and nucleic acids either separately or in combination with protein components and may be useful even in the presence of adjuvants and delivery system components. It is not our purpose here to discuss this issue in any depth, but rather to simply point out that analyses of vaccines is hopefully moving toward the situation that currently exists with purified proteins; namely, a combination of methods that characterize an extensive variety of structural features of the major vaccine components. To what extent such methods can increase our confidence in the quality of vaccine entities remains to be seen, although total elimination of more biologically based assays for vaccines seems only a remote possibility in the immediate future.

References

Arnott D, Shabanowitz J and Hunt DF (1993) Mass spectrometry of proteins and peptides: sensitive and accurate mass measurement and sequence analysis. Clin. Chem. 39: 2005–2010.

Bax A (1991) Experimental NMR techniques for studies of biopolymers. Curr. Op. Struct. Biol. 1: 1030–1035.

Billeci TM and Stults JT (1993) Tryptic mapping of recombinant proteins by matrix-assisted laser desorption/ionization mass spectrometry. Anal. Chem. 65: 1709–1716.

Burstein EA, Vedenkina NS and Ivkova MN (1973) Fluorescence and the location of tryptophan residues in protein molecules. Photochem. Photobiol. 18: 263–279.

Carr SA, Hemling ME, Bean MF and Roberts GD (1991) Integration of mass spectrometry in analytical biochemistry. Anal. Chem. 63: 2802–2824.

Chait BT and Kent SBH (1992) Weighing naked proteins: practical, high-accuracy mass measurement of peptides and proteins. Science 257: 1885–1894.

Claes P, Fowell S, Woollin C and Kenney A (1990) On-line molecular size determination for protein chromatography. Am. Lab. 22: 58–62.

Clore GM and Gronenborn AM (1994) Structures of larger proteins, protein-ligand and protein-DNA complexes by multidimensional heteronuclear NMR. Protein Sci. 3: 372–390.

Dollinger G, Cunico B, Kunitani M, Johnson D and Jones R (1992) Practical on-line determination of biopolymer molecular weight by high-performance liquid chromatography with classical light-scattering detection. J. Chromatogr. 592: 215–228.

Ealick SE and Walter RL (1993) Synchroton beamlines for macromolecular crystallography. Curr. Op. Struct. Biol. 3: 725–736.

Faure P, Micu A, Perahia D, Doucet J, Smith JC and Benoit JP (1994) Correlated intramolecular motions and diffuse X-ray scattering in lysozyme. Struct. Biol. 1: 124–128.

Granzow R and Reed R (1992) Interactions in the fourth dimension. Bio/Technology 10: 390–393.

Feng R and Konishi Y (1993) Stepwise refolding of acid-denatured myoglobin: Evidence from electrospray mass spectrometry. J. Am. Soc. Mass Spectrom. 4: 638–645.

Johnson Jr. WC (1990) Protein secondary structure and circular dichroism: a practical guide. Proteins: Struct., Func., Genet. 7: 205–214.

Jones AJS (1993) Analysis of polypeptides and proteins. Adv. Drug Delivery Rev. 10: 29–90.

Lasky SJ and Buttry DA (1990) Development of a real-time glucose biosensor by enzyme immobilization on the quartz crystal microbalance. Am. Biotech. Lab 2: 8–16.

Mach H and Middaugh CR (1994) Simultaneous Monitoring of the environment of of tryptophan, tyrosine and phenylalanine residues in proteins by near-UV second derivative absorption spectroscopy. Anal. Biochem. (in press).

McConnell HM, Owicki JC, Parce JW, Miller DL, Baxter GT, Wada HG and Pitchford S (1992) The cytosensor microphysiometer: biological applications of silicon technology. Science 257: 1906–1912.

Middaugh CR (1990) Biophysical approaches to the pharmaceutical development of proteins. Drug Develop. Ind. Pharm. 16: 2635–2654.

Mirza UA, Cohen SL and Chait BT (1993) Heat-induced conformational changes in proteins studied by electrospray ionization mass spectrometry. Anal. Chem. 65: 1–6.

Pflugrath JW (1992) Developments in X-ray detectors. Curr. Op. Struct. Biol. 2: 811–X15.

Surewicz WK, Mantsch HH and Chapman D (1993) Determination of protein secondary structure by fourier transform infrared spectroscopy: a critical assessment. Biochemistry 32: 389–394.

Volkin DB, Burke C, Marfia K, Middaugh R, Oswald B, Hennessey J, Orella C, Hagen A, Sitrin R and Oliver C (1994) Biophysical characterization of hepatitis A virus (HAV). Abstracts of Papers, 207th ACS National Meeting, March 13–17, 1994, American Chemical Society, Washington, DC, BIOT 10.

Wyatt PJ and Papazian LA (1993) The interdetector volume in modern light scattering and high performance size-exclusion chromatography. LC-GC 11: 862–872.

Address for offprints: C.R. Middaugh, Department of Pharmaceutical Research, Merck Research Laboratories, WP78-302, West Point, PA 19486.

Cytotechnology **15:** 195–208, 1994.

The effect of protein synthesis inhibitors on the glycosylation site occupancy of recombinant human prolactin

Marc Shelikoff[1], A. J. Sinskey[2] and Gregory Stephanopoulos[1]
Departments of Chemical Engineering[1] and Biology[2], and the Biotechnology Process Engineering Center, 18 Vassar St., 20A–207, Massachusetts Institute of Technology, Cambridge, Massachusetts, 02139–4308

Key words: Cycloheximide, glycosylation, inhibition of protein synthesis, post-translational processing, prolactin

Abstract

The relationship between synthesis and N-linked glycosylation site occupancy of recombinant human prolactin produced from C127 cells was studied with the aid of a battery of protein synthesis inhibitors. Non-lethal concentrations of sodium fluoride, gougerotin, puromycin, anisomycin, and emetine did not alter site occupancy, but low concentrations (<10 μg ml^{-1}) of cycloheximide increased the fraction of secreted prolactin bearing oligosaccharide from 20% to 80% of the total. Cycloheximide is an inhibitor of the elongation step of protein synthesis. The observed increase in glycosylation site occupancy upon addition of cycloheximide is consistent with the current opinion that the initial glycosylation event occurs cotranslationally during a limited time period. Cycloheximide may extend this time period by reducing elongation rate. However, the absence of any effect from treatment with other inhibitors of elongation suggests that cycloheximide is unique in its behavior on this system.

Abbreviations: clp-PRL = clipped form of prolactin, DMEM/F12 = 1:1 Dulbecco's Modified Eagle's Medium/Ham's nutrient mixture F12, G-PRL = glycosylated (N-linked) fraction of prolactin, NG-PRL = prolactin fraction without N-linked glycosylation, PMSF = phenylmethylsulfonylfluoride

Introduction

Glycosylation site occupancy

The general pathway of asparagine-linked glycosylation has been known for some time (Kornfeld, 1985). A completed oligosaccharyldolichol precursor provides the sugar substrate for the most critical reaction in N-linked glycoprotein formation (Fig. 1). This is the transfer of oligosaccharide from dolichol to a polypeptide at the asparagine of its Asn-X-Ser/Thr consensus sequon. It is often the case, both in cultured cells and *in vivo*, that this reaction does not occur for every molecule of a given protein. Two protein populations, one bearing and one lacking oligosaccharide, may be produced for each potential glycosylation site on a secretory protein. The variability in the *presence* of oligosaccharide has been called 'glycosylation site occupancy heterogeneity', 'core glycosylation heterogeneity', or 'glycosylation macroheterogeneity'. It is conceptually and physically distinct from variability in the *identity* of oligosaccharide, which has been referred to as 'microheterogeneity'. Several reviews of glycoprotein production with a biotechnology emphasis have recently been published (Goochee, 1990; Cumming, 1991; Welply, 1991; Goochee, 1992; Jenkins, 1994).

Oligosaccharyltransferase, a multimeric, membrane-bound protein complex (Kelleher, 1992), catalyzes asparagine-linked glycosylation. It has been known for some time that this enzyme usually acts on the protein while it is still being translated (Rothman, 1977; Bergman, 1978). The spatial and temporal relationship among translation, glycosylation, and folding have been investigated previously by several groups.

A relationship between glycosylation and folding was first indicated when the normally unglycosylated potential site of ovalbumin was able to become gly-

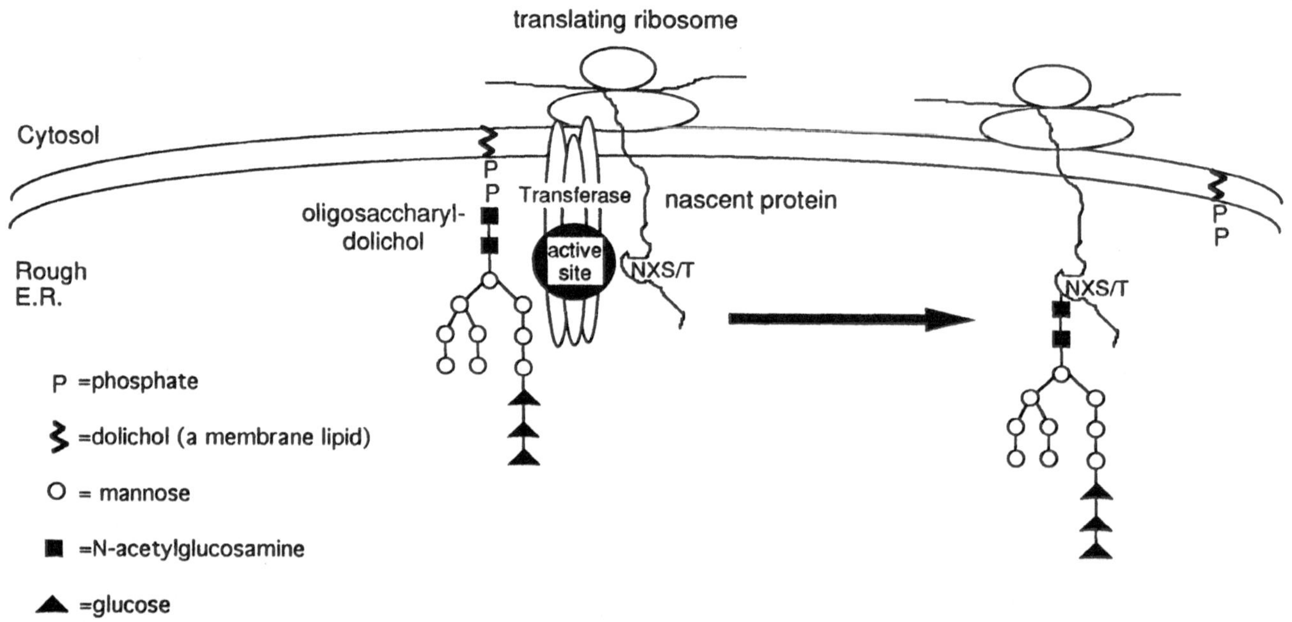

Fig. 1. Asparagine-linked oligosaccharide transfer.

cosylated only after denaturation (Pless, 1977). The partial glycosylation site occupancy of a non-standard Asn-X-Cys sequon in protein C (Miletich, 1990) was believed to result from destruction of the site by disulfide bond formation. These authors speculated that a disulfide bond with this cysteine removes the hydrogen necessary for oligosaccharide transfer only if the bond forms before glycosylation occurs. Direct evidence of an inverse relationship between disulfide bond formation and core glycosylation of a standard site in mutated t-PA was demonstrated by Bulleid (1992) by altering the redox potential of an *in vitro* translation system. These results were interpreted by the above authors as demonstrating that cotranslational folding events prevent oligosaccharide transfer by burying the site within the interior of the protein.

A relationship between glycosylation and translation was suggested by Lau (1983) who noted that only a brief moment in time may exist when the potential glycosylation site on a nascent polypeptide is near the region of space where the transferase active site resides. The region occupied by the active site is known to begin roughly 4 nanometers from the E.R. membrane (Nilsson, 1993). Potential glycosylation sites close to the N-terminus of a protein would be expected to have a high glycosylation site occupancy because they spend more time near the transferase than sites close to the C-terminus. Two pieces of evidence are consistent with this view. A statistical analysis of glycoproteins found in nature showed that sites near the C-terminus are more likely to be completely unoccupied (Gavel, 1990). Site-directed mutagenesis has also been used to introduce premature termination codons near a potential glycosylation site of rabies virus glycoprotein (Shakin-Eshleman, 1993). Site occupancy was decreased by moving the C-terminus of the protein closer to the potential site.

In addition to the complex relationship among translation, glycosylation, and folding, the identity of amino acids near or within the sequon and the availability of transferase and oligosaccharide precursor relative to protein production have been suggested as potential determinants of glycosylation site occupancy. With regards to the amino acid sequence, there is both experimental (Bause, 1981; Roitsch, 1989) and statistical evidence from sequenced glycoproteins (Gavel, 1990), that threonine in the third position is more likely to result in glycosylation than serine (with cysteine an unlikely third possibility), and that a proline residue at position X or at a position one amino acid C-terminal to the sequon strongly disrupts glycosylation. Correlations between the solution conformation and glycosylation efficiency of different tripeptide substrates (Imperiali, 1991) suggest other parts of the amino acid sequence may determine site occupancy, but these relationships must be subtle because they do not eliminate the site occupancy of sequenced glycoproteins (Gavel, 1990).

Insufficient oligosaccharyltransferase expression or activity may also determine glycosylation site occupancy. The ability of a soluble peptide substrate containing a glycosylation sequon to compete for transferase with a synthesized protein (Lau, 1983) provides some evidence for the importance of transferase expression. However, recognition that transferase complexes are present in roughly 1:1 stoichiometry with membrane bound ribosomes and that transferase contains a ribosome-binding protein (Kelleher, 1992) has led to the hypothesis that one transferase complex may be positioned at every nascent polypeptide by first binding to the ribosome. Saturation of transferase due to high protein production is impossible if this hypothesis is correct. A decline in transferase activity in regenerating liver (Oda-Tamai, 1985) causes decreased core glycosylation. This effect may be mediated by the phosphatidylcholine composition of the E.R. membrane which is thought to modulate site occupancy by altering transferase specific activity (Chalifour, 1988).

Oligosaccharyldolichol precursor availability is believed to be a key regulatory mechanism for glycosylation site occupancy. Insufficient precursor may result from not enough formation or from overabundant degradation of the complete oligosaccharyldolichol (Spiro, 1991). The glycosylation site occupancy of recombinant gamma-interferon decreases in cell culture with increasing glucose limitation (Hayter, 1992). This effect was believed to be due to insufficient precursor because glucose starved cells are known to contain less completed oligosaccharyldolichol (Chapman, 1988). It was recently determined (Konrad, 1994) that stimulation of core glycosylation by the second messenger cAMP is due to an increased level of oligosaccharyldolichol. A cAMP mediated increase in precursor may explain the effect of various hormones on core glycosylation in several cell types (Konrad, 1994).

It has been known for some time that inhibitors of protein synthesis cause a decrease in the formation of completed oligosaccharyldolichol (Schmitt, 1979; Hubbard, 1980), presumably because less precursor is required by the cell. A drop in protein synthesis due to detachment of anchorage dependent Chinese hamster ovary cells caused a similar decline in oligosaccharyldolichol (Cacan, 1993). This decrease does not occur in glucose-starved cells, but, in fact, an increase in oligosaccharyldolichol synthesis upon inhibition of protein synthesis is observed (Chapman, 1988). One hypothesis explaining this behavior is that a decline in protein synthesis in glucose-fed cells causes a high GTP level which acts as a feedback signal to inhibit enzymes of the oligosaccharyldolichol synthesis pathway (Grant, 1983; Spiro, 1986). In glucose-starved cells, high GTP levels may not be possible, thus destroying the feedback response to protein synthesis inhibition (Chapman, 1988).

Glycosylated prolactin

This paper describes the effect of inhibitors of protein synthesis on the glycosylation site occupancy of recombinant human prolactin. Prolactin contains only one potential asparagine-linked glycosylation site (Asn^{31}-Leu-Ser-Ser) which is partially occupied in protein derived both from serum and from culture fluid of cells transfected with the prolactin gene (Cole, 1991). Prolactin is an ideal model protein to use for these studies because it exhibits the simplest type of glycosylation macroheterogeneity possible: one protein population produced with and one without a single N-linked oligosaccharide.

Prolactin is a 199 amino acid hormone with a wide variety of effects on the body (reviewed in Cooke, 1989), most notably stimulation of lactation post partum. Prolactin is secreted into the bloodstream by mammotropes (specialized prolactin-producing cells) in the pituitary via a regulated pathway. Secretion may be triggered by stress, eating, or nursing. Sleep and pregnancy increase the quantity of prolactin released during each episode and increase the frequency of these secretion events. Decidual cells of the endometrium also secrete prolactin, but via a constitutive pathway into the amniotic fluid during pregnancy.

The glycosylated form of human prolactin was first reported less than ten years ago (Lewis, 1985). About 15% of prolactin in human pituitary (Lewis, 1989) and 30% of prolactin from decidual cells (Lee, 1986) is N-glycosylated. The site occupancy of prolactin in the serum of human subjects has been shown to decrease under the same conditions that induce regulated secretion from the pituitary (Hashim, 1990; Brue, 1992). This decrease is explained by the observation that NG-PRL is stored and secreted more efficiently along the regulated pathway of mammotropes than G-PRL (Pellegrini, 1990). Prolactin found in serum and in amniotic fluid exhibits considerable molecular weight heterogeneity due not only to glycosylation, but to proteolysis and the formation of dimers and aggregates (Fukuoka, 1991). A hypothesis has evolved (Lewis, 1984; Fonseca, 1991) that different patterns of prolactin heterogeneity control a balanced physiological response to the event which triggered the prolactin release.

G-PRL and NG-PRL exhibit different biological activities on several systems. Standard bioassays reveal that G-PRL has a lower specific activity than NG-PRL (Pellegrini, 1988; Lewis, 1989). This is most likely a result of the lower binding affinity of G-PRL towards prolactin receptor (Pellegrini, 1988; Haro, 1990). Due to its microheterogeneity, some glycoforms of G-PRL may be cleared from the body more slowly and some more rapidly than NG-PRL (Sinha, 1991). G-PRL and NG-PRL also exhibit different binding affinities in various immunoassays (Pellegrini, 1988), thus altering the accuracy of diagnostic prolactin determinations (Haro, 1990).

Protein synthesis inhibitors

A large number of antibiotic and chemical inhibitors of eukaryotic protein synthesis have been found. The site of action with respect to the translation cycle of the six inhibitors used in these experiments is known with some certainty (Fig. 6), but the biochemical mechanism of inhibition of only puromycin has been determined in detail (Vazquez, 1979; Gale, 1981; Ballesta, 1991). It is also known that some of these inhibitors have effects on the cell which are unrelated or only indirectly related to their effect on translation. In addition to oligosaccharyldolichol synthesis, previous work has also determined the effect of different inhibitors of protein synthesis on thermotolerance (Lee, 1987) and induction of the immediate early response genes (Edwards, 1992), both of which may be related to the initial glycosylation event (Borrelli, 1991; Henle, 1993).

Materials and methods

Cell line and tissue culture

A mouse C127 cell line transfected with human prolactin cDNA in a bovine papilloma virus-metallothionein vector (Hsiung, 1984; Cole, 1991) was a gift of Genzyme Inc., Framingham, MA. C127 is an anchorage dependent transformed adenocarcinoma derived from murine mammary tissue. Cells were thawed into T-flasks containing DMEM/F12 (JRH Biosciences) supplemented with 10% fetal bovine serum (Sigma). Cultures were maintained in a humidified incubator at 37 °C in the presence of 5% CO_2. Medium with serum was replaced after 2 days, and cultures were split four fold when confluent – four days after thawing. After two such passages, cells were seeded into wells used in the experiments. Medium with serum was replenished after 2 days. Four days after passage, when cells were confluent, cultures were rinsed twice in phosphate buffered saline to remove serum, and protein-free DMEM/F12 without serum was added to the cultures. While no further growth has been observed in the absence of serum, confluent cultures secreting prolactin have been maintained for three weeks by replacing medium every other day.

Prolactin heterogeneity

The presence of N-linked oligosaccharide on prolactin in raw supernatant was determined by removal with recombinant N-Glycanase (Boehringer Mannheim and Genzyme). The protocol included with the Genzyme material was followed.

Proteolytic activity in raw supernatant was determined by incubation at 37 °C for 5 days of supernatant from serum-free cultures at various pH with and without 1 mM PMSF (Sigma), a serine protease inhibitor (Gold, 1967; Scopes, 1987).

Protein synthesis inhibition

Medium was removed two days after switching to serum-free conditions. Cells were preincubated in fresh DMEM/F12 with protein synthesis inhibitor for 40 minutes. This medium was then removed, and cells were incubated for one to two days in fresh DMEM/F12 containing identical concentrations of inhibitor. Medium containing prolactin secreted in the presence of inhibitor was then collected. One negative control using identical medium exchanges in the absence of inhibitor was performed for all six inhibitor experiments. Viability was determined by cell counting using exclusion from trypan blue (Sigma).

Inhibitors used were cycloheximide, anisomycin, gougerotin, puromycin, (all from Sigma), emetine (ICN Biochemicals), and sodium fluoride (Mallinckrodt). A range of concentrations of each inhibitor were used which have previously been reported to cause a partial inhibition in protein synthesis (Contreras, 1978; Mankovitz, 1978). These concentrations are shown in the legend of Fig. 3.

Electrophoresis and western blots

Raw supernatant from cells incubated in the presence of inhibitors was subjected to Laemmli's method of SDS/PAGE (Hames, 1990) followed by electrophoretic transfer onto nitrocellulose (Schleicher & Schuell) and western blotting (Garfin, 1989). Precast 12% acrylamide gels (Bio-rad or Jule) were used in Bio-Rad's Mini-Protean II electrophoresis cell. Prelabelled molecular weight standards were from Gibco-BRL. Purified human prolactin standard was a gift from Genzyme Inc., Framingham, MA.

Polyclonal rabbit antisera to human prolactin was purchased from Ventrex Laboratories. Goat-anti rabbit alkaline phosphatase conjugate was purchased from Bio-rad. Goat-anti-rabbit-horseradish peroxidase conjugate was purchased from Cappel. Bio-rad's system for alkaline phosphatase or horseradish peroxidase color development was used.

Densitometry

Air dried nitrocellulose blots were analyzed on a Hoefer GS300 scanning densitometer. Peak area was integrated using Hoefer's GS370 data analysis system on a Macintosh IIsi.

Results

Prolactin heterogeneity detected by SDS/PAGE

Prolactin from cell culture and from purified standard appear on western blots as 2, 3, or 4 major bands. Comparison with standards indicates that the four bands have apparent molecular weights between 23 and 29 kilodaltons. Digestion with N-Glycanase removes the two highest molecular weight bands (Fig. 2a). This indicates that band 1 and band 2 correspond to prolactin bearing an N-linked oligosaccharide.

Incubation of cell free supernatant for 5 days resulted in an increase in band 4 and decrease in band 1. This change was inhibited by acidic conditions and by the presence of PMSF (Fig. 2b), suggesting that band 4 represents a form that is clipped by an extracellular serine protease. A possible substrate site for the action of this protease is Arg-10/Cys-ll which was shown by Cole *et al.* (1991) to be cleaved in recombinant baboon PRL produced in C127. The position of band 2 relative to band 1 in figure 2a suggests that it represents a glycosylated proteolysis product. We have assigned the names G-PRL, G-clp-PRL, NG-PRL, and NG-clp-PRL to bands 1 through 4 respectively. Clipped forms are often faint or difficult to resolve from complete forms.

Protein synthesis inhibition

Western blots of prolactin secreted during incubation with inhibitor are shown in Fig. 3. For anisomycin, gougerotin, puromycin, sodium fluoride, and emetine (Fig. 3b–f), no significant effect on glycosylation site occupancy of secreted prolactin was noted. Glycosylation site occupancy increased substantially with cycloheximide concentration, however (Fig. 3a). In medium from cells incubated 17 hours in the absence of inhibitor (lane 9), the lower band is the major form, but in medium isolated from cells incubated in the presence of 0.6 $\mu g\ ml^{-1}$ cycloheximide (lane 5), the top band is darker. Similar results were obtained from medium withdrawn after 2 and after 44 hours of incubation (Fig. 4), indicating that this is not a time-dependent phenomenon.

Viability was determined at a single high inhibitor concentration for each experiment (Table 1). In every case tested, cultures were found to be greater than 70% viable.

Densitometry

Optical scanning allows us to present the protein synthesis inhibition data of Fig. 3 in a compressed form. When prolactin concentration varies by less than an order of magnitude on a single blot, a linear response of densitometer peak area to prolactin standard concentration has been found for concentrations ranging from 2 to 280 $\mu g\ ml^{-1}$ (results not shown). Figure 5a describes the response of total prolactin production to each inhibitor as determined by optical scanning. Each data point represents one lane from Fig. 3 containing one culture sample from cells incubated in the presence of inhibitor. The sum of the densitometer peak areas corresponding to prolactin (all bands found between 20 and 30 kilodaltons) produced in the presence of inhibitor was divided by the sum of the densitometer peak areas corresponding to prolactin in the negative control from the same experiment depicted on the same blot. This ratio is the ordinate of Fig. 5a. An average value of the two negative control lanes for sodium fluoride was used. These inhibition curves are similar to those observed previously for total cellular protein (Contreras, 1978; Mankovitz, 1978) with the exception

Table 1. Viability of cultures exposed to high inhibitor concentrations

Inhibitor	Concentration range (μg ml^{-1}) used to partially inhibit synthesis[1,2]	Concentration (μg ml^{-1}) used to determine viability	Viability
no inhibitor (n=4)			81±11%
Cycloheximide	0.02–20	60	88%
Puromycin	0.4–10	80	70%
Anisomycin	0.02–20	60	70%
Emetine	0.3–200	200	74%
Gougerotin	0.3–300	50	71%
Sodium fluoride	20–85	60	80%

[1] Contreras, 1978; [2] Mankovitz, 1978

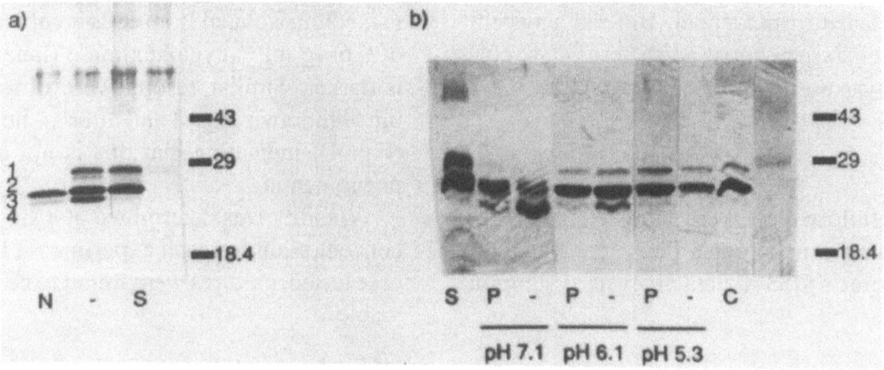

Fig. 2. Prolactin heterogeneity detected by western blotting. a) Digestion with N-glycanase – N-glycanase digested cell supernatant (N) and undigested control (–) are compared with prolactin standard (S); b) Proteolysis inhibition – Cell-free supernatant incubated for five days in the presence of PMSF (P) and with no PMFS (–) are compared at pH 7.1, 6.2, and 5.3. An additional control (C) consists of cell supernatant which was stored at –20 °C for the same five day period. Lane S contains prolactin standard.

of emetine where we required a higher concentration to inhibit prolactin production.

The effect of protein synthesis inhibitors on prolactin glycosylation site occupancy is shown in Fig. 5b. The ratio of glycosylated to total prolactin was estimated from the blots shown in Fig. 3 by dividing the densitometer peak area of band 1 (band 2 is not resolved from band 1 in any of the blots shown in Fig. 3) by the peak area of all bands found between 20 and 30 kilodaltons. This ratio is the ordinate of Fig. 5b. The combined effect of variability in measured site occupancy due to both measurement imprecisions and to experimental conditions may be found by comparing the G-PRL / total- PRL ratio obtained in the negative controls for all six experiments. An average value for glycosylated/total prolactin of 0. 13 is obtained for the negative controls with a standard deviation (n=6) of 0.04.

The effect of protein synthesis inhibitors on both production and site occupancy of prolactin is found in Fig. 5c where the ordinates from Fig. 5a and 5b are plotted against each other. Linear curve fits shown in Fig. 5c were obtained from all data with prolactin production greater than 10% of the uninhibited value (method of least squares: n=6 for cycloheximide, n=34 for all others). We suspect that data obtained when prolactin production drops below 10% of the control is unreliable due to both the extreme faintness of these bands and to the lack of a linear response as described above.

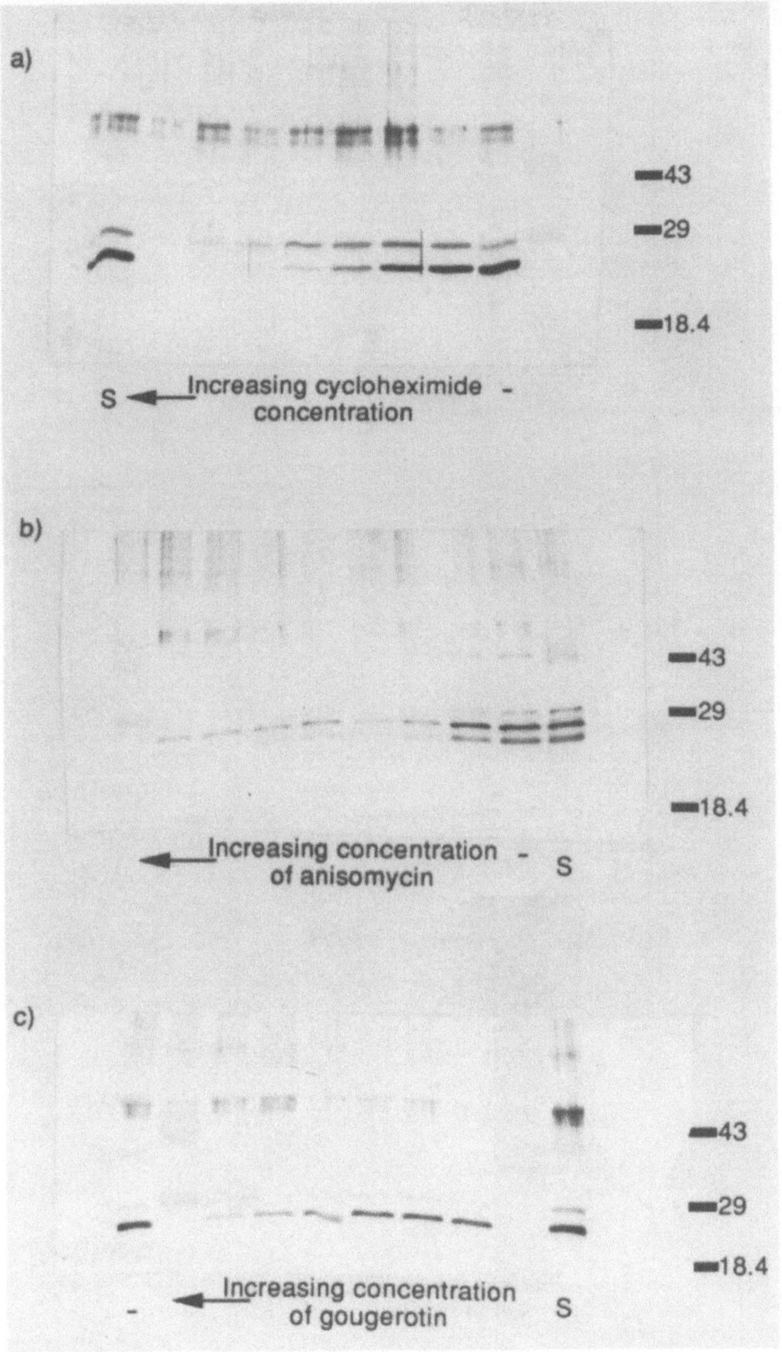

Fig. 3. Effect of protein synthesis inhibitors on prolactin glycosylation. Confluent cell cultures were pre-incubated with inhibitor for 40 minutes, and incubated at the same inhibitor concentrations for one day. Lanes marked S contained prolactin standard. Lanes marked – were incubated without inhibitor. a) Effect of cycloheximide. Lane 1: prolactin standard. Lane 2: incubated with 20μg ml^{-1} cycloheximide, lane 3: 6 μg ml^{-1}, lane 4: 2 μg ml^{-1}, lane 5: 0.6 μg ml^{-1}, lane 6: 0.2 μg ml^{-1}, lane 7: 0.06 μg ml^{-1}, lane 8: 0.02 μg ml^{-1}. Lane 9: incubated without inhibitor. Lane 10: molecular weight standards; b) Effect of anisomycin. Lane 1: molecular weight standards. Lane 2: incubated with 20μg ml^{-1} anisomycin, lane 3: 6 μg ml^{-1}, lane 4: 2 μg ml^{-1}, lane 5: 0.6 μg ml^{-1}, lane 6: 0.2 μg ml^{-1}, lane 7: 0.06 μg ml^{-1}, lane 8: 0.02 μg ml^{-1}. Lane 9: incubated without inhibitor. Lane 10: prolactin standard; c) Effect of gougerotin. Lane 1: incubated without inhibitor. Lane 2: molecular weight standards. Lane 3: incubated with 300μg ml^{-1} gougerotin, lane 4: 150 μg ml^{-1}, lane 5: 50 μg ml^{-1}, lane 6: 12 μg ml^{-1}, lane 7: 3 μg ml^{-1}, lane 8: 0.3 μg ml^{-1}. Lane 9: blank. Lane 10: prolactin standard.

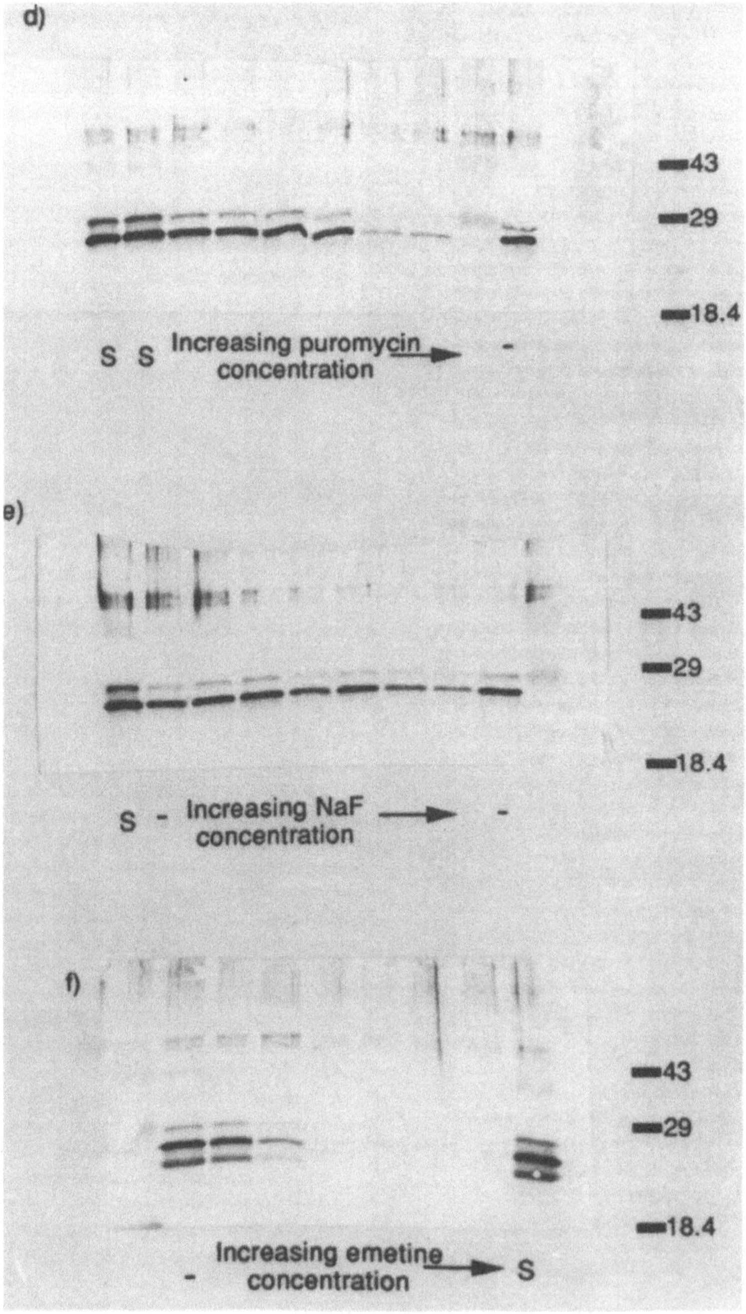

Fig. 3. d) Effect of puromycin. Lanes 1 and 2: prolactin standard. Lane 3: incubated with 0.4μg ml^{-1} puromycin, lane 4: 0.75 μg ml^{-1}, lane 5: 1.5 μg ml^{-1}, lane 6: 4 μg ml^{-1}, lane 7: 7.5 μg ml^{-1}, lane 8: 10 μg ml^{-1}. Lane 9: molecular weight standards. Lane 10: incubated without inhibitor; e) Effect of sodium fluoride. Lane 1: prolactin standard. Lane 2: incubated without inhibitor. Lane 3: incubated with 20μg ml^{-1} sodium fluoride, lane 4: 40 μg ml^{-1}, lane 5: 50 μg ml^{-1}, lane 6: 60 μg ml^{-1}, lane 7: 75 μg ml^{-1}, lane 8: 85 μg ml^{-1}. Lane 9: incubated without inhibitor. Lane 10: molecular weight standards; f) Effect of emetine. Lane 1 (not shown): incubated with 4 μg ml^{-1} emetine. Lane 2: molecular weight standards. Lane 3: incubated without inhibitor. Lane 4: incubated with 0.3μg ml^{-1} emetine, lane 5: 1 μg ml^{-1}, lane 6: 20 μg ml^{-1}, lane 7: 60 μg ml^{-1}, lane 8: 120 μg ml^{-1}, lane 9: 200 μg ml^{-1}. Lane 10: prolactin standard.

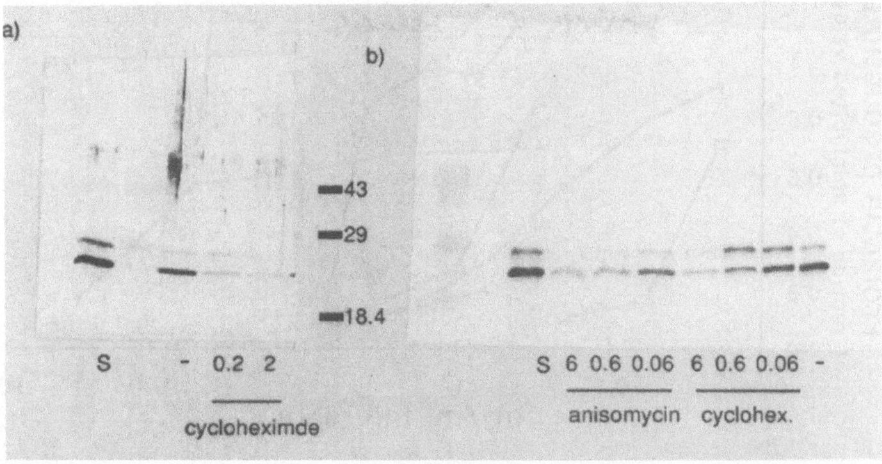

Fig. 4. The effect of cycloheximide on prolactin site occupancy is maintained from 2 to 44 hours of incubation. Confluent cell cultures were pre-incubated with inhibitor for 40 minutes, and incubated at identical inhibitor concentrations. Numbers indicate inhibitor concentration in $\mu g\ ml^{-1}$. Lanes marked S contained prolactin standard. Lanes marked – were incubated without inhibitor. a) Effect of cycloheximide-total synthesized after 2 hours. b) Effect of cycloheximide and anisomycin-total synthesized after 44 hours.

Discussion

Figure 6 describes current opinion about the effect on the translation cycle of the protein synthesis inhibitors used in this experiment. With the exception of puromycin, the detailed biochemical mode of action of these inhibitors is not yet known. Puromycin inhibits peptide bond formation by acting as an aminoacyl-tRNA analogue and causing premature termination. Both puromycin and sodium fluoride, an inhibitor of initiation, would be expected to reduce protein synthesis without slowing the elongation rate of individual ribosomes.

Anisomycin and gougerotin are known to inhibit peptide bond formation, and cycloheximide and emetine inhibit ribosome movement along an mRNA from one codon to the next. All four are thought to decrease the elongation rate of individual ribosomes. (Vazquez, 1979; Gale, 1981; Ballesta, 1991).

Cycloheximide may increase prolactin glycosylation site occupancy by selective induction of proteins which alter the glycosylation process. A specific cycloheximide-induced pattern of protein induction is known to occur (Lee, 1987) and may be responsible for altered availability of oligosaccharyldolichol (Chapman, 1988), protection of cells from heat killing (Lee, 1987), and expression of the immediate early response genes (Edwards, 1992). The effects of cycloheximide on thermotolerance and oligosaccharyldolichol were found to occur after incubation with puromycin. A similar pattern of protein induction was also determined to occur after treatment with puromycin (Lee, 1987). Cycloheximide and anisomycin both induce expression of transcription factors *c-fos* and *c-jun* (Edwards, 1992) which trigger the immediate early response to growth factors and phorbol esters. These results are summarized in Table 2. The fact that only cycloheximide altered prolactin glycosylation in our system suggests that a different mechanism may be at work than has been described previously.

Cycloheximide may increase site occupancy by its effect on the elongation step of translation. Cycloheximide reduces the elongation rate (in codons per second) of ribosomes along the mRNA coding for prolactin. It is believed that the kinetics of translocation are very rapid when compared with translation (Simon, 1992), so cycloheximide would also be expected to decrease the rate (in amino acids per second) at which individual nascent polypeptides of prolactin are translocated through the endoplasmic reticulum membrane. It was recently determined that the active site of oligosaccharyltransferase is positioned in the ER lumen at a minimum of 3-4 nm from the membrane (Nilsson, 1993), but the length of the region where oligosaccharyltransferase may act has yet to be determined. The potential glycosylation site on a single prolactin polypeptide would be expected to pass through this region of oligosaccharyltransferase activity at a rate (in nanometers per second) proportional to the local elongation rate of mRNA undergoing transla-

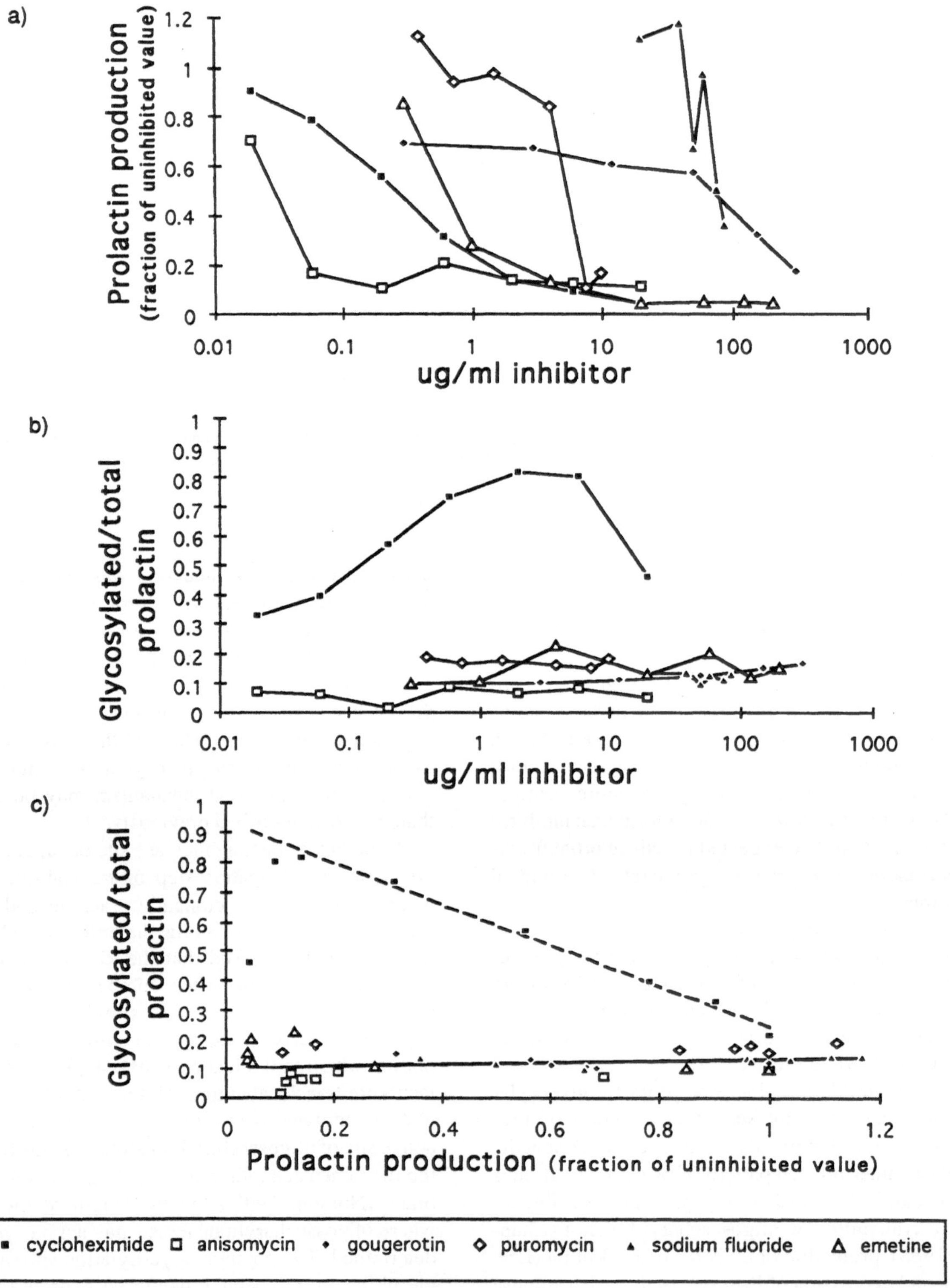

Fig. 5. Effect of protein synthesis inhibitors-densitometry. Prolactin production and glycosylation site occupancy were determined as described in Results section. a) Effect of protein synthesis inhibitors on prolactin production. b) Effect of protein synthesis inhibitors on glycosylation site occupancy. c) Glycosylation site occupancy as a function of prolactin production.

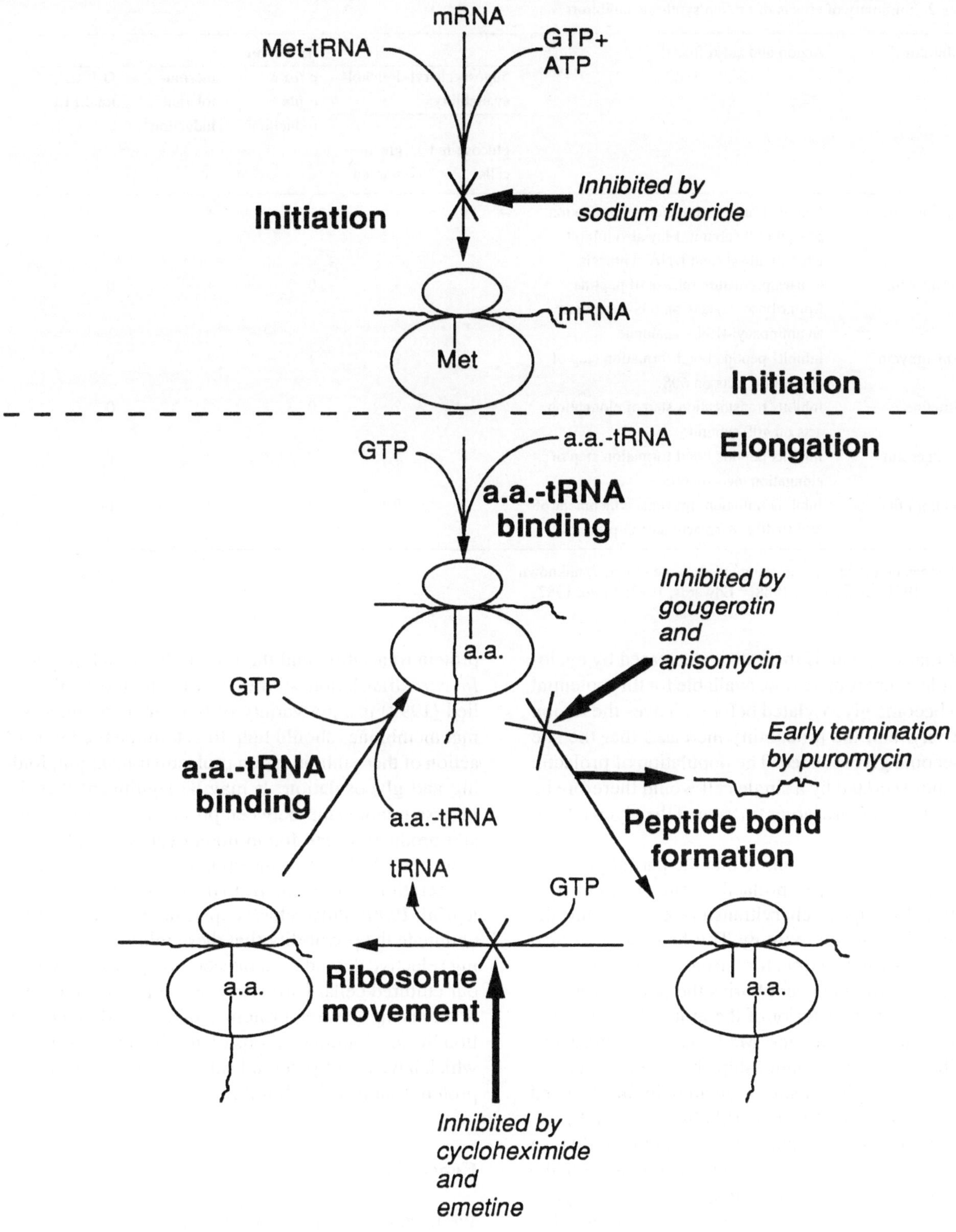

Fig. 6. Mode of action of protein synthesis inhibitors (Vazquez, 1979).

Table 2. Summary of effects of protein synthesis inhibitors

Inhibitor	Action and side effects[1]	Effect on				
		oligosaccharyl-dolichol availability[2]		*c-fos* & *c-jun* induction[3]	thermo-tolerance induction[4]	G-PRL/ total PRL
		glucose-fed cells	glucose-starved			
Cycloheximide	Inhibits translocation step of elongation acts on 60S subunit. May also inhibit RNA synthesis and DNA synthesis.	–	+	+	+	+
Puromycin	Causes premature release of peptide from ribosome, acts on 60S subunit as an aminoacyl-tRNA analogue.	–	+	0	+	0
Anisomycin	Inhibits peptide bond formation step of elongation, acts on 60S.	?	?	+	?	0
Emetine	Inhibits translocation step of elongation. acts on 40S subunit.	?	?	0	?	0
Gougerotin	Inhibits peptide bond formation step of elongation, acts on 60S.	?	?	?	?	0
Sodium fluoride	Inhibits initiation, prevents attachment of 60S to 40S. Also activates G-proteins.	?	?	?	?	0

–, decreasing effect; +, increasing effect; 0, no effect; ?, unknown
[1] Gale, 1981; [2] Chapman, 1988; [3] Edwards, 1992; [4] Lee, 1987

tion at that moment. If this rate is decreased by cycloheximide, there is more time available for this potential site to become glycosylated before it leaves the region of activity, and the probability increases that the site will become glycosylated. The population of prolactin molecules secreted by a single cell would therefore be expected to contain a larger fraction of the glycosylated protein.

The above model assumes that the partial glycosylation site occupancy of prolactin results from insufficient time for oligosaccharyltransferase to act while the potential site is in its proximity. This behavior may also be due to insufficient time for the enzyme to act before a competing folding event buries the potential glycosylation site in the interior of the molecule where it is inaccessible to the enzyme. Mathematical expressions describing expected relationships between elongation rate and glycosylation site occupancy in an idealized system are derived elsewhere (Shelikoff, unpubl.).

Both models presented above would predict an increase in site occupancy upon treatment with the other three inhibitors of elongation (anisomycin, gougerotin, and emetine) which was not observed in our laboratory. Our results emphasize the need for further experimental study of the relationship between protein translation and the initial glycosylation event. *In vitro* translation studies similar to those of Bullied (1992) using a variety of treatments on microsomal membranes should help to determine the mode of action of these inhibitors on prolactin translation, folding and glycosylation. It may be significant that the inverse relationship between prolactin site occupancy and production rate found upon cycloheximide treatment of C127 is also found upon stimulation of regulated secretion of prolactin from the pituitary *in vivo* (Pellegrini, 1990). Pulse-chase experiments are required to eliminate the possibility that these inhibitors are altering selective degradation or secretion pathways within our cultured cells. Ultimately, we hope to explore the relationship between protein synthesis and glycosylation by developing methods of altering elongation rate which have fewer potential side-effects than the use of protein synthesis inhibitors.

Conclusions

We have determined that out of six inhibitors of protein synthesis, cycloheximide and only cycloheximide dramatically altered the glycosylation site occupancy of recombinant human prolactin. The inability of other inhibitors of elongation to alter site occupancy suggests that a model explaining the effect of cycloheximide as resulting solely from a change in elongation rate, total protein synthesis, or oligosaccharyldolichol may be too simplistic. We suspect that cycloheximide may be acting in an unknown manner to alter oligosaccharyltransferase activity or to interfere with a competing folding event.

Acknowledgements

This work was supported by the National Science Foundation, Grant No. EEC8803014 through the MIT Biotechnology Process Engineering Center. Additional financial support from the National Science Foundation in the form of a graduate fellowship to M.S. is gratefully acknowledged. The authors are grateful to Susan Richards of Genzyme Inc., Framingham, MA. for the gift of prolactin standard and the C127 cell line transfected with the prolactin gene. The authors also wish to thank Naftali Cohn, Sara Howe, and Christine Lin for excellent technical assistance.

References

Ballesta JPG (1991) Inhibitors of eukaryotic protein synthesis. In H. Trachsel (ed.) Translation in Eukaryotes. CRC Press, Boca Raton.

Bause E and Legler G (1981) The role of the hydroxy amino acid in the triplet sequence Asn-Xaa-Thr(Ser) for the N-glycosylation step during glycoprotein biosynthesis. Biochem. J. 195: 639–644.

Bergman LW and Kuehl WM (1978) Temporal relationship of translation and glycosylation of immunoglobulin heavy and light chains. Biochemistry 17: 5174-5180.

Borrelli MJ, Lee YJ, Frey HE, Ofenstein JP and Lepock JR (1991) Cycloheximide increases the thermostability of proteins in Chinese hamster ovary cells. Biochem. Biophys. Res. Commun. 177: 575–581.

Brue T, Caruso E, Morange I, Hoffmann T, Evrin M, Gunz G, Benkirane M and Jaquet P (1992) Immunoradiometric analysis of circulating human glycosylated and nonglycosylated prolactin forms: spontaneous and stimulated secretions. J. Clin. Endocrinol. Metab. 75: 1338–1344.

Bullied NJ, Bassel-Duby RS, Freedman RB, Sambrook JF and Gething MH (1992) Cell-free synthesis of enzymatically active tissue-type plasminogen activator: protein folding determines the extent of N-linked glycosylation. Biochem. J. 286: 275–280.

Cacan, R, Labiau, O, Mir, A and Verbert, A (1993) Effect of cell attachment and growth on the synthesis and fate of dolichol-linked oligosaccharides in Chinese hamster ovary cells. Eur. J. Biochem. 215: 873–881.

Chalifour RJ and Spiro RG (1988) Effect of phospholipids on thyroid oligosaccharyltransferase activity and orientation: evaluation of structural determinants for stimulation of N-glycosylation. J. Biol. Chem. 263: 15673–15680.

Chapman AE and Calhoun IV JC (1988) Effects of glucose starvation and puromycin treatment on lipid-linked oligosaccharide precursors and biosynthetic enzymes in Chinese Hamster Ovary cells in vivo and in vitro. Arch. Biochem. Biophys. 260: 320–333.

Cole ES, Nichols EH, Lauziere K, Edmunds T and McPherson JM (1991) Characterization of the microheterogeneity of recombinant primate prolactin: implications for posttranslational modifications of the hormone in vivo. Endocrinology 129: 2639–2646.

Contreras A, Vazquez D and Carrasco L (1978) Inhibition, by selected antibiotics, of protein synthesis in cells growing in tissue culture. J. Antibiotics 31: 598–602.

Cooke NE (1989) Prolactin: normal synthesis, regulation, and actions. In: L. J. DeGroot (ed.) Endocrinology 2nd Ed. (pp. 384–407) W.B. Saunders Co., Philadelphia.

Cumming DA (1991) Glycosylation of recombinant protein therapeutics: control and functional implcations. Glycobiology 1: 115–130.

Edwards DR and Mahadevan LC (1992) Protein synthesis inhibitors differentially superinduce c-fos and c-jun by three distincs mechanisms: lack of evidence for labile repressors. EMBO J. 11: 2415–2424.

Fonseca ME, Ochoa R, Moran C and Zarate A (1991) Variations in the molecular forms of prolactin during the menstrual cycle, pregnancy, and lactation. J. Endocrinol. Invest. 14: 907–912.

Fukuoka H, Hamamoto R and Higurashi M (1991) Heterogeneity of serum and amniotic fluid prolactin in humans. Horm. Res. 35 (suppl. 1): 58–63.

Gale EF, Cundliffe E, Reynolds PE, Richmond MH and Waring MJ (1981) The Molecular Basis of Antibiotic Action 2nd Ed. John Wiley and Sons, London.

Garfin DE and Bers G (1989) Basic aspects of protein blotting. In: B.A. Baldo and E.R. Tovey (eds.) Protein Blotting: Methodology, Research, and Diagnostic Applications. (pp. 5–42) Karger, Basel.

Gavel Y and von Heijne G (1990) Sequence differences between glycosylated and non-glycosylated Asn-X-Thr/Ser aceptor site: implications for protein engineering. Protein Eng. 3: 433–442.

Gold AM (1967) Sulfonylation with sulfonyl halides. Methods Enzymol. 11: 706–711.

Goochee CF (1990) Environmental effects on protein glycosylation. Bio/technology 8: 421–427.

Goochee CF, Gramer MJ, Andersen DC, Bahr JB and Rasmussen (1992) The oligosaccharides of glycoproteins: factors affecting their synthesis and their influence on glycoprotein properties. In: P. Todd, S. K. Sikdar and M. Bier (eds.). Frontiers in Bioprocessing II. (pp. 199–240) American Chemical Society, Washington D.C.

Grant SR and Lennarz WJ (1983) Relationship between oligosaccharide-lipid synthesis and protein synthesis in mouse LM cells. Eur. J. Biochem. 134: 575–583.

Hames BD (1990) One-dimensional polyacrylamide gel electrophoresis. In: B.D. Hames and D. Rickwood (eds.) Gel Electrophoresis of Proteins: A Practical Approach 2nd Ed. (pp. 1–147) Oxford University Press, Oxford.

Haro LS, Lee DW, Singh RNP, Bee G, Markoff E and Lewis UJ (1990) Glycosylated human prolactin: alterations in glycosylation pattern modify affinity for lactogen receptor and values in prolactin radioimmunoassay. J. Clin. Endocrinol. Metab. 71: 379-383.

Hashim IA, Aston R, Butler J, McGregor AM, Smith CR and Norman, M (1990) The proportion of glycosylated prolactin in serum is decreased in hyperprolactinemic states. J. Clin. Endocrinol. Metab. 71: 111-115.

Hayter PM, Curling EMA, Baines AJ, Jenkins N, Salmon I, Strange PG, Tong JM and Bull AT (1992) Glucose-limited chemostat culture of Chinese Hamster Ovary cells producing recombinant human interferon-g. Biotechnol. Bioeng. 39: 327-335.

Henle KJ, Kaushal GP, Nagle WA and Nolen GT (1993) Prompt protein glycosylation during acute heat stress. Exp. Cell Res. 207: 245–251.

Hsiung N, Fitts R, Wilson S, Milne A and Hamer D (1984) Efficient production of hepatitis B surface antigen using a bovine papilloma virus-metallothionein vector. J. Mol. Appl. Gen. 2: 497–506.

Hubbard SC (1980) Synthesis of the N-linked oligosaccharides of glycoproteins: assembly of the lipid-linked precursor oligosaccharide and its relation to protein synthesis in vivo. J. Biol. Chem. 255: 11782–11793.

Imperiali B and Shannon KL (1991) Differences between Asn-Xaa-Thr-containing peptides: a comparison of solution conformation and substrate behavior with oligosaccharyltransferase. Biochemistry 30: 4374–4380.

Jenkins N and Curling EMA (1994) Glycosylation of recombinant proteins: problems and prospects. Enzyme Microb. Technol. 16: 354–364.

Kelleher DJ, Kreibich G and Gilmore R (1992) Oligosacchalyltransferase activity is associated with a protein complex composed of ribophorins I and II and a 48 kD protein. Cell 69: 55–65.

Konrad M and Merz WE (1994) Regulation of N-glycosylation: long term effect of cyclic AMP mediates enhanced synthesis of the dolichol pyrophosphate core oligosaccharide. J. Biol. Chem. 269: 8659–8666.

Kornfeld R and Kornfeld S (1985) Assembly of asparagine-linked oligosaccharides. Ann. Rev. Biochem. 54: 631–664.

Lau JTY, Welply JK, Shenbagamurthi P, Naider F and Lennarz WJ (1983) Substrate recognition by oligosaccharyl transferase: inhibition of co-translational glycosylation by acceptor peptides. J. Biol. Chem. 258: 15255–15260.

Lee DW and Markoff E (1986) Synthesis and release of glycosylated prolactin by human decidua in vitro. J. Clin. Endocrinol. Metab. 62: 990–994.

Lee YJ, Dewey WC and Li GC (1987) Protection of Chinese hamster ovary cells from heat killing by treatment with cycloheximide or puromycin: involvement of HSPs? Radiat. Res. 111: 237–253.

Lewis UJ, Singh RNP, Lewis LJ, Seavey BK and Sinha YN (1984) Glycosylated ovine prolactin. Proc. Natl. Acad. Sci. USA 81: 385–389.

Lewis UJ, Singh RNP, Sinha YN and Vanderlaan WP (1985) Glycosylated human prolactin. Endocrinology 116: 359–363.

Lewis UJ, Singh RNP and Lewis LJ (1989) Two forms of glycosylated human prolactin have different pigeon crop sac-stimulating activities. Endocrinology 124: 1558-1563.

Mankovitz R, Kisilevsky R and Florian M (1978) Chinese hamster ovary cell lines resistant to the cytotoxic action of fluoride. Can. J. Genet. Cytol. 20:71–84.

Miletich JP and Broze Jr. GJ (1990) b protein C is not glycosylated at asparagine 329. J. Biol. Chem. 265: 11397–11404.

Nilsson I and von Heijne G (1993) Determination of the distance between the oligosaccharyltransferase active site and the endoplasmic reticulum membrane. J. Biol. Chem. 268: 5798–5801.

Oda-Tamai S, Kato S, Hara S and Akamatsu N (1985) Decreased transfer of oligosaccharide from oligosaccharide-lipid to protein acceptors in regenerating rat liver. J. Biol. Chem. 260: 57–63.

Pellegrini I, Gunz G, Ronin C, Fenouillet E, Peyrat J, Delori P and Jaquet P (1988) Polymorphism of prolactin secreted by human prolactinoma cells: immunological, receptor binding, and biological properties of the glycosylated and nonglycosylated forms. Endocrinology 122: 2667–2674.

Pellegrini I, Gunz G, Grisoli F and Jaquet P (1990) Different pathways of secretion for glycosylated and nonglycosylated human prolactin. Endocrinology 126: 1087–1095.

Pless DD and Lennarz WJ (1977) Enzymatic conversion of proteins to glycoproteins. Proc. Natl. Acad. Sci. USA 74: 134–138.

Roitsch T and Lehle L (1989) Structural requirements for protein N-glycosylation: influence of acceptor peptides on cotranslational glycosylation of yeast invertase and site-directed mutagenesis around a sequon sequence. Eur. J. Biochem. 181: 525–529.

Rothman JE and Lodish HF (1977) Synchronised transmembrane insertion and glycosylation of a nascent membrane protein. Nature 269: 775–780.

Schmitt JW and Elbein AD (1979) Inhibition of protein synthesis also inhibits synthesis of lipid-linked oligosaccharides. J. Biol. Chem. 254: 12291–12294.

Scopes RK (1987) Protein Purification 2nd Ed. Springer-Verlag, Berlin.

Shakin-Eshleman SH, Wunner WH and Spitalnik SL (1993) Efficiency of N-linked core glycosylation at asparagine-319 of rabies virus glycoprotein is altered by deletions C-terminal to the glycosylation sequon. Biochemistry 32: 9465–9472.

Simon SM, Peskin CS and Oster GF (1992) What drives the translocation of proteins? Proc. Natl. Acad. Sci. USA 89: 3770–3774.

Sinha YN, DePaolo LV, Haro LS, Singh RNP, Jacobsen BP, Scott KE and Lewis UJ (1991) Isolation and biochemical properties of four forms of glycosylated porcine prolactin. Mol. Cell. Endocrinol. 80: 203–213.

Spiro MJ and Spiro RG (1986) Control of N-linked carbohydrate unit synthesis in thyroid endoplasmic reticulum by membrane organization and dolichol phosphate availability. J. Biol. Chem. 261: 14725–14732.

Spiro MJ and Spiro RG (1991) Potential regulation of N-glycosylation precursor through oligosaccharide-lipid hydrolase action and glucosyltransferase-glucosidase shuttle. J. Biol. Chem. 266: 5311–5317.

Vazquez D (1979) Inhibitors of Protein Biosynthesis. Springer-Verlag, Berlin.

Welply, JK (1991) Protein glycosylation: function and factors that regulate oligosaccharide structure. In: C. S. Ho and D. I. C. Wang (eds.) Animal Cell Bioreactors. (pp. 59–72) Butterworth-Heinemann, Boston.

Address for offprints: M. Shelikoff, Massachusetts Institute of Technology, Department of Chemical Engineering, 18 Vassar Street, 20A–207, Cambridge, MA, 02139 U.S.A.

Cytotechnology **15**: 209–215, 1994.

Effect of lipid supplements on the production and glycosylation of recombinant interferon-γ expressed in CHO cells

Nigel Jenkins, Paula Castro, Sunitha Menon, Andrew Ison[1] and Alan Bull
Research School of Biosciences, University of Kent, Canterbury, Kent CT2 7NJ, U.K; [1] *SERC Advanced Centre for Biochemical Engineering, University College, London University, Torrington Place, London WC1E 7JE, U.K.*

Key words: Glycosylation, recombinant, interferon, CHO cells, lipids

Abstract

The effects of lipids on the glycosylation of recombinant human interferon-γ expressed in a Chinese Hamster Ovary cell line were investigated in batch culture. Lipids form an essential part of the N-glycosylation pathway, and have been shown to improve cell viability. In control (serum-free) medium the proportion of fully-glycosylated interferon-γ deteriorated reproducibly with time in batch culture, but the lipoprotein supplement ExCyte was shown to minimise this trend. Partially substituting the bovine serum albumin content of the medium with a fatty-acid free preparation also improved interferon-γ glycosylation, possibly indicating that oxidised lipids carried on Cohn fraction V albumin may damage the glycosylation process.

Abbreviations: BSA – bovine serum albumin; CHO – chinese hamster ovary; DHFR-dihydrofolate reductase; FCS – foetal calf serum; IFN-γ – human interferon-gamma; q_{IFN} – specific interferon production rate; μ – specific growth rate; 2N – doubly-gycosylated; 1N – singly-glycosylated; ON – non-glycosylated

Introduction

Glycosylation is the most extensive of all the post-translational modications undergone by nascent peptides prior to secretion from animal cells. The precise role of the oligosaccharide chains varies in each glycoprotein, however their innuence on plasma half-life, protein antigenicity, and protection from protease attack has been demonstrated in many glycoproteins (Cumming, 1991). Protein structure, host cell type and cell culture conditions all influence the complex N-glycosylation process, resulting in considerable heterogeneity of the product (Goochee *et al.*, 1991; Jenkins and Curling, 1994). Thus, control of glycoprotein heterogeneity is a major goal of the biotechnology industry.

We have studied the control of N-glycosylation in a Chinese Hamster Ovary (CHO) cell line producing human recombinant interferon-γ (IFN-γ). This protein can exist in three glycoforms: doubly-glycosylated (at Asn_{25} and Asn_{97}; 2N), singly glycosylated (at Asn_{25} only; 1N) or non-glycosylated (0N; Jenkins *et al.*, 1993). Glycosylation deteriorates reproducibly with time in batch culture, with 0N IFN-γ reaching up to 25% of the total product by 140h and concomitant decreases in both 2N and 1N glycoforms (Curling *et al.*, 1990). Simply adding extra nutrients such as glucose and glutamine at the start of batch culture does not overcome this limitation (Curling *et al.*, 1990; Hayter *et al.*, 1991) but the glycoform proportions can be held constant in steady-state gucose-limited chemostat cultures (Hayter *et al.*, 1992; 1993).

Lipids such as dolichol form an essential part of the N-glycosylation pathway (Kaiden and Krag, 1992; Rosenwald *et al.*, 1990), and in this paper we report on the effects of lipid supplements on CHO cell growth, IFN-γ productivity and its glycosylation.

Materials and methods

Cells

The CHO 320 cell line expresses human recombinant IFN-γ under the control of an early SV40 promoter, with the IFN-γ gene copy number co-amplified using the DHFR-methotrexate selection system (Curling *et al.*, 1990). Seed cultures for all experiments were made within ten generations of a working cell bank (held in liquid nitrogen), passaged during the exponential growth phase of culture by centrifugation (1,000rpm, 10 minutes at room temperature) and resuspending in fresh medium. Replicate shake-flasks containing 100ml of cells in media were seeded at 0.18–0.20$\times 10^6$ cells.ml^{-1}, and cultured under an atmosphere of 95% air, 5% CO_2 at 100rpm and 37°C. During each experiment 6ml samples were taken daily for analysis.

Media supplements

Cells were adapted to grow suspended our in-house serum-free media in 1988 (Hayter *et al.*, 1991; Jenkins, 1991), and this medium was the base for the experiments described in this paper. In this medium RPM1-1640 (Gibco Ltd. Paisley, U.K.) was supplemented with glutamine (294mg.l^{-1}), insulin (5mg.l^{-1}), transferrin (5mg.l^{-1}), Cohn fraction V BSA (5g.l^{-1}, Pentex grade from Bayer Diagnostics) and trace elements (Jenkins, 1991; Hayter *et al.*, 1991).

Two different lipoprotein supplements derived from adult bovine serum were tested: Supplement A, used at 0.1% (v/v) from Sigma Chemicals (Poole, U.K,) contained cholesterol (11.4g.l^{-1}; 97% complexed with HDL) phospholipids (11.3g.l^{-1}; composed of 65% phosphatidyl choline, 10% lysophosphatidyl choline, 25% sphingomyelin), fatty acids (12.5g.l^{-1}, 51% linoleic, 2% arachidonic, 47% polyunsaturated) and 23g.l^{-1} total protein. Supplement B was Miles ExCyte Very Low Endotoxin Lipoprotein mixture used at 0.5% (v/v) containing cholesterol (9.7g.l^{-1}) and protein (15.4g.l^{-1}). Both supplements were used at the concentrations within those recommended by the manufacturer. Another series of flasks contained 8% (v/v) foetal calf serum (FCS; Gibco Ltd. Paisley, U.K.) in standard RPMI-1640 medium with no further supplements except for glutamine (294mg.l^{-1}).

IFN-γ analysis

Recombinant IFN-γ in cell culture medium was purified using an anti IFN-γ immunoafinity matrix described previously, yielding >98% pure IFN-γ (Curling *et al.*, 1990; Hayter *et al.*, 1993). The proportion of IFN-γ glycoforms (2N, 1N and 0N) was determined by running the immunoprecipitated IFN-γ on 14% SDS-polyacrylamide gels, followed by staining with silver nitrate and quantitation by scanning densitometry (Curling *et al.*, 1990; Hayter *et al.*, 1993). Total IFN-γ levels were determined using a double antibody ELISA (Hayter *et al.*, 1993) calibrated against an international reference preparation supplied by the National Institute of Biological Standards and Control, U.K.

Results and discussion

The following experiments were designed to assess the effects of lipids on the growth, productivity and glycosylation properties of the CHO-320 cell line. Two lipid supplements from different manufacturers (A and B) and FCS were tested. The BSA derived from bovine serum by Cohn fractionation (Finlayson, 1980) and used in the standard serum-free medium at 5g.l^{-1} contains up to 5mg.g^{-1} of total lipids (i.e. approximately 25mg.l^{-1} lipids are already in the standard medium). Therefore, the effect of reducing these BSA-associated lipids was also investigated. However, the CHO-320 cell line could not grow in the total absence of Cohn-derived BSA, but cell growth could be restored using 1g.l^{-1} Cohn-derived BSA, 1g.l^{-1} fatty-acid free BSA, and 0.1% of the Sigma lipoprotein mixture. The copolymer Pluronic F68 (0.1% v/v) was also used to protect against shear damage at low levels of BSA.

Effects of lipids on cell growth and IFN-γ production

FCS (8% v/v) and lipid supplement A (0.1% v/v) both improved cell growth (mean specific growth rates, μ of 0.026±0.002 and 0.026±0.001 respectively) resulting in a higher peak cell densities (14.8$\times 10^5$ cells.ml^{-1} and 10.2x10^5 cells.ml^{-1} respectively) compared to the standard serum free medium (μ=0.021±0.002, peak density of 8.8$\times 10^5$ cells.ml^{-1}) as shown in Fig. 1. Cells with fatty acid free BSA plus supplement A showed growth a similar profile to control cultures (μ=0.019±0.002, peak density of 7.9$\times 10^5$ cells.ml^{-1}). However, the IFN-γ production was decreased by all these treatments (by 60% in FCS, 35% with sup-

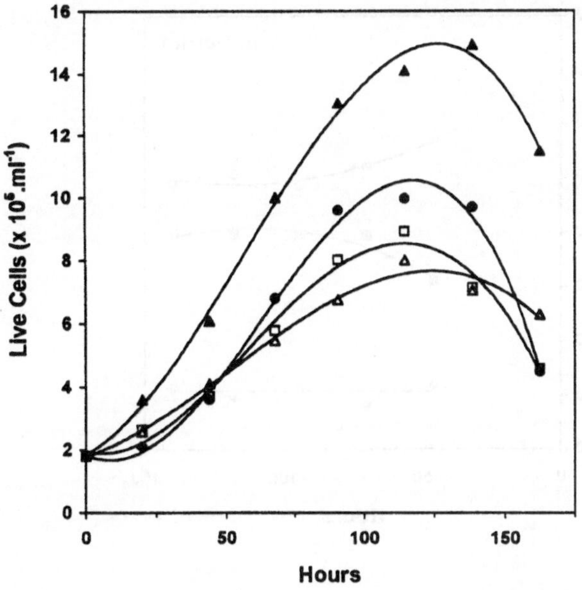

Fig. 1. CHO-320 viable cell counts following supplementation of the standard serum-free medium at 0 hours with 8% foetal calf serum (▲), lipid supplement A (•), or lipid supplement A in fatty acid reduced BSA medium (△), compared to control serum-free cultures (□). Mean values from duplicate flasks are shown.

plement A, and 42% with fatty acid free BSA pus supplement A) compared to control cells (peak concentration of 8,350±450 $IU.m^{-1}$). This was reflected in lower specific production rates (q_{IFN}) with FCS (95 $IU.10^6$ $cells^{-1}.day^{-1}$) supplement A (99 $IU.10^6$ $cells^{-1}.day^{-1}$) and fatty acid free BSA pus supplement A (117 $IU.10^6$ $cells^{-1}.day^{-1}$) compared to control cultures (200 $IU.10^6$ $cells^{-1}.day^{-1}$). The negative effects of FCS on IFN-γ production may be related to its endogenous lipid content since supplement A also suppressed IFN-γ production, but serum also contains factors that undoubtedly stimulate CHO cell growth. Cell lines vary in their lipid requirements (Kovar and Franek, 1986) and no single combination of lipids can support the growth of all mammalian cell lines.

Effects of lipids on IFN-γ glyucosylation

Cells grown in 8% FCS showed poor IFN-γ glycosylation in the early phase of batch culture (Fig.2b) but this did not deteriorate further, in contrast to cells grown in the control serum free medium (Fig. 2a). Immunoglobulin glycosylation has also been shown to differ between cells grown in FCS and serum-free conditions (Patel *et al*, 1992; Lund *et al.*, 1993; Maiorella *et al.*, 1993). IFN-γ glycosylation was not affected by lipid supplement A alone (Fig.2c), even though IFN-γ production was inhibited by 35%. Previous data gained on the same cell line in chemostat culture also show that marked changes in q_{IFN} do not influence its glycosylation site occupancy (Hayter *et al.*, 1993). Glycosylation of tPA is also independent of its production rate *in vitro* (Bulleid *et al.*, 1992) and *in vivo* (Lin *et al.*, 1993).

In contrast, cells treated with fatty acid free BSA plus supplement A showed a marked improvement in IFN-γ glycosylation (60–65% 2N by the end of culture; Fig.2d) compared to control cells or cells given supplement A alone (Figs.2a and 2c). We postulate that a lipid component of the Cohn-derived BSA may inhibit IFN-γ glycosylation, possibly arising from the oxidation of fatty acids in this supplement during the extensive heat treatment used in its preparation (Finlayson, 1980). The positive effect of fatty acid free BSA (Fig.2c) may thus derive from a lack of toxic lipid components or provision of alternative lipid sources to the cell. Previous studies have shown that albumin can be substituted by a complex of α-cyclodextrin, linoleic acid and cholesterol for the production of interferon by lymphoblastoid cells (Minamoto *et al.*, 1991).

Finally, a longer time course experiment (240h) was run using a different lipoprotein mixture (B; Miles ExCyte VLE) to supplement the standard medium. The most significant effect was seen when lipid supplement B was added during the late log phase of culture (45h), when it improved peak cell density and lengthened the growth phase of the culture (Fig.3a). This supplement did not signicantly reduce IFN-γ yields (peak concentration of 8,130±490 $IU.ml^{-1}$), in contrast to supplement A or FCS. A similar improvement in both cell growth and productivity has been shown using ExCyte on A431 cells producing interleukin-1 (Hewlett *et al.*, 1989). The glycosylation profile of control cells showed a substantial decline in 2N IFN-γ (Fig.3b), whereas supplement B reduced this decline, maintaining 50% 2N IFN-γ by the end of culture (Fig.3c). This improved glycosylation arose from a shift in the proportion of 2N and 1N glycoforms, rather than a reduction in 0N IFN-γ which increased substantially in both cultures over the final 50–100h of culture. Furthermore, because our previous mass spectrometry analysis has shown that Asn_{25} is the predominant site glycosylated in 1N IFN-γ (Jenkins *et al.*, 1993) we can surmise that supplement B specifically improved glycosylation at Asn_{97}. Note that the nomenclature for the amino acid sequence of human IFN-γ has recently changed (Farrar and Schreiber, 1993): Asn_{28} in the old

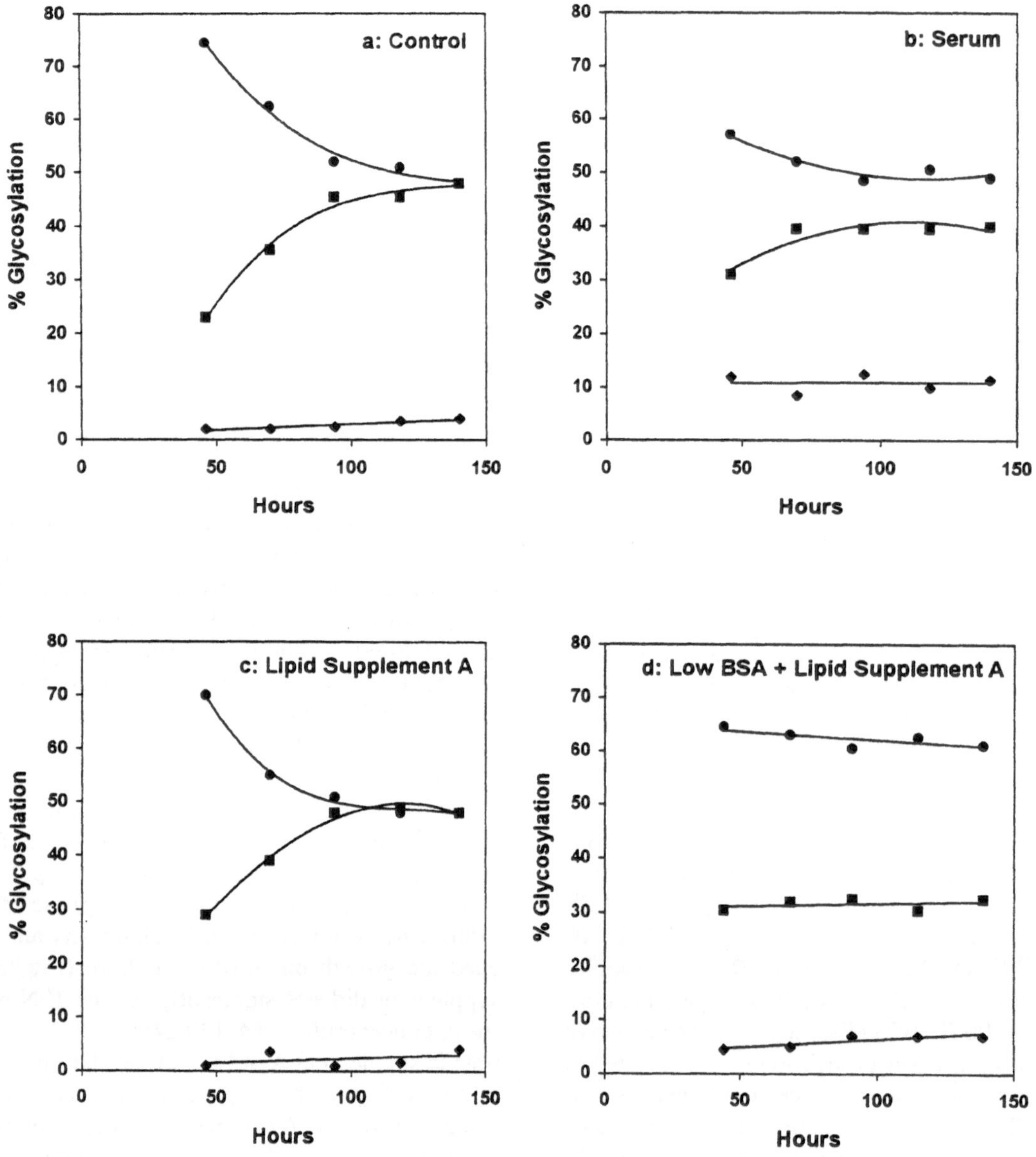

Fig. 2. Glycosylation pattern of recombinant IFN-γ following supplementation of the standard serum-free medium at 0 hours with 8% foetal calf serum (b), lipid supplement A (c), or lipid supplement A in fatty acid reduced BSA medium (d), compared to control serum-free cultures (a). Mean percentages of doubly-glycosylated (•) singly-glycosylated (■) and non-glycosylated IFN-γ (♦) were determined by scanning densitometry following gel electrophoresis of IFN-γ immunoprecipitated from duplicate flasks.

numbering sequence now becomes Asn_{25}, and Asn_{100} becomes Asn_{97}.

Several explanations are possible for the N-glycosylation preference of Asn_{25} over Asn_{97}, Firstly, the target glycosylation sequence Asn-X-Thr (found at Asn_{25} in IFN-γ) has been found to be N-glycosylated forty times more rapidly by oligosaccharyl transferase *in vitro* than the Asn-X-Ser sequence found at Asn_{97} of IFN-γ (Bause, 1984). Secondly, an extensive analysis of glycosylation site occupancy (Gavel and von Heijne, 1990) has revealed that glycosylation of potential target sequences is more likely to occur near the N terminus than the C terminus. Thirdly, X-crystallographic data on *E.coli*-derived IFN-γ (Ealick *et al.*, 1991) and NMR proton assignments (Grzesiek *et al.*, 1992) have shown that the Asn_{25} site lies in the middle of a random coil structure. In contrast, Asn_{97} is the last residue of an α-helix which may impose restrictions on the

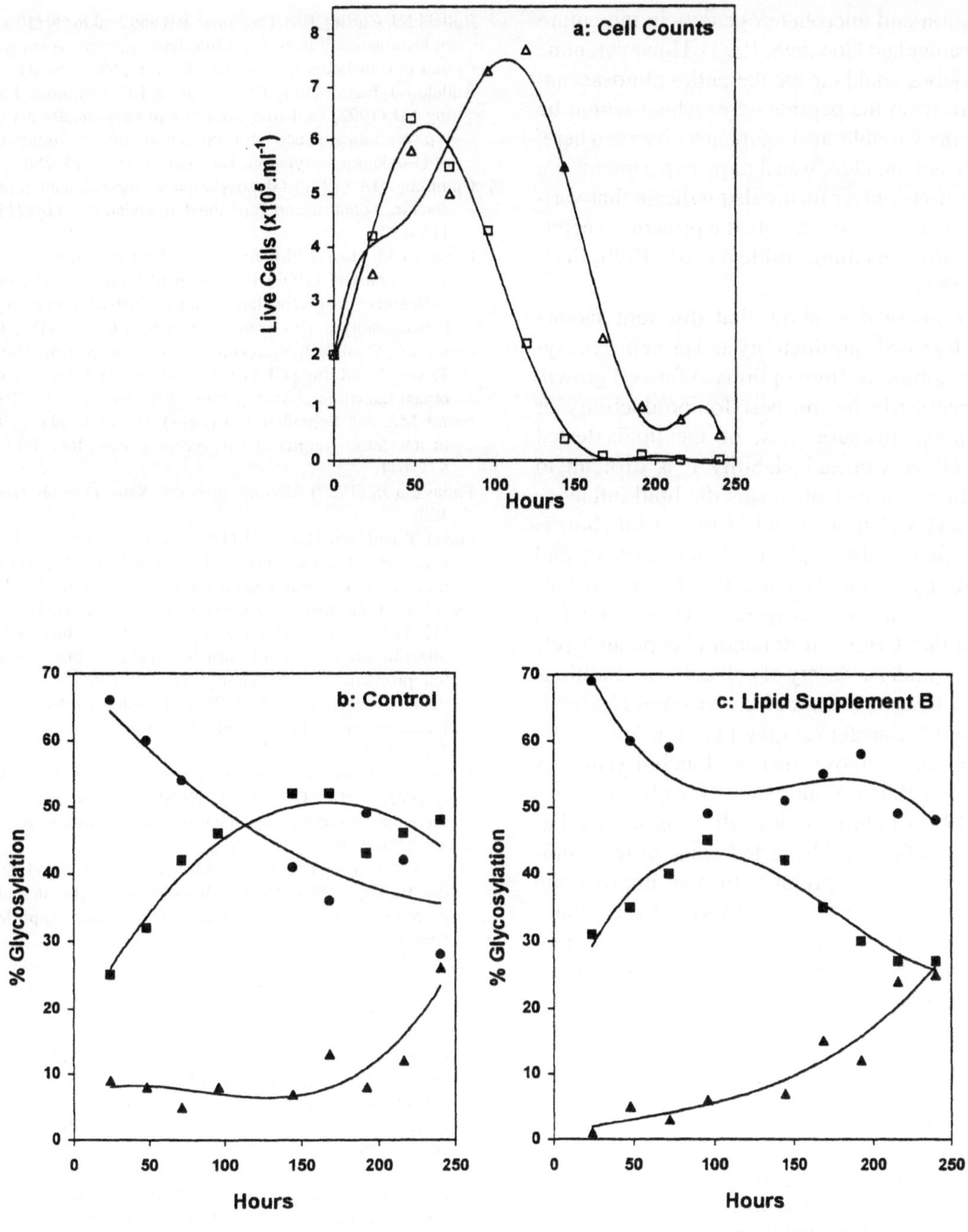

Fig. 3. CHO-320 viable cell counts (a), and IFN-γ glycosylation profiles (b and c), following supplementation of the standard serum-free medium at 48h with lipid supplement B (△ and c), compared to control serum-free cultures (□ and b). Mean percentages of doubly-glycosylated (•) singly-glycosylated (■) and non-glycosylated IFN-γ (▲) were determined by scanning densitometry following gel electrophoresis of IFN-γ immunoprecipitated from individual flasks.

access of not only oligosaccharyl transferase (leading to variable site occupancy) but also Golgi based glycosyltransferases such as fucosyltransferase, galactosyltransferase and sialyltransferases (leading to differences in microheterogeneity).

CHO cells have been shown to produce a variety of glycosidases that could conceivably lead to glycan degradation and microheterogeneity in the cuture medium (Gramer and Goochee, 1993). However, none of these enzymes could cleave the entire oligosaccharide structure from the peptide core, which would be required for the variable site occupancy observed here. Pulse-chase and *in vitro* translation experiments on intracellular IFN-γ glycoforms also indicate that variable site occupancy occurs before the protein is secreted into cell culture medium (Bulleid *et al.*, 1990; Curling *et al.*, 1990).

Together, these data show that different factors influence cell growth, productivity and protein gycosylation, and a cuture medium optimised for cell growth may not necessarily be the best for productivity or product quality. Because most of the lipids tested improved cell growth and viability it is difficult to dissociate these effects from a specific lipid influence on protein glycosylation. A lipid carrier (dolichol) is an essential part of the N-glycosylation process, and lipid availability might influence this process resulting in glycoform macroheterogeneity. However, it has been shown that CHO cell dolichol phosphate levels do not change under a variety of cell culture conditions (Kaiden and Krag, 1992), and the enzyme N-acetylglucosamine-l-P transferase may play a more central role in regulating oligosaccharide-dolichol synthesis in response to cellular requirements for glycoproteins (Lehrman, 1991). Future studies will focus on the influence of more defined lipids, tested alone or in combination. Work is also in progress to both improve the rapid monitoring of IFN-γ glycoforms by capillary electrophoresis (James *et al.*, 1994) and to characterise fully the changes in oligosaccharide structures that occur during batch culture of recombinant CHO cells producing IFN-γ.

Acknowledgements

This work was supported by the SERC Biotechnology Directorate, and PC was awarded a studentship from the Portuguese National Foundation for Scientific and Technological Research (JNICT). The authors are grateful to the Wellcome Foundation for supplying the CHO-320 cell line.

References

Bause E (1984) Model studies on N-glycosylation of proteins. Biochem. Soc. Trans. 12: 514–517.

Bulleid NJ, Curling EM, Freedman RB and Jenkins N (1990) Source of heterogeneity in secreted interferon-gamma. A study on products of translation in vitro. Biochem. J, 268: 777–781.

Bulleid NJ, Bassel-Duby RS, Freedman RB, Sambrook J and Gething MJ (1992) Cell-free synthesis of enzymically active tissue-type plasminogen activator. Protein folding determines the extent of N-linked gycosylation. Biochem. J. 286: 275–280.

Cumming DA (1991) Glycosylation of recombinant protein therapeutics: control and functional implications. Glycobiology 1: 115–130.

Curling EM, Hayter PM, Baines AJ, Bull AT, Gull K, Strange PG and Jenkins N (1990) Recombinant human interferon-gamma. Differences in glycosylation and proteolytic processing lead to heterogeneity in batch culture. Biochem J, 272: 333–337.

Ealick SE, Cook WJ, Vijaykumar S, Carson M, Nagabhushan TL, Trotta PP and Bugg CE (1991) 3-dimensional structure of recombinant human interferon-gamma. Science 252: 698–702.

Farrar MA and Schreiber RD (1993) The molecular cell biology of interferon-gamma and its receptor. Ann. Rev. Immunol. 11: 571–611.

Finlayson JS (1980) Albumin products. Sem. Thromb. Hem. 6: 85–120.

Gavel Y and von Heijne G (1990) Sequence differences between glycosylated and non-glycosylated Asn-X-Thr/Ser acceptor sites: implication for protein engineering. Protein Eng. 3: 433–442.

Goochee CF, Gramer MJ, Andersen DC, Bahr JB and Rasmussen JR (1991) The oligosaccharides of glycoproteins – bioprocess factors affecting oligosaccharide structure and their effect on glycoprotein properties. Bio/technology 9: 1347–1355.

Gramer MJ and Goochee CF (1993) Glycosidase activities in chinese hamster ovary cell lysate and cell culture supernatant. Biotechnol. Prog. 9: 366–373.

Cizesiek S, Dobeli H, Gentz R, Garotta G, Labhardt AM and Bax A (1992) ^{1}H, ^{13}C, and ^{15}N NMR backbone assignments and secondary structure of human interferon-gamma. Biochemistry 31: 8180–8190.

Hayer PM, Curling EM, Baines AJ, Jenkins N, Salmon I, Strange PG and Bull AT (1991) Chinese hamster ovary growth and interferon production kinetics in stirred batch culture. Appl. Microbiol. Biotechnol. 34: 559–564.

Hayter PM, Curling EM, Baines AJ, Jenkins N, Salmon I, Strange PG, Tong JM and Bull AT (1992) Glucose-limited chemostat culture of chinese hamster ovary cells producing recombinant human interferon-gamma. Biotechnol. Bioeng. 39: 327–335.

Hayter PM, Curling EM, Gould ML, Baines AJ, Jenkins N, Salmon I, Strange PG and Bull AT (1993) The effect of dilution rate on CHO cell physiology and recombinant interferon-gamma production in glucose-limited chemostat cultures. Biotechnol. Bioeng. 42: 1077–1085.

Hewlett G, Duvinski MS and Montalto JG (1989) Pentex ExCyte growth enhancement media supplement as a lipoprotein additive for mammalian cell culture. Miles Science Journal 11: 9—14.

James DC, Freedman RB, Hoare M and Jenkins N (1994) High resolution separation of recombinant human interferon-gamma by micellar electrokinetic capillary chromatography. Anal. Biochem. (in press).

Jenkins N (1991) Growth Factors. In: Butler M. (ed.) Mammalian Cell Biotechnology: A Practical Approach. (pp. 39–55) Oxford University Press, Oxford, U.K.

Jenkins N, Wingrove C, Strange PG, Baines AJ, Curling EM, Freedman RB and Pucci P (1993) Changes in the glycosylation pattern of interferon-gamma during batch culture. In: Kaminogawa S, Ametani A and Hachimura S. (ed.) Animal Cell Technology:

Basic and Applied Aspects Vol. 5. (pp. 231–235) Kluwer Academic Publishers, Dordecht, The Netherlands.

Jenkins N and Curling EM (1994) Glycosylation of recombinant proteins: problems and prospects. Enzyme Microb. Technol. 16: 354–364.

Kaiden A and Krag SS (1992) Dolichol metabolism in Chinese hamster ovary cells. Biochem. Cell. Biol. 70: 385–389.

Kovar J and Franek F (1986) Serum-free medium for hybridoma and parental myeloma cell cultivation. Methods Enzymol. 121: 277–292.

Lehrman MA (1991) Biosynthesis of N-acetylglucosamine-P-P-dolichol, the committed step of asparagine-linked oligosaccharide assembly. Glycobiology 1: 553–562.

Lin AA, Kimura R and Miller WM (1993) Production of tPA in recombinant cho cells under oxygen-limited conditions. Biotechnol. Bioeng. 42: 339–350.

Lund JT, Takahashi N, Hindley SA, Tyler R, Goodall M and Jefferis R (1993) Glycosylation of human IgG subclass and mouse IgG2b heavy chains secreted by mouse J558L transfectoma cell lines as chimeric antibodies. Human Antibodies and Hybridomas 4: 20–25.

Maiorella BL, Winkelhake J, Young J, Moyer B, Bauer R, Hora M, Andya J, Thomson J, Patel T and Parekh RB (1993) Effect of culture conditions on IgM antibody structure, pharmacokinetics and activity. Bio/technology 11: 387–392.

Minamoto Y, Ogawa K, Abe H, Iochi Y and Mitsugi K (1991) Development of a serum-free and heat sterilizable medium and continuous high density cell culture. Cytotechnology 5: 35–51.

Patel TP, Parekh RB, Moellering BJ and Prior CP (1992) Different culture methods lead to diferences in glycosylation of a murine IgG monoclonal antibody. Biochem. J. 285: 839–845.

Rosenwald AG, Stoll J and Krag SS (1990) Regulation of glycosylation. Three enzymes compete for a common pool of dolichyl phosphate in vivo. J. Biol. Chem. 265: 14544–14553.

Address for offprints: N. Jenkins, Research School of Biosciences, University of Kent, Canterbury, Kent CT2 7NJ, U.K.

Cytotechnology **15:** 217–221, 1994.

Role of environmental conditions on the expression levels, glycoform pattern and levels of sialyltransferase for hFSH produced by recombinant CHO cells

W. Chotigeat, Y. Watanapokasin, S. Mahler and P.P. Gray
Department of Biotechnology, University of New South Wales, Sydney, Australia

Key words: β actin, CHO, hFSH, perfusion cultures, sialyltransferase

Abstract

A recombinant CHO cell line in which the expresison of human follicle stimulating hormone (hFSH) was under the control of the β actin promoter was maintained in steady state perfusion cultures on a protein free medium. The level of expression of the hFSH was controlled by varying the steady state level of dissolved oxygen (10–90% of air saturation) and of sodium butyrate (0–1.5mM). Under these conditions, the specific productivity of hFSH (q_{FSH}) varied from 0.7 to 4.8 ng hFSH/10^6 cells/h. As the specific productivity of hFSH increased, there was a shift in the FSH isoforms to the lower pI fractions, corresponding to increased sialic acid content. As the specific productivity of hFSH increased, shifting the isoform distribution towards the lower pI isoforms, that the sialyltransferase enzymic activity also increased.

Introduction

There has been considerable conjecture to date as to whether the rate of expression of a glycoprotein by a recombinant mammalian cell affects the glycoform pattern of the secreted protein. The role of the host cell and the environmental conditions on the glycoform distribution has been described (e.g. Goochee and Monica, 1990). However, the question as to whether with the same cell line under basically the same environmental conditions, the distribution of isoforms is a function of the rate of recombinant glycoprotein expression has, to date, been unanswered.

In this paper, the role of expression rate on the glycoform distribution of human follicle stimulating hormone (hFSH) being produced by recombinant CHO cells is reported. Follicle stimulating hormone (hFSH) is a heterodimeric molecule with 2 N-glycosylation sites on each peptide chain. hFSH is secreted by the anterior pituitary gland and is a member of the gonadotropin class of peptide hormones, sharing a common alpha subunit with hCG, hTSH and hLH (Pierce and Parsons, 1981; Chappel *et al.*, 1983). Recombinant CHO and B cell lines producing hFSH have been used in our group as model systems to study heterologous glycoprotein production. (Gebert and Gray, 1994; Gray *et al.*, 1992).

In this study, CHO cells in which the r-hFSH expression was under the control of the actin promoter were grown in steady state perfusion cultures. By varying the steady state level of dissolved oxygen in the cultures, it was possible to vary the specific productivity (q_{FSH}) of the hFSH. Further increases in q_{FSH} were obtained by adding sodium butyrate to the perfusion medium. As the expression levels increased, there was an increase in the proportion of acidic isoforms in the expressed hFSH.

There is a close relationship between the isoform pI and the sialic acid content of hFSH. (Ulloa – Aguirre *et al.*, 1988). Accordingly the levels of sialyl transferase, the enzymes responsible for sialic acid addition in the terminal stages of hFSH synthesis were studied as a function of the rate of expression. Data are presented which shows that sialyl transferase activity increased as the specific productivity of hFSH increased.

Materials and methods

Cell line

The Darren cell line used in this study had the c-DNA coding for hFSH alpha and beta subunits under the control of the β actin promoter (Gebert and Gray, 1994). The cells were grown until confluent on Cytodex 2 microcarriers in DMEM:F12 medium containing 10% FCS. Once confluent, the medium was replaced with DMEM: F12 protein free production medium. All other growth conditions and assays were as previously described (Gray *et al.*, 1990).

Bioreactor and associated equipment

The bioreactor was a 3 litre vessel (Applikon) with an operating volume of 1.5 litres. The vessel was controlled by an FC-4 computer linked biocontroller (Real Time Engineering). The cells growing on microcarriers were retained in the bioreactor by a rotating stainless steel screen with a 75 micron mesh size. Fresh sterile medium was added to the bioreactor at the rate of 0.3 volumes per 10^6 cells per day; the exit pump was set at a slightly higher flow rate than the inlet pump and scavenged medium from inside the spin filter down to the level set by the height of the exit pipe.

The level of dissolved oxygen in the bioreactor was controlled by the addition of oxygen to the vessel, and carbon dioxide addition was used to control pH. For each steady state, three to four volumes were allowed to pass through the bioreactor, then samples were taken to ensure steady state conditions existed. Several samples were taken over the next day and these data points used as the steady state data.

Preparation of microsomal membrane fraction

Microsomal membrane preparations of the CHO cells were prepared and used in the determination of sialyl transferase activity. Approximately $1\text{–}2 \times 10^8$ CHO cells were collected by centrifugation at 1,000 rpm for 10 mins, washed with phosphate buffer saline and recentrifuged. Cells were suspended in homogenization buffer (0.1MTris-HC*l*, 0.25mM sucrose, 1mM EDTA, 10mM mercaptoethanol, 0.5mM PMSF) homogenized by a Potter Elvehjem homogenizer, then centrifuged at 10,000 rpm for 30 mins. The supernatant was collected and centrifuged at 30,000 rpm for 60 mins. The microsomal pellet obtained from the second centrifugation was dissolved in solubilization buffer (Levart *et al.*, 1990; Kaplan *et al.*, 1988). Protein concentrations were determined by the Bradford assay.

Sialyltransferase activity determination

Total sialyltransferase activity in the microsomal preparation was assayed using asiolofetuin as an exogenous acceptor. The reaction mixture contained 800 μg asialofetuin, 12.5 μmol sodium phosphate buffer pH 6.8, 0.5% triton X-100, 55 nmol CMP-(^{14}C) and enzyme preparation (100 μg protein) to a final volume of 100μL. Incubation was carried out for 1 hr at 37°C and terminated by the addition of 0.5 ml of 15% trichloroacetic acid containing 5% phosphotungstic acid. Precipitates were filtered under suction through 2.4 cm glass fibre filters (Whatman 934-AH) and washed with 10 ml 5% trichloroacetic acid three times. Radioactivity in the discs was determined in 10 ml Ecolite with a Beckman LS-250 Scintillation Counter. Replicates of reactants without the added acceptor were used as blanks (Chu and Walker, 1986).

Purification and quantification of hFSH

Samples were prepared for assay and characterisation using the following methods. The harvested medium was loaded onto a Protein G affinity column (1.5 × 2.0 cm). The Protein G Sepharose was bound with anti-FSH (78/24) according to the method of Harlow and Lane (1989). The FSH bound on the column was eluted with 0.1M Glycine pH 2.5. The FSH in the sample and after purification was quantified by Enzyme Linked Immunosorbent Assay (ELISA).

The ELISA used in this study was performed by using purified hFSH(80/1) as a standard. The monoclonal anti-FSH(79/7) was coated onto the plate, which was then exposed to the sample containing the FSH which then bound to the antibody. The bound FSH was detected by the conjugated biotin-anti FSH(78/24) and Streptavidin-horse radish peroxidase (HRP). The colour was developed by ABTS(2,3-azino-bis-3-ethylbenzthiazoline-6-sulfonic acid diammonium salt) in 0.04M citric acid/0.06M phosphate buffer pH 4 and determined at 410 nm.

Isoelectric focusing (IEF) gel and Blotting

Immunoaffinity purified samples 0.5μg obtained under the different conditions were run on IEF gels, pI 3–9 (Phastgel System, Pharmacia). The gel was then

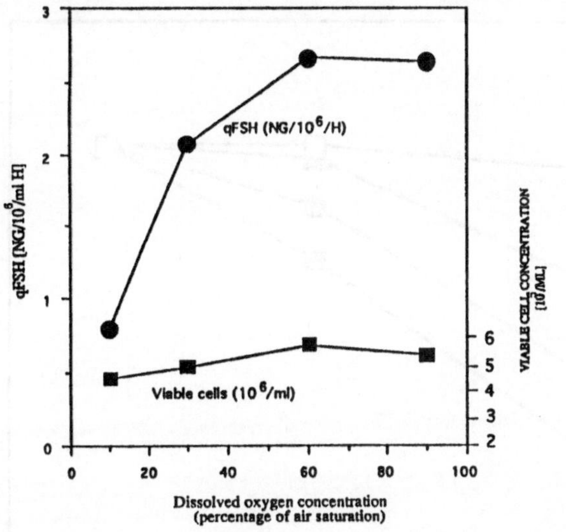

Fig. 1. Steady-state specific productivity of hFSH (q_{FSH}) and viable cell concentration as functions of the dissolved oxygen concentration.

Table 1.

Conditions	Specific acctivity of Sialyl transferase (n mole/mg protein)/hour
Level of Dissolved Oxygen	
10%	1.0
30%	1.2
60%	2.05
90%	4.9
Concentration of Sodium Butyrate	
0.5 mM	2.8
1.0 mM	2.9
1.5 mM	3.85

transferred onto a polyvinylidene difluoride membrane (PVDF Millipore). The rhFSH on the PVDF membrane was detected by rabbit anti-FSH and mouse anti-rabbit conjugated with alkaline phosphatase, respectively. The colour was developed by 15mg of 5-bromo-4 chloroindolylphosphate and 30 mg of nitroblue tetrazolium in 100ml of 0.1M carbonate buffer/ 1mM $MgCl_2$ pH 9.8, and the isoform patterns on the membrane were scanned and quantified by the Bio-image gel scanner (Millipore).

Results

The results of steady-state perfusion cultures at a range of different dissolved oxygen levels are shown in Figure 1. Increasing the level of dissolved oxygen increased the specific productivity of FSH (q_{FSH}), from a value of 0.7 at 10% DO to a value of 2.6 ng/10^6cells/h at 90% DO. The number of viable cells was relatively constant, ranging from 4.5–5.7 $\times$ 10^6 cells/ml over the dissolved oxygen levels studied. Figure 2 shows the variation in isoform distribution within a pI range of about 4.5 to 8.6 at the selected DO levels. The data is presented as the percentage of isoforms (determined by the percentage of integrated intensity for the bands on an IEF gel) which have a pI less than the pI value plotted on the abscissae. At the lowest level of dissolved oxygen studied (DO of 10%), 65% of the FSH isoforms had a pI of 4.5 or lower. As the level of dissolved oxygen increased, the percentage of FSH isoforms at the lower pI's increased, with the highest level of dissolved oxygen studied, 90%, having 86% of the FSH bands occurring at a pI of 4.5 or lower.

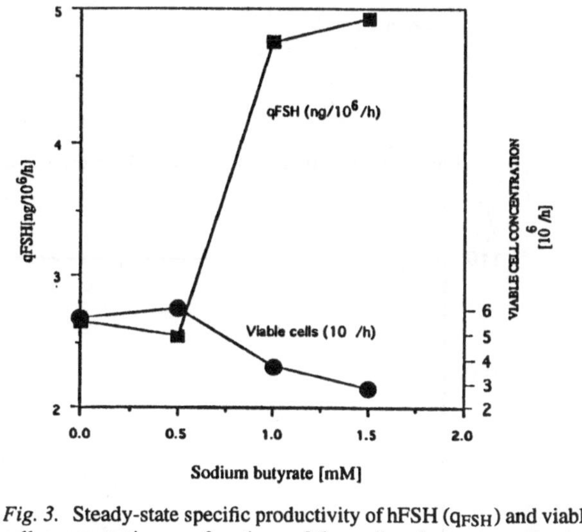

Fig. 3. Steady-state specific productivity of hFSH (q_{FSH}) and viable cell concentrations as functions of the sodium butyrate concentration.

The levels of the sialyltransferase activity observed at the different levels of dissolved oxygen are shown in Table 1. The sialyl transferase activity determined correlated positively with the dissolved oxygen level.

In Fig. 3, the effect of increasing levels of sodium butyrate in the medium on the q_{FSH} and on viable

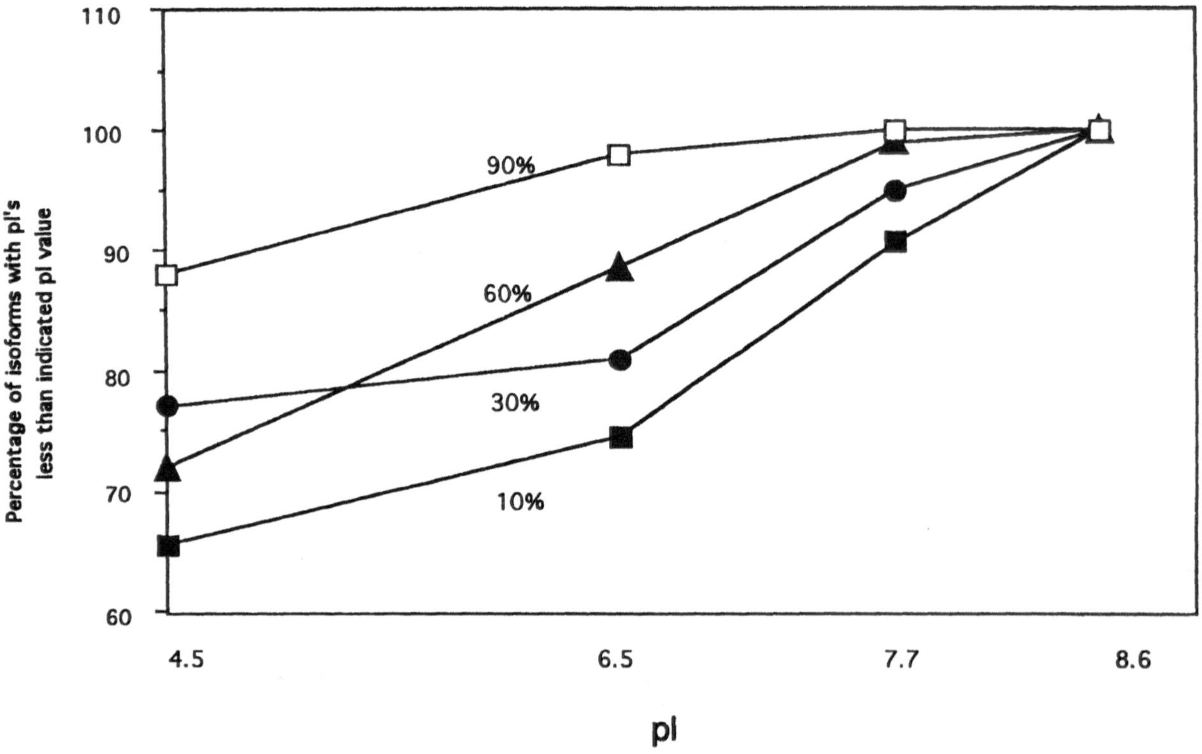

Fig. 2. Distribution of pI's of r-hFSH isoforms as a function of the steady-state dissolved oxygen concentration.

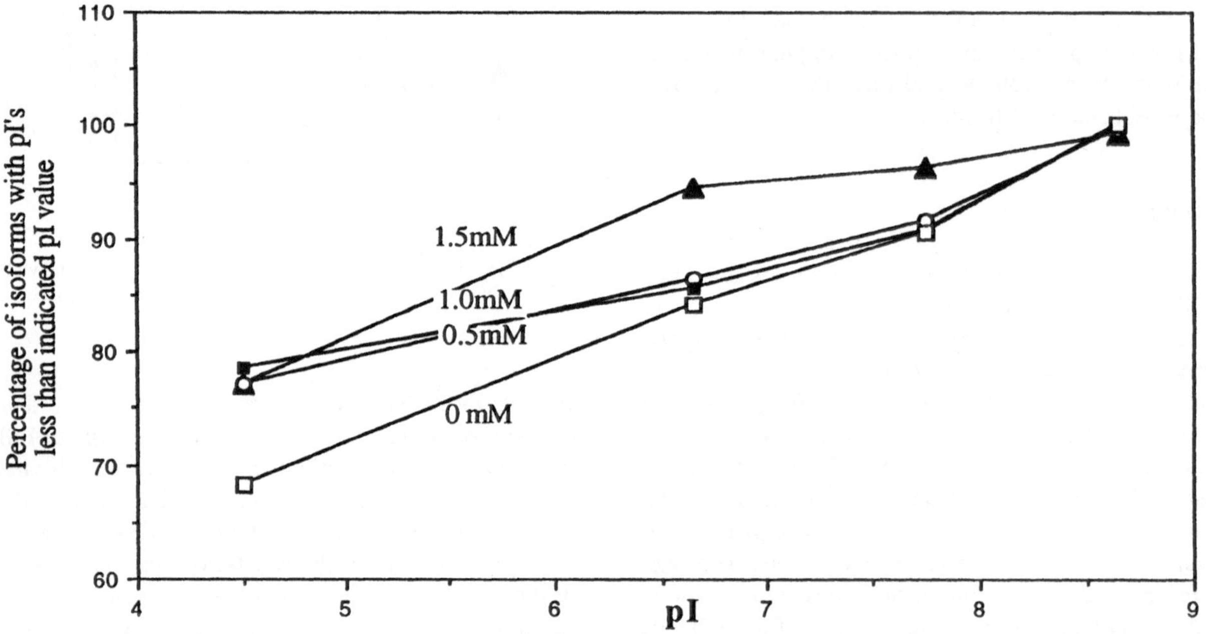

Fig. 4. Distribution of pI's of r-hFSH isoforms as a function of the steady-state sodium butyrate concentration (□, 0mM; ■, 0.5mM; ○, 1.0mM; ▲, 1.5mM)

cell numbers is shown. These experiments were carried out at a dissolved oxygen concentration of 60%. The addition of sodium butyrate resulted in further increase in the specific productivity, from 2.6 up to 4.8 ng/10^6cells/h at 1.5mM sodium butyrate. Higher levels of sodium butyrate are toxic to the cells, and although the levels used in these experiments were not high enough to have a major impact on viability, there was some reduction in cell numbers, from 5.7×10^6 cells/ml down to 2.8×10^6 cells/ml at the 1.5mM level. Figure 4 shows that the increases in specific productivity of FSH caused by the added butyrate were accompanied by further shifts in the isoform profile to lower pI's.

Table 1 shows that there were modest increases in the sialyl transferase activity in the experiments with added butyrate.

Conclusions

By altering the concentrations of dissolved oxygen and sodium butyurate in a perfusion culture of a recombinant CHO cell line expressing hFSH, it was possible to vary the specific productivity, q_{FSH}, over a seven fold range. At the lowest value of q_{FSH}, 65% of the FSH isoforms had a pI of 4.5 or less; at the highest value of q_{FSH}, 90% of the FSH isoforms had a pI of 4.5 or less. It was also observed that as the q_{FSH} increased there was an increase of approximately four fold in the total sialyltransferase activity. The observation that the q_{FSH} had higher content of sialic acid at high expression rates, means that as the expression rate increased that at least one enzyme on the glycosylating pathway, the sialyl transferase, did not become limiting, in fact the inverse occurred.

The results obtained in this study clearly demonstrated that the rate of expression of a glycoprotein by a recombinant mammalian cell affects the glycoform pattern of the secreted protein, in this case FSH. Steady state perfusion cultures were used to ensure that the only environmental variation the cells were experiencing was the change in the single parameter viz dissolved oxygen level or the concentration of sodium butyrate used to regulate expression levels. Different environmental conditions and modes of bioreactor operation (batch, fed batch etc) will produce glycoproteins with differing sets of glycoform distribution (e.g. Watson *et al.*, 1994). Steady state continuous operation offers the possibility of maintaining constant environmental conditions to produce glycoproteins of constant glycoform distribution.

Acknowledgements

The cell lines used in the work were supplied by Dr. Peter Schofield, of the Garvan Institute of Medical Research. The monoclonal antiFSH clone 78/24 and 79/7 was generous gift from Peter Anderson of Bioquest Pty Ltd. Supported under the Australian Government's Cooperative Research Centre Program.

References

Chappel SC, Ulloa-Aguirre A, Coutifaris C (1983) Biosynthesis and secretion of FSH. Endo. Rev. 4: 179–211.

CHU SW and Walker WA (1986) Developmental changes in the activities of sialyl and fucosyltransferases in rat small intestine. Biochem. Biophys. Acta. 883: 496–50.

Gebert C and Gray PP (1994) Expression of FSH in CHO cells I. Comparison of promoter types and effects of their respective inducers. Cytotechnology 14: 39–45.

Goochee CF and Monica T (1990) Environmental effects on protein glycosylation. Bio/Technology. 8: 421–427.

Gray PP, Crowley JM and Marsden WL (1990) Growth of a recombinant CHO cell line and high level expression in protein free medium. In: Murakami H (ed.) Trends in Animal Cell Technology, VCH I Kodansha, Tokyo: 265–270.

Gray PP, Jirasripongpun K and Gebert C (1992) Maximising mammalian cell expression of heterologous proteins in continuous flow bioreactors. In: Ladisch MR and Bose A (eds.) Harnessing Biotechnology for the 21st Century, ACS, Washington: 197–201.

Kaplan HA, Naider F and Lennarz WJ (1988) Partial characterisation and purification of the glycosylation site recognition of oligosaccharyltransferase J. Biol. Chem. 263: 7814–7820.

Levart C, Ardail D and Louisot P (1990) Comparative study of the N-glycoprotein synthesis through dolichol intermediates in mitochondria, golgi apparatus and endoplasmic reticulum – rich fraction. Int. J. Biochem. 22: 287–293.

Ulloa-Aguirre A, espinoza R, Damian-Matsumwa P and Chappel SC (1988) Immunological and biological potencies of the different molecular species of gonadotrophins. Human Reproduction 3(4): 491–501.

Watson E, Shah B, Leiderman L, Hsu Y-R, Karkare, S, Lu H d and Lin F-K (1994) Comparison of N-Linked Oligosaccharides of Recombinant Human Tissue Kallikrein Produced by Chinese Hamster Ovary Cells on Microcarrier Beads and in Serum-Free Suspension Culture. Biotechnol. Prog. 10: 39–44.

Weinstein J, de Souza-e-Silva U and Paulson JC (1982) Sialylation of glycoprotein oligosaccharides N-linked to asparagine. J. Biol. Chem. 257: 13845–13853.

Address for correspondence: P.P. Gray, Department of Biotechnology, University of New South Wales, P.O. Box 1, 2033 Kensington, NSW, Australia.

Cytotechnology **15:** 223–228, 1994.

Glycosylation and functional activity of anti-D secreted by two human lymphoblastoid cell lines

Diane Cant, John Barford, Colin Harbour[1], Anne Fletcher[2], Nicole Packer[3] and Andrew Gooley[3]

Departments of Chemical Engineering and [1] Infectious Diseases, The University of Sydney, NSW 2006, Australia; [2] NSW Red Cross Blood Transfusion Service, 153 Clarence St., Sydney, NSW 2000, Australia; [3] Macquarie University Centre for Analytical Biotechnology, Balaclava Road, North Ryde, NSW 2113, Australia

Key words: Anti-D; effector function; glycosylation; immunoglobulin G; monoclonal antibody

Abstract

Cell lines BTSN4 and BTSN5 were produced by the Epstein-Barr Virus (EBV) transformation of B-lymphocytes from the same human donor. Both secrete an anti-D monoclonal of the IgG1 subclass but these antibodies display vastly different effector activities. Specifically, anti-D from BTSN4 has a far greater activity in both monocyte- and lymphocyte-mediated ADCC reactions and causes a higher percentage of rosettes to be formed with monocyte-like U937 cells. This variation in functional activity is shown to coincide with changes in the structure of the sugar chains attached to the asparagine-297 site on the immunoglobulin heavy chain.

Abbreviations: ADCC – antibody dependent cellular cytotoxicity; EBV – Epstein-Barr virus; GlcNAc – glucosamine; Gal – Galactose; Man – Mannose; Fuc – Fucose; SA – Sialic acid.

Introduction

The routine administration of anti-D immunoglobulin to pregnant Rh negative women has successfully reduced both the incidence of Rh alloimmunisation following childbirth and the occurrence of haemolytic disease of the newborn (Bowman, 1988). At present, the anti-D used is a serum-derived IgG. Predominant sources of this are either Rh negative women who have been sensitised by a previous pregnancy or Rh negative male donors who are deliberately immunised and boosted to secrete high levels of anti-D. Each source presents its own problems and so in recent years there has been a push to replace the therapeutic polyclonal immunoglobulin with a cell culture-derived human monoclonal anti-D.

When producing anti-D *in vitro*, we are aiming for an antibody of consistent quality that has a reproducible high activity in functional assays. Functional activity has previously been shown to depend on the structure of the C_H2 doman in the antibody heavy chain (Jefferis *et al.*, 1990). The tertiary structure of this region is stabilised by an oligosaccharide chain attached to asparagine-297 and the presence of this sugar chain is essential for the maintainence of antibody functional activity (Nose and Wigzell, 1983; Leatherbarrow *et al.*, 1985; Leader *et al.*, 1991). From limited studies that have been performed elsewhere, it appears that the structure of these sugar chains is affected by cell type and, to a lesser extent, by culture conditions (Goochee and Monica, 1990). Several studies have been published comparing the functional activity of anti-Ds from different sources. Most (Hadley and Kumpel, 1989; Hadley *et al.*, 1989; Merry *et al.*, 1989; Kumpel, 1990) discuss differences between antibodies of different subclasses (specifically between IgG1 and IgG3) but largely neglect variation within a particular subclass. Kumpel *et al.*, 1989, investigated variations in a panel of 23 IgG1 anti-Ds in their ability to mediate red cell lysis by lymphocyte-mediated ADCC. The variation observed is attributed to changes in the Fc structure of the antibodies and is suspected to be post-translational as the activity of the anti-D from some cell lines changes depending on how it is produced. Similarly, Armstrong *et al.*, 1987, noted differences in both lymphocyte-mediated ADCC and macrophage binding activity between three IgG1 anti-Ds. However, they offer little explanation as to how these differences arise. In this study, we investigate two monoclonal IgG1 anti-Ds, both of which have been produced by EBV transformation of B-lymphocytes from the same human donor. Activity in functional assays is compared and this is related to changes in the sugar structure as the asparagine-297 site on the antibody heavy chain.

Materials and methods

Cell lines

Cell lines BTSN4 and BTSN5 were produced by the EBV transformation of B-lymphocytes from the same immune boosted human donor. Both secrete anti-D of the IgG1 subclass and secretion was stabilised by cloning and rosetting of the cell lines with O group, D positive erythrocytes (Fletcher, 1992). Culture conditions for the two cell lines were identical. The medium used was IMDM (Gibco BRL) pH 6.8 supplemented with 10% heat inactivated, bovine IgG-free foetal calf serum (Commonwealth Serum Laboratories), 2mM glutamine, 60iu/ml penicillin and 40μg/ml gentamicin. Cultivation was in 250ml or 500ml spinner vessels (Corning) agitated at a constant 35 rpm.

Haemagglutination

Haemagglutination of papain-treated O^+ (R_1R_2) erythrocytes was performed according to the method of Whitson (1985). Anti-D-containing supernatant was serially diluted on U-well microtitre plates (NUNC) and the highest dilution capable of causing agglutination noted.

IgG ELISA

Flat-bottomed microtitre plates (NUNC) were coated with a 1/800 dilution of affinity isolated goat anti-human IgG (Sigma). After blocking with 1% BSA in PBS, serially diluted samples were applied and incubated for 90 minutes at 37 °C. The plates were then washed prior to application of 1/4000 dilution of goat-anti-human IgG-peroxidase conjugate (Sigma). Reaction with a substrate of ABTS was catalysed by hydrogen peroxide (Kirkegaard and Perry Laboratories) and absorbance read at 405mm. Antibody concentration was calculated by comparison with a human IgG standard (Sigma) present on each plate.

Antibody purificiation

Cell free supernatant was centrifuged at 7000g for 15 minutes and then filtered through an 0.2μM Acrodisc membrane filter (Gelman Sciences) to remove particulates. Purification was by affinity chromatography using a protein G column attached to a FPLC® system (Pharmacia). Antibody was eluted using 0.1M glycine-HCl pH 2.7 and the eluted antibody was immediately restored to a slightly alkaline environment (~pH 7.5) using 1.0M Tris-HCl pH 9.0 The purifed anti-D was shown to be free of contamination using SDS-PAGE electrophoresis under both reducing and non-reducing conditions.

Antibody-Dependent Cellular Cytotoxicity (ADCC)

The release of ^{51}Cr by monocytes or lymphocytes from antibody sensitised erythrocytes was monitored according to the method of Kirkwood *et al.* (1993).

Rosetting using monocyte-like U937 cells

A 5% suspension of washed O^+ (R_1R_2) erythrocytes was maximally sensitised by incubation with purified anti-D. The sensitised cells were then rewashed and contacted with U937 monocyte-like cells in an effector cell: erythrocyte ratio of 1:30. After incubation, 100 effector cells were visually scored for rosette formation (a rosette is defined as an effector cell with 3 or more attached red blood cells). Percentage rosette formation was corrected for effector cell viability (as determined by trypan blue dye exclusion) and each anti-D sample was assayed in quadruplicate.

Deglycosylation of Anti-D

Intact oligosaccharide moietes were released from the purified antibody by treatment with Glycopeptidase F (Boehringer Mannheim) in a reducing environment. Purified antibody was suspended in a 12.5mM Sodium Phosphate buffer (pH 7.0) and supplemented with 12.5mM EDTA, 0.25% Triton-X100, 0.1% SDS and 0.5% 2-mercaptoethanol. A small aliquot was removed to serve as a control and 20 units/ml Glycopeptidase F

(Boehringer Mannheim) were added. Digest and control were incubated overnight at 37 °C and then stored at –20 °C until required.

Monosaccharide quantitation

Purified anti-D was heated at 100 °C for 4 h in 2M trifluoroacetic acid to hydrolyse the oligosaccharide residues present into monosaccharides. The hydrolysed product was separated by HPLC using a CarboPac PA1 anion exchange column (Dionex) linked to a pulsed electrochemical detector (Waters). Elution was by 15mM sodium hydroxide and the resultant peaks were quantitated against an internal standard of 2-deoxyglucose. Sialic acid was detected by the thiobarbituric assay (Skoza and Mohos, 1976).

Results and discussion

The first step in this study was to compare antibody production by the cell lines in batch culture. In each case, production appeared to be growth associated with the maximal rate occurring during the exponential growth phase. However, while BTSN5 cells secrete anti-D at a mean rate of 4.2 ± 0.1 ng/10^5 viable cells/h, the secretion rate by BTSN4 cells is approximately seven times greater than this at 28.6 ± 5.2 ng/10^5 viable cells/h. Similarly, the maximum total antibody produced by a typical BTSN5 batch culture is of the order 3.3 ± 0.2 μg anti-D/ml compared to 36.5 ± 3.8 μg/ml for BTSN4. Maximum cell densities and mean doubling times for the two cell lines do not show the same disparity.

As can be seen from table 1, supernatant from both BTSN4 and BTSN5 batch cultures causes strong haemagglutination of papain-treated O^+ (R_1R_2) red blood cells. In fact, a minimum of 1.97 ng of BTSN5-derived anti-D is required to agglutinate 25μl of 2% red blood cells compared to only 0.027ng of BTSN4-derived anti-D. This 75-fold difference probably results from a combination of effects. The avidity with which BTSN5-derived anti-D binds red blood cells may be significantly lower than anti-D from BTSN4. Alternatively, some of the antibody secreted by BTSN5 may be inactive. These issues will be addressed in future work.

Anti-D from BTSN4 cells displays consistently high activity in functional assays. Table 1 shows responses in monocyte- and lymphocyte-mediated ADCC experiments and in rosette formation with

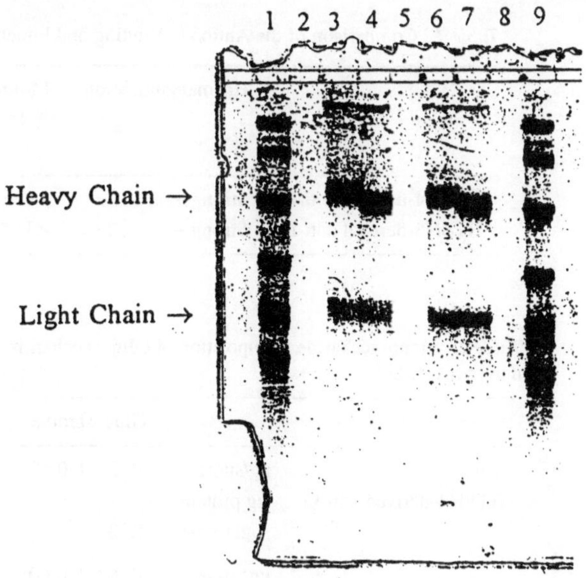

Fig. 1. 10–15% SDS-PAGE gel (reducing conditions) showing the effect of Glycopeptidase F on Anti-D from BTSN4 and BTSN5. Lane 1, Molecular Weight Standards; Lane 2, Sample Buffer Only; Lane 3, BTSN5 Anti-D Control; Lane 4, BTSN5 Anti-D Glycopeptidase Digest; Lane 5, Sample Buffer Only; Lane 6, BTSN4 Anti-D Control; Lane 7, BTSN4 Anti-D Glycopeptidase Digest; Lane 8, Sample Buffer Only; 9, Molecular Weight Standards.

monocyte-like U937 cells. In each case, the red blood cells are incubated with a vast excess of anti-D to eliminate differences between the cell lines. However, in spite of this, anti-D from BTSN4 cells shows significantly higher acxtivity than anti-D from BTSN5 cells in all the assays. At saturating antibody concentrations, as we have here, activity in these assays is solely dependent on the structure of the antibody Fc region and, in particular, the structure of the sugar residue attached to the asparagine-297 site. It is this that we now consider.

Figure 1 shows the effect of Glycopeptidase F digestion on the light and heavy chains of anti-D from BTSN4 and BTSN5. The shift in the heavy chain molecular weights compared to the controls indicates that N-linked sugar residues have been removed. This is expected as the conserved site of glycosylation on the IgG molecule is at the asparagine-297 site in the Fc region of the heavy chain (Rademacher *et al.*, 1986). By comparing the migration shift with the controls, the decrease in heavy chain molecular weight is around 2900 for both BTSN-4 and BTSN5-derived anti-D. As the anti-D light chains do not move on Glycopeptidase F treatment, they can be assumed to be devoid of sugar chains in their native form.

Table 1. Comparison of the Antibody Binding and Functional Activity of Anti-D from BTSN4 and BTSN5

	Haemagglutination	Monocyte ADCC % Red Cell Lysis	Lymphocyte ADCC % Red Cell Lysis	Rosetting (% U937 forming rosettes)
BTSN4-derived anti-D	Strong +ve	44 ± 3.8%	65%	35.9 ± 2.5%
BTSN5-derived anti-D	Strong +ve	<33%	28%	24.1 ± 1.4%

Table 2. Monosaccharide Composition of Oligosaccharide Residues attached to Asparagine-297 of BTSN4- and BTSN5-derived Anti-D

		Glucosamine	Galactose	Mannose	Fucose	Sialic Acid
BTSN4-derived anti-D	ng/sugar μg protein	10.25 ± 0.48	3.44 ± 0.24	5.89 ± 0.97	1.58 ± 0.20	1.29
	sugar ratio	5.22	1.75	3	0.80	0.66
BTSN5-derived anti-D	ng/sugar μg protein	12.68 ± 0.60	3.91 ± 0.50	13.06 ± 0.30	1.90	1.58
	sugar ratio	2.91	0.90	3	0.44	0.36

Next, the composition of the sugar chains from BTSN4- and BTSN5-derived anti-D was compared. Table 2 shows the result of neutral sugar and sialic acid quantitation assays for each antibody. In absolute terms, expressed on a basis of nanograms of sugar per microgram of protein, each antibody contains a similar amount of glucosamine, galactose, fucose and sialic acid. However, the mannose content of BTSN5-derived anti-D is consistently higher than in BTSN4-derived anti-D. In attempting to relate these monosaccharides to possible sugar structures on the IgG molecules two factors should be taken into consideration. Firstly, not all the antibody molecules from a particular cell line will have the same sugar structure. Rather, different glycoforms exist differing by maybe one or two sugar residues (Mizuochi *et al.*, 1982). In the case of BTSN4 and BTSN5-derived anti-D, each antibody displays seven glycoforms (results not shown). Because of this, any overall sugar content determination can, at best, only give an indication of the average glycosylation present. Secondly, N-linked sugars can take several possible forms. The fact that the glycosylation is removed by Glycopeptidase F (see above) establishes that the sugars are N-linked and that there are certain core residues present (Fig. 2). However, substitutions onto this core structure are variable and the overall sugar may have a high mannose, bi-/triantennary or hybrid structure.

When the monosaccharide compositions given in table 2 are considered in terms of ratios, we get some clue as to the types of structure that may be present. For example, in BTSN4-derived anti-D, the mannose: glucosamine : galactose ratio is 3 : 5.22 : 1.75. This could suggest that we have a biantennary structure, probably with a bisecting glucosamine (refer to Fig. 2) and that this structure is almost completely substituted with galactose. The presence of a high proportion of galactose in the high activity BTSN4-derived anti-D agrees with the results of Kumpel *et al.* (1993) who report that Fc receptor interactions are increased in anti-Ds that are highly substituted with galactose. Further analysis of the monosaccharide ratios for BTSN4-derived anti-D reveals that most (80%) of the glycoforms contain fucose and there is some terminal sialic acid substitution. The monosaccharide ratios for BTSN5-derived anti-D are less revealing. The fucose, galactose and sialic acid content rules out the presence of only high mannose glycoforms, but the ratio of mannose : glucosamine is too high to be the product of solely bi- or triantennary structures. Either there is gross variation between different BTSN5-derived anti-D glycoforms, with some having a high mannose structure and others being bi- or triantennary, or there are some hybrid chains present. This could be further investigated using lectins specific for certain sugar-sugar linkages.

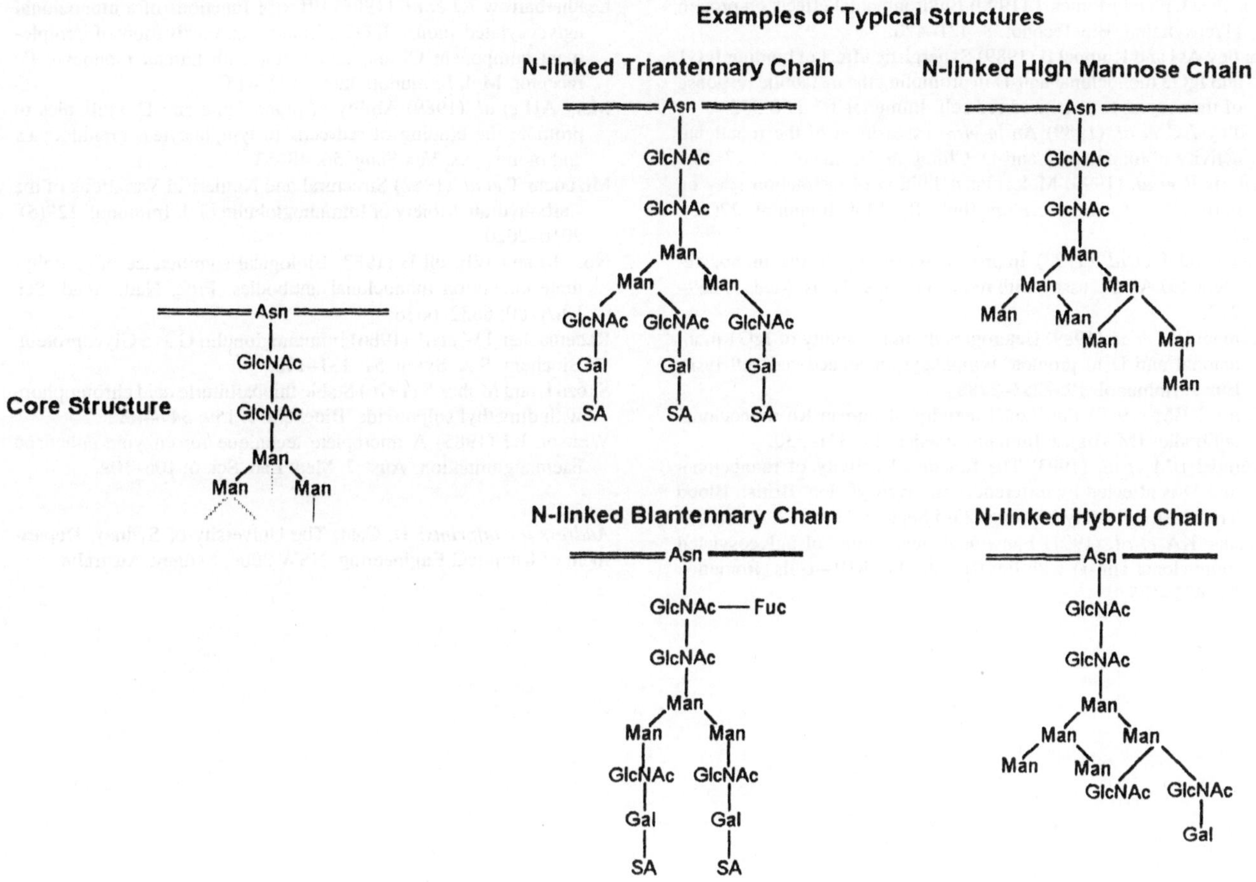

Fig. 2. Typical N-linked Sugar Structure (Alberts *et al.*, 1989: 'Glycan Analysis Pathway', Boehringer Mannheim).

Conclusions

The production of cell lines from the same source material (ie. a single human donor) using the same method, does not necessarily mean that the cell lines produced nor the product secreted will be the same. In the case studied here, antibody from two cell lines so produced differs widely in functional interactions with both FcγRI and FcγRIII receptors. The high activity of one antibody may be related to increased substitution of the core N-linked sugar chains by galactose. Stearic effects due to the presence of bulky high mannose and/or hybrid sugar residues could account for the reduction in activity of the other antibody.

Acknowledgements

This work was funded by an Australian Research Council (ARC) grant. Cell lines BTSN4 and BTSN5 were supplied by Dr Anne Fletcher of the NSW Red Cross Blood Transfusion Service.

References

Alberts B *et al.* Molecular Biology of the Cell. 2nd Edn. Garland Publishing Inc.

Armstrong SS *et al.* (1987) Heterogeneity of IgGl monoclonal anti-Rh(D): an investigation using ADCC and macrophage binding assays. B. J. Haematol. 66: 257–262.

Bowman JM(1988) The prevention of Rh immunization. Trans. Med. Rev. 2: 129–150.

Fletcher A (1992) The production and characterisation of lymphoblastoid cell lines secreting human monoclonal anti-D antibody. PhD Thesis, The University of Sidney.

Glycan Analysis Pathway. Boehringer Mannheim Technical Publication No. 1437950.

Goochee CF and Monica T (1990) Environmental effects on protein glycosylation. Bio/Technol. 8: 421–426.
Hadley AG and Kumpel B (1989) Synergistic effect of binding IgG1 and IgG3 monoclonal anti-D in promoting the metabolic response of monocytes to sensitised red cell. Immunol. 67: 550-552.
Hadley AG *et al.* (1989) An *in vitro* assessment of the functional activity of monoclonal anti-D. Clin. Lab. Haematol. 11: 47–54.
Jefferis R *et al.* (1990) Molecular definition of interaction sites on human IgG for Fe receptors (huFγR). Mol. Immunol. 27(12): 1237–1240.
Kirkwood J *et al.* (1993) Improved sensitivity in the monocyte-mediated ADCC assay with red cell targets. Trans. Med. 3: 269–273.
Kumpel BM *et al.* (1989) Heterogeneity in the ability of IgG1 monoclonal anti-D to promote lymphocyte-mediated red cell lysis. Eur. J. Immunol. 19: 2283–2288.
Kumpel BM (1990) Functional activity of human Rh monoclonal antibodies (MABs). J. Immunogenetics. 17: 321–330.
Kumpel BM *et al.* (1993) The functional activity of monoclonal anti-D is affected by differences in glycosylation. British Blood Transfusion Society Meeting. 23rd Sept. 1993.
Leader KA *et al.* (1991) Functional interactions of aglycosylated monoclonal anti-D with FcγRI+ and FcγRIII+ cells. Immunol. 72: 481–485.
Leatherbarrow RJ *et al.* (1985) Effector functions of a monoclonal aglycosylated mouse IgG2a: binding and activation of complement component Cl and interaction with human monocyte Fc receptor. Mol. Immunol. 22(4): 407–415.
Mery AH *et al.* (1989) Ability of monoclonal anti-D antibodies to promote the binding of red cells to lymphocytes, granulocytes and monocytes. Vox Sang. 56: 48–53.
Mizuochi T *et al.* (1982) Structural and Numerical Variations of the Carbohydrate Moiety of Immunoglobulin G. J. Immunol. 129(5): 2016–2020.
Nose M and WIgzell H (1983) Biological significance of carbohydrate chains on monoclonal antibodies. Proc. Natl. Acad. Sci. USA. 80: 6632–6636.
Rademacher TW *et al.* (1986) Immunoglobulin G as a Glycoprotein. Biochem. Soc. Symp. 51: 131–148.
Skoza L and Mohos S (1976) Stable thiobarbituric acid chromophore with dimethyl sulphoxide. Biochem. J. 159: 547–462.
Whitson KJ (1985) A microplate technique for enzyme enhanced haemagglutination. Aust. J. Med. Lab. Sci. 6: 106–109.

Address for offprints: D. Cant, The University of Sydney, Department of Chemical Engineering, NSW 2006, Sydney, Australia.

Cytotechnology **15:** 229–242, 1994.

Interaction of cell culture with downstream purification: a case study

Wolf Berthold and Ralph Kempken
Process Development, Department of Biotechnological Production, Thomae / Boehringer Ingelheim GmbH, Birkendorfer Str. 65, 88397 Biberach at the Riss, Germany

Key words: Cell harvest, biomass removal, clarification, large scale cell centrifugation, large scale tangential flow filtration, particles, cell debris

Abstract

Separation of product from secreting mammalian cells in the culture broth means the transition from product generation to product isolation. This interface within a biotech production process has to perform a proper solid / liquid phase separation of the cell suspension to make the product containing fluid amenable for further purification. These subsequent steps require fluid with low occurence of contaminants in order to function properly. The goal of this study was to evaluate some economic and fast cell separation methods for the preparation of a product fluid ready for use in further ultrafiltration and chromatographic processes. We have performed experiments to test the usefulness of disc stack centrifuges and tangential flow microfiltration units at large scale. Both systems revealed outstanding prospects with regard to throughput and scale up properties. However, the centrificgation did not lead to a fluid sufficiently free of particles for direct ultrafiltration or chromatography. Thus, an additional filtration step was necessary. On the other hand microfiltration led to an acceptable quality of process fluid directly. By optimisation of process parameters an effective, reproducible and robust cell separation can be obtained. However, our experience has been that such optimal conditions are somewhat specific for a narrow range. Thus, even the equipment functioning well with one type of cell would possibly not perform as well with another cell or even with the same cell under conditions slightly different to the usual situation.

Introduction

Even today a biotech production represents in concept a procedure akin to agriculture, which is probably the most significant and fundamental technology in the evolution of our civilisation.

The production process has the goal to obtain a desired product by use of highly selected natural (and today man modified micro-) organisms and plants in an artificial containment controlled by man. The product may be the entire culture, as in the case of hey and yoghurt, or it may be part of it as for wheat and beer. In the latter case a separation technique has to achieve the isolation and enrichment to a useful concentration of the desired product. Dependent on the nature of the product the technique used exploits different physical or chemical properties of the product.

This general concept still applies to today's very highly sophisticated biotechnological production, which experiences a rapid development due to the novel possibility of man modified microorganisms and cells.

Some of the technically most demanding processes are those, where mammalian cells are the source of the desired product. The cells are comparatively fragile and 'defenseless' in culture because they are cultivated outside any functional immune system of their original animal. In addition they require a rather complex nutrient medium for 'homeostasis'. The cells proliferate very slowly, especially compared to invading 'self-contained' microorganisms with their inbuild defense mechanisms. Still the unique capability of mammalian cells to synthesize even the largest and most complex structures that the human body could require as medicine, make them an indispensable source for a new class of pharmaceutical products.

Today the most common types of products from mammalian cells are secreted proteins. The isolation of the protein therefore requires the removal of particles and molecules of very different sizes from cell aggregates to soluble proteins and smaller molecules.

However, even in mammalian cell culture the product could be the cells themselves (e.g. in the new and exciting perspective of *in vitro* organ generation e.g. for skin substitution). The cells can also harbor the product intracellularly, like in some virus production schemes, or the product may be associated with cellular membranes like for membrane anchored receptors. Thus, for the purpose of product isolation there are different starting conditions and targets and usually they can not be achieved in one step, but in a sequence of steps. The most common is the removal of cells. This topic will be addressed in this paper in more detail.

Techniques using either differences in density or size have found wide applications for the removal of cells.

The use of gravitational settlers for cell retention on a larger scale and slow operation is inconvenient because of the large size of settling tank relative to the bioreactor. Recently, some novel types of more compact settlers have been designed (Thompson *et al.*, 1994; Tokashiki and Arai, 1989; Stevens *et al.*, 1994; Searles *et al.*, 1994; Brown *et al.*, 1991). These systems allow gentle handling of the cells during separation, and can hereby seperate viable cells from dead cells, because viable cells have a greater settling velocity than non-viable cells that shrink upon death. On the other hand, scalability of these settler devices is limited and typical throughputs are in the range of 1 VVD (volume per volume and day) (Thompson *et al.*, 1994; Stevens *et al.*, 1994; Searles *et al.*, 1994). Thus, the sedimentation techniques are most appropriate for a selective removal of non-viable cells in continuous cultivation systems (Hülscher *et al.*, 1992) or typical small cell isolation. For rapid and quantitative removal of both viable and non-viable cells plus debris, other techniques should be superior to the sedimentation of cells.

By centrifugation, the density difference between solid and liquid is amplified through the application of centrifugal force by rotating the suspension at high speed. For large scale, continuous tubular bowl and disc stack centrifuges are required to handle the volumes in appropriate time (Mahar, 1993). These devices are much less efficient than laboratory centrifuges because lower g forces and short residence times are applied.

Using a tubular bowl centrifuge, the inside wall becomes covered with precipitate and the efficiency of the centrifuge is lowered gradually by reduction of the effective radius (Lee, 1989). The precipitate inside the bowl cannot be removed without stopping the centrifuge. However, the machine has to be dismantled for cleaning. Due to these limitations and to the long residence time of 'pelleted' cells inside the rotating bowl at high centrifugal forces, disc stack centrifuges should be more adequate for the removal of sensitive mammalian cells.

The disc stack centrifuge is run at relatively low speed (e.g. 5,000·g) but has nevertheless a high capacity for separating particles, because the settling distance to the sediment is small for suspended particles (Ball, 1985). Discharge of sediment is possible during the run of the centrifuge, and cleaning procedures (CIP and SIP) can be performed without dismantling (Westfalia, 1991). Such centrifuges have been used very successfully in the food industry (Hemfort, 1983) and in the biotechnology of yeasts and bacteria (Hemfort and Kohlstette, 1988; Bell *et al.*, 1983; Mannweiler and Hoare, 1992; Jin *et al.*, 1994, see below). However, there is very little experience with animal cells handled with disc stack centrifugation (Jäger, 1992; Wang *et al.*, 1968).

In contrast to centrifugation techniques, filtration exploits the difference in particle size. The better the uniformity of pore size the better defined is the cut off point of particles retained. If the effective pore size of membranes can remain uniform and unifluenced by the operation an excellent standardization of particle separation can be achieved in a wide range of sizes from particles to molecules (Cheryan, 1986; Tutunjian and Sewin, 1984). The benefit and application of membrane processes in the fields of chemical engineering, for chemistry, waste water treatment, desalination, food industry and medicine has been reviewed e.g. by Rautenbach and Albrecht (1982), Chmiel *et al.* (1983), Strathmann and Chmiel (1985), Belfort (1984), Rippberger (1988), Belfort (1989), Winzeler (1990). Microfiltration techniques take advantage of depth absorption, zeta potential and sieve retention (Ball *et al.*, 1985).

Depth filtration filters have a high capacity for debris removal compared to screen-type filters. On the other hand, flow rate and differential pressures may vary during the filtration process. Particles may pass through the filter when increasing the pressure to maintain the desired flow rate all, 1985). Precoating the filter may extend the filtration performance significantly but cannot prevent the biomass from clogging the filter (Lee, 1989).

A more uniform filtration of large 'precipitates' of particles can be achieved by tangential flow filtration (TFF) instead of dead end filtration. A cross-flow of feed solution over the filtration surface prevents cells from being deposited on the membrane during filtration (Grabosch, 1987; Bowen, 1993). Careful fluid management is necessary to avoid clogging and to minimise the boundary layer. Exceptional challenges of TFF processes for sensitive mammalian cells are both the intactness of viable and non-viable cells and the presence of debris and 'sticky' molecules like proteins, lipids, DNA, etc. Characteristics and applications of TFF membranes for cell harvesting were investigated e.g. by Hanisch (1986), van Reis *et al.* (1991), Maiorella *et al.* (1991), Sarry and Sucker (1992), Tanaka *et al.* (1993). Some effects of adsorption, shear and pressure on microfiltration membrane properties were evaluateed by Bowen and Gan (1991, 1992). Characteristics of different membrane materials were compared by Strathmann (1979) and the characteristics of different module configurations by Mackay and Salusbury (1988). Besides the chemistry of the filtration membrane, the size and uniformity of pores and the module design, the success of tration will highly depend on a proper process engineering (Maiorella *et al.*, 1991, van Reis *et al.*, 1991).

This paper will describe attempts to test these diferent techniques in their present technological state of perfection for the use in large scale mammalian cell culture.

Materials and methods

All experiments on fluid clarification reported in this paper were performed using a Westfalia CSA-1 high speed stainless steel disc stack centrifuge (Westfalia Separator AG, Oelde, FRG). A schematic cross section of this separator and its operating mode is given by Hemfort (1983). The prototype makes discharge of sediment possible either by the conventional mode (discontinuously by activating an intermittent discharge mechanism) or by a modified output mechanism (continuous discharge via a special separation disc) that allowed shorter residence times of the sedimented cells in the bowl. The sediment containing phase was about 3% of the fed culture fluid in order to get high concentration factors and to minimize product loss accompanying the sediment containing phase.

All tagential flow microfiltration experiments described in this paper were performed with Enka Microdyn® hollow fiber modules (Enka AG, Wuppertal, FRG) with polypropylen membranes of 0.2 μm pore size. Those membrane cartridges are commercially available and are amenable to linear scale up without changes in flow geometry or fluid dynamics. Scale up was performed by combining the TFF cartridges in parallel order in stainless steel housings.

Fresh cell suspension of a mouse hybridoma cell line from fed batch fermentations at 80 liter, 400 liter and 2000 liter scale were used as preharvest suspension. The fermentation runs were performed according to GLP rules in the process development area of Thomae/BTP according to Werner *et al.* (1993) and Werner and Noé (1992). Virtually the same raw materials, cultivation modes, process design, bioreactor and utility systems were used in all experiments. Thus, representative starting conditions were attained for the following experiments. As product either recombinant proteins or murine monoclonal antibodies were secreted into the culture medium and monitored by ELISA. Cell concentration and viability were analysed microscopically by trypan blue exclusion. Wet cell biomass was determined by weighing the sample of centrifuged pellet (37000 g, 20 minutes).

Results and discussion

The biotech production can be divided into two major production phases with very different goals. The first phase consists of the proliferation of cells and/or stimulation of the cell synthesis of the product. Thus the goals is generation of product as much as possible by facilitating the biological synthesis of product in and by cells. The second phase has the goal to obtain the product from the primary broth with as high a yield as possible and transfer it into a form most useful for the intended purpose. Thus the product should be purified with as little loss as possible.

The interface between synthesis and isolation is usually called harvest and can involve many different techniques. All are dually dependent on the mode of cell fermentation in combination with the properties of the product. Because of this interrelated dependency the harvest methods constitute a very difficult operation in contrast to subsequent purification steps. Especially in the production of biopharmaceuticals the entire process is further constrained by the requirement for reproducibility of the process and consistency of the product.

There are two major forms of fermentation: a continuous cell expansion alternating with a continuous or repeated harvesting and batch fermentation resulting in a single harvest. The demands on the efficiency can be very different according to the mode of operation and optional cell retention (Table 1). The product containing cell suspension may vary from an almost clarified fluid (e.g. immobilised systems, continuous perfusion with spinfilter or micro-filtration modules) via suspensions of mediocre cell density (e.g. batch, chemostat) to suspensions of very high cell density (fed batch, hollow fiber reactors with harvest of product and cells in the extracapilluy space).

This paper describes attempts to obtain a reliable, fast and effective separation of mammalian cells from large preharvest volumes that are typically obtained in batch and fed batch fermentation processes. General requirements for mammalian cell separation are listed in Table 2. With preference, a product containing fluid should be obtained that is amenable for further downstream processing. Special attention was directed towards disc stack centrifugation and tangential flow microfiltration because both methods have been the most efficient in performance and scale up prospects for industrial scale applications in biotechnology (van Reis *et al.*, 1991; Hemfort and Kohlstette, 1988).

Case I: Disc Stack Centrifugation (DSC)

In many biotechnological processes centrifugal clarifiers are used extensively in technical and industrial scale. Production processes with bacteria and yeast cells use various types of disc stack centrifuges at large scale, e.g. to concentrate single cell protein and baker's yeast, to separate the cells at citric acid and glutamate production, to recover intracellular enzymes, to perform counter-current extraction of antibiotics, to perform contained cell separation at the vaccine production, etc.. Nevertheless the centrifugal clarification of mammalian cell suspensions requires appropriated equipment (see below) with some advantageous features. An extensively standardized and automated mode of operation combined with automatic tunning of validated CIP and SIP may reduce interference by or attention of staff and may allow nexible use of the same equipment for several production processes (i. e. no dedication to a product in case of campaign or concurrent manufacturing in a multi-use biotech facility). In addition fouling and ageing effects of filtration processes could be omitted.

These aspects and the separation efficiency of DSC were tested with cell suspensions originated by a hybridoma (cell line A). Using the intermittent discharge system evety time 30 liter clarified liquid had been processed, a clarification capacity of 100% was obtained. Cells were concentrated approximately 20 fold in the discharged sediment and the mode of operation was simple and reliable. Although about 20% of the cells were destroyed in the DSC and lots of subcellular particles were present in the clarified liquid phase (visible by microscope as 'sandy' ground, detectable by Coulter counter and laser light diffraction analyses and by subsequent dead-end filtration procedures). Thus, DSC runs with continuous sediment ejection were performed to reduce the residence time of the mammalian cells inside the bowl from ca. 30 minutes to approximately 1 – 2 minutes.

Figures la and 1b reveal different outcomes of such experiments with continuous sediment removal depending on process parameters. An optimal run is shown in Fig. 1a. The sediment was discharged continuously and rapidly via the separation disc. Remaining cell concentration in the clarified liquid phase was measured as about 1% with respect to the feed concentration (Fig. la). At low cell densities in the feed stock the cell sediment accumulated in the sediment holding space and was discharged lately when some packed sediment was mobilized by chance or by pressure fluctuations (Fig. 1b). This effect was enhanced when cell density in the feed was decreased.

In contrast the DSC did not cope with the cell discharge at very high cell densities (data not shown here). The sediment holding space filled with cells and large aggregates clogged the effluent stream via separation disc. Then cells overflowed and were discharged in the 'clarified' liquid phase. Hence, the continuous discharge system operated well only at a narrow range. In addition significant fluctuation of pressure and fluid flow occured at this mode of operation. The same amount of cells were destroyed inside the centrifuge as reported for the intermittent discharge system.

Product recovery in the clarified liquid phase was excellent in all DSC experiments. No significant loss was detected in the liquid phase during the entire operation (Fig. 2) and no accumulation was found in the sediment containing phase. Product in the sedimed was not attempted to be recovered and approximately 3% of the quantity of the cell suspension was lost, therefore.

Table 1. Harvest modes for mammalian cells dependent on the cultivation method

Fermentation Mode	Separation Mode	Cell Culture Fluid	Consequences
attached growing cells			
• immobilised systems	repeated harvesting	cell retention without special separation device	withdrawal of almost cell-free supernatant
• hollow fiber modules	repeated harvest of the extracapillary space	cell centrifugation in small scale	very dense cell populatio
	continuous perfusion of the intracacapillary space	UF membranes	cellfree ultrafiltrate, very little cell debris
suspended cells or microcarrier			
• continuous perfusion	spinfilter systems	rotating sieve	little cell bleed
	slow centrifugation	disc stack centrifuge	very little cell bleed
	sedimentation	settling devices	removal of dead cells
• batch and fed batch	single or repeated batch harvest	MF membranes	cellfree microfiltrate, very little cell debris
	ditto	I high speed disc stack centrifuge	I very little cell bleed

Table 2. Requirements for mammalian cell separation processes

Desirable parameters	goal
single step procedure	ease of operation
contained system	process control
CIP and SIP feasibilities	cleaning
design of process and equipment	validatible, GMP
defined cut-off of particle clearance	transformation of a suspension to a solution
low particulate and chemical fouling	low contribution of surface layer on separation performance
reproducibility of process	GMP
short process time	limited exposure to contaminants
numerous re-use	economics
short downtime	economics
low generation of cell debris	limited exposure to contaminants
quantitative removal of cells	handshake with downstream processing (ultrafiltration,chromatography)
clearance of subcellular particles	definite solution as starting point for purification
low shear stress for the cells	limited exposure to contaminants
low pressure exposure of the cells	gentle treatment of the sensitive cells
short residence time of the cells in the system	ditto
linear scale up	flexibility of use

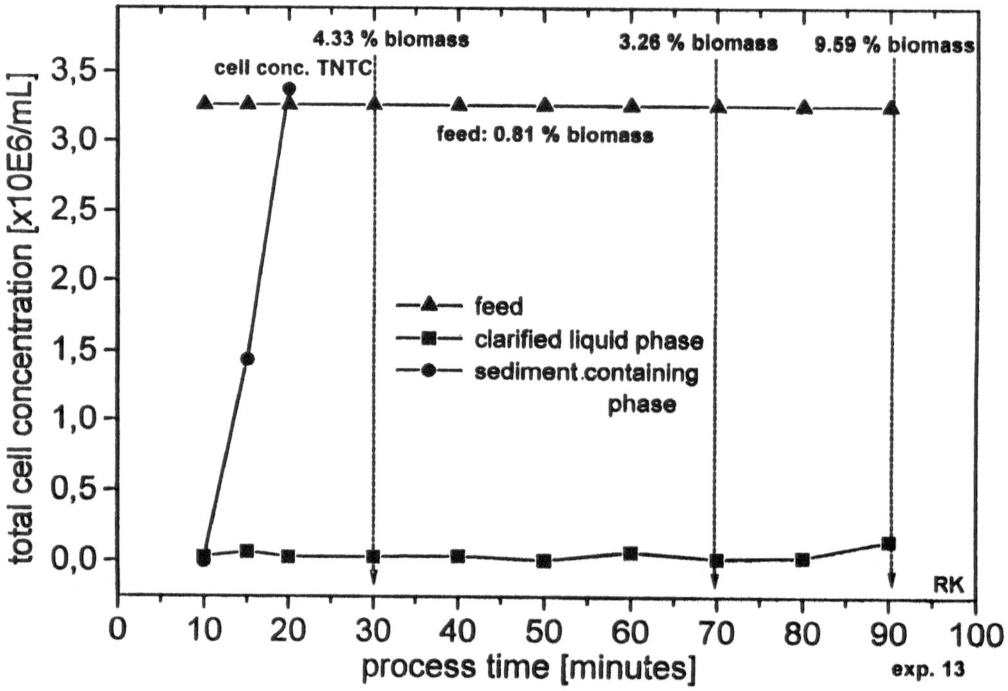

Fig. 1a.

Fig. 1(a). Optimal DSC run with continuous removal of the sediment containing phase.

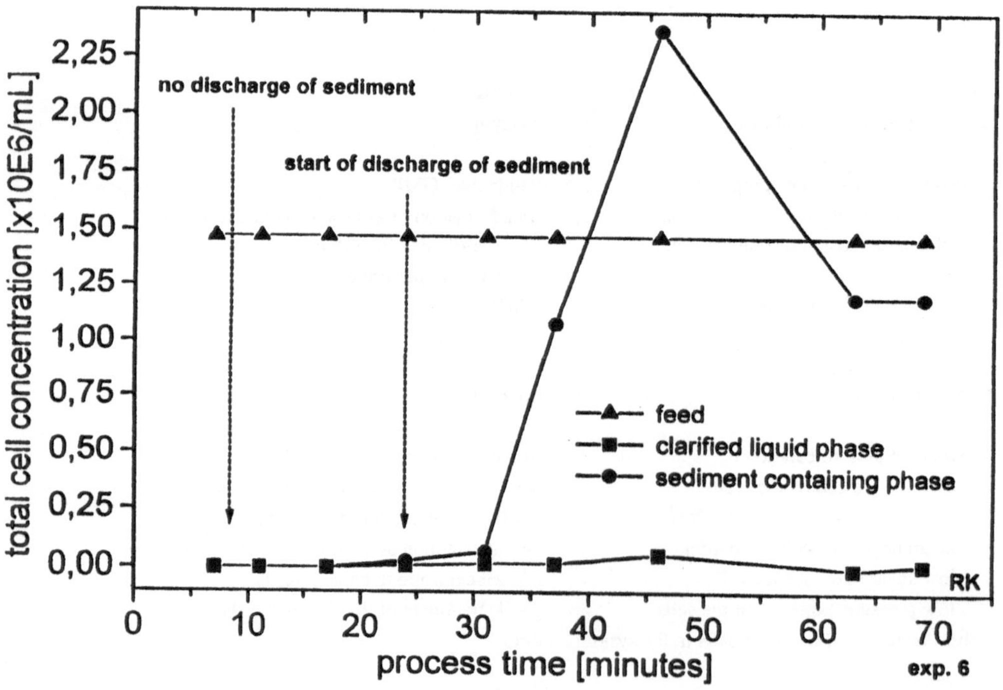

Fig. 1b.

Fig. 1(b). Suboptimal DSC run with continuous removal of the sediment containing phase at low cell concentration in the feed.

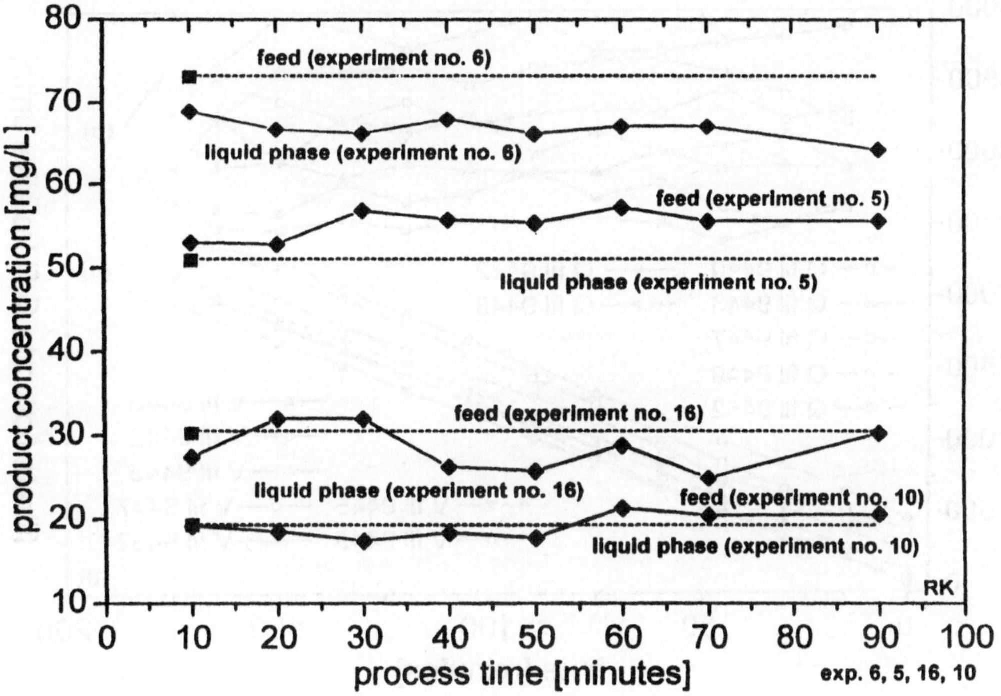

Fig. 2.

Fig. 2. Course of product titers in feed and liquid phase during some DSC experiments.

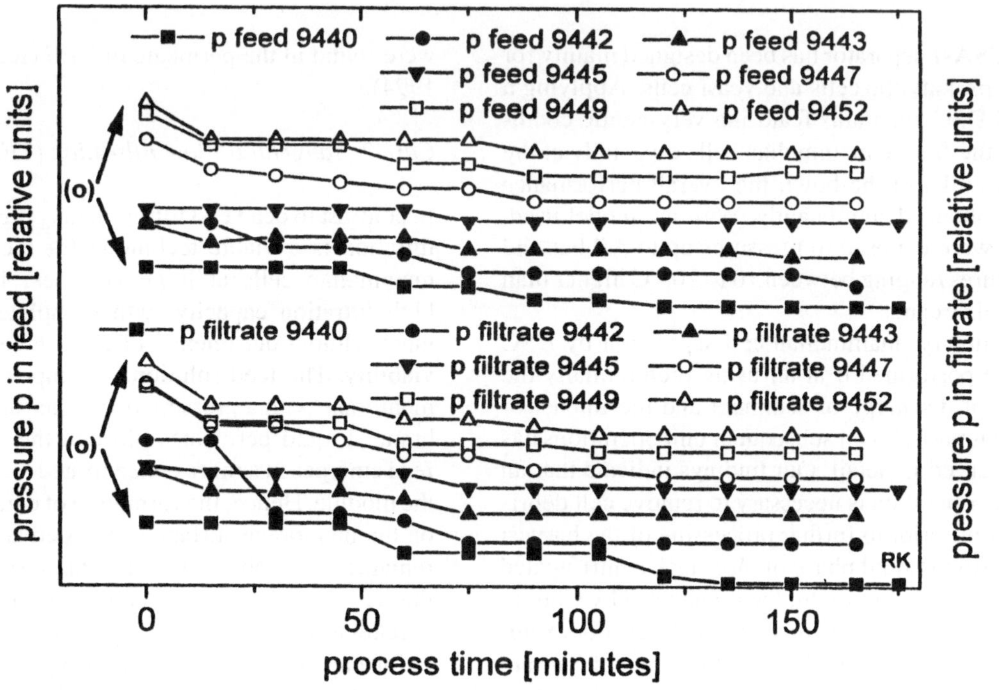

Fig. 3a.

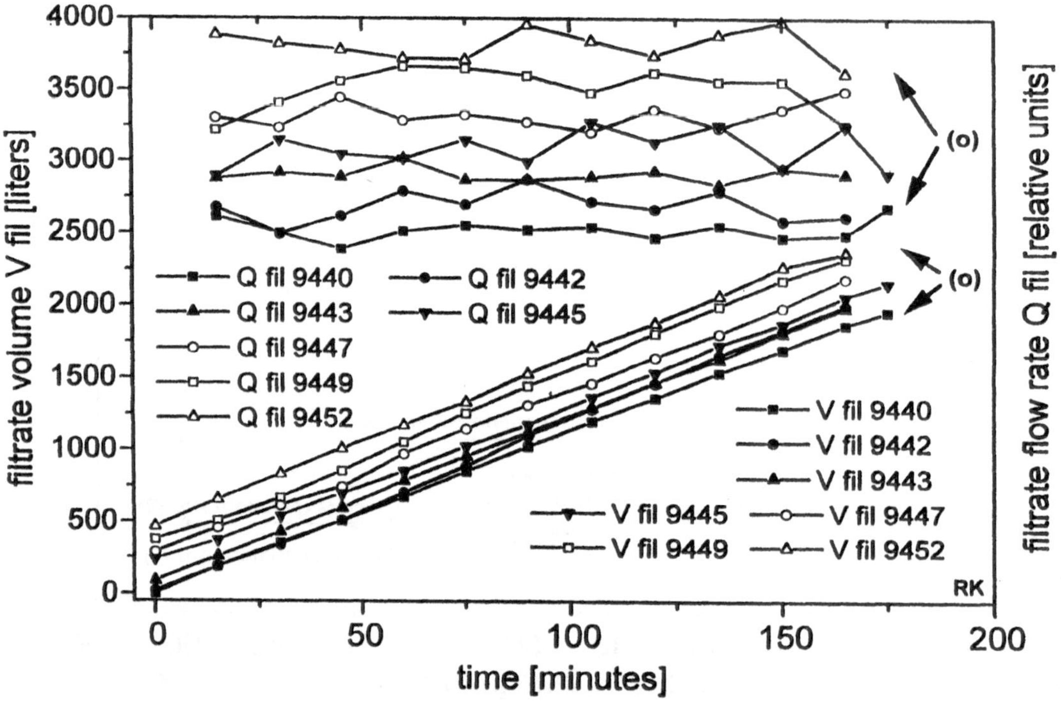

Fig. 3b.

Figs. 3(a)–(b). Consistency and reproducibility of TFF procedures at the 2000 liter scale (representation in displayed layers); a) course of pressure in feed and filtrate; b) course of filtrate flow rate and accumulation of filtrate fluid; (o): these curves are at identical levels, they have been displaced to distinguish the invididual data spots.

The CSA-1 separator has been designed mainly for use with prokatyotic cells and yeast cells. Applying a modified DSC apparatus featuring very gentle conditions for the fragile mammalian cells especially at the inlet and outlet of the bowl, the overall performance may be enhanced significantly. With the model used, the cells were exposed to pressures up to 3.0 bar and temperatures ranging between 10 to 20 °C higher than in the feed stream.

In summary, mammalian cell separation by DSC should be performed if at all using preferentially the intermittent discharge of sediment and the impact of subcellular particles on subsequent unit operations has to be observed in detail. Our findings indicate that an additional operation is necessary to remove cell debris and particles prior to further processing of the harvest fluid. Clarified liquid phase of disc stack centrifugated mammalian cell culture fluid was analysed by Coulter counter and by laser light diaction methods. For example, many particles were found in the range of 5 μm to 1.5 μm with a steady increase of particle numbers with decreasing size. At the same range almost no particles were found in the permeate of TEF (Kempkens *et al.*, 1994).

Case 2: Tangential Flow Filtration (TFF)

In contrast to dead end filtration tangential flow microfiltration is a viable technique for the separation of mammalian cells at large volumes. It can provide high filtration capacity with complete cell containment wlthout detrimental effects on cell number and viability. The feed solution is pumped parallel to the membrane with a pressure difference across the membrane. Liquid permeates through the membrane and feed emerges in a more concentrated form at the exit of the module. Hence, the formation of extensive deposits on the membrane surface is avoided and the retentate remains in a mobile form. For large scale application the TFF membranes are configured in series and in parallel to get an arrangement with the required separation capability, e.g. to process approximately 10,000 to 15,000 liters of cell culture fluid in a few hours (van Reis *et al.*, 1991, Werner *et al.*, 1992). Howev-

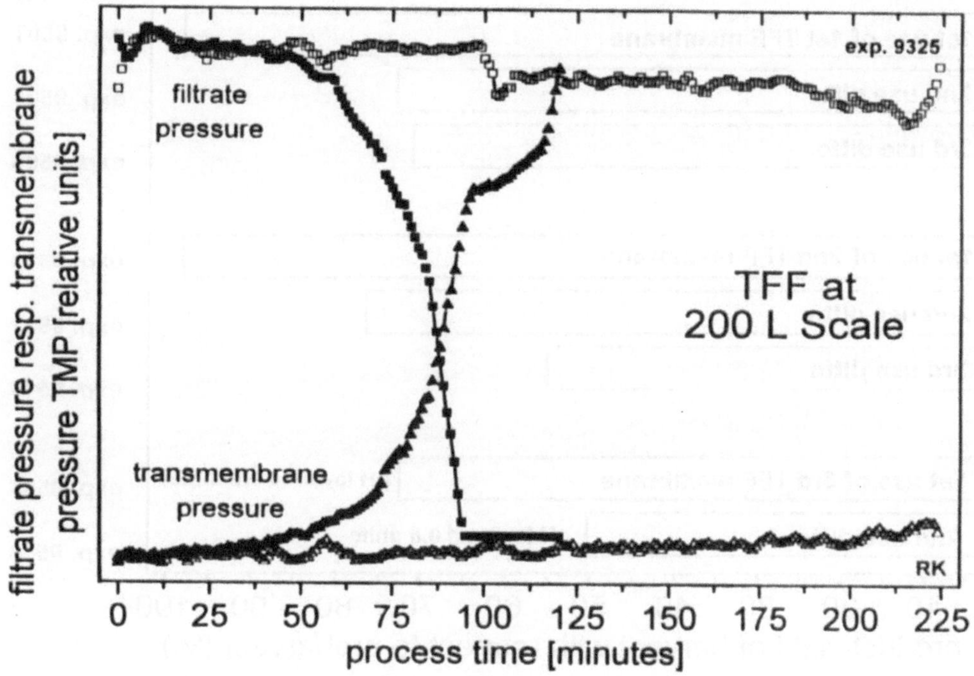

Fig. 4. Effect of slight variation of the filtrate flow rate on TFF performance; see text for more details.

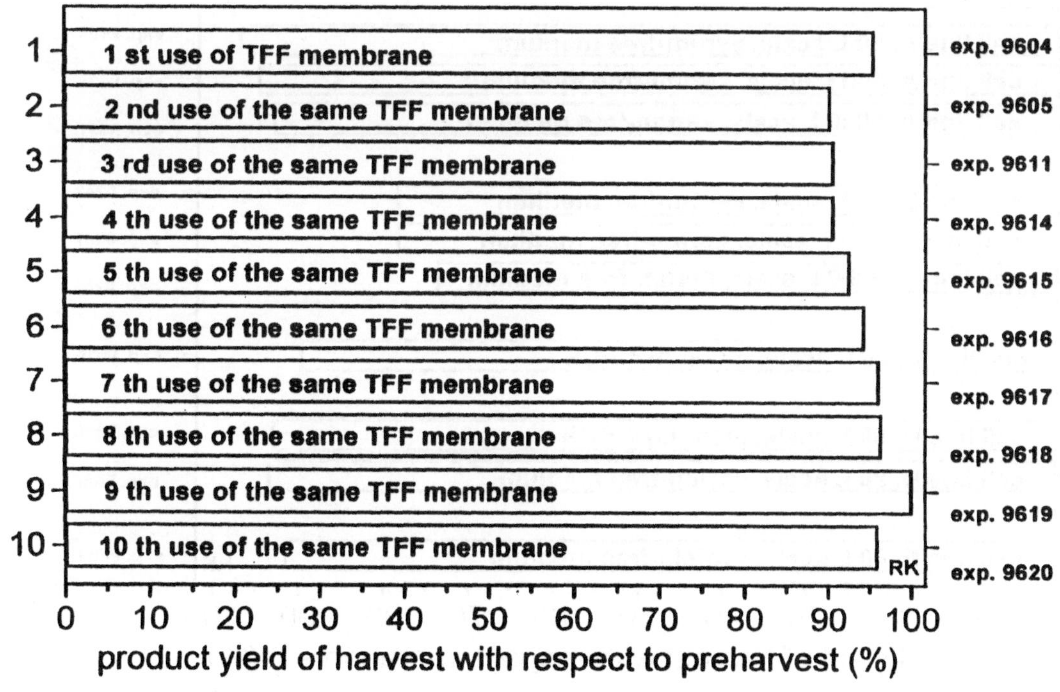

Fig. 5 Efficiency and reproducibility of TFF cell harvest with multiple re-use of the membrane.

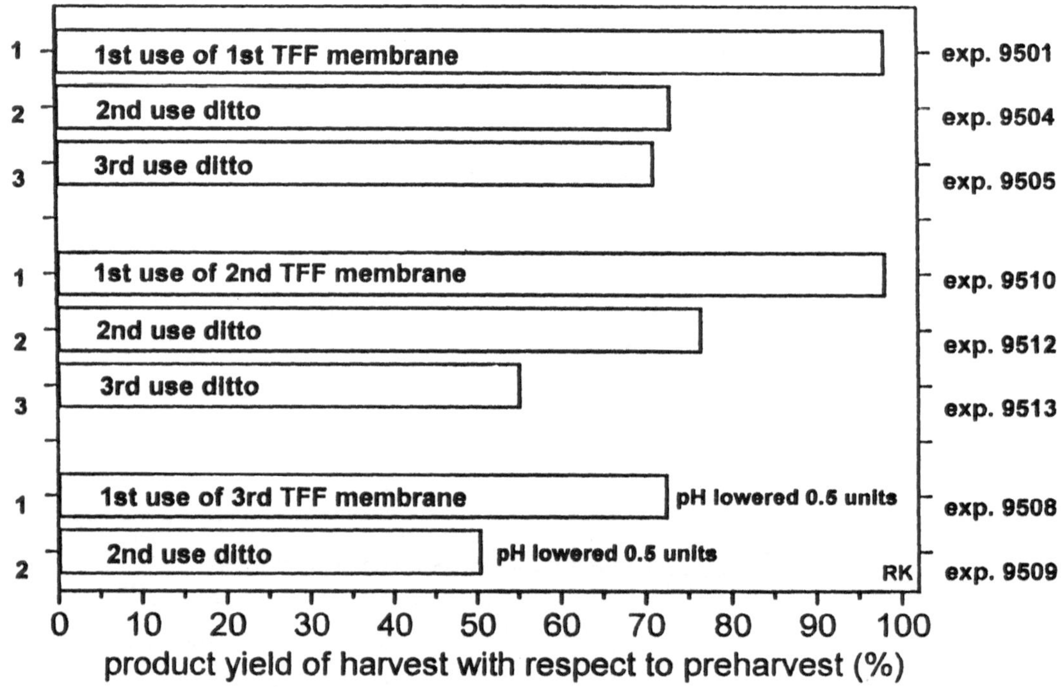

Fig. 6. Suboptimal TFF performance when re-using the membrane or when altering the process conditions (here: slight decrease of pH).

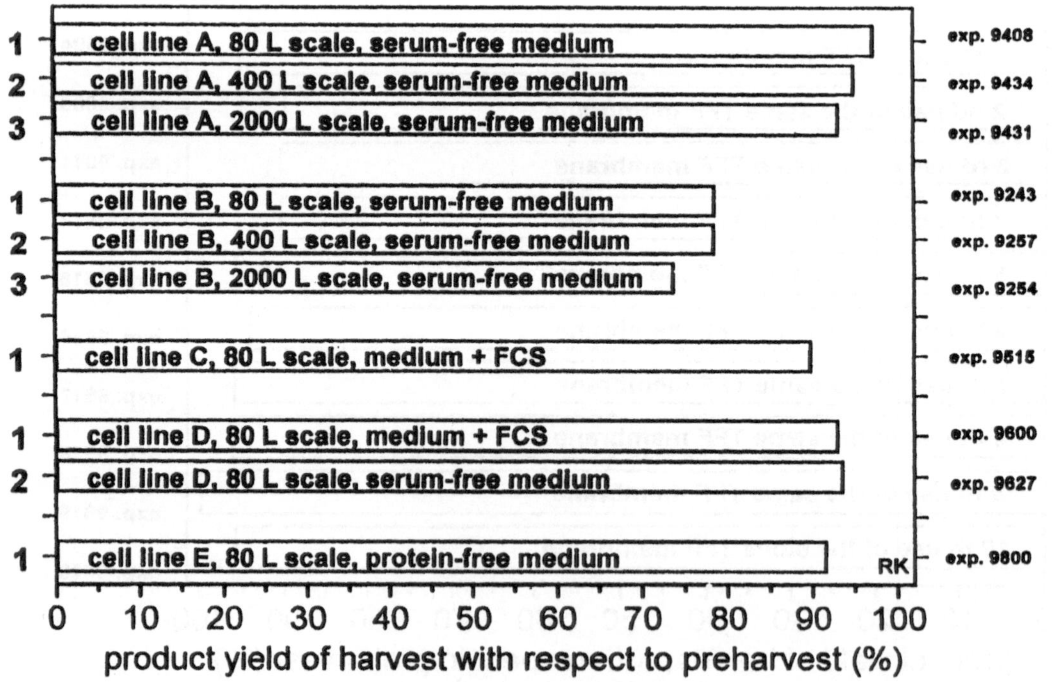

Fig. 7. Applicability of one optimised TFF process for various cell lines, production scales and culture media.

er, this well-established technique is not completely trouble-free for every cell line or process and critical parameters of TFF are listed below. Some experiments of industtial scale TF runs demonstrate however its consistency and reproducibility and sensitivity as well as its limits of efficiency.

In general, a uniform cell separation performance can be realised by process optimisation. Fig. 3a reveales the equal time course of pressure in feed and filtrate respectively at seven TFF harvests in the 2,000 liter scale. For a comparative visualisation the curves were plotted in displaced layers. The same coordinates are applied in Fig. 3b for the course of filtrate flow and the accumulating filtrate fluid at the same experiments. While constant physical parameters are obtained during the entire run, no decrease of filtrate permeation is found and product yield is about 90%. The experiments were performed with identical hardware and process conditions and with equal cell suspension.

Slight variation of process parameters might interfere severely with the outcome of the separation process. Fig. 4 shows the filtrate pressure and transmembrane pressure (TMP) profiles of two experiments in 200 liter scale with identical cell suspension (cell line F), the same hardware and equal process parameters except for the filtrate flow rate. At the standard filtrate flow rate (120 minutes process time, solid symbols in Fig. 4) a rapid fouling of the membrane was found and confirmed by a sharp drop of filtrate pressure and a contrary incline of TMP respectively. This experiment resulted in a modest product yield of 86%. When reducing the filtrate to half the flow rate, the process time was considerably longer but an excellent and constant filtrate pressure and TMP was maintained (hollow symbols in Fig. 4). The product yield of this run was excellent at 98%.

These findings illustrate that a fine tuning of process parameters can improve the separation performance considerably even if defined and invariable hardware configuration has to be maintained.

Efficient cell harvest processes excel by high product yields at low process costs. Thereby, a multiple re-use of the TFF modules with consistently good performance is advantageous particularly since they have to be used in a product specific mode for GMP production and to avoid cross contamination. The membranes have to be stored in an appropriate manner in between different campaigns of production. Figure 5 demonstrates a consistently high product yield greater than 90 % with an optimised cell harvest procedure for cell cine D. A single TFF module was used ten times for cell harvests in the 80 liter scale with the same cell suspension, the same harvest unit and the same process parameters. During the runs pressures and flow rates were as uniform as shown in Fig. 3a,b for cell line A (data not shown here).

Nevertheless, not for all cell lines and products optimal results for cell harvest can be expected when using one and the same hardware and method. For example Fig. 6 indicates considerable losses in product yield when re-using the same TFF membranes whereas the first use revealed very good results greater than 95% (TFF in the 80 liter scale using cell line C, equal cell suspension, same harvest units and process parameters). Probably the cleaning procedure was not compatible with cell line and product in this special combination, since other biological and technical parameters than fermentation process, culture medium, harvest system and parameters were compatible to those applied for the experiments of Fig. 5.

Hence, some experiences with certain combinations of cell harvest system and parameters, cell line and product, composition of preharvest will not allow to predict the performance of the same combination when using another product from another production cell line. Yet after suitable optimisation a cell harvest system can be used successfully for different cell lines and products, for various cell densities, product titers and culture media in the technical scale and in the production scale. For example the product yields of cell harvest experiments in the 80 liter scale are summarised in Fig. 7 for five different cell lines (A – E) and products. The cell density varied from 2 to 20 million cells per ml and the medium composition from 0.2% FCS via serum-free to protein-free media. The cell separation method was suboptimal merely for cell line B, otherwise product yields greater than 90% were obtained in all other cases.

At the scale up of the harvest process from technical scale to production scale the TFF modules were combined in parallel order to a stainless steel piped harvest unit. Both processes could be performed with approximately the same yield in the 80 liter, 400 liter and 2000 liter scale.

All preharvest cell culture suspensions that were used in this study were fresh culture broths from fed batch fermentations with low viabilities at the day of harvest (ca. 30% to 5% living cells). Thus, those suspension fluids represented a worst case for membrane separation processes with mammalian cell cultures. Nonetheless a 0.1 or 0.2 μm dead-end filtration was sufficient in all experiments to process the cell-free permeate of the single step TFF procedure appropriately for a trouble-free performance of subsequent downstream procedures using ultrafiltration and chromatography steps (Berthold and Walter, 1994).

Conclusion

At the interface between fermentation and product recovery, two scalable unit operations were tested for mammalian cell separation from the product containing culture broth. Regarding the removal of whole cells the DSC (disc stack centrifuge) with discontinuous sediment removal could almost achieve a quantitative cell removal of 0.2 μm TFF (tangential flow filtration) systems. Since DSC is not liable to fouling or ageing, effects of TFF systems and since operation and cleaning of the DSC apparatus can be automated partially, the centrifugation should theoretically be the superior system. On the other hand shear and pressure forces on the cells are much higher when using DSC. For example a balance of fed cells to discharged and sedimented cells at the DSC runs of this study indicated that approximately 20% of the overall cells were destroyed. This is likely to lead to a release of proteolytic enzymes from organelles and lead to subsequent interference with the product.

Using TFF procedures the destruction of cells could be avoided by appropriate process parameters. With help of suitable chemical and physical characteristics of the TFF membrane and the modules (charge, surface properties, size and uniformity of pores, robustness, module design) and optimised parameters for operation and cleaning (flow and pressure profiles, shear forces, low adhesion of solids, high transmission of product) the disadvantages of the TFF principle can be reduced very much for a great variety of cells and products. Consequently the advantages of the TFF method are more relevant, e.g. the definite cut-off at a low and certified pore size, the low exposure of cells to shear and pressure and the single step operation mode. The resulting fluid could be dead-end filtered by 0.1 or 0.2 μm filters and was ready for use in subsequent ultrafiltration or chromatography steps.

Using the DSC the cell culture was exposed to high shear stress (at the inlet and outlet) and high pressures (at the outward parts of the bowl) at the entire range of operation that enabled a reasonable clarification performance. Many subcellular particles could not be removed and presumably were even generated by the DSC process.

The situation did not improve when applying a continuous discharge of sediment with a considerably reduced residence time of the cells inside the bowl. Dead-end filtration of the clarified liquid phase was ditcult because several filters of 0.1 to 1.2 μm pore size were clogged up when less than 10 liters of filtrate were passed through.

When the clarified liquid phase was ultrafiltrated directly, the ultrafiltration membranes blocked up and were unable to regenerate. Thus, an additional process step (probably a small TFF procedure) has to be included before the clarified liquid phase of DSC can be processed in downstream steps such as ultrafiltration or chromatography.

Possibly some improvements in mechanical engineering can lead to a more refined equipment for the centrifugal clarification of mammalian cell suspensions. Nevertheless some aspects of this mode of separation is basically inconsistent for mammalian cells: a) the density difference between mammalian cells and their culture medium is very small; b) mammalian cells are much more sensitive than microorganisms when exposed either to high levels or to fluctuations of shear force, pressure and temperature; c) cell lysis may result in the exposure of product to lytic enzymes if they are released from their cellular compartments; d) mammalian cells generate a considerable amount of cell debris and the number and size of subcellular particles dependent on the fermentation mode.

Separation experiments with various hybridoma and recombinant CHO cell lines in our hands indicate that TFF procedures are most promising to meet the requirements for a robust and rapid batchwise harvest of big quantities of mammalian cells: a) the products have to be recovered by sophisticated downstream process steps with a pure and essentially particle free clarified liquid as a prerequisite; b) the cell-free harvest fluid ought to be sterile filtered by dead-end devices prior to subsequent processing; c) contents of contaminating protein, endotoxin and nucleic acid levels have to be equal or less than those in preharvest; d) cell number and viability should not be affected by the harvest procedure; e) high processing capacities should be obtained at economically feasible costs.

Recommendation

In summary the following recommendations can be expressed for the cell separation procedures evaluated in this study. In case of cell associated proteins the

DSC may allow a rapid and efficient collection of cells in biomass suitable for further extraction of the protein product. In case of cell retention with modest flow rates a continuous perfusion system can be established based on slow speed DSC and used in combination with continuous recirculation of the sediment and continuous removal of the product containing liquid phase (Jäger, 1992). In case of single or repeated harvests of big preharvest batches the quality of the permeate after TFF is superior to the clarified liquid phase after disc stack centrifugation. Besides, the economics of a single step TFF are expected to be superior to a two step procedure using disc stack centrifugation and additional filtration process. Finally, the technical limitations of TFF are tolerable and the TFF procedures can be adopted for large scale operation. This TFF technology has been used for many years in the manufacture of kilogram quantities of pharmaceutical products derived from mammalian cells (van Reis *et al.*, 1991, Werner *et al.*, 1992).

Acknowledgement

Special thanks go to Jörg Schäfer and Anja Preißmann for conception and performance of experiments with DSC. The technical skill and in depth discussions of Hubert Rechtsteiner, Uwe Karnapp, Uwe Katz, Bärbel Skopp and Herbert Hertz are greatfully acknowledged. The authors are indebted to Westfalia Separator AG company (59302 Oelde, FRG) for the loan of a prototype CSA-1 disc stack centrifuge and for constructive discussions.

References

Ball GD (1985) Clarification and Sterilisatrion. In: Spier RE, Griffiths (eds.) Animal Cell Biotechnology 2, Academic Press Inc. pp. 87–127.

Bell DJ, Hoarel M, Dunnill P (1983) The Formation of Protein Precipitates and Their Centrifugal Revovery. In: Fichter A (ed.) Advances in Biochemical Engineering / Biotechnology 26, Springer Verlag: 1–72.

Belfort G (1984) Synthetic Membrane Processes Fundamentals and Water Applications. Academic Press, Inc.

Belfort G (1988) Membranes and Bioreactors: A Technical Challenge in Biotechnology and Bioengineering 33: 1047–1066. Bioreactors: A Technical Challenge in Biotechnology.

Berthold W, Walter J (1994) Protein Purification: Aspects of Processes for Pharmaceutical Products. Biologicals 22: 135–150.

Bowen R, Gan Q (1991) Properties of Microfiltration Membranes: FLux Loss During Constant Pressure Permeation of Bovine Serum Albumin. Biotechnology and Bioengineering 38: 688–696.

Bowen R, Gan Q (1991) Properties of Microfiltration Membranes: The Effects of Adsorption and Shear on the Recovery of an Enzyme. Biotechnology and Bioengineering 40: 491–497.

Bowen R (1993) Understanding Flux Patterns in Membrane Processing of Protein Solutions and Suspensions. Trends in Biotechnology 11: 451–460.

Brown PC, Lanel JC, Wininger MT, Chow R (1991) On-line Removal of Cells from Continuous Suspension Cultuires. In: Meignier , Spier RE, Griffiths GB (eds.) Production of Biologicals from Animal Cells in Culture, Butterworth-Heinemann Ltd., pp. 240–242.

Cheryan M (1986) Ultrafiltration Handbook. Technomic Publishing Company Inc.

Chmiel H, Strathmann H, Streicher E, Schneider H (1983) Membranes in Medicinal Process Enneering. Chemie-Ingenieur-Technik 55: 4282–292.

Grabosch M (1986) Cross-flow Filtration Technique in Biotechnology. BioEngineering 1: 62–68.

Hanisch W (1986) Cell Harvesting. In: McGregor WC (ed.) Membrane Separations in Biotechnology, Marcel Dekker, Inc. pp. 61–88.

Hemfort H (1983) Centrifugal Separators for the Food Industry. Technical Scientific Documentation No. 3 (2nd ed.) of Westfalia Separator AG, 59302 Oelde, FRG,

Hemfort H (1988) Separators. Technical Scientific Documentation No. 1 (booklet No. 9997-7133-010/0483) of Westfalia Separator AG, 59302 Oelde, FRG.

Hemfort H, Kohlstette W (1988) Centrifugal Clarifiers and Decanters for Biotechnology. Technical Scientific Documentation No. 5 (booklet No. 9997-8203-000/0688) of Westfalia Separator AG, 59302 Oelde, FRG .

Hülscher H, Scheibler U, Onken U (1992) Selective Recycle of Viable Cells by Coupling of Airlift Reactor and Cell Settler. Biotechnology and Bioengineering 30: 442–446.

Jäger V (1992) A Novel Perfusion System for the Large-scale Cultivation of Animal Cells Based on a Continuous Flow Centrifuge. In: Spier RE, Griffiths JB, MacDonald C (eds.) Animal Cell Technology: Developments, Processes and Products, pp. 397–4502, Butterworth-Heinemann Ltd.

Jin K, Thomas ORT, Dunnill P (1994) Monitoring Recombinant Inclusion Body Recovery in an Industrial Disc Stack Centrifuge. Biotechnology and Bioengineering 43: 455–460.

Kempken R, Preißmann A, Berthold W (1994) Assessment of an Industrial Disc Stack Centrifuge for Use in Mammalian Cell Separation. Paper submitted to Biotechnology and Bioengineering.

Lee S-M (1989) The Primary Stages of Protein Recovery. Journal of Biotechnology 11: 103–118.

Mackay D, Salusbury T (1988) Choosing Between Centrifugation and Cross-flow Microfiltration. Chemical Enneering 447: 45–50.

Mahar JT (1993) Scale-up and Validation of Sedimentation Centrifuges. Part I: Scale-up. BioPharm (September, 1993) 42–51.

Maiorella B, Dorin G, Carion A, Harano D (1991) Crossflow Microfiltration of Animal Cells. Biotechnology and Bioengineering 37: 121–126.

Mannweiler K, Hoare M (1992) The Scale-down of an Industrial Disc Stack Centrifuge. Bioprocess Engineering 8: 19–25.

Ratenbach R, Albrecht R (1982) Separation by Membranes. Chemie-Ingenieur-Technik 54 (3): 229–240.

Rippberger S (1988) Approaches to Microfiltration of Process Solutions and its Application. Chemie-Ingenieur-Technik 60 (3): 155–161.

Sarry C, Sucker H (1992) Adsorption of Proteins on Microporous Membrane Filters: Part I. Pharmaceutical Technology International (October, 1992) 16–23.

Searles JA, Todd P, Kompala DS (1994) Perfusion Culture of Suspended CHO Cells Employing Inclined Sedimentation. In: Spier RE, Griffiths GB, Berthold W (eds.) Animal Cell Technology: Products of Today, Prospects for Tommorow, pp. 240–242. Butterworth-Heinemann Ltd.

Stevens J, Eickel S, Onken U (1994) Lamellar Clarifier – A New Device for Animal Cell Retention in Perfusion Culture Systems. In: Spier RE, Griffiths GB, Berthold W (eds.) Animal Cell Technology: Products of Today, Prospects for Tommorow, pp. 234–239. Butterworth-Heinemann Ltd.

Strathmann H (1979) Trennung von molekularen Mischungen mit Hilfe synthetischer Membranen. Steinkopff Verlag, Darmstadt.

Strathmann H, Chmiel H (1985) Membranes in Chemical Engineering. Chemie-Ingenieur-Technik 57 (7): 581–596.

Tanaka T, Kamimura R, Itoh K, Nakanishi K (1993) Factors Affecting the Performance of Crossflow Filtration of Yeast Cell Suspension. Biotechnology and Bioengineering 41: 617–624.

Thompson KJ, Wilson JS (1994) A Compact Gravitational Settling Device for Cell Retention. In: Spier RE, Griffiths GB, Berthold W (eds.) Animal Cell Technology: Products of Today, Prospects for Tommorow, pp. 227-229. Butterworth-Heinemann Ltd.

Tokashiki M, Arai T (1989) High Density Culture of Mouse-Human Hybridoma Using Perfusion Apparatus with Multi-settling Zones to Separate Cells from the Culture. Cytotechnology 2: 5–8.

Tutunjian RS, Sweing R (1984) Hollow-fiber-ultrafiltration in Biotechnology. Biotech-Forum 1 (3/4): 1–8.

van Reis R, Leonard LC, Hsu CC, Builder SE (1991) Industrial Scale Harvest of Proteins from Mammalian Cell Culture by Tangential Flow Filtration. Biotechnology and Bioengineering 38: 413–422.

Wang DIC, Sinskey JJ, Gerner RE, De Filippi RP (1968) Effect of Centrifugation on the Viability of Burkitt Lymphoma Cells. Biotechnology and Bioengineering 10: 641–649.

Werner RG, Walz F, Noé W, Konrad A (1992) Safety and Economic Aspects of Continuous Mammalian Cell Culture. Journal of Biotechnology 22: 51–68.

Werner RG, Noé W (1993) Mammalian Cell Cultures. Part II: Genetic Engineering, Protein Glycosylation, Fermentation and Process Control. Arzneimittel-Forschung/Drug Research 43 (II): 1242–1249.

Westfalia Separator Report. Introduction to Westfalia Separator AG and Some of the Fields of Application of the Machines, Installations and Processes. Booklet no. 9997-4776-010/0189 (1991) of Westfalia Separator AG, 59302 Oelde, FRG.

Winzeler HB (1990) Membrane Filtration with High Separation Performance and Minimal Power Demand. Chimia 44: 288–291.

Address for offprints: W. Berthold, Department of Biotechnological Production, Thomae / Boehringer Ingelheim GmbH, Birkendorfer Str. 65, 88397 Biberach at the Riss, Germany.

Cytotechnology **15:** 243–251, 1994.

Evaluation of membranes for use in on-line cell separation during mammalian cell perfusion processes

Heino Büntemeyer, Christoph Böhme and Jürgen Lehmann
Institute for Cell Culture Technology, University of Bielefeld, P.O. Box 100131, 33501 Bielefeld, Germany

Key words: Hybridomate, monoclonal antibody, perfusion, microfiltration, membrane, fouling

Abstract

In this study two microporous hollow fibre membranes were evaluated for their use as cell retention device in continuous perfusion systems. A chemically modified permanent hydrophillic PTFE membrane and a hydrophilized PP membrane were tested. To investigate the filtration characteristic under process conditions each membrane was tested during a long term perfusion cultivation of a hybridoma cell line. In both cultivations the conditions influencing membrane filtration (e.g. transmembrane flux) were kept constant. Filtration behaviour was investigated by monitoring transmembrane pressure and protein permeability. Transmembrane pressure was measured on-line with an autoclavable piezo-resistive pressure sensor. Protein permeability was determined by quantitative evaluation of unreduced, Coomassie stained SDS-PAGE. The membrane fouling process influences the filtration characteristic of both membranes in a different way. After fermentation the PP membrane was blocked by a thick gel layer located in the big outer pores of the asymmetric membrane structure. The hydraulic resistance was higher but the protein permeability was slightly better than of the PTFE membrane. For this reason the PP membrane should be preferred. On the other hand, transmembrane pressure decreases slower when the PTFE membrane is used, which favours this membrane for long term cultivations, especially when low molecular weight proteins (<30 KD) are produced.

Abbreviations: PP – Polypropylene; PTFE – Polytetrafluoroethylene

Introduction

For the cultivation of mammalian cells and the production of biopharmaceuticals (recombinant proteins, monoclonal antibodies) a wide variety of bioreactor systems were developed. Because of the low growth rate and high nutrient demands of mammalian cells continuous systems are often used. To obtain high cell densities cell retention devices have been employed. These devices can be placed into the bioreactor or in an external loop. As internal systems spin filters (rotating sieves) (Fenge *et al.*, 1993) and membrane devices proved to be successful (Büntemeyer *et al.*, 1987). Synthetic membranes can be used as hollow fibres or flat membranes in an arrangement of static or cross flow elements.

The main problems using membranes in static or cross flow mode occur from fouling processes which affect transmembrane flow rate and filtration characteristic. Synthetic membranes in bioreactors used for cell retention are of microfiltration type with pore sizes of 0.1 – 5 μm. In most cases microporous membranes are used. Because of their special structure cells, cell debris and macromolecules (proteins, DNA, etc.) can attach to the membrane and cause the fouling. They fill the pores and form a layer which, in addition to the synthetic membrane, behaves as a secondary membrane (Le and Gollan 1989; Meireles *et al.*, 1991). The hydraulic resistance rises. The secondary membrane, mainly consisting of a gel formed by the macromolecules, changes the filtration type from microfiltration to ultrafiltration. The active pore size decreases from micropores (0.1 – 5 μ) to nanopores (1 – 20 nm).

At this stage a filtration phenomenon, the concentration polarization, plays a major role. Macromolecules are now retained by the secondary membrane and the filtrate concentration decreases whereas the concentration in front of the secondary membrane rises. When the saturation concentration of those substances is reached they precipitate to the gel layer and enlarge it further (Flaschel *et al.*, 1983). The consequence is a significant increase in transmembrane pressure drop if the flux through the membrane is kept constant.

The PTFE membrane used in this study was a symmetric microporous membrane (pore size is the same on both sides) while the PP membrane was of asymmetric structure with the smallest pores on the inner surface.

The aim of this study was the evaluation of the two hollow fibre membranes for their use in continuous perfusion processes of mammalian cell cultivations. Transmembrane pressure drop and protein permeability are used as characteristic quantities.

Table 1. Summary of the characteristics of the PP and PTFE membranes studied in this investigation

	PP	PTFE
Material	Polypropylene (Accurel) hydrophilized (with ethanol)	Polytetrafluoroethylene (PTFE), modified permanent hydrophilic
pore size	0.3 μm	3 μm
wall thickness	0.4 mm	0.5 mm
specific surface		
outside	81.7 $cm^2\ m^{-1}$	94.3 $cm^2\ m^{-1}$
inside	56.5 $cm^2\ m^{-1}$	62.8 $cm^2\ m^{-1}$
porosity	75%	40–80 %
bubble point (in water)	0.25 bar	1.55 bar

Materials and methods

Bioreactor

The bioreactor system used in this study was a 2 litre bench scale perfusion system based on a modified BIOSTAT BF2 bioreactor (B. Braun Biotech International, Melsungen, Germany). The reactor was equipped with the double membrane stirrer on which aeration and perfusion membranes were fixed (Büntemeyer *et al.*, 1987).

The bioreactor was aerated with a hydrophobic hollow fibre membrane ($3\ m\ l^{-1} = 245\ cm^2\ l^{-1}$) (Lehmann *et al.*, 1988) connected to a 4-channel gas supply for air, O_2, N_2 and CO_2 controlled by a digital control unit (DCU, B. Braun International) depending on set-points of pO_2 and pH. Temperature was set to 37 °C, stirrer speed to 35 rpm, pH to 7.2 and pO_2 to 40% air saturation.

Furthermore, the stirrer was equipped with internal hollow fibre microfiltration membranes. For cell-free continuous medium exchange during perfusion two different hollow fibre membranes were tested (Table 1). In each case the same filtration area of approx. 377 cm^2 membrane was mounted onto the stirrer. Because of the different outer surface area of both membranes, it was necessary to use 4 m of the PTFE membrane and 4.62 m of the PP membrane, respectively. The membranes were connected on one end via a peristaltic pump (Watson-Marlow 501UR) to the medium reservoir vessel and on the other end via a second pump to the harvest vessel (Fig. 1). Both pumps were controlled by a level sensor and ran alternatingly in a special manner. First, the feeding pump fills the reactor with a transmembrane flux of $0.075\ l\ m^{-2}\ min^{-1}$ until maximal level is reached. Then the harvesting pump withdraws the supernatant with a lower transmembrane flux of $0.059\ l\ m^{-2}\ min^{-1}$ for a defined time interval (30 min). Next, the reactor is filled again and the whole procedure is repeated (Büntemeyer *et al.*, 1987). For the measurement of the pressure drop during the harvesting step a sterilizable, piezo-resistive pressure sensor (type 4045 A5, Kistler, Winterthur, Switzerland) was fitted into the harvesting stream very close to the bioreactor outlet (Fig. 1). Above feed and harvest flux were chosen to provoke a more rapid membrane fouling. Optimal flux values are described earlier (Büntemeyer *et al.*, 1987). Additionally, a bleed stream was connected direct to the bioreactor to control cell density and maintain steady state conditions.

Cells

The cell line used in this study was the hybridoma HB 58 (ATCC). This cell is a rat mouse hybridoma which produces rat antibodies type IgG_1, specific for mouse κ light chain. Cell numbers were determined microscopically by trypan blue exclusion.

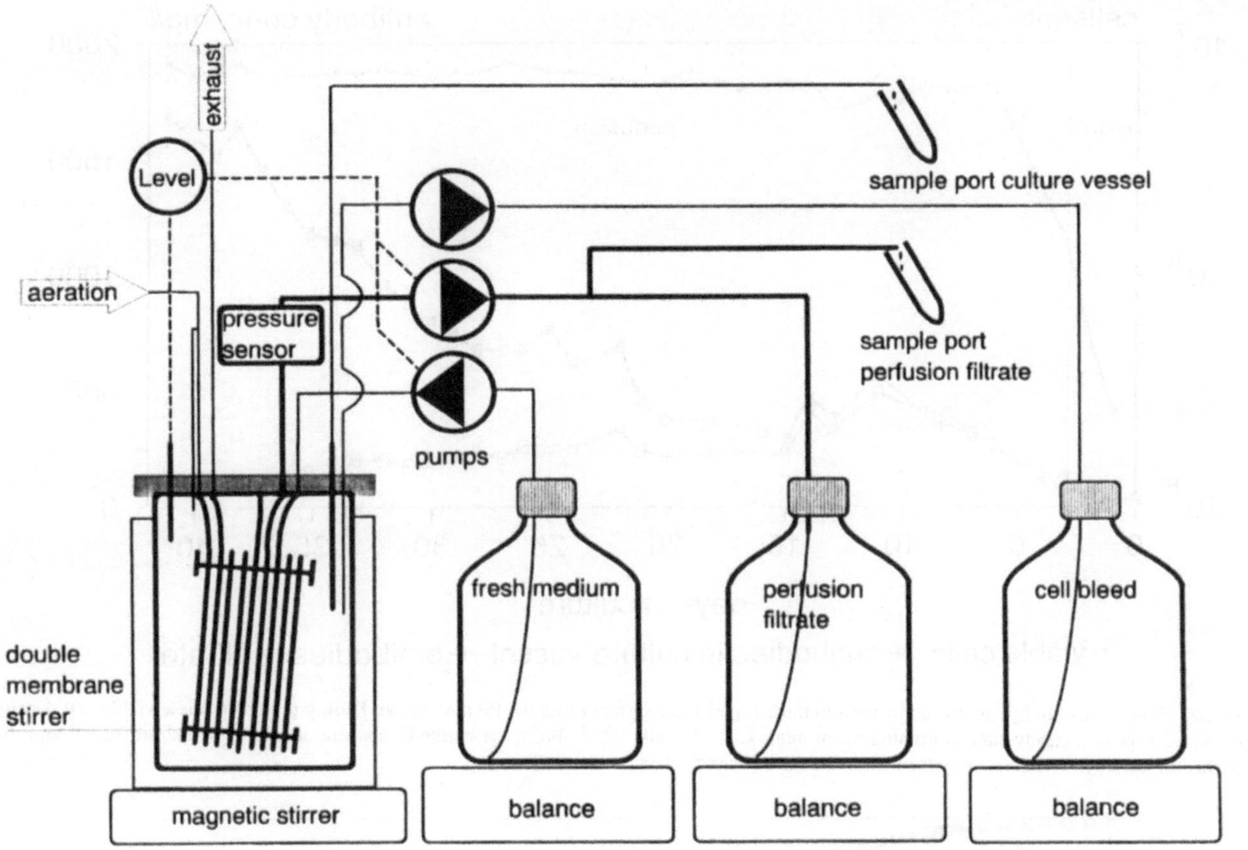

Fig. 1. Experimental setup of continuous perfusion system. A 2 L bench scale bioreactor was used. The double membrane stirrer was equipped with a hydrophobic PP membrane for aeration and the investigated membranes for perfusion.

Medium

The medium used for both experiments was a serum-free low protein medium (SF-medium) consisting of a 1:1 mixture of DMEM and F12 (Gibco, Eggenstein, Germany). The following supplements were used per litre medium: 10 mg human transferrin (Fe saturated), 10 mg bovine insulin, 50 μmol ethanolamine and 2 ml lipoprotein solution (ExCyte I, Bayer Diagnostics, München, Germany). To obtain high cell density in perfusion mode the serumfree medium was supplemented with glucose, sodium pyruvate and various concentrations of amino acids and glutamine accordingly to the demand of the cells in high cell density culture.

Analytical methods

Antibody concentrations in the supernatant were analyzed by a kinetic sandwich ELISA method (Enzyme: Peroxidase; Substrate: o-Phenylenediamine) as described previously (Büntemeyer *et al.*, 1991). For the determination of protein concentrations unreduced SDS-PAGE electrophoresis was done. The gels were stained with Coomassie Brilliant Blue using the Sensiquant method (Bülles *et al.*, 1990). Quantitative evaluation of the gels was carried out by scanning with an Epson flat bed scanner GT6000 and analysing with Pharmacia's Gel Image software package 1DEVA. The quantitative results for antibody concentration obtained by gel scanning and ELISA were compared. They showed nearly the same accuracy and error margins. Comparable results for the other proteins should be expected. For optimal nutrient supply of the culture during perfusion concentrations of glucose and amino acids were analyzed by methods described previously (Büntemeyer *et al.*, 1991).

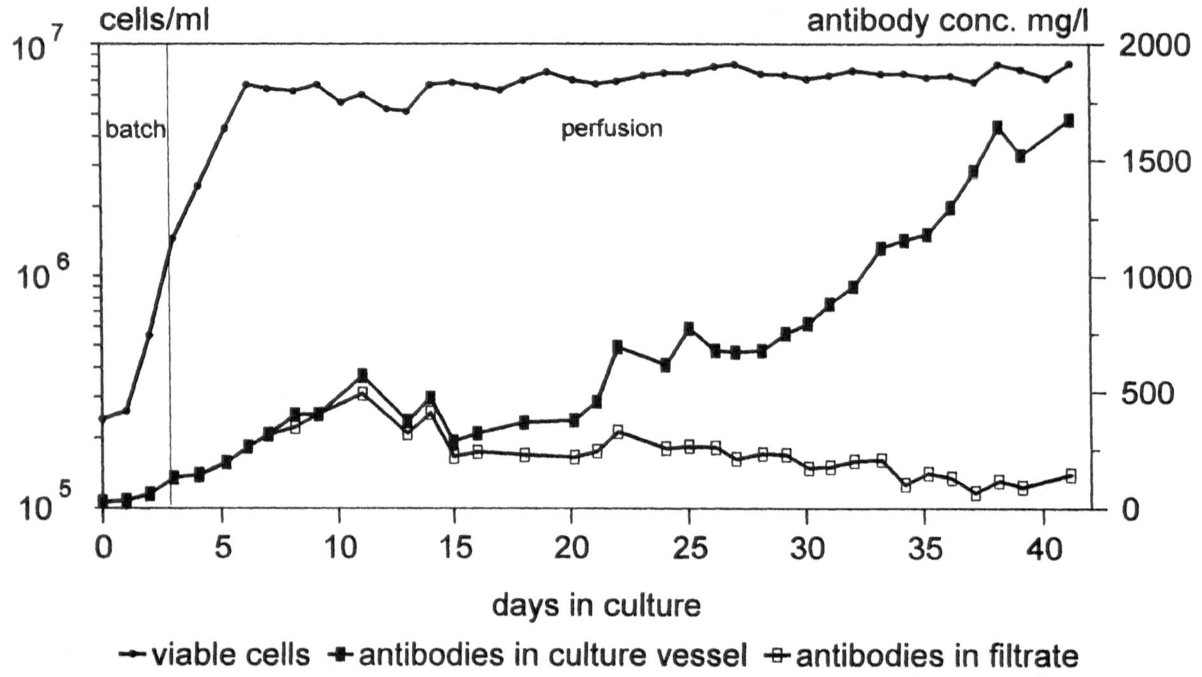

Fig. 2. Time course of hybridoma cultivation in the 2 L perfusion system using the PP membrane. During perfusion phase viable cell density could be kept at a steady state concentration of approx. $7 \cdot 10^6$ cells ml^{-1}. From the antibody concentrations in the culture vessel and the perfusion filtrate the retainment by the membrane clearly can be seen.

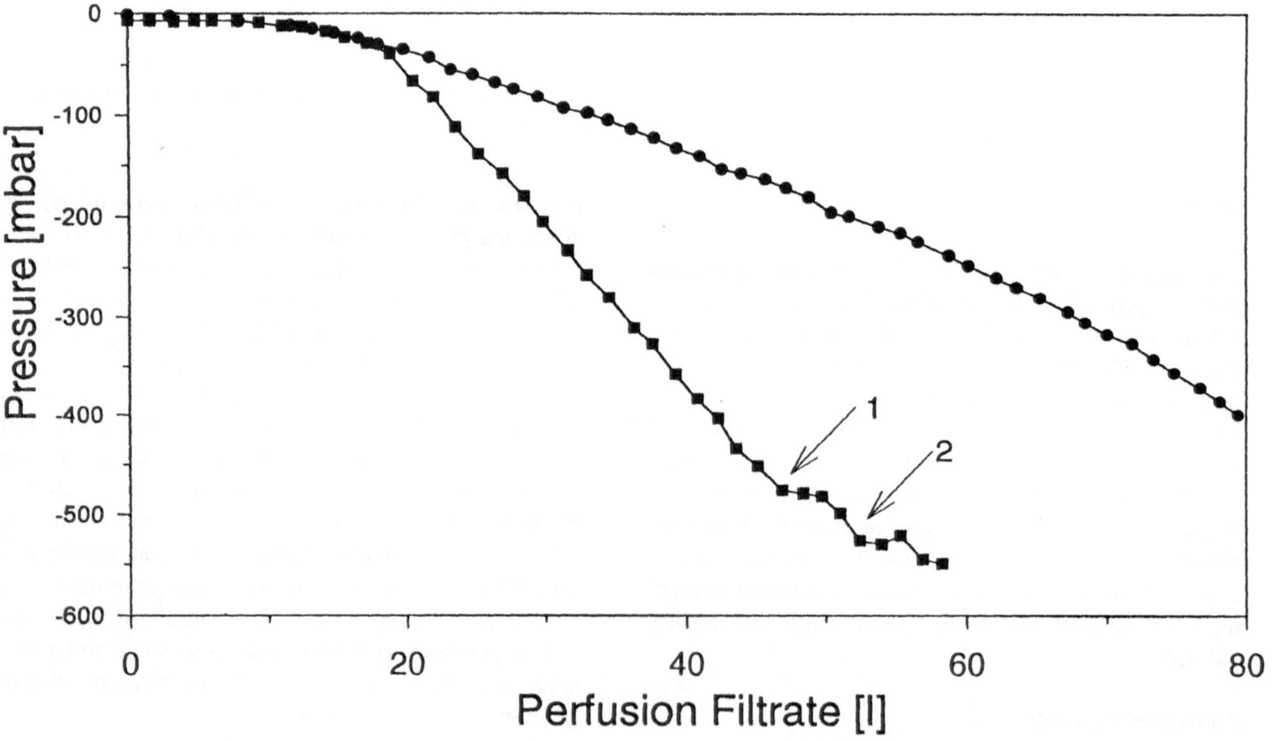

Fig. 3. Transmembrane pressure drop in the perfusion membranes during fermentation shown as function of filtrated volume. –■– PP membrane, –●– PTFE membrane. 1) Increase of backflushing rate from 0.075 l m^{-2} min^{-1} to 0.15 l m^{-2} min^{-1}; 2) further increase to 0.32 l m^{-2} min^{-1}

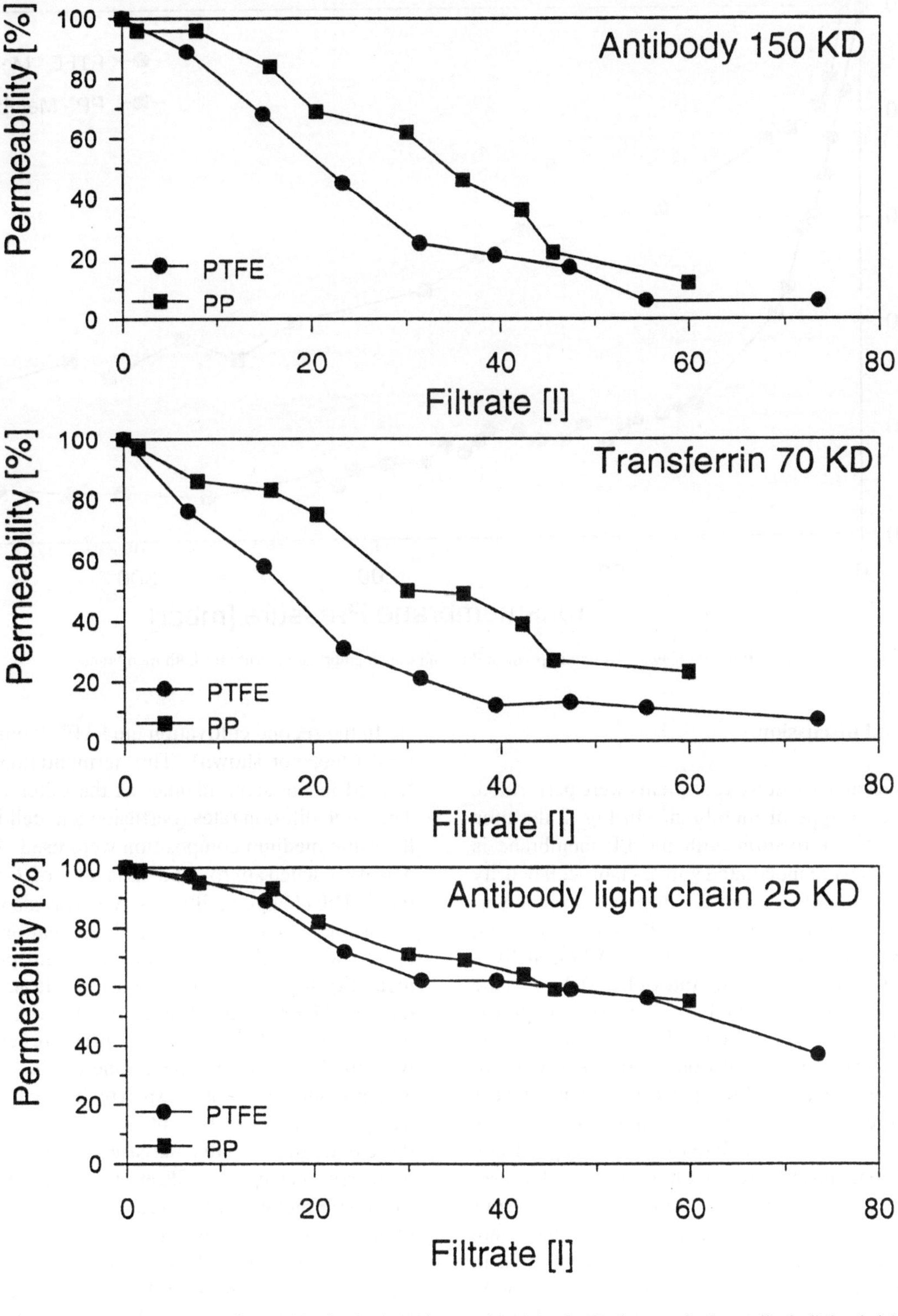

Fig. 4. Protein permeability of the 3 major proteins in the hybridoma cultivations (antibody, transferrin, antibody light chain) shown as function of filtrated volume. The permeability is the percentage of the ratio of the concentration in the filtrate (permeate) and in the fermenter (concentrate) of each protein.

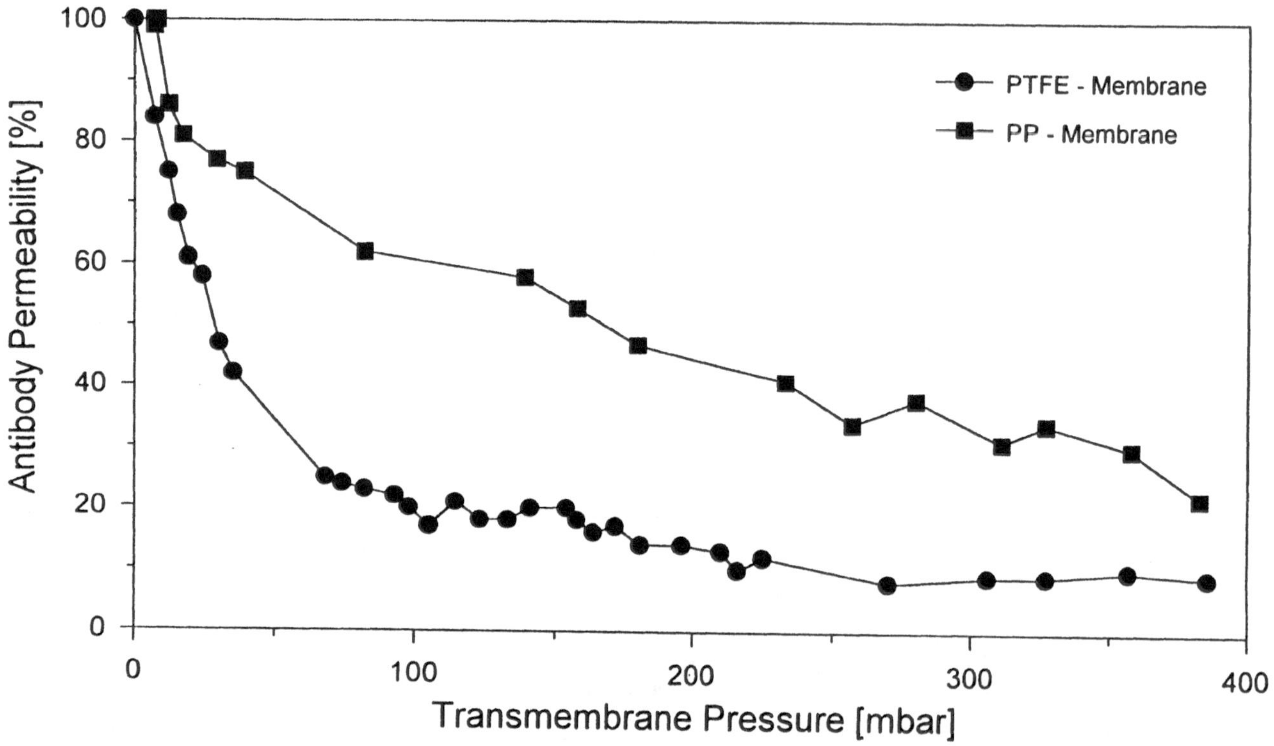

Fig. 5. Relation between antibody permeability and transmembrane pressure for both membranes.

Results and discussion

Two long term perfusion experiments were performed, one with each type of membrane. In Fig. 2 the time course of the cultivation with the PP membrane is shown. Cells were inoculated with a viable cell density of $2{,}4 \cdot 10^5$ cells ml^{-1}. After a 3 days batch mode perfusion was activated with a rate of $D=0.8\ d^{-1}$ and kept constant for the rest of the cultivation. Additionally, a cell bleed was activated with a rate of $D_B=0{,}2\ d^{-1}$ at the same time. After day 8 a viable cell density of approx. $7 \cdot 10^6$ cells ml^{-1} was reached and could be maintained during the rest of the cultivation (35 days). The perfusion phase with cell bleed was performed under glucose limitation (data not shown) to maintain a stable steady state in respect to viable cell density. During perfusion antibody concentration was monitored in the bioreactor and the perfusion filtrate. From Fig. 2 it is obvious that, beginning with day 15, antibody molecules were retained in the reactor, whereas the concentration in the perfusion filtrate decreased. The type of filtration changed from microfiltration to ultrafiltration. For this reason the antibody concentration reached a level of $1.6\ g\ l^{-1}$ at the end of the fermentation.

In the second cultivation the PTFE membrane was used (data not shown). This fermentation was performed in the same manner as the other cultivation. The same dilution rates (perfusion and cell bleed) and the same medium composition were used. Viable cell density reached approx. the same level of $7\text{–}8 \cdot 10^6$ cells ml^{-1}. The change in filtration type was also observed.

Special attention during both cultivations was given to transmembrane pressure drop and protein permeability to investigate changes of filtration characteristic. The transmembrane pressure was monitored with a sterilizable piezoresistive pressure sensor which was fitted close to the membrane outlet (see Fig. 1). The pressure difference Δp which is caused by the pump during harvesting phase was calculated from these data. In Fig. 3 the increase in pressure drop with increasing perfusion (=filtration) volume is depicted. The PTFE membrane shows a slower pressure increase than the PP membrane. Two attempts to decelerate pressure rise in the PP membrane by increasing the backflushing rate during filling step failed. The pressure rise could not be retarded significantly. Contrary to the findings concerning the pressure drop the PP membrane shows a better protein permeability for large molecules (see Fig. 4). For antibody and transferrin

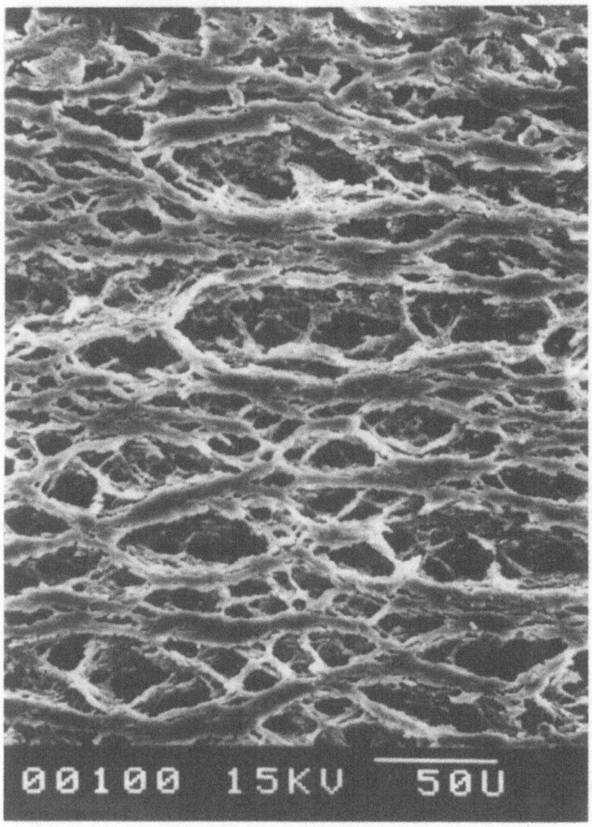

Fig. 6a) PP membrane, unused.

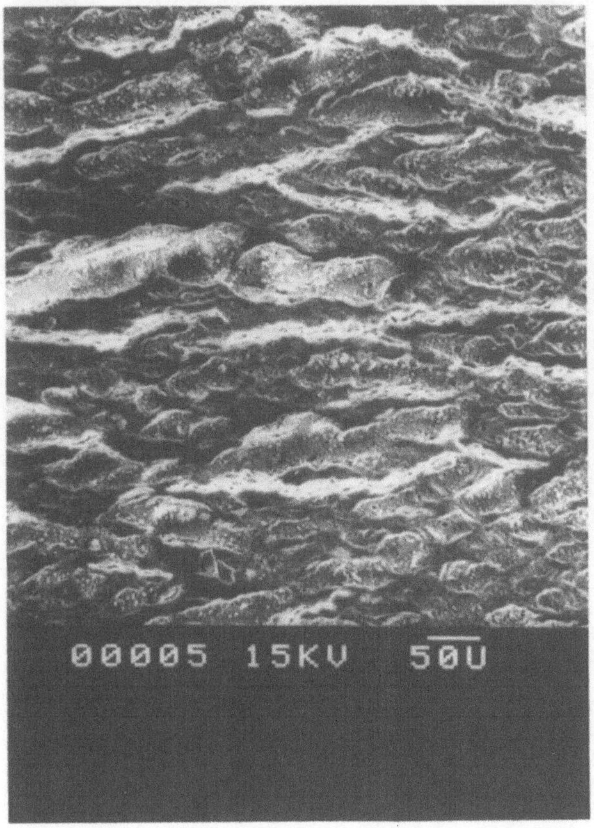

Fig. 6b) PP membrane after long term perfusion culture.

molecules the permeability of the PP membrane did not decrease with increasing filtration volume to such an extent as it occurred when the PTFE membrane was used. For smaller molecules like the antibody light chain the behaviour of both membranes is nearly the same. These findings lead to the conclusion that the fouling process has a different quality for each membrane. In case of the PP membrane fouling causes the formation of a gel layer with greater pores and higher hydrodynamic resistance. To solve thist apparent contradiction it is necessary to look at the filtration theory (Flaschel *et al.*, 1983).

In the pore model ideal cylindric pores are assumed. According to *Hagen-Poiseuille's* law (1) the flux **J** is proportional to the transmembrane pressure drop $\Delta \mathbf{p}$. The hydraulic resistance **W** (2) is dependent on the square of the pore diameter **d**, the membrane porosity ε and inverse proportional to the dynamic viscosity η and the pore length **l**. The total hydraulic resistance is a sum of the resistance of the synthetic membrane itself and the secondary membrane formed during fouling ($W_{total} = W_{membrane} + W_{layer}$).

$$J = \frac{1}{W} \cdot \Delta p \qquad (1)$$

$$\frac{1}{W} \sim \varepsilon \cdot \frac{d^2}{\eta \cdot l} \qquad (2)$$

It can be assumed that the resistance of the synthetic membrane is constant during the whole process. But, the equations have to be applied also to the gel layer formed as the secondary membrane since the transmembrane pressure drop is mainly caused by it. It can be assumed that the porosity of the gel and the viscosity of the culture broth are of same quality in both experiments. Since the apparent pore diameter **d** (see Fig. 4) and transmembrane pressure drop $\Delta \mathbf{p}$ (see Fig. 3) of the gel layer on the PP membrane are higher, also the value of the pore length **l** must be higher under the given condition of a constant flux. A greater pore length also means a thicker gel layer. Therefore, the fouling process caused the formation of a substantial thicker gel layer on the PP membrane than on the PTFE membrane. Fig. 5 presents the relation-

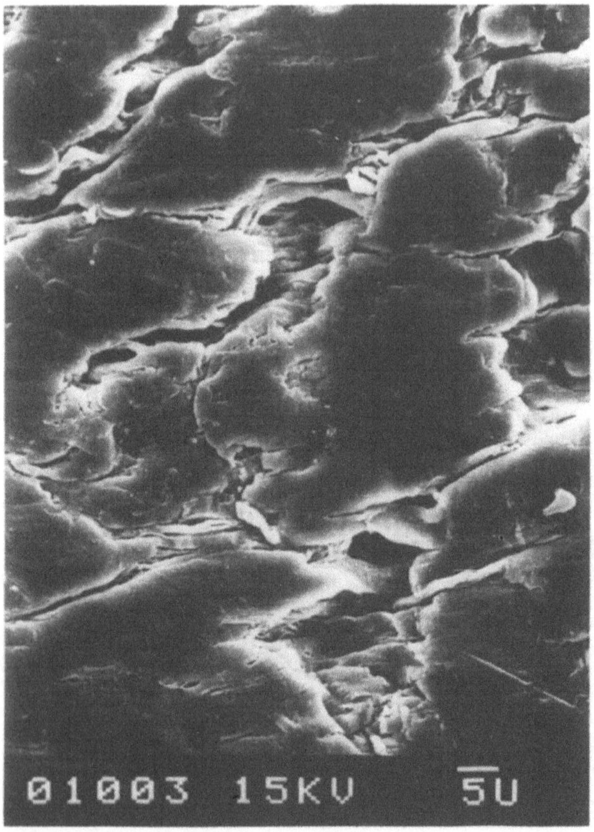

Fig. 6c) PTFE membrane, unused.

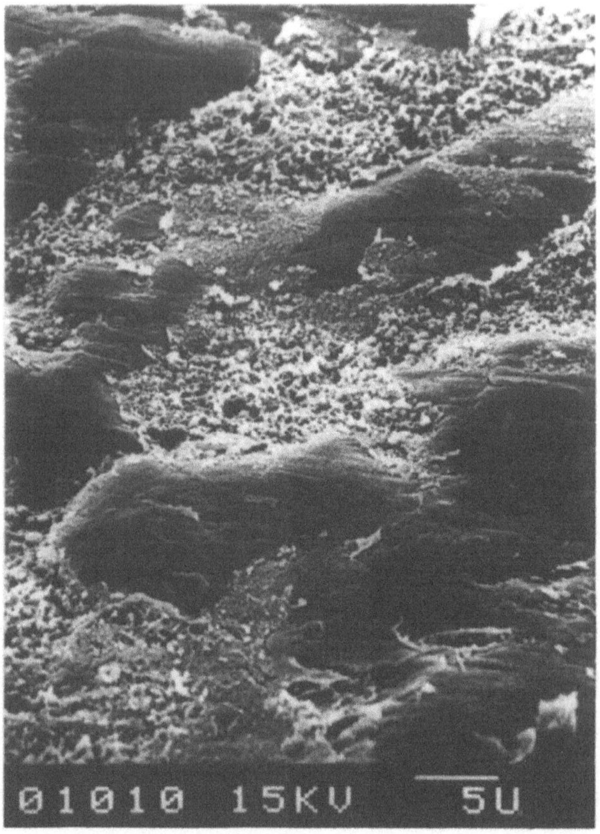

Fig. 6d) PTFE membrane after long term perfusion culture.

Figs. 6(a)–(6d). REM – micrographs of outer surfaces of both membranes used for microfiltration.

ship between transmembrane pressure and antibody permeability. In case of the PTFE membrane the antibody permeability already drops to about 20% when a transmembrane pressure of 100 mbar is reached. A permeability of 10% seems to be a stable situation even when the transmembrane pressure increases. For the PP membrane a decreased permeability of 20% is reached at 400 mbar, at the earliest. A further increase of transmembrane pressure will probably lower antibody permeability. The antibody permeability of the PP membrane is in any case about twice as high as the permeability of the PTFE membrane.

On the REM micrographs (Fig. 6a–d) the different structures of the membranes can be seen. The PTFE membrane has those pores, which are relevant for filtration, on the outer surface, while the PP membrane is very asymmetric with great pores on the outer surface and with the filtration pores on the inner surface. After the fermentation these large pores on the outer surface of the PP membrane are filled and blocked by the protein gel. That confirms the findings above which predict a longer pore length of the gel.

Conclusion

Both investigated microfiltration perfusion membranes can be used for internal cell retention if the correct filtration procedure is used. In both cases the filtration type changes during perfusion from microfiltration to ultrafiltration by forming a protein layer (Blasey, 1990; Blasey *et al.*, 1991). The membrane fouling process influences the filtration characteristic of both membranes in a different way. The permeability of the PP membrane is slightly better than that of the PTFE membrane and therefore this type of membrane should be preferred. On the other hand, fouling cause a slower decrease of transmembrane pressure when the PTFE membrane is used, which favours this membrane for long term cultivations, especially when low molecular weight proteins (<30 KD) are produced.

Acknowledgmenets

We like to thank B. Braun International (Dr. C. Fenge) for providing the PTFE membranes and Dr. P. Heimann, Faculty of Biology, University of Bielefeld, for preparing the REM micrographs. This work was supported in part by the project 'Development of a procedure and a plant for the recirculation of nutrient media for animal cell culture' (BMFT ref No. 0319346A) of the German Ministry of Research.

References

Blasey HD (1990) Untersuchung zur Optimierung eines Perfusionsreaktorsystems für die Kultivierung von Animalzellen. Ph.D. Thesis, University of Hannover, Germany.

Blasey HD and Jäger V (1991) Strategies to increase the efficiency of membrane aerated and perfused animal cell bioreactors by an improved medium perfusion. In: Animal Cell Culture and Production of Biologicals, Sasaki R and Ikura K (eds.), Kluwer, Dordrecht, 61–73.

Bülles J, Barziuk H, Klossom RJ, Schickle HP and Gronau S (1990) *Pharmacia Application Paper*, A 49 6/90.

Büntemeyer H, Bödeker BGD and Lehmann J (1987) Membrane-stirrer-reactor for bubble free aeration and perfusion. In: Modern Approaches to Animal Cell Technology, Spier RE and Griffiths JB (eds), Butterworth, London, 411–419.

Büntemeyer H, Lütkemeyer D and Lehmann J (1991) Optimization of serum-free processes for antibody production. Cytotechnology 5: 57–67.

Fenge C, Klein C, Heuer C, Siegel U and Fraune E (1993) Agitation, aeration and perfusion modules for cell culture bioreactors. Cytotechnology 11: 233–244.

Flaschel E, Wandrey C and Kula M-R (1983) Ultrafiltration for the separation of biocatalysis. In: Advances in Biochemical Engineering/Biotechnology, Fiechter A (ed), Springer, Berlin, Vol. 26: 73–142 .

Meireles M (1991) Effects of protein fouling on the apparent pore size distribution of sieving membranes. Journal of Membrane Science 56: 13–28.

Le MS and Gollan KL (1989) Fouling of microporous membranes in biological applications. Journal of Membrane Science 40: 231-242.

Lehmann J, Vorlop J and Büntemeyer H (1988) Bubble free reactors and their development for continuous culture with cell recycle. In: Animal Cell Biotechnology 3, Spier RE and Griffiths JB (eds), Academic Press, London, 221-237.

Address for offprints: Heino Büntemeyer, University of Bielefeld, Technical Faculty, Institute for Cell Culture Technology, P.O. Box 100131, 33501 Bielefeld, Germany..

Cytotechnology **15:** 253–258, 1994.

Vortex flow filtration of mammalian and insect cells

Steven J. Hawrylik, David J. Wasilko, Joann S. Pillar, John B. Cheng and S. Edward Lee
Pfizer Central Research, Eastern Point Road, Groton, CT 06340 U.S.A.

Key words: Insect cells, mammalian cells, vortex flow filtration

Abstract

The use of vortex flow filtration for harvesting cells or conditioned medium from large scale bioreactors has proven to be an efficient, low shear method of cell concentration and conditioned medium clarification. Several 8–10 L batches of the human histiocytic lymphoma U-937 cell line (ATCC CRL 1593) were concentrated to less than 1 L by vortex flow filtration through a 3.0 μm membrane. An aggressive filtration regimen caused a 17% loss of cell viability and a 32% loss of IL-4 receptor binding capacity when compared to a batch centrifuged control. A reduction of the rotor speed from 1500 to 500 RPM and reduction of system back pressure from 10 to 0 PSIG resulted in cell viability and IL-4 binding capacity comparable to the control. Several 10 L batches of baculovirus infected Sf-9 cells were also concentrated to less than 1 L by vortex flow filtration through a 3.0 μm membrane. SDS-PAGE analysis of filtrate samples showed that aggressive filtration caused cell damage which led to contamination of the process stream by cellular lysate. When rotor speed was reduced to 500 RPM and system back pressure was reduced to 0 PSIG, the amount of contaminating lysate proteins in filtrate samples was comparable to a batch centrifuged control

Introduction

Traditionally, continuous centrifugation has been used for concentration of cells and clarification of conditioned medium from large scale bioreactors. As an alternative to continuous centrifugation, microporous tangential flow filtration can also be used (Maiorella *et al.*, 1991). In microporous tangential flow filtration, the filter membrane is swept clean by the retentate at high flow rates. The high flow rate required to prevent membrane clogging often results in excessive shear force which leads to cell damage and contamination of the process stream by unwanted cellular protein. Vortex flow filtration has also been shown to be an effective, low shear method of cell concentration and broth clarification for microbial fermentations (Kroner *et al.*, 1987; Kroner *et al.*, 1988) and mammalian cell cultures (Rebsamen *et al.*, 1987). Unlike microporous tangential flow filtration, the filter membrane in vortex flow filtration is swept clean by the Taylor vortices created by a spinning rotor in close proximity to the membrane surface (Taylor, 1923; Liegberherr, 1979). In vortex flow filtration, the membrane cleaning action of the Taylor vortices can be accomplished at relatively low retentate circulation rates, thus, decreasing the amount of shear force exerted on the cells and products (Parnham *et al.*, 1993).

Here, we report results on mammalian cell concentration and baculovirus infected Sf-9 conditioned medium clarification using vortex flow filtration. The quality of U-937 human histiocytic lymphoma cell concentrates was evaluated by monitoring cell viability and IL-4 receptor binding capacity. The quality of clarified broth from baculovirus infected Sf-9 insect cells was analyzed by SDS-PAGE. Several filtration control parameters, rotor speed, retentate flow rate and system back pressure, were optimized to reduce shear damage to cells and minimize the accumulation of contaminating cellular proteins in the process stream. The use of vortex flow filtration for harvesting mammalian cells or conditioned medium from large scale bioreactors has proven to be an efficient, low shear method of cell concentration and conditioned medium clarification.

Materials and methods

The U-937 human histiocytic lymphoma cell line (ATCC CRL 1593), which expresses IL-4 receptors on its cell surface, was cultured in a 1:1 mixture of RPMI 1640 (Gibco Life Technologies #11875) and Dulbecco's Modified Eagle Medium (DMEM) (Gibco #11965). The medium was supplemented with 5% fetal bovine serum (Hyclone A-1111L), 10 mL L^{-1} 200 mM L-glutamine solution (Gibco #25030) and 10 mL L^{-1} 10,000 U mL^{-1} – 10,000 ug mL^{-1} Penicillin-Streptomycin solution (Gibco 15140). A 1-L inoculum culture was prepared in a spinner flask agitated at 90 RPM in a 37 °C incubator with 5% CO_2. A 10-L (working volume) bioreactor (Biostat EC 10, B. Braun Biotech) was inoculated and run in semi-continuous batch mode. Dissolved oxygen (D.O.) was controlled above 50% of saturation by head space aeration at 0.5 LPM and by sparging O_2 on demand. pH was controlled at 7.3 by sparging CO_2 and temperature was maintained at 37 °C. U-937 cells grown in the 10-L bioreactor typically had a doubling time of 20–24 hours and were harvested at a density of $0.5–1.2 \times 10^6$ cells ml^{-1} with viability greater than 95%.

The *Spodoptera frugiperda* (Sf-9) insect cell line was cultured at 27 °C in Sf-900 II SFM (Gibco #11902) supplemented with 0.5 mL L^{-1} gentamicin solution (Gibco #15710). The inoculum for 10-L (working volume) bioreactor runs was prepared in four 500 mL shake flasks (Corning), each containing 200 mL medium and incubated on a rotary shaker at 130 RPM (Innova 2100, New Brunswick Scientific). When the inoculum cell density reached $2.5–3.0 \times 10^6$ cells mL^{-1}, the four shake flask cultures were pooled and used to inoculate the bioreactor.

The 10-L insect cell cultures were run in stirred tank bioreactors from Applikon Dependable Instruments. All process parameters were monitored and controlled by an ML-4100 controller (New Brunswick Scientific). D.O. was controlled at 50% of saturation by continuous aeration of the vessel head space and by sparging pure oxygen on demand. pH ranged from 6.0 to 6.4. The agitation shaft was fitted with a pitched blade vortexing impeller at the bottom of the shaft and a six blade Rushton impeller at about the 10-L liquid level. Agitation was controlled at 75 RPM. The bioreactor runs were inoculated at a density of $2–2.5 \times 10^5$ cells ml^{-1} and infected when the cell density reached $2–3 \times 10^6$ cells ml^{-1} with greater than 98% cell viability. The 10-L cultures were infected at an MOI of 1 with a recombinant baculovirus (obtained from Dr. D. Auperin, Dept. of Molecular Genetics, Pfizer Central Research) which expresses human procollagenase. The BaculoGold expression kit (PharMingen) was used in the construction and selection of the recombinant baculovirus. A maximal level of recombinant protein was secreted into the medium at a cell density of $3–5 \times 10^6$ mL^{-1} at 48 hours post infection. Cell viability ranged from 79–92% at the time of harvest.

Mini Pacesetter (Membrex) vortex flow filtration system was fitted with a 3 μm, 400 cm^2 re-usable stainless steel filter membrane. A schematic of the Membrex Mini-Pacesetter flow-path is shown in Fig. 1. Cell suspension was circulated through the filtration unit and concentrated cell retentate was returned to the reservoir. The rotor speed and the filtrate rate were controlled as indicated. Back pressure was controlled by the back pressure regulator valve and the differential pressure was determined by monitoring the difference between the filtrate pressure gauge and the back pressure gauge. Prior to each run, the rotor speed, circulation loop and filtrate flow rates, and back pressure were checked using deionized water. The deionized water was then drained, and for U-937 cell runs, the unit was primed with Dulbecco's phosphate buffered saline (Gibco 14190). During filtration, the Mini Pacesetter rotor speed was maintained at 1500, 1200 or 500 RPM. The whole broth recirculation rate was controlled at 2 L min^{-1}, 1 L min^{-1} or 0.6 L min^{-1} using a peristaltic pump (Watson-Marlow, model 604U). The filtrate flow rate was maintained at 300 ml min^{-1} using a Masterflex pump (Cole Parmer). The system back pressure was set at 10 PSIG, 3 PSIG or 0 PSIG. Under these filtration conditions, a 10-L batch was processed in about 30 minutes. After each filtration, the unit was flushed with deionized water to remove residual cells/medium and disinfected with 0.5% NaOH solution. The unit was then rinsed with deionized water.

Several 8–10 L batches of U-937 cells were concentrated to less than 1 L using the Mini Pacesetter. Whole broth filtration was initiated and cell free filtrate was collected at a rate of 300 ml min^{-1}. Cell concentrate samples were taken at several points during the run to monitor cell viability. A non-filtered control sample was taken before each run and centrifuged for 10 minutes at 2200 g in a J2-21 centrifuge (Beckman). Cell number and viability of the filter concentrated and

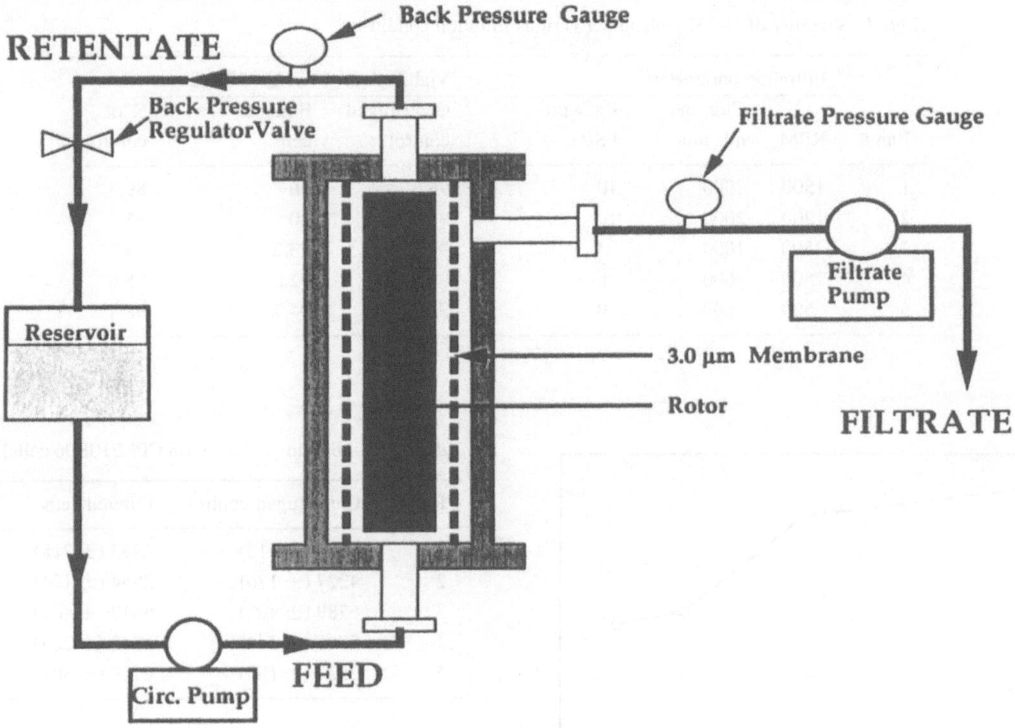

Fig. 1. A schematic of the Membrex Mini-Pacesetter flow-path

the control U937 cells were determined by trypan blue staining. Cell density of the filter concentrated and the control cells was adjusted to 1.0×10^7 mL^{-1} and IL-4 receptor binding capacity was determined by an ^{125}I radiolabeled IL-4 binding assay (Pillar *et al.*, 1992).

Baculovirus infected insect cells were concentrated to less than 1 L using the Mini Pacesetter following a similar protocol. Samples were taken from the filtrate stream at the beginning of the filtration, when the filtrate volume reached 5 L, 7.5 L and 9 L, and at the completion of the filtration. A control sample was prepared before each run by centrifuging unfiltered whole broth for 10 minutes at 1400 g using a GPR centrifuge (Beckman). The filtrate samples and the centrifuged control were analyzed by SDS-PAGE to monitor expression of recombinant human procollagenase and cell lysis during filtration. A 10% Mini Plus precast gel (Integrated Separation Systems #SG22100N) was used, and the electrophoresis was run at 150 volts for 2.5 hours. Protein bands were detected by silver stain (Daiichi #SE140000).

Results and discussion

Concentration of U-937 cells

Five 8–10 L batches of U-937 cells were concentrated to less than 1 liter by vortex flow filtration (Table 1). In runs 1 and 2 an aggressive filtration regimen with high rotor speed and retentate circulation rate was used to minimize clogging and build-up of differential pressure across the 3.0 μm membrane. When it became apparent that membrane clogging was not a problem, the filtration variables (rotor speed, retentate circulation rate and system back pressure) were modified to minimize possible damage to the cells. In runs 4 and 5 where the gentlest filtration conditions were applied, there was still no evidence of clogging or differential pressure build up. In all of the filtration runs, the filtrate flow rate was maintained at 300 ml min^{-1} and there was no measurable differential pressure.

Each of the 5 runs started with a cell concentration ranging from 0.5–1.2×10^6 cells ml^{-1} with a cell viability of 96.2–97.9%. In runs 1 and 2, the viability of the cells after filtration was 86 and 80% respectively (Table 1). In these two runs, the viability of the filtered

Table 1. Viability of U-937 cells under various filtration conditions

	Filtration parameters			Viability (in %) of U-937 cells		
Run #	RPM	Circ. rate mL min^{-1}	Back pres. PSIG	Centrifuged control	Filtered cells	% of control
1	1500	2000	10	96.3	86	89.3
2	1200	2000	10	96.4	80	83
3	500	1000	3	97.9	93.2	95.2
4	500	600	0	96.5	92.2	95.6
5	500	600	0	96.2	94.7	98.4

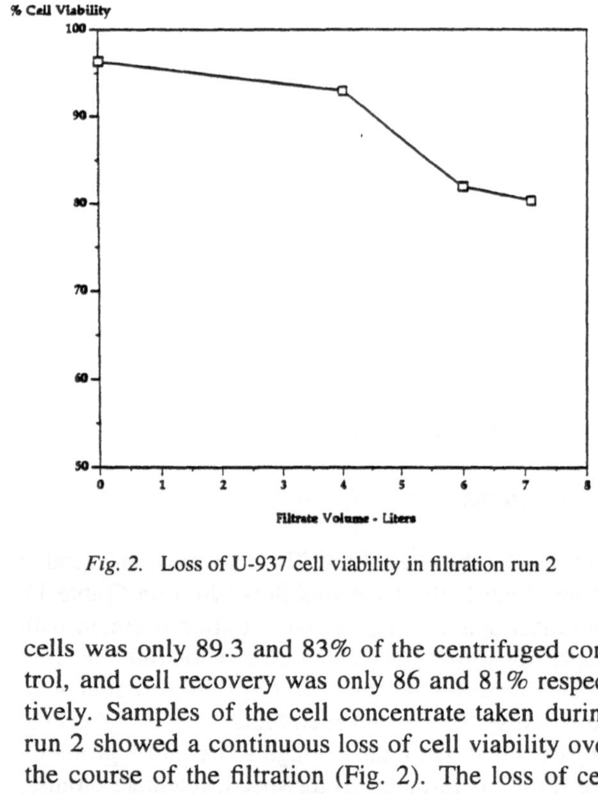

Fig. 2. Loss of U-937 cell viability in filtration run 2

Table 2. IL-4 binding capacity (in CPM/10E06 cells) of U-937 cells

Run #	Centrifuged control	Filtered cells	% of control
1	3882 (± 313)	2793 (± 718)	71.9
2	4227 (± 176)	2854 (± 174)	67.5
3	6789 (± 402)	5617 (± 402)	82.7
4	5085 (± 244)	4842 (± 285)	95.2
5	2524 (± 180)	2729 (± 501)	108.1

cells was only 89.3 and 83% of the centrifuged control, and cell recovery was only 86 and 81% respectively. Samples of the cell concentrate taken during run 2 showed a continuous loss of cell viability over the course of the filtration (Fig. 2). The loss of cell viability in runs 1 and 2 was apparently due to shear within the filtration unit. When the filtration control parameters were modified in runs 3, 4 and 5, the viability of the recovered cells ranged from 92.2–94.7%. The viability of the cells recovered under the gentlest filtration conditions (runs 4 and 5) averaged 97% of the centrifuged control with essentially a full recovery of the cells in the filter retentate.

The U-937 human histiocytic lymphoma cell line expresses IL-4 receptors on its surface. A radiolabeled IL-4 binding assay was used to quantitate the receptor binding capacity and to assess the quality of the filter concentrated cells. The aggressive filtration conditions used in runs 1 and 2 resulted in a considerable loss of IL-4 binding capacity (Table 2). The IL-4 binding capacity of U-937 cells recovered from runs 1 and 2 was only 71.9 and 67.5% of the centrifuged controls respectively. The modified filtration conditions used in run 3 resulted in improved cell viability, however, the binding capacity of recovered cells was only 82.3% of the centrifuged control. The gentlest filtration conditions used in runs 4 and 5 resulted in an average IL-4 binding capacity that was 102% of the centrifuged control. The minimal loss of cell viability and complete recovery of cell surface IL-4 binding capacity suggests that the cells remained intact under the gentle filtration conditions. The standard deviation of the binding assay results was between 4-25%, and averaged 10%.

Clarification of baculovirus infected insect cell broth

The whole broth from four 10-L baculovirus infected Sf-9 insect cell runs was clarified by vortex flow filtration. In run 1 (Table 3), an aggressive filtration regimen was chosen to minimize clogging and build-up of differential pressure across the membrane. In subsequent experiments, several filtration parameters (rotor speed, retentate circulation rate and system back pressure) were modified to minimize cell damage and

Table 3. Filtration conditions and baculovirus infected Sf-9 cell concentration and viability before filtration

Run #	RPM	Circ. rate ml min^{-1}	Back pres. PSIG	Total cells # mL^{-1}	Cell viability %
1	1500	2000	10	3.4 × 10E06	79
2	500	1000	3	4.6 × 10E06	91
3	500	1000	2	3.9 × 10E06	91.6
4	500	600	0	3.0 × 10E06	88.7

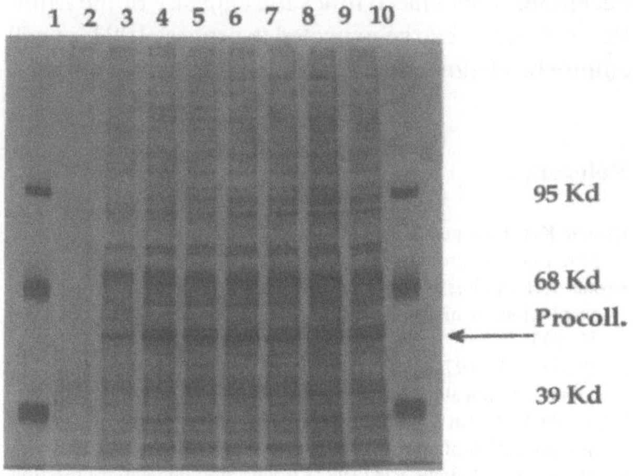

Fig. 3. SDS-PAGE of filtrate stream samples from run 1 (For lane description, see text.)

improve the quality of the filtrate. In run 4, where the gentlest filtration conditions were used, no evidence of clogging or differential pressure build-up was observed. In all of the filtration runs the filtrate flow rate was maintained at 300 ml min^{-1} and there was no measurable pressure change across the membrane.

Baculovirus infected Sf-9 insect cells were harvested at 48 hours post infection. The cell density of the four batches ranged from 3.0 to 4.6 × 10^6 cells ml^{-1} and the viability from 79 to 92% (Table 3). The expression of recombinant human procollagenase and the release of contaminating cellular proteins during filtration was monitored by SDS-PAGE. The results from run 1 are shown in Fig. 3. Molecular weight markers are shown in lanes 1 and 11. Lane 2 shows a sample of broth supernatant from the bioreactor before infection, and the 24 and 48 hour post infection samples are shown in lanes 3 and 4 respectively. Lanes 5 through 8 show samples of the filtrate stream collected during the filtration run when the filtrate volume was 0 L, 5 L, 7.5 L and 9 L respectively and a final sample from the filtrate stream is shown in lane 9. As the filtration progressed, the amount of cellular proteins in the filtrate stream steadily increased. Lane 10 shows a pooled filtrate sample with a large amount of cellular protein when compared to the 48 hour post infection sample (lane 4) or the initial filtrate sample (lane 5). Shear force within the filtration unit run under the aggressive conditions apparently caused excessive cell lysis and protein release.

In runs 2 and 3, the filtration control parameters were modified to reduce cellular protein contamination of the broth filtrate. SDS-PAGE analysis (data not presented) showed a reduction of the amount of contaminating cellular protein in the filtrate stream. However, when samples from the pooled filtrate were compared to the 48 hour post infection samples from the bioreactors, it was evident that shear force within the filtration unit still had an adverse effect on cell integrity and caused the accumulation of contaminating cellular protein.

The filtration control parameters were further modified, in run 4, to improve the quality of the broth filtrate. Unlike the previous filtration runs, in run 4, there was only a minimal increase in contaminating cellular proteins. Samples of the insect cell filtrate stream were taken during run 4 and analyzed by SDS-PACE (Fig. 4). Molecular weight markers are shown in lanes 1 and 10. Lane 2 shows a sample of broth supernatant from the bioreactor before infection, and the 24 and 48 hour post infection samples are shown in lanes 3 and 4 respectively. Lanes 5 through 7 show samples of the filtrate stream collected during the run when the filtrate volume was 0 L, 5 L and 7.5 L respectively and a final sample from the filtrate stream is shown in lane 8. A sample of the pooled filtrate is shown in lane 9. Only a slight increase in contaminating cellular pro-

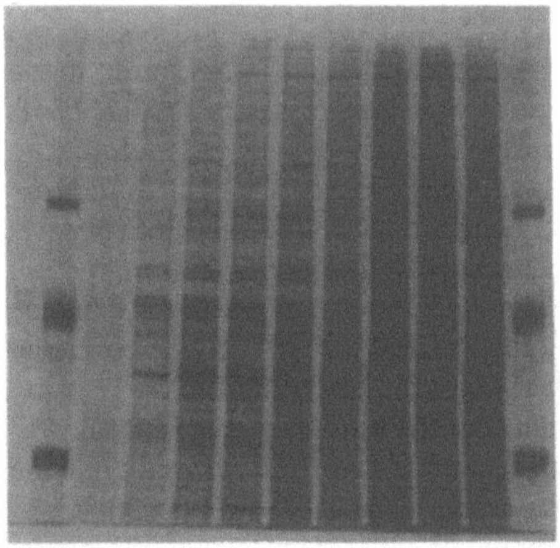

Fig. 4. SDS-PAGE of filtrate stream samples from run 4 (For lane description, see text.)

teins is observed in the final filtrate stream sample (lane 8). Contaminating protein accumulation in the filtrate pool (lane 9) appears to be minimal when compared to the 48 hour post infection sample (lane 4).

Conclusion

Efficient concentration of U-937 human histiocytic lymphoma cells by vortex flow filtration with full recovery of intact cells was demonstrated. Filtration conditions (rotor speed, retentate circulation rate and systemback pressure) were varied to improve the quality of the cell concentrate. Under the gentlest filtration conditions tested, no evidence of filter clogging or of differential pressure build-up across the filter was observed. Cell viability and IL-4 binding capacity of cells recovered under the gentlest filtration conditions were comparable to the centrifuged controls.

Whole broth from baculovirus infected insect cells was clarified using vortex flow filtration. The filtration conditions were varied to minimize the accumulation of contaminating cellular proteins. SDS-PAGE showed a minimal amount of contaminating cellular proteins in filtrate samples from the runs using the gentlest filtration conditions.

The data presented here show that the optimal operating conditions of the MiniPacesetter are with a retentate circulation rate of 600 ml min^{-1}, a rotor speed of 500 RPM, and no system back pressure. The use of vortex flow filtration to concentrate mammalian cells and to clarify baculovirus infected insect cells has been shown to be an effective alternative to other, more traditional cell concentration and broth clarification methods. This method of filtration can be scaled up by lengthening the filter module. For example, the Pacesetter model has 6 times the capacity of the MiniPacesetter, and can be expected to process 100 L of cell culture broth down to a few liters in about 60 minutes.

References

Kroner KH, Nissinen V and Ziegler H (1987) Improved dynamic filtration of microbial suspensions. Bio/Technology 5: 921–925.

Kronet KH and Nissinen V (1988) Dynamic filtration of microbial suspensions using an axially rotating filter. J. Membr. Sci. 36: 85–100.

Liegberherr J (1979) Hydrodynamics of the annular gap flow between permeable cylinder walls. Escher Wyss New 2: 24–30.

Maiorella B, Dorin G, Carion A and Harano D (1991) Crossflow microfiltration of animal cells. Biotechnol. Bioeng. 37: 121–126.

Parnham CS and Davis RH (1993) Rotary and tangential crossflow microfiltration for protein recovery from cell debris. Adv. Filtration Sep. Technol. 7: 349–352.

Pillar TS, Hawrylik SJ, Wasillco DJ Lee SE and Cheng TB (1992) Characterization and regulation of interleukin-4 receptor sites on U-937 cells. Twenty-first annual meeting of New England pharmacologists, Portland Maine U.S.A., February 7.

Rebsamen E, Goldinger W, Scheirer W, Merten O-W and Palfi GE (1987) Use of a dynamic filtration method for separation of animal cells. Develop Biol. Standard 66: 273–277.

Taylor GI (1923) Stability of a viscous liquid contained between two rotating cylinders. Phil. Trans. Royal Soc. 223: 289–343.

Address for offprints: S. E. Lee, Pfizer Central Research, Eastern Point Road, Groton, CT 06340 U.S.A.

Cytotechnology **15:** 259–269, 1994.

Automated imunoanalysis systems for monitoring mammalian cell cultivation processes

Birgitt Schulze, Cornelia Middendorf, Martin Reinecke, Thomas Scheper, Wolfgang Noé[1] & Michael Howaldt[1]
Institut für Biochemie, Universität Müster, Wilhelm-Klemm-Str. 2, 41489 Münster, Germany; Dr. Karl Thomae GmbH, Birkendorfer Str. 66, 88400 Biberach an der Riß, Germany

Key words: Automated immunoanalysis, flow-injection analysis, heterogeneous assay, turbidimetric assay, process analysis

Abstract

Two different automated immunoanalysis systems are presented. Both are based on the principles of flow-injection analysis and were developed to provide reliable, rapid monitoring of relevant proteins in animal cell cultivation processes. One system uses a turbidimetric analysis, and the other employs a heterogeneous chemistry with immobilized immunocomponents. For both systems, the analysis time is in the range of a few minutes, and a complete analysis cycle, including triplicate analyses and various washing steps, is in the range of 20–30 minutes. Samples from cultivation processes can be analyzed directly without dilution. Quantitation of proteins such as rt-PA or monoclonal antibodies can be performed over an analyte concentration range of 1–1000 mg/L. Both systems were compared to conventional ELISA assays on microtiter plates. The turbidimetric analysis system also included a biosensor for simultaneous glucose determination.

Introduction

Monitoring of protein products, especially in mammalian cell cultivation processes, is of immense interest for process documentation and optimization. Currently, time consuming conventional ELISA assays carried out in microtiter plates are the most widely used method. These assays provide data for bioprocess control with a long time delay, and are labor intensive and difficult to automate. Additionally, high dilution factors are necessary to decrease the analyte concentration to fall within an acceptable range (about 1,000- to 10,000-fold dilution). Thus, data can be obtained from a cultivation sample with a long time delay and with high standard deviations. For direct control, faster and more reliable analytical systems are needed to control the bioprocess in a timely manner.

Alternatives to off-line ELISA methods exist. To date, at least five different automated analysis systems for direct bioprocess monitoring have been described. These have been based on measurements of changes in the flow potential (Miyabayashia and Mattiasson, 1990), the turbidity during immunocomplex formation (Freitag *et al.*, 1991a; Middendorf *et al.*, 1993), and spectroscopic effects during competitive heterogeneous assays (Stöcklein *et al.*, 1992, Freitag *et al.*, 1991b). Another system uses calorimetric immunoanalysis, and even flow ELISA systems (Mattiasson and Hakanson, 1992; Brandes *et al.*, 1993; Scheper *et al.*, 1993; Danielsson and Larson, 1990) have been described.

In this paper, two automated assay systems for the fast, accurate and inexpensive immunoanalysis of proteins are presented. One system is based on a turbidimetric assay in which the turbidity of immunocomplexes formed is measured, and the other uses a heterogeneous chemistry with immobilized antibodies. Both are based on the principles of flow injection analysis (FIA) (Ruzicka and Hansen, 1988; Nilsson *et al.*, 1991)), and were optimized with model media and in simulated cultivation processes for analysis of rt-PA (recombinant tissue-type plasminogen activator),

ATIII (antithrombine III), and different antibodies of the IgG type. Additionally, both systems were later applied to direct bioprocess monitoring of the production of rt-PA and a monoclonal antibody of the IgG-type.

Materials and methods

Materials

Turbidimetric assay
A 0.01 M sodium phosphate buffer (pH = 7.4, 0.077 M NaC1, 3% polyethylene glycol 6000) was used as the carrier (turbidimetric immuno assay (TIA) buffer). The reaction coil was thermostatted at 37 °C. Antibodies were diluted in TIA buffer.

An enzyme cartridge with immobilized glucose oxidase was used for glucose analysis (Anasyscon, Germany). A phosphate buffer (0.1 M potassium phosphate buffer, pH = 7.4) was used as carrier buffer. The glucose concentration was monitored via the oxygen consumption during the enzymatic reaction via an amperometric oxygen electrode (Anasyscon, Germany) (Dullau and Schügerl, 1991). A spectrophotometric enzyme assay (Merck, Germany) was used as reference method for glucose analysis.

Heterogeneous assay
The buffer systems used had the following composition: injection-washing-elution buffer: 0.1 M potassium phosphate (pH = 7.4); alkaline elution buffer: 0.1 M potassium phosphate (pH = 12.3), acidic elution buffer: 0.1 M glycine (pH = 2.0). The sample injection volume was 40 μl, unless otherwise mentioned. The dispersion coefficient in the FIA system was 20 as measured by the methods of Ruzicka and Hansen (1988).

Three different support materials were used for the experiments: Sepharose Cl4B (Pharmacia, Germany), VA-Epoxy Biosynth (Riedel de Haen, Germany), and Eurocell-ONB (Knauer, Germany). For each immobilization procedure, 200 μg of the antibody or protein G were dissolved in 800 μl of injection buffer. This solution was mixed with 270 mg of the Sepharose or VA-Epoxy Biosynth matrix or with 1 g of the Eurocell-ONB. After vigorous mixing over a 30-min period, the samples were incubated for 16 hours at 4 °C. After this incubation time, the matrix was washed and loaded into the flowthrough cartridges (Mobitec, Germany).

Analytes
The monoclonal and polyclonal antibodies for the detection of rt-PA (recombinant tissue plasminogen activator) and anti-A-Mab (a monoclonal antibody of the IgG type), as well as the analytes themselves were supplied by Dr. Karl Thomae GmbH; AT III (Antithrombin III) and antibodies for AT III were purchased from Behringwerke AG.

Turbidimetric assay method

In the turbidimetric assay system, 50 μl of the antigen-containing sample is injected into the FIA system at the same time as an aliqout of antibody solution (Fig. 1). The streams converge and mix and immunocomplexes form within 100 s in the reaction coil. Turbidity of this mixture is measured at 340 nm and can be used to determine the antigen concentration with a calibration curve. The medium blank absorbance is monitored via a reference channel to obtain the true absorbance of the immuno complexes. A sample can be analyzed every 6 min., and a complete analysis cycle (three analyses and a washing step) can be run within 20 min.

An analysis cycle consists of the following steps: Injection of the sample (15 s), incubation for immunocomplex formation (20 s for anti-A-Mab, 90 s for rt-PA), detection (120 s), and reequilibration of system (240 s). The computer-controlled system can be used for on-line bioprocess monitoring or can be used in combination with a sample injector. Sampling systems are necessary for on-line bioprocess monitoring to provide the analyzer with a cell-free sample stream (e.g., BioPEM, BBI, Germany; tubular in situ probe, ABC and Eppendorf, Germany; flat membrane cross-flow module, A-Sep^{TM}, Applikon, The Netherlands, or Sartorius, Germany).

Glucose analysis via a biosensor channel (enzyme cartridge and oxygen electrode (Anasyscon, Hannover, Germany) was incorporated into the system to obtain more process information. During a complete immunoanalysis cycle (three analyses of one sample), the sample was also injected into the glucose channel. The glucose was converted into gluconic acid and hydrogen peroxide by the enzyme glucose oxidase with consumption of oxygen. Oxygen consumption is proportional to the sample glucose concentration (Dullau and Schügerl, 1991).

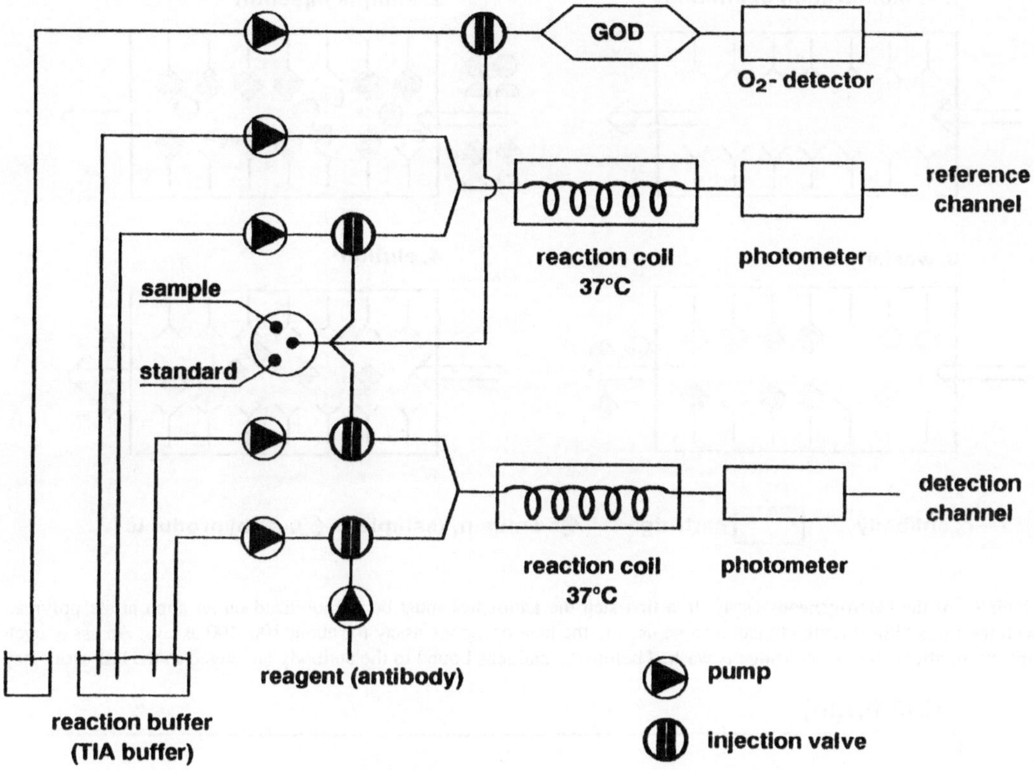

Fig. 1. Schematic setup of the automated turbidimetric immunoassay (including second channel for glucose monitoring).

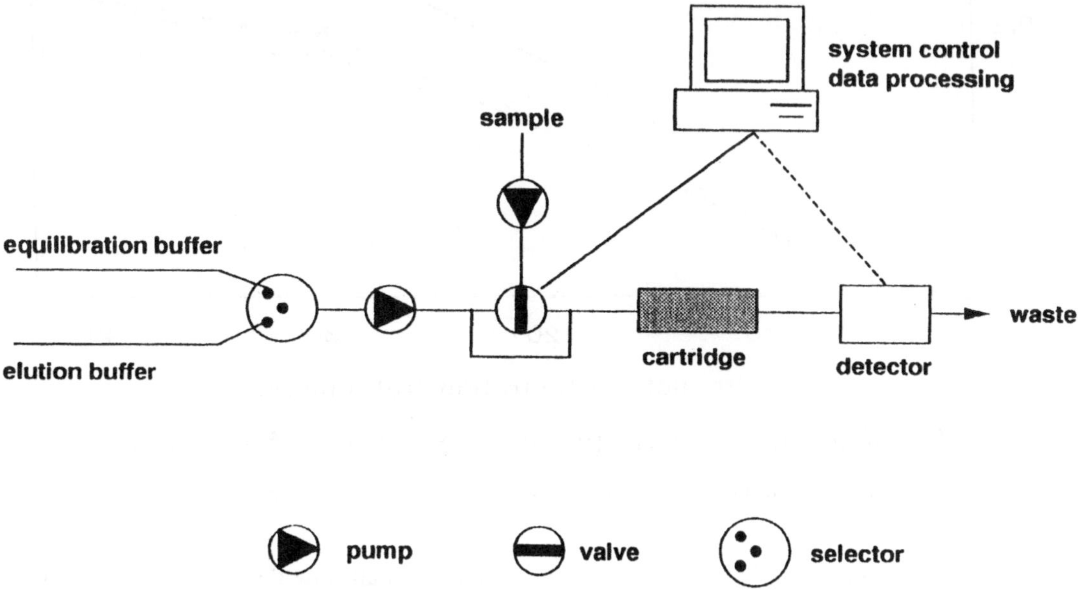

Fig. 2. Schematic setup of the automated heterogeneous assay.

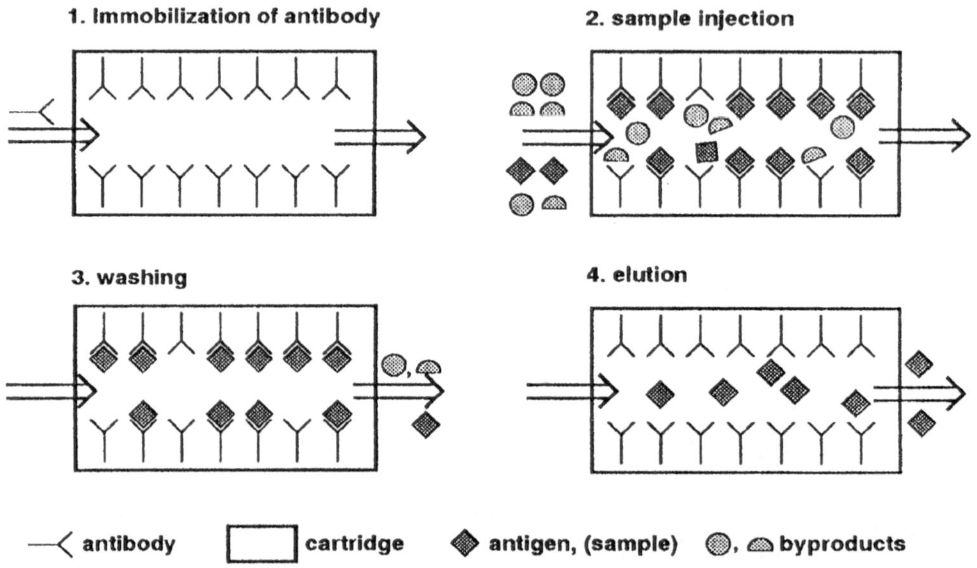

Fig. 3. Principles of the heterogeneous assay. In a first step the antibodies must be immobilized on an appropriate polymer support. The cartridges with the immobilized antibody can then be used in the heterogeneous assay for about 100–400 assays. An assay cycle is started by the sample injection, afterwards the cartridge is washed before the antigens bound to the antibody are eluted in the last assay step.

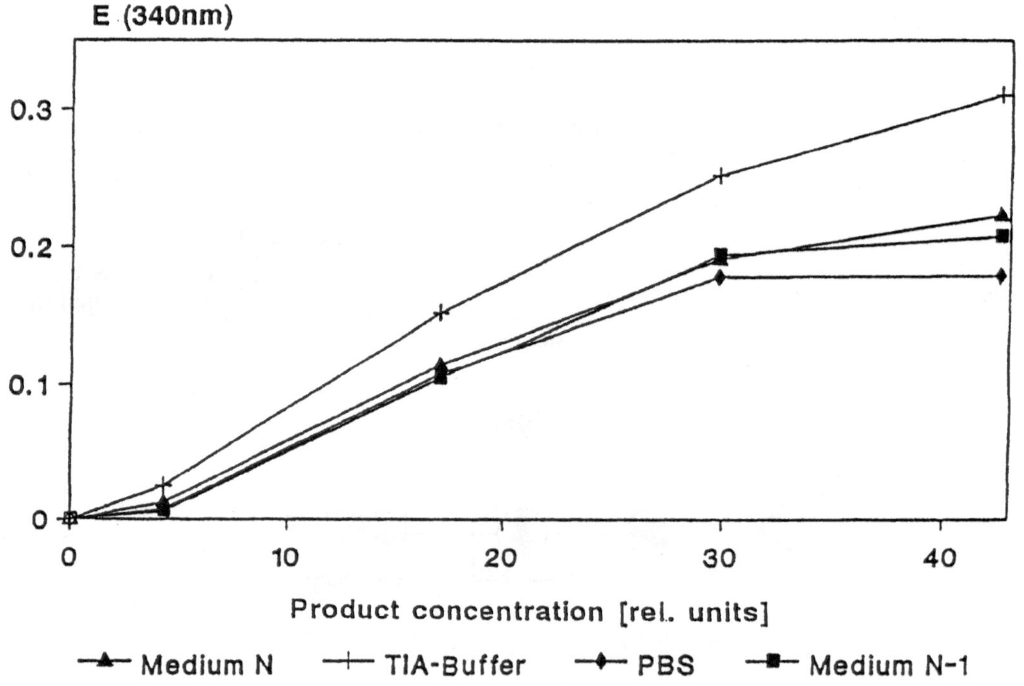

Fig. 4. Calibration of the system in different media, especially in production media.

Heterogeneous assay method

In the heterogeneous assay system, one immunocomponent is immobilized on a polymer that is loaded into a small flow-through cartridge. Antibodies were immobilized for the specific analysis of proteins, and protein G was immobilized for the analysis of antibodies of the IgG type. The cartridge is then used in

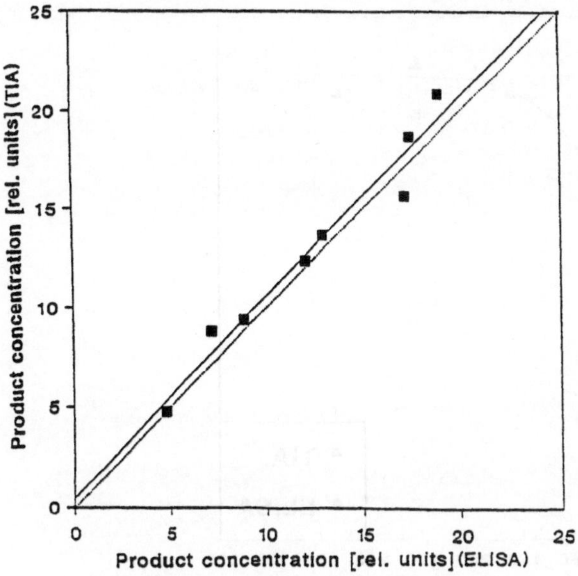

Fig. 5. Comparison of the ELISA data with the data obtained from the turbidimetric assay (solid line: calculated graph from the experimental data: $y = 1.1 + 1.01x$; regression coefficient: $r = 0.98$.

the FIA system shown in Fig. 2. The analyte protein-containing sample is injected into the carrier stream and is pumped through the cartridge. All target proteins bind to the antibodies or protein G and can be eluted after a washing step. The eluent fluorescence is monitored via a spectrofluorometer (Merck-Hitachi, Model F1050) in an HPLC flow through cell (excitation: 273 nm, emission: 340 nm). An analysis cycle is composed of the following steps: sample injection (about 10 s), washing (about 170 s), elution and detection (150 s), and re-equilibration (30 s). No sample dilution is necessary, and an assay can be run every 6 minutes. The sensitivity of the analytical system can be varied by changing the injected sample volume (at low concentrations a larger volume is injected and vice versa). The major advantages of this technique are that the immobilized immuno component can be used several times, interfering substances can be eluted in the washing step (Fig. 3), and the analysis range can be adjusted easily via the volume of the injected sample.

Results and discussion

Turbidimetric assay

Optimization of the turbidimetric assay

Several analysis variables, such as reaction temperature, injection and reaction time, ratio of antigen to antibody concentrations and the influence of medium components must be carefully studied to optimize the turbidimetric assay. The optimal incubation time (period required for immunocomplex formation) was found to be in the range of 180 s. The influence of additives in production media on the analysis of rt-PA is shown in Fig. 4. The calibration graphs from the rt-PA analysis (rt-PA was dissolved in each medium) show that the best results were obtained for the optimized TIA buffer. However, analysis could be performed with high accuracy even in production media (mean deviation of data less than 5%). A comparison between the conventional microtiter plate ELISA and the optimized rt-PA assays is plotted in Fig. 5. The correlation coefficient calculated for the curve is 0.98. Both analysis techniques thus provide equivalent results.

Results from the application of the automated immunoanalysis system optimized for rt-PA to monitoring a 'simulated cultivation' is shown in Fig. 6. Here, production medium from the completed cultivation process (high rt-PA concentration) was mixed continuously with rt-PA-free medium from the beginning of a cultivation process continuously in a gradient mixing chamber. The effluent was analyzed on line with the immunoanalysis system. This procedure can be used for studying the analysis equipment under pseudo-process conditions, with particular attention to lifetime, stability and possible problems with interfering byproducts. The good correlation to the ELISA data is obvious.

Validation of the turbidimetric assay

A validation of the immunoassay was carried out before the analytical equipment was used for bioprocess control during a mammalian cell cultivation. The validation provided information about the quality of the analytical data. Four people took part in this test as analysis operators. Three of them had had no previous experience with this analytical method before the test was started. The immunoassay was validated for the antigen/antibody-system of anti-A-Mab and polyclonal antibodies versus anti-A-Mab. The process of

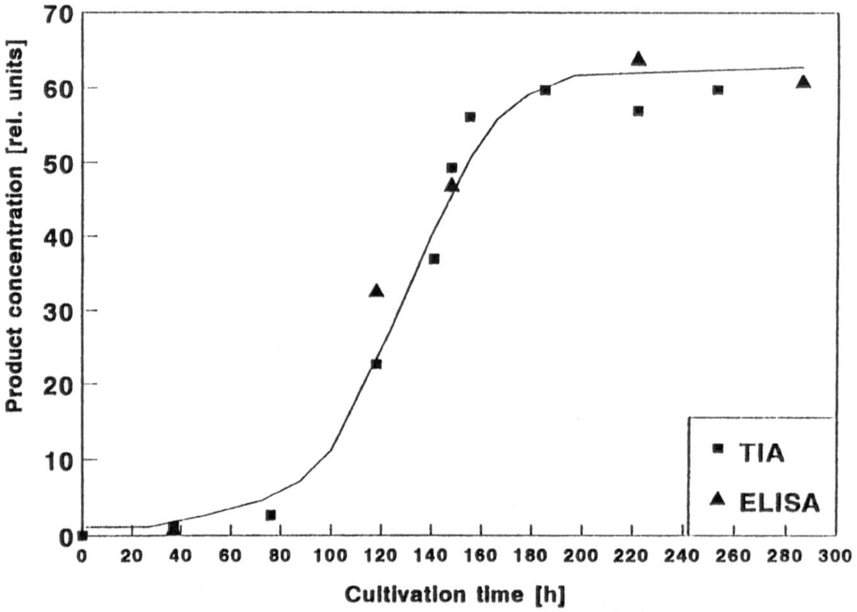

Fig. 6. Comparison of ELISA and on-line turbidimetric immunoanalysis data for a simulated cultivation.

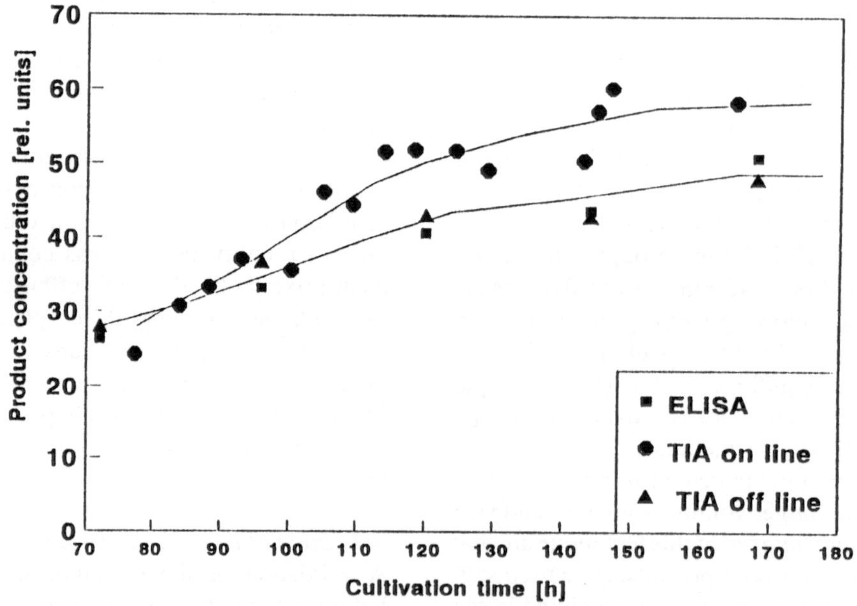

Fig. 7. On-line application of the immunoanalysis system for a direct bioprocess monitoring. The automated system was used to analyze the filtrate side every four hours with a complete analysis cycle (three single assays).

validation consisted of the following statistical analyses:

(i.) Intra-assay variation
(ii.) Inter-assay variation
(iii.) Inter-analyst variation
(vi.) Specifity

To determine the intra-assay variation, the antigen content of two different samples (A and B) was determined by four people. Each sample was analyzed in three independent tests by one analyst. Each single test contained three measurements, the mean value, mean square deviation and percent error were determined.

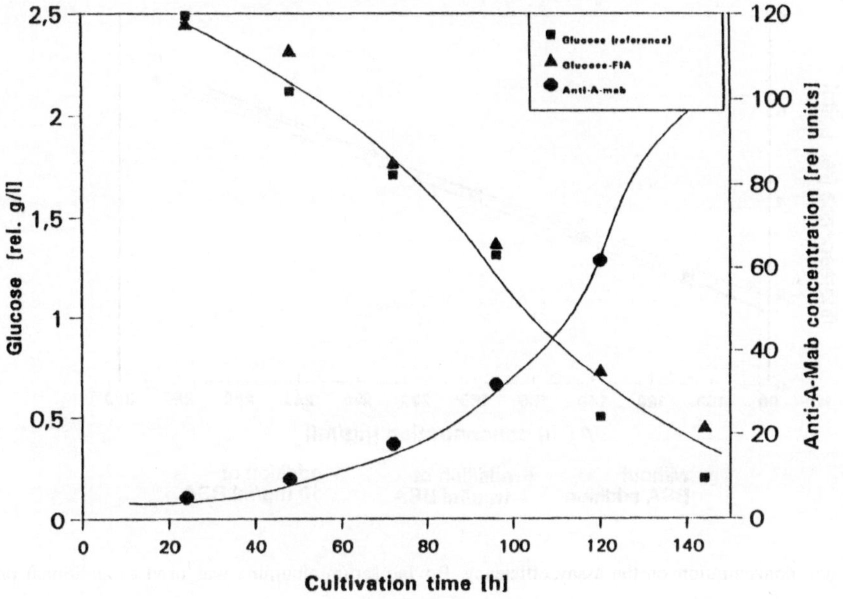

Fig. 8. Simultaneous analysis of glucose and protein concentration with the extended analysis system. A sample could be analyzed every 25 min (three single immunoassays, three glucose assays).

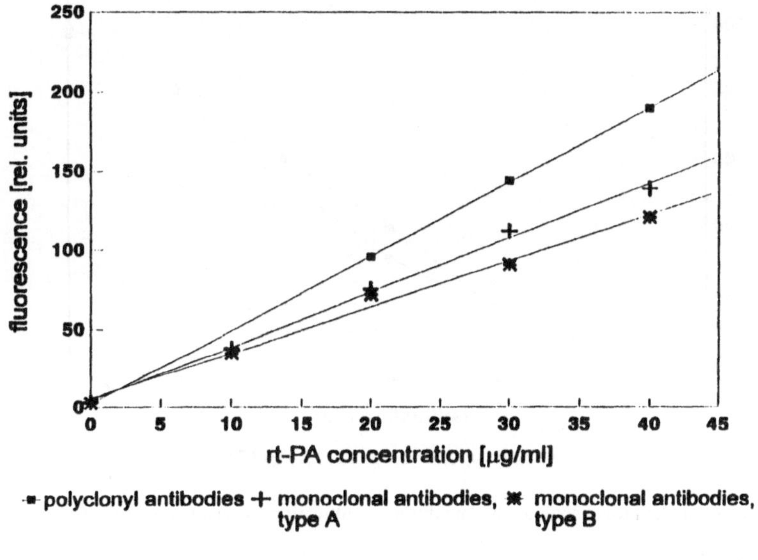

Fig. 9. Calibration curves for the rt-PA analysis with different types of antibodies.

The samples A and B were analyzed on four successive days. The mean value of the results varied by about 6%. The percent error ranges from 1–13%.

To assess inter-assay variation, six samples were analyzed by four analysts on four successive days. Each sample was analyzed in a single test. The antigen was dissolved in buffer or in cultivation medium. The percent error was in the range between 3 and 11%. The mean value of the results changed about 6%.

Inter-analyst variation was measured by having four analysts determine the antigen content of six samples on four successive days. Each day, the results of the analysts for one identical sample were compared. The mean value of the percent error of the results was about 7.7%.

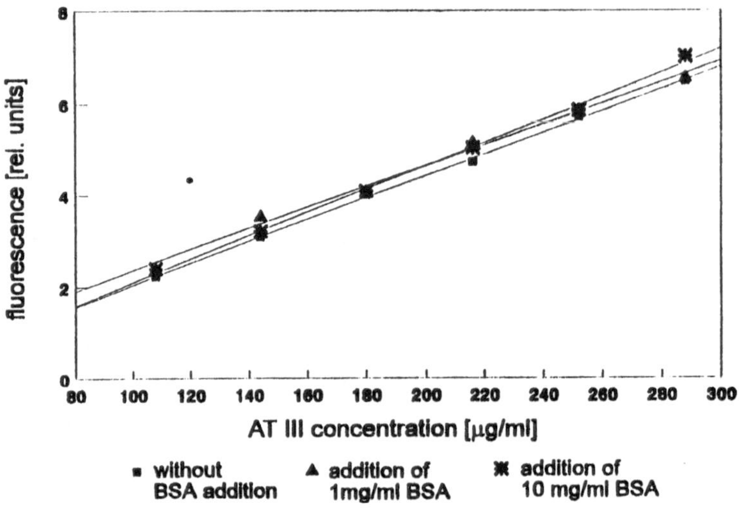

Fig. 10. Influence of protein concentration on the assay efficiency. Bovine serum albumine was used as additional protein marker in the experiments.

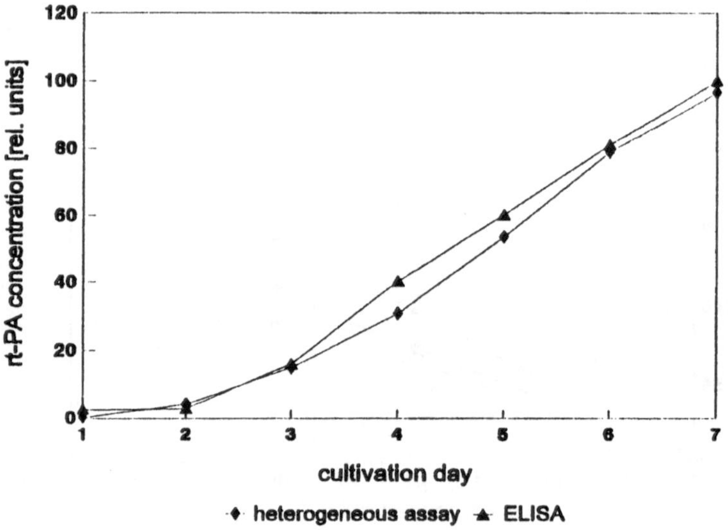

Fig. 11. Comparison of ELISA and heterogeneous assay data during a cultivation experiment of a CHO cell for the production of rt-PA.

Finally, the specificity of the assay system was determined by performing the test in different media (buffer systems, production media with and without additives). The percent error of one antigen concentration measured in different media was between 0.5 to 0.8%.

Application for cultivation monitoring
The application of the complete turbidimetric immunoanalysis system for the analysis of a batch mammalian cell cultivation process is shown in Fig. 7. A cell-free sample was withdrawn via a Sartocon mini-filtration unit (Sartorius, Germany). For this purpose, cell-containing medium was pumped through the filtration units at high volumetric rates. The anti-A-Mab concentration was monitored not only on-line via the automated analysis system but also via measurements on samples withdrawn from the bioreactor which were centrifuged and frozen. These samples were analyzed off-line using a conventional ELISA and with the auto-

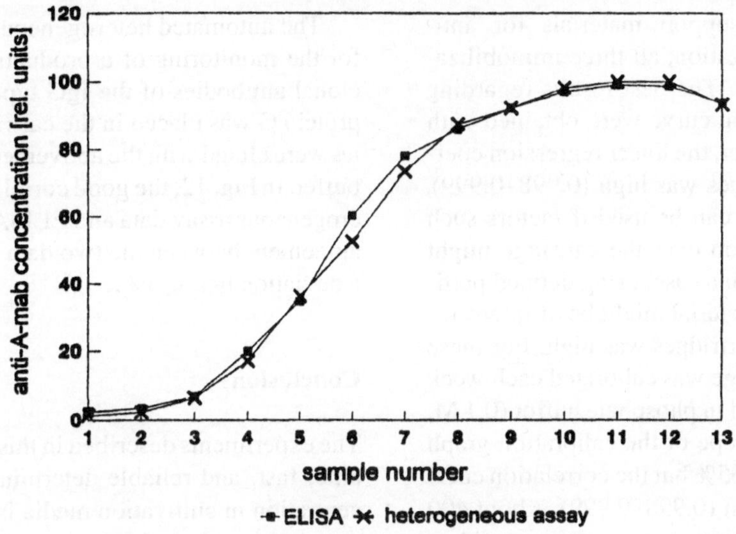

Fig. 12. Application of a heterogeneous assay system with immobilized protein G for the determination of an antibody of an IgG type during cultivation experiments.

mated immunoanalysis system after the end of the cultivation. Examination of the data in Fig. 7 reveals that the deviations in the on-line analysis are due to problems in continuous sampling. The correlation between the off-line data is good, but these data do not correlate with the antibody concentrations in the filtrate. The filtrate side of the sampling device was activated every four hours, ten minutes prior to analysis. Due to several effects such as nonspecific adsorption, the concentration appears to be higher in the filtrate and the mean deviation of the turbidimetric data fluctuated between 1 and 8%. These problems indicate again the most crucial problem of on-line monitoring of bioprocesses: Continuous sampling and provision of a reliable (or more representative) cell-free sample.

Extended system for simultaneous immuno- and glucose analysis
The extended analysis system was used to get additional data about the bioprocess. In general, the system can easily be extended for the analysis of additional variables such as lactate concentration, glutamine concentrations, etc. Such biosensor systems are described in detail elsewhere (Scheper and Reardon, 1991). In Fig. 8, the analysis of samples of a mammalian cell cultivation process for the production of anti-A-Mab are shown. The data obtained from the analysis system are compared to those analyzed with conventional off-line techniques. A sample was analyzed every 25 min (three analyses for anti-A-Mab and one analysis for glucose) (These investigations will be published in detail in a future publication).

Heterogeneous assay

Optimization of the heterogeneous assay
The optimization of the heterogeneous assay will be presented for the analysis of rt-PA and AT III. Polyclonal or monoclonal antibodies were used for the experiments. Fig. 9 shows the use of different antibodies for the analysis of rt-PA. Polyclonal antibodies gave the best results, however, the monoclonals might be used if the availability and the price are advantageous (Middendorf *et al.*, 1993). Since the concentration of the analyte protein such as rt-PA is measured via its fluorescence (excitation: 273 nm, emission: 340 nm), the effect of the different buffer systems on the analyte fluorescence must be optimized. It was verfied in preliminary tests that rt-PA could be eluted efficiently with phosphate buffer in the pH range of 10 to 14 and with only a small effect on the conformation of the antibodies. The use of a buffer system at pH 12.3 results in the highest fluorescence intensities. Thus, this buffer was used for elution. No fluctuations in the spectrofluorometer reading were observed during changes from pure injection-washing-equilibration buffer to pure elution buffer, since all of the buffers had the same molarity. Another advantage of this buffer system is that no contamination was observed during experiments due to the high pH.

To obtain the best support materials for antibody/protein G immobilization, all three immobilization matrices were tested. The best results regarding the slope of the calibration curve were obtained with the Sepharose 4b. However, the linear regression coefficient for all three matrices was high (0.998–0.999), and so the other systems can be used if factors such as increasing pressure drop over the cartridge might become a problem. In such a case, using defined particle sizes of the support material might be of interest.

The stability of the cartridges was high. For these measurements, the cartridge was calibrated each week for one day and was stored in phosphate buffer (0.1 M, pH = 7.4) at 4 °C. The slope of the calibration graph decreased down to about 65% but the correlation coefficients were still excellent (0.998–0.999). About 400 assays can be performed with one immuno cartridge. To obtain reliable data, calibration was performed at least once each day.

To study the influence of other proteins on the accuracy of the analysis, the effect of high concentrations of bovine serum albumine (BSA) in ATIII-containing samples was investigated. For this purpose AT III or BSA were labelled with fluorescein isothiocyanate (FITC) to distinguish between the analyte protein and other proteins. As can be seen from Fig. 10, even the addition of high concentrations of BSA did not influence the accuracy of the analysis to a large degree. Furthermore, the BSA was washed out of the immuno cartridge efficiently during the washing step. The standard deviation of the optimized assays was in the range of 1.5% even when cultivation media were used.

Application for cultivation monitoring

The entire automated immunoanalysis system was used for the monitoring of rt-PA production processes. It can be interfaced directly to a cultivation process when an appropriate sampling system is available, or individual samples can be analysed via a sample injector.

A cultivation process of recombinant CHO cells for the production of rt-PA was monitored via the conventional microtiter plate ELISA assay and the automated heterogeneous assay. The correlation between each assay was excellent during the entire process, as can be seen in Fig. 11. The samples were analyzed directly during the process within a few minutes and thus the data could be used for direct bioprocess control. However, it also would have been possible to analyze all of the samples in a batch at a later time.

The automated heterogeneous assay was also used for the monitoring of a production process of monoclonal antibodies of the IgG type. Here, immobilized protein G was placed in the cartridge, and the antibodies were eluted with the abovementioned acidic glycine buffer. In Fig. 12, the good correlation between the heterogeneous assay data and ELISA data are shown. The agreement between the two data sets is excellent, with a deviation below 5%.

Conclusions

The experiments described in this report show that specific, fast, and reliable determination of protein concentration in cultivation media is possible. These systems produced results that correlate well with the conventional microtiter plate ELISA. Moreover, the automated assays offer tremendous advantage of analysis within a very short time and can thus be used for online monitoring. In addition, other FIA-based sensor systems such as biosensors can be combined with the automated computer controlled systems to give more detailed insight into the bioprocess. Using these systems, effective, direct bioprocess control becomes possible. However, a limiting factor for on-line bioprocess monitoring, especially of high molecular weight components, is the appropriate sampling device. Future work should focus especially on this problem.

Acknowledgement

The authors would like to thank the BMFT-Verbund 'Entwicklung und Anwendung von Biosensoren in der Biotechnologie' for financial support of this work.

References

Brandes W, Maschke HE, Scheper T (1993) Specific flow injection sandwich binding assay for IgG using protein a and a fusion protein, Anal. Chem. 65: 3368–3371.

Danielsson B, Larson PO (1990) Specific monitoring of chromatographic procedures, Trend Anal. Chem. 91: 223–227.

Dullau T, Schügerl K (1991) Process analysis and control by enzyme-FIA-systems, in: Flow injection analysis (FIA) based on enzymes or antibodies (ed.: R.D: Schmid), GBF Monographs, Vol. 14, Weinheim: 27–39.

Freitag R, Scheper T, Schügerl K (1991) Development of a turbidimetric immunoassay for on-line monitoring of proteins in cultivation processes, Enzyme Microb. Technol., 13: 969–975.

Freitag R, Scheper T, Spreinat A, Antranikian G (1991b) On-line monitoring of pullulanase production during continuous culture of Clostridium thermosulfurogenes. Appl. Microbiol. Biotechnol. 35: 471–476.

Mattiasson B, Hakanson H (1992) Immunochemically based assays for process control, in: Modern Biochemical Engineering (ed A. Fiechter) Advances in Biochem. Eng. Biotechnol., Vol. 46, Springer Verlag Berlin, pp: 81–101.

Middendorf C, Schulze B, Freitag R, Scheper T, Howaldt M, Hoffmann H (1993) On-line immunoanalysis for bioprocess control, J. Biotechnol. 3: 395–403. Miyabayashi A, Mattiasson B (1990) A dual streaming potential device used as an affinity sensor for monitoring hybridoma cell cultivation, Anal. Biochem. 184: 165–171,

Nilsson M, Hankanson H, Mattiasson B (1991) Flow-injection ELISA for process monitoring and control, Anal. Chim. Acta 249: 163–168.

Ruzicka J, Hansen EH (1988) Flow injection analysis (2nd edition), Wiley & Sons, New York.

Scheper T, Reardon KF (1991) Sensors in Biotechnology in: Sensors Vol. 2(2) (eds.: W Göpel, J Hesse, JN Memel) VCH Weinheim: 1024–1046 .

Scheper T, Brandes W, Maschkel H, Plötz F, Muller C (1993) Two FIA-based biosensor systems studied for bioprocess monitoring, J. Biotechnol. 31: 345–356.

Stöcklein W, Jäger V, Schmid RD (1992) Monitoring of mouse immunoglobulin G by flow injection analytical aftinity chromatography, Anal. Chim. Acta 245(1): 1–6.

Address for offprints: T. Scheper, Universität Münster, Institut für Biochemie, Wilhelm-Klemm-Straße 2, 41489 Münster, Deutschland.

Cytotechnology **15:** 271–279, 1994.

A direct computer control concept for mammalian cell fermentation processes

Heino Büntemeyer, Rainer Marzahl and Jürgen Lehmann
Institute of Cell Culture Technology, University of Bielefeld, P.O. Box 100131, 33501 Bielefeld, Germany

Key words: Bioreactor, control, computer, fermentation, perfusion

Abstract

In the last 10 years, new assignments and the special demands of mammalian cells to the culture conditions caused the develoepment of complex small scale fermentation setups. The use of continuous fermentation and cell retention devices requires appropriate process control systems. An arrangement for control and data-acquisition of complex laboratory-scale bioreactors is presented. The fundamental idea was the usage of a standard personal computer, which is connected to pumps, valves and sensors via ADA-transformation. The possibility of free programming allowed the development of user-oriented software, especially designed for the far-reaching requirements of a university laboratory in the field of animal cell culture. Control of aeration, pumps, data-acquisition and data-storage are combined within one program, which allows the automation of standard operations like measurement of k_La- or OTR-values. Pump control algorithms for all common fermentation strategies (batch, fed batch, chemostat, perfusion) are included and can be selected any time during cultivation. Oxygen partial pressure and pH are controlled via direct digital control (ddc), providing simple adaption of control parameters and set points to current fermentation conditions.

Introduction

Starting conditions

The comparison between eucaryotic and prokaryotic cells shows, that the former are much more pretentious referred to the fermentation conditions. Adapted to the lower growth rate, the higher susceptibility to shear stress and the higher substrate requirements, special fermentation processes have been developed in the last years. They allow continuous cultivation and include internal or external cell retention systems, like spin filters, continuous centrifuges or hollow fibre membranes. In these sophisticated setups, controlling of pumps for the various medium streams is more complex compared to fed batch or chemostat fermentations.

A process control system (PCS) for common laboratory scale fermenter setups like those, is usually available as a combination of several independent components. Subsystems, controlling aeration, pumps and data acquisition must be combined to provide the possibility of an integrated process control. Information exchange and manipulation of the running process need to take place at different modules of the fermentation setup. This decreases the effectiveness and increases the working effort of mammalian cell cultivation processes.

Demands to a process control system

There are several aspects that are important for the development of a PCS. First, it must control all process relevant variables, like pH, oxygen partial pressure, liquid flows, etc. Second, the possibility of rational user information and interaction with the system must be provided. Finally, the data acquisition should be performed in such way, that the stored data can be processed by word processing or spreadsheet software.

Common PCS have been laid out for a wide spectrum of even prokaryotic fermentations. Thus, only a few functions, specifically used for mammalian cell cultivation, are implemented. Others must be added.

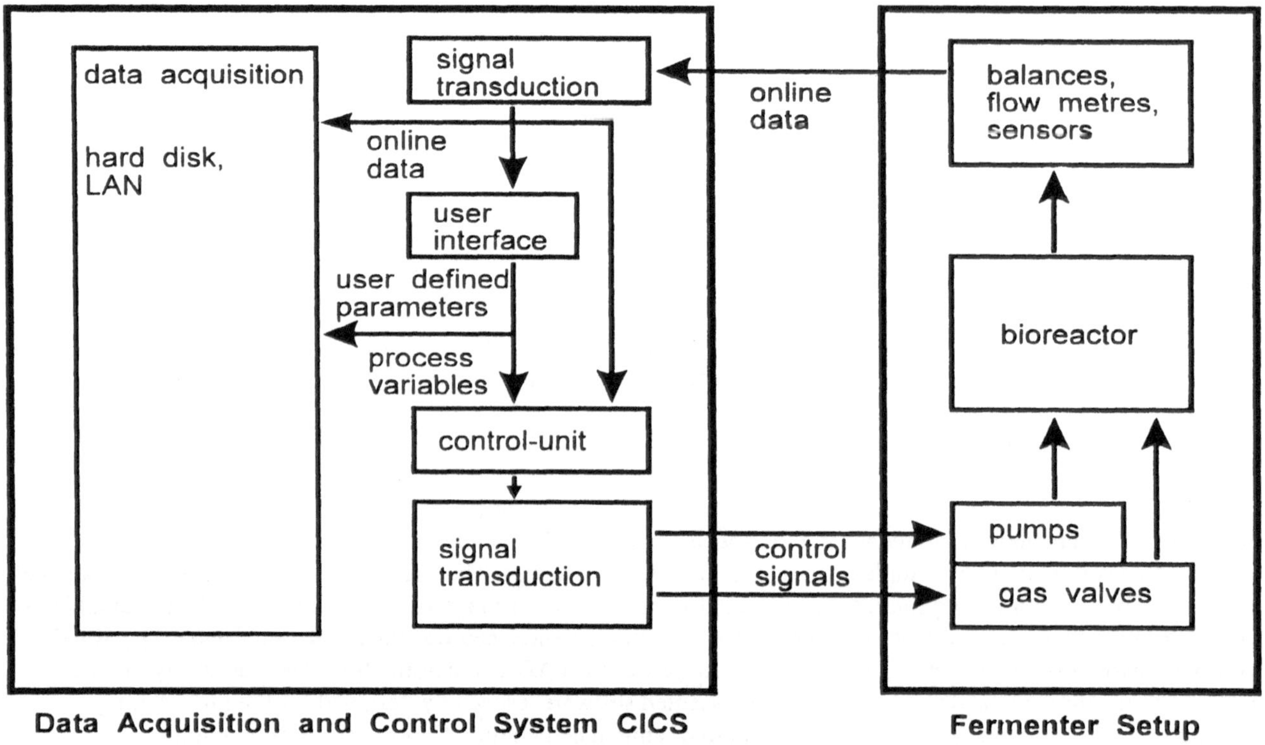

Fig. 1. The configuration of the computer integrated control system CICS is shown schematically.

Control units for pumps, aeration and data filing often are discrete modules. Manual control has to take place at different locations in the experimental setup and the medium flow rates are defined via revolutions per minute of the peristaltic pump rotor. Consequently, calculation and adjustment of feed, harvesting and dilution rates are very difficult and complicated. Since the amount of data accumulated during a continuous fermentation of 60 days and more may be voluminous, a suitable compression for on-line data storage is necessary to reduce the archive file length.

There is a tremendous need for a PCS, which facilitates the culture process handling by unifying data acquisition and control components.

Material and methods

The CICS – data acquisition and control system

The final configuration of the computer integrated control system (CICS), which is schematically shown in Figure 1, was developed as a compromise between the technical possibilies and the demands that will be made of this system. The primary features of the control set-up include the following:

1. Central user interface (terminal/monitor); information about all process relevant values; allowing manual operations necessary during a running process.
2. Reception of on-line data like pH, pO_2, filling level of fermenter and medium vessels.
3. pH and oxygen partial pressure control by regulating the mass flows of air, oxygen, nitrogen and carbon dioxide.
4. Reception of actual gas flow, possible by using thermal mass flow controllers (0–10 Nl h^{-1}, HI-TEC, Ruurlo, NL).
5. Complete control of up to four pumps (Watson-Marlow 501 U, Watson-Marlow, Fallmouth, UK) dependant on the current fermentation mode.
6. Providing data collection and filing of on-line and used defined data.

Hardware

All on-line data (pH, pO_2, balances and aeration) are connected to the computer integrated ADA (ADA 16,

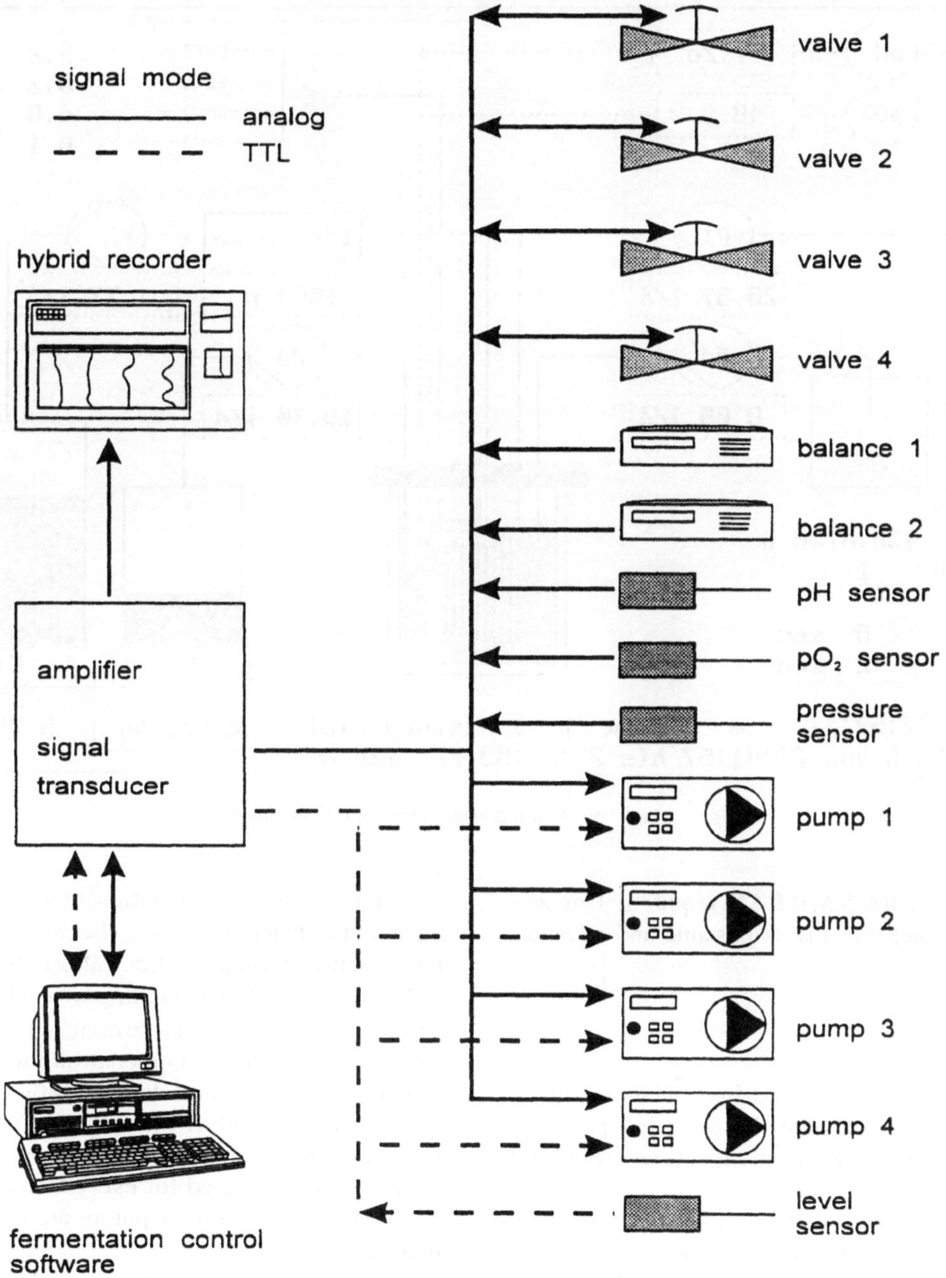

Fig. 2. Summary of possible hardware connections. Computer output signals are transduced, if required, and directly control the fermenter periphery. Input data are collected, transduced if necessary, processed and stored.

1989a+b) and TTL (Fischer, 1990) – interface cards (see Fig. 2). A transducer changes the voltage of this input signals into 0–2V or 0–10V, respectively, needed for the interface cards. This renders a central processing of all on-line data from probes, and measuring devices like balances or mass flow controllers. If the voltage signals are compatible, the program can be adapted to various other sensors. The transduced output controls aeration valves, feeding and harvesting pumps.

Minimal requirement for running the process control system is an AT386 computer, 25 MHz, equipped with 2 MB RAM and 20 MB hard disk under a multitasking operation system like MultiUserDos 5.1 (Nov-

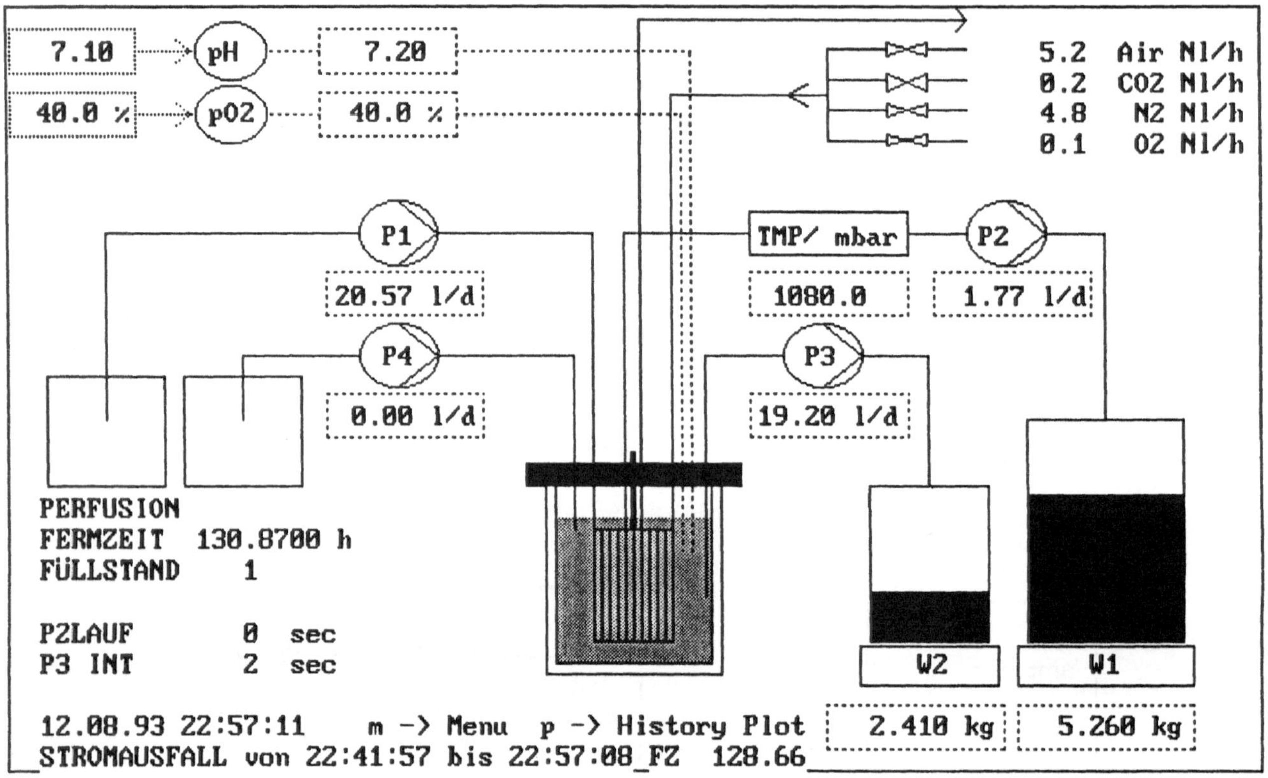

Fig. 3. Hardcopy of on-line main screen of CICS software.

ell) or OS/2 2.x (IBM, 5 MB RAM required). Connection to a computer network is possible and advantegeous.

Software

The central processing of all on-line and used defined data is done by a TurboBASIC compiled program. Controlling (ddc), data acquisition, data recording and user interface are totally integrated.

The status of the fermentation process is shown on a main screen (Fig. 3), which provides information about all relevant process and user defined variables like pH, pO_2, set points and flow rates of liquids and gases. The signal of the level sensor is shown directly: 1 – maximum level, 0 – level beyond maximum.

As far as data acquisition, pumps, controllers and aeration are concerned, all manual interventions during fermentation happen in menus at the terminal. Menus are as secure as possible against nonsense inputs. Senseless data like inputs that provoke a division by zero, pump rates that are higher than the maximum output or negative time intervals will be rejected.

A menu called 'fermentation mode' permits switching between batch, fed batch, chemostat and perfusion fermentation formats. After calibration of the peristaltic pumps at the beginning of each fermentation period, all relevant inputs are done in the units litre per day. A menu called 'pump control/pump parameters' provides a direct control device for the flow rates and intervals. The input of primary parameters like perfusion, harvest and dilution rate is possible in special control menus, offered for every fermentation mode. The control signals for the pumps are figured out automatically.

If required, a direct control of the gas flow controller is also possible. The corresponding menu allows set points between zero and the maximum mass flow for each valve.

During the fermentation pH and pO_2 are controlled by software simulated PID controllers (ddc). Depending on the oxygen requirements of the culture the controller combines air and nitrogen or air and oxygen producing a constant 10 Nl h^{-1} gas flow. Control of pH can take place supplementary with up to 10 Nl h^{-1} carbon dioxide. Control parameters as set point,

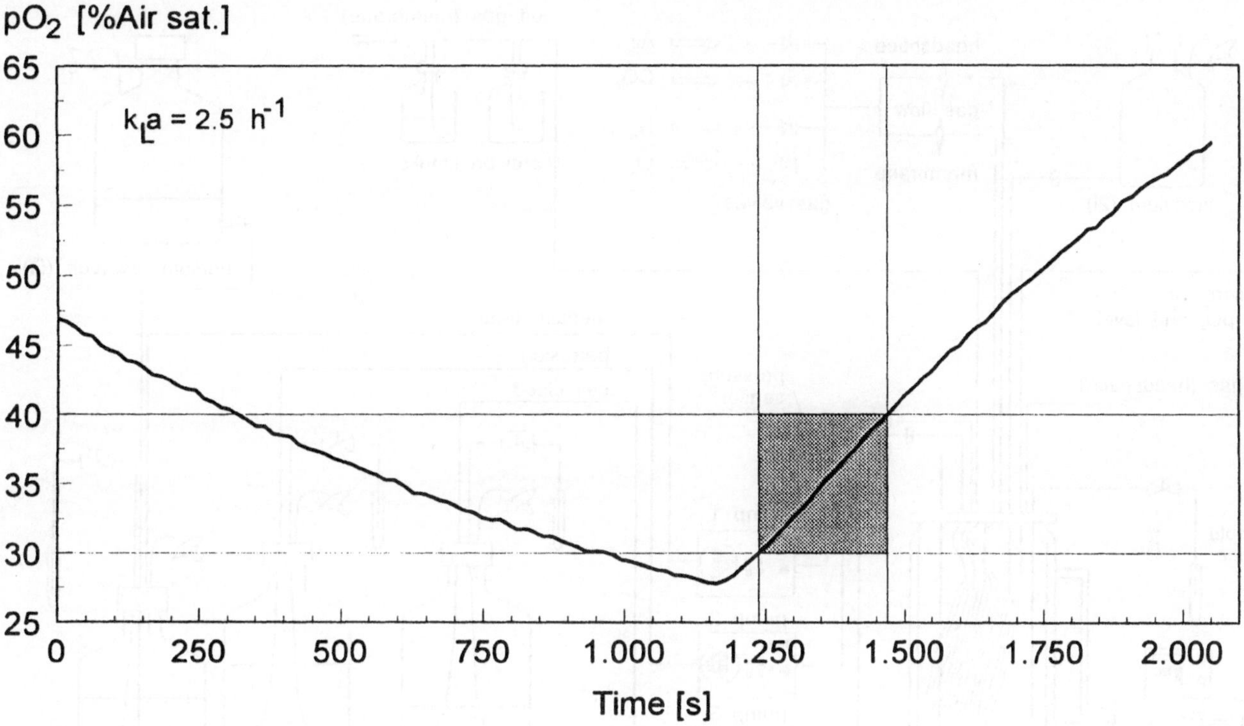

Fig. 4. Time course for pO_2, generated by the PCS for the measurement of k_La-value. The k_La is calculated automatically from the slope of the $pO_2(t)$ curve in the range between 30 and 40% air saturation (marked box).

derivative and integration constant can easily be adapted to the current process conditions.

Data acquisition is realized in several files. The intervals between the single data points are use defined and can be changed every time during the fermentation. Each data set consists of every on-line measured value, date, time and fermentation duration.

The short term filing is done both in a RAM-matrix and in a short term archive file on hard disk, including the last 540 data sets. It can be displayed on screen as a 'history plot' and if the default value for the short term interval (60 seconds) is used the file will represent the last nine hours of fermentation.

A second, long term archive file consists of arithmetic mean values of the short term data received in the long term interval of the course of the whole fermentation. Both files are stored in a compressed format, which reduces the file length. A complete long term file, including twelve hundred data sets, needs only about 80 KB memory space on hard disk. This is equivalent to 30 days fermentation duration if the default value for the long term interval (60 minutes) is used. A third initialization file stores all current user defined data like control parameters, pump rates, tara weights of balances, calibration constants, etc. The usage of a computer network or a multitasking operating system provides a permanent grip on the stored data. After power or software failure, CICS is able to restart the fermentation in the most recent state by loading all required data from archive files. During the failure, no action will take place, because both PCS and fermenter setup are parts of the same circuit. The duration of software failures is at most one minute until the system works properly again. A main screen message shows time and duration of the power failure. After repeated error occurrence the system runs automatically in 'secure' batch mode to prevent pump malfunction.

Further on, the program provides standard operations like pump and balance calibration or OTR and k_La value determination. The latter is done by the dynamic method (Vorlop 1990). For this the control program produces a pO_2-profile like it is shown in Fig. 4.

All in all, results become more reproducible and the procedures cause no additional working effort. The accumulated data are stored in special files for later manipulation in word processing or spreadsheet software.

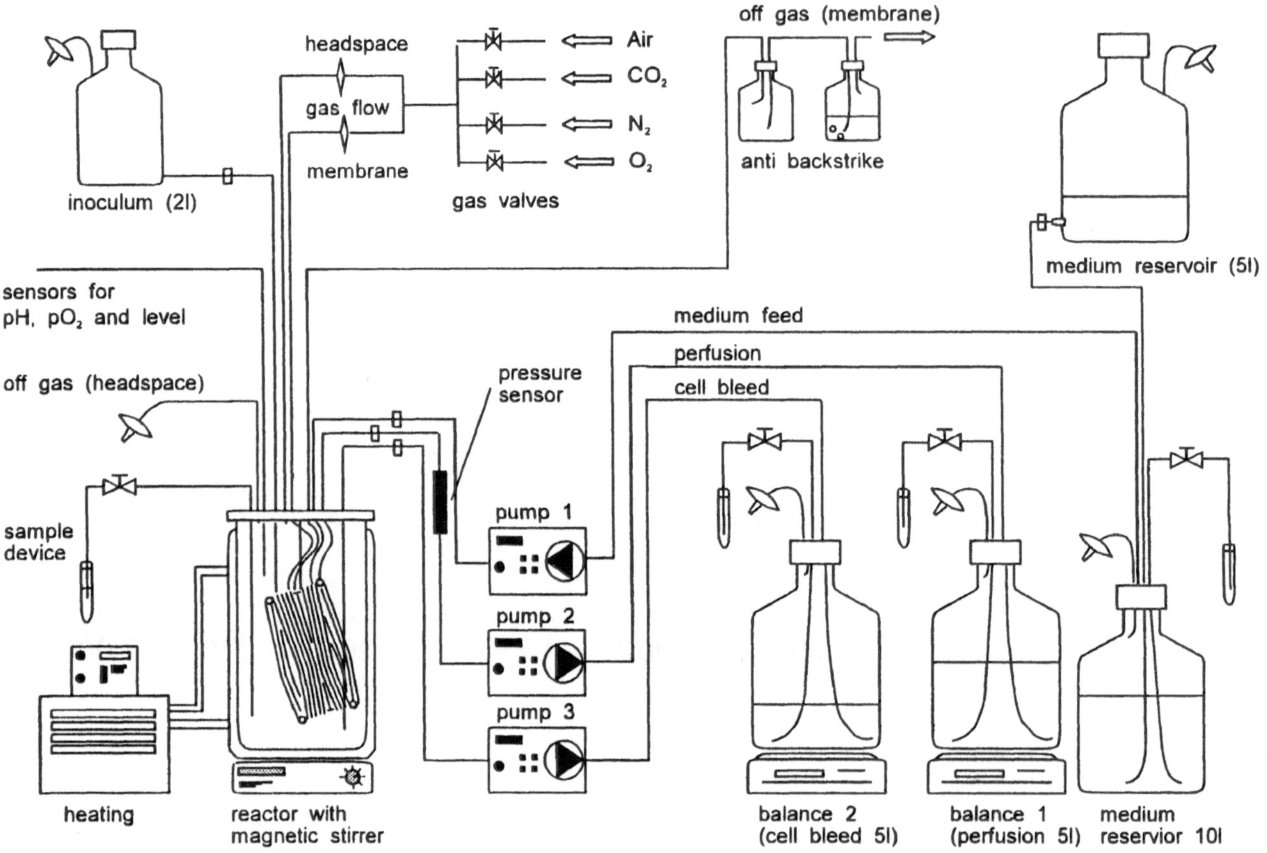

Fig. 5. Scheme of bioreactor and periphery setup controlled by CICS for the fermentation described in this study.

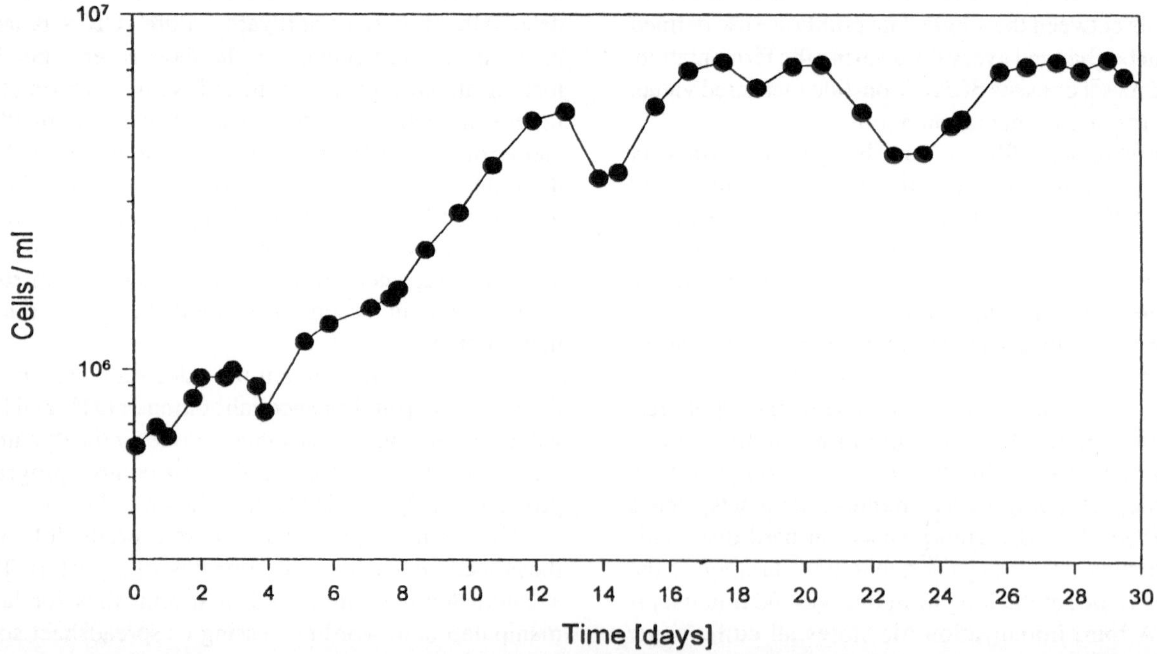

Fig. 6. Time course of total cell density of CHO fermentation.

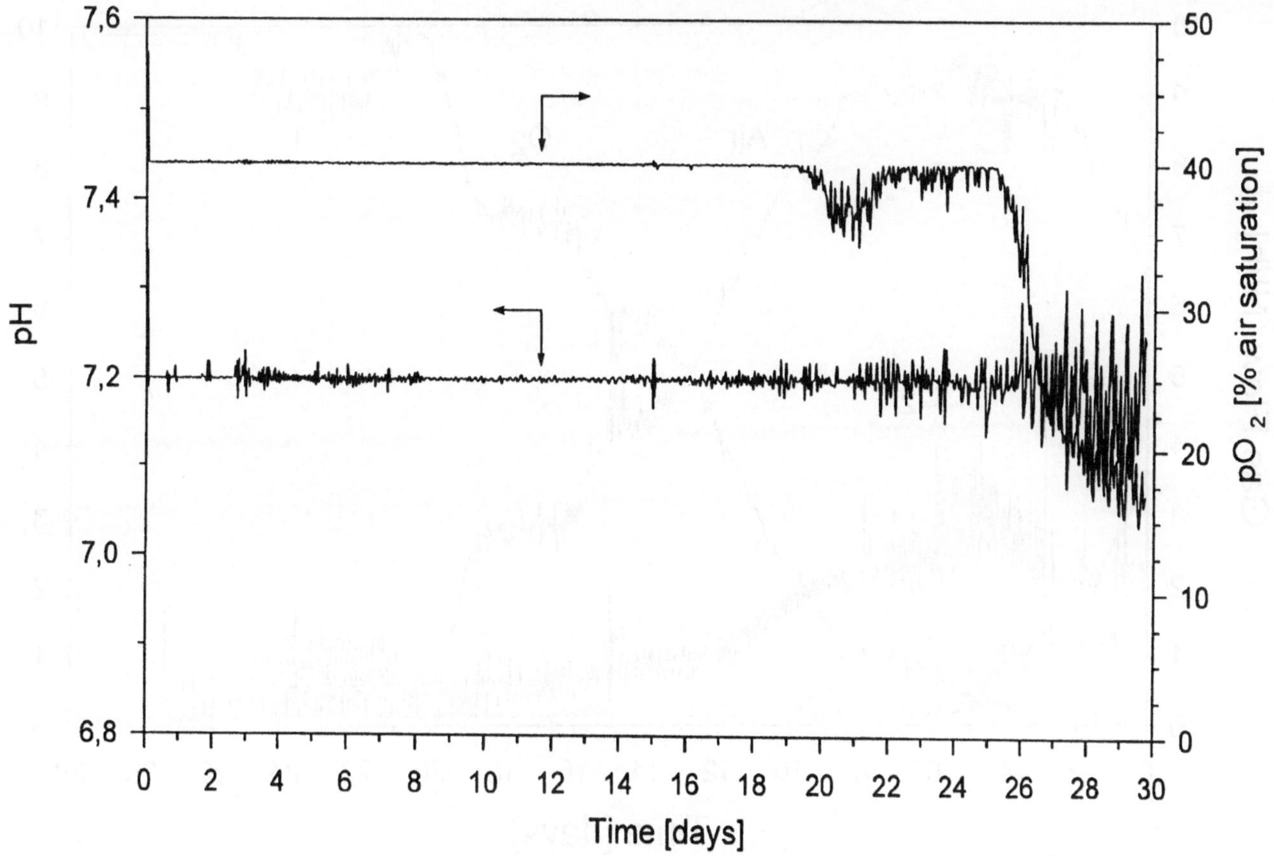

Fig. 7. History plot of pH and pO_2 on-line data during CHO fermentation.

Reactor setup and cell culture

A fermentation of recombinant CHO cells (Zettlmeissl *et al.* 1987) (Behringwerke AG, Marburg, FRG) was carried out in a 2 L BIOSTAT BF2 bioreactor (B. Braun Biotech International, Melsungen, FRG) with bubble-free aeration (Lehmann *et al.*, 1988) and internal cell retention via microporous membranes (Büntemeyer *et al.*, 1987) (Fig. 5) (polypropylene for aeration, polysulfone for perfusion, Microdyn, Wuppertal, FRG). A serum-free, low protein medium DMEM/F12 1:1 (Gibco, Paisley, UK) as described in detail previously (Riese *et al.*, 1993) was used. Counts of cell nuclei were figured out microscopically by crystal violet staining. A trypan blue stain allowed the determination of vital cells.

Results

The control software was installed on an AT386 computer, 25 MHz with 5 MB RAM under the multitasking operating system MultiUserDos 5.1 (Novell) integrated in a local area network (Netware 386 V 3.11). All software features were adapted to the flow scheme shown in Fig. 5. Four gas mass flow controllers (Air, O_2, N_2, CO_2) were used to provide an appropriate gas mix at each state of the culture, depending on data from pH and pO_2 sensors. Two pumps were controlled to perform a continuous perfusion mode with internal cell retention (microfiltration membrane) in a special manner to avoid early clogging of the membrane. Backflushing with fresh medium is integrated in this procedure (described in detail in Büntemeyer *et al.*, 1987). Additionally, a third pump is controlled to carry out cell removal for maintaining steady state cell concentrations. Since CHO-cells form aggregates while growing in suspension culture, the cell bleed must be carried out discontinuously to prevent cell

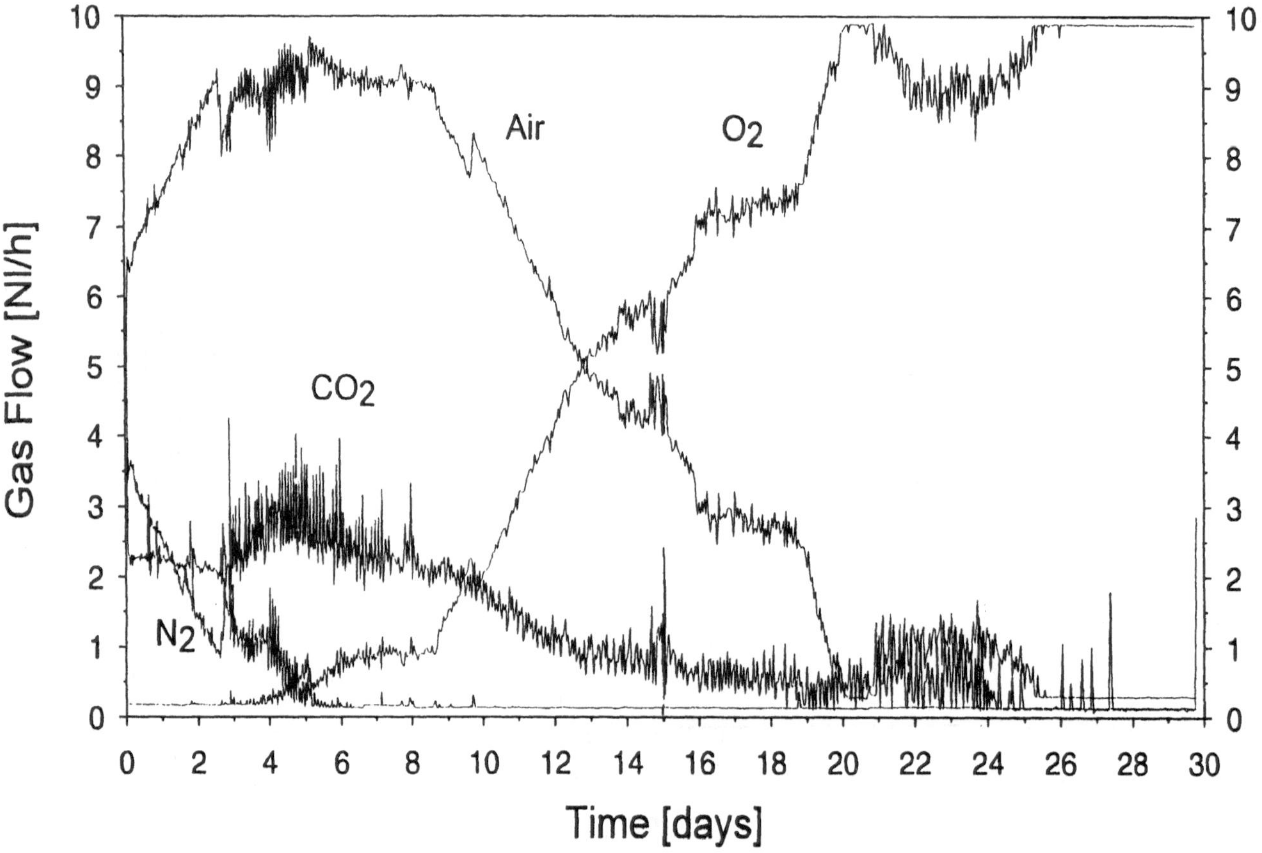

Fig. 8. History plot of the flow of the four controlled gases (air, O_2, N_2, CO_2) during CHO fermentation depending on pH and pO_2 data.

sedimentation in the sampling tube. It is easy to adapt the control program in a way that the bleed pump works only for short intervals with high flow rates. The flow rates were calculated automatically after input of bleed rate and interval.

Data from two balances were also collected on-line to calculate actual flow rates for the verification of all pump rates. An on-line signal from a pressure sensor was used to monitor the fouling status of the perfusion membrane. In Fig. 6 the changes in cell concentration during a 30 day CHO cultivation are shown. After a four days batch, mode status was switched to perfusion mode. Additionally, the cell bleed stream was started at day 12 to maintain a stable cell density. The cells grew slowly (typical for this CHO cell line), reaching a final steady state density of about 8×10^6 total cells per ml. From Fig. 7 it can be seen that for the first 26 days both pH and pO_2 could be kept very close to their set point values of pH = 7.2 and pO_2 = 40% air saturation, respectively. Only for the last four days pO_2 could not be maintained at 40% air saturation and pH showed a higher fluctuation. This can be explained by following the oxygen gas flow in Fig. 8. The instability in pO_2 and pH occurred when the oxygen valve is completely opened but the demand of the culture was not satisfied. In this case stable values of pH and pO_2 could not be controlled sufficiently. The drop in pO_2 beginning at day 26 is caused by a decrease in oxygen transfer capacity of the membrane aeration system due to membrane fouling effects of the hydrophobic polypropylene membrane. Higher oxygen transfer rates can be obtained by using a more hydrophobic membrane or by using a higher oxygen gas flow rate (requires a mass flow controller with more than a maximum of 10 Nl h^{-1}) or by employing a back pressure controller at the membrane outlet.

Conclusion

The presented 'computer integrated control system (CICS)' joins components of aeration and pump control, data acquisition and central user interface for an easy to use and integrated detailed fermentation control system. In contrast to other PCS commonly used for small scale fermentations, like MICRO-MFCS (B. Braun International, Melsungen, FRG) or UBICON (EDS GmbH, Hannover, FRG) (Gollmer *et al.*, 1992), this system was developed especially for application to cell culture bioreactors. Routine functions and interactive menus for the determination of parameters like dilution or bleed rate are embedded. Additionally user defined software controller for pH and pO_2 are implemented in the process control program. Detailed data acquisition of all on-line data in user defined intervals is available in this system. A further processing of the data in spreadsheet programs is possible. Security features to overcome minor and major problems including power failure hardware and software errors are implemented in the program to guarantee correct fermentation performance and to minimize the loss of important cultures. An adaptation to all commercially available bioreactor systems, including amplifiers of sensors and probes, pumps and valves is easily possible if analog signals for data sampling and device control are available.

Acknowledgements

This work was partly supported by the project 'Development of a procedure and a plant for the recirculation of nutrient media for animal cell culture' (BMFT ref. no 0319346A) of the German Ministry of Science and Technology. We like to thank the Beringwerke AG, Marburg, FRG, for kindly supplying the CHO cell line and Microdyn, Wuppertal, FRG, for supplying the polysulfone membrane.

References

ADA 16: 8–16 Bit A/D-D/A-Wandlerkarte für PCs (Part 1 ELV Journal 2/89 (1989a).

ADA 16: 8–16 Bit A/D-D/A-Wandlerkarte für PCs (Part 2 ELV Journal 3/89 (1989b).

Büntmeyer H, Bödeker BGD and Lehmann J (1987) Membrane-stirrer-reactor for bubble free aeration and perfusion. In: Modern Aproaches to Animal Cell Technology, Spier RE and Griffiths JB (eds), Butterworth, London, 411–419.

Fischer J, Werges U (1990) MSR-Karten für IBM-Kompatible. Elector, 9 (1990), 24.

Gollmer K, Gäbel T, Nothnagel J and Posten C (1992) UBICON – a universal bioprocess control system. In: DECHEMA Biotechnology Conferences 5, VCH Verlagsgesellschaft, Weinheim.

Lehmann J, Vorlop J and Büntemeyer H (1988) Bubble free reactors and their development for continuous culture with cell recycle. In: Animal Cell Biotechnology 3, Spier RE and Griffiths JB (eds), Academic Press, London, 221–237.

Riese U, Heidemann R, Lütkemeyer D and Lehmann J (1993) Partial cell culture medium recirculation: Experience in pilot scale. In: Animal Cell Technology: Basic and Applied Aspects, Kaminogawa S, Ametani A and Hachimura S (eds), Kluwer, Dordrecht, 443–446.

Vorlop J (1990) Entwicklung eines Membranrührers zur Blasenfreien Begasung und Durchmischung von Zellkulturen im Pilotmaßstab. PhD Thesis, Technical University of Braunschweig.

Zettlmeissl G, Ragg H and KLarges HE (1987) Expression of biologically active human antithrombin III in Chinese hamster ovary cells. Biotechnology 5: 720–725.

Address for correspondence: Heino Büntmeyer, University of Bielefeld, Technical Faculty, Institute of Cell Culture Technology, P.O. Box 100131, 33501 Bielefeld, Germany.

Cytotechnology **15:** 281–289, 1994.

Automated monitoring of cell concentration and viability using an image analysis system

Fumio Maruhashi[1], Sei Murakami[2] & Kenji Baba[1]
[1] *Hitachi Research Laboratory, Hitachi Ltd., 7-1-1 Omika-cho, Hitachi, Ibaraki 319-12, Japan;* [2] *Kasado Works, Hitachi Ltd., 794 Higashitoyoi, Kudamatsu, Yamaguchi 744, Japan*

Key words: Animal cell culture, cell debris, cell viability, image analysis

Abstract

In order to automate measurements of cell concentration and viability in a suspended animal cell culture, we have developed an *in situ* microscopic image analysis system with an effective cell recognition algorithm. With a small amount of sample, this system can measure the cell density rapidly and aseptically. In addition, it can measure a cell size histogram including cell debris small particle distribution. These small particles have been found to be related to the viability of the mouse-mouse hybridoma STK1 cell line. By using cell debris small particle density as an indicator of cell viability, the developed system provides non-destructive viability monitoring without trypan blue staining.

Introduction

Industrial applications of cell culture for producing cytokines, monoclonal antibodies, *etc.* have already started to be established, especially in the pharmaceutical industry. In these applications, culture reproducibility and consistency are strongly required for process validation. Unfortunately, reliable on-line cell culture monitoring is limited to pH, dissolved oxygen, and temperature measurements. On the other hand, microscopic cell observation is indispensable for monitoring cultured cell conditions. This type observation has been often performed by the human eye and brain because of the associated morphological ionformation recognition and classifying ability. However, human eyes are not suitable for measuring quantitative values, such as size, number, density, and area. To overcome this disadvantage, many automated image analysis methods for acquiring numerical information from cell images have been reported (Bradbury, 1979; Goebel *et al.*, 1990; Tucker *et al.*, 1994). Scanning electronic microscopy, SEM, is often used for detailed cell structure analysis (Bradbury, 1979; Pons *et al.*, 1992). Although this method provides numerous morphological information acquisition abilities, the time consuming sample preparation (especially freeze drying of the cell preparation) makes SEM difficult to use for real time cell culture measurement. To automate the time consuming hemacytometer cell number counting, a flow cytometric measurement method with electric conductivity and light scattering is often used to rapirly measure the cell size distribution and many other cell characteristics (Broise *et al.*, 1991; Brooks *et al.*, 1985; Chuck *et al.*, 1992; Frame *et al.*, 1990; Goebel *et al.*, 1990; Livne *et al.*, 1987; Miller *et al.*, 1986; Rudt *et al.*, 1992; Sen *et al.*, 1989; Shields *et al.*, 1978; Wheatley *et al.*, 1987). However, residues of electrolytes from the culture medium interfere with the flow cytometric conductivity measurement, and cell clumps may block the flow cytometer capillaries (Rudt *et al.*, 1992). Since the flow cytometric light scattering method cannot directly measure cell size, the measured cell size is less precise than that of direct measurement methods and requires standardized particle calibration for each sample.

A cell concentration and viability measurement system for on-line real-time monitoring should have following characteristics:

- no sample pre treatment (staining, dilution, etc.)
- rapid measurement

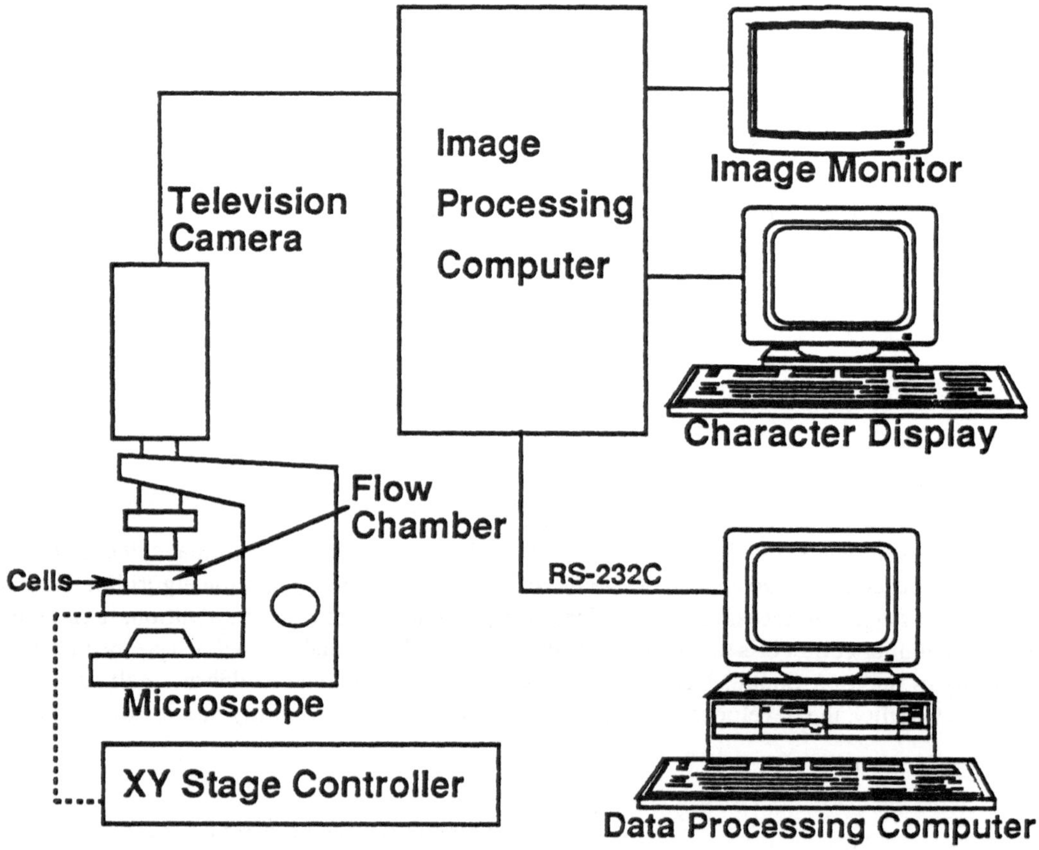

Fig. 1. Experimental apparatus

- small sample volume for avoiding cell loss
- negligible cell damage for sample cell recycle
- aseptic handling

In this paper, an image analysis system suitable for on-line real-time cell concentration measurement is described. Some experimental data encouraging the application of non-destructive cell viability measurement with cell debris particle counting are also presented.

Materials and methods

Equipment

The configuration of the experimental apparatus is represented in Fig. 1. The apparatus is constructed of a flow chamber, a conventional optical microscope (Nikon Optiphoto) with automatic XY stage, a television camera (Hitachi, HV-730), an image processing computer (Hitachi IP-21) and a data processing computer (Hitachi B-16EXII). A small amount of suspended culture broth was sampled intermittently into a flow chamber, in which a microscopic image was taken by the television camera. The flow chamber contains 15 micro-liters of the culture broth in a detection room of 0.1 mm thickness. The flow chamber was designed to be capable of autoclave sterilization. The image of the culture broth captured in 0.016 seconds by the television camera is memorized in the image processing computer, then processed for cell concentration and viability measurement. The flow chamber on the XY automatic stage is moved by the stage controller. The captured image of the culture broth is divided into 256 pixels horizontally and 240 pixels vertically as shown in Fig. 2. The brightness of each pixel is converted into 128 gray scale levels, then the intensity value of each pixel is numbered as g(x, y) as shown on Fig. 2. The size of each pixel is equivalent to 1 micrometer of the culture broth. This image magnitude and resolution

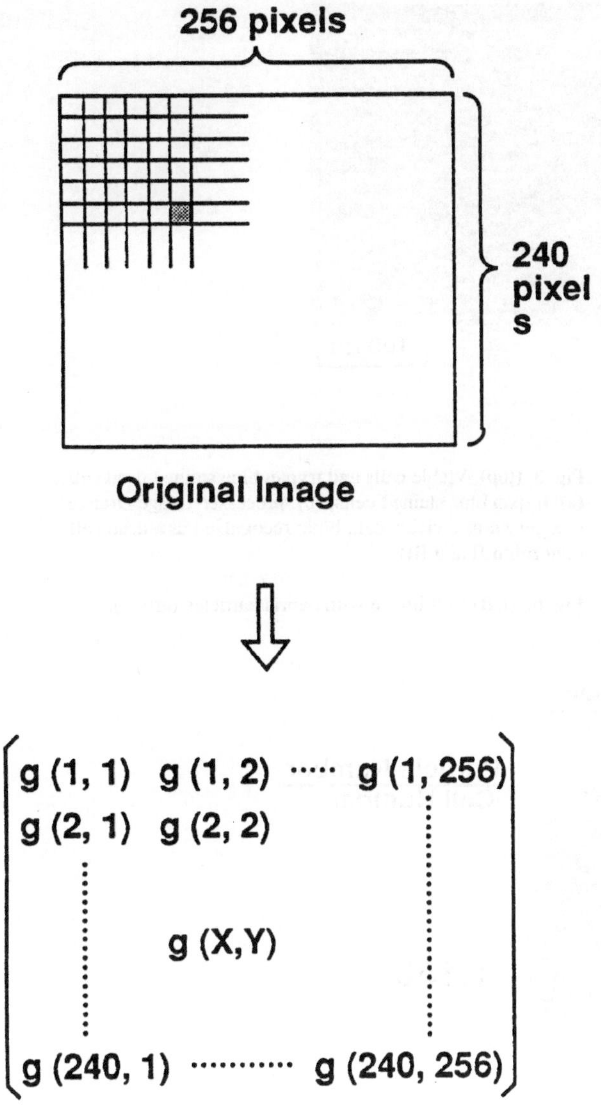

Fig. 2. Image conversion to gray scale levels.

were determined in order to retain image resolution high enough to recognize the cell, and capture a sufficient number of cells in one field for reliable cell number counting. The signal g(x, y) was processed using the image processing computer, in which the cells or particles were recognized from background (culture liquid) based on g(x, y). Viable cells, dead cells and cell debris particles were distinguished by using an image analysis algorithm described later. The results are presented on the display with quasi-color. The analyzed data set is transmitted to the data processing computer through an RS-232C cable. The cell and particle size distribution and the cell viability are

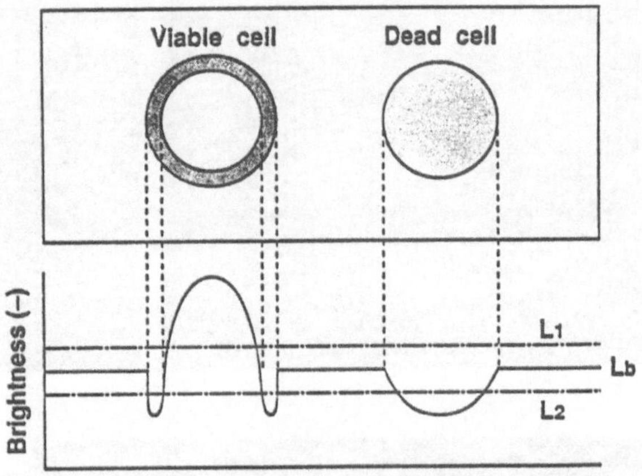

Fig. 4. Brightness distribution model of viable and dead cell.

calculated statistically in the data processing computer, and the results are presented on the display.

Image analysis algorithm

Two types if image analysis methodologies were developed. One is destructive measurement using trypan blue stained cells and the other one is non-destructive measurement without staining cells but making use of cell debris particle measurement. In both measurements, the same recognition algorithm for viable cells was developed, but different algorithms for cell viability measurement were developed for the destructive and the non-destructive method. Below are the image analysis algorithms tried during the course of this research.

I Binarization method (destructive)
Viable cells and trypan blue stained dead cells are observed in the gray scale image of culture broth as shown on Fig. 3(a). A brightness distribution model of the viable and trypan blue stained dead cell image is illustrated in Fig. 4. The preliminary experimental results of viable cell brightness distribution are shown in Fig. 5. At first, the most simple image recognition was carried out based on brightness of each pixel. For instance, the upper part over the first threshold level L1, is recognized as viable cell, on the other hand, the lower part of the second threshold level L2 is recognized as dead cell as shown on Fig. 4. This algorithm is called a binarization method. However, the outline of the viable cell is recognized as a dead cell due to its low brightness. Furthermore, the background brightness Lb includes noise signals and changes easily depending on the lighting and medium conditions.

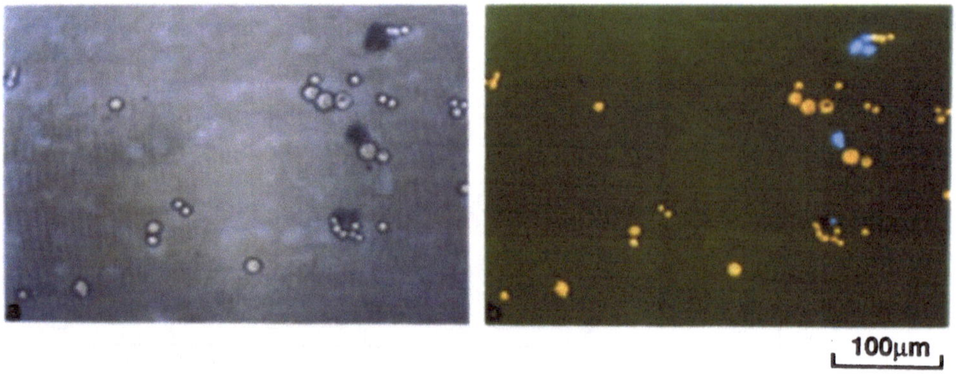

Fig. 3.

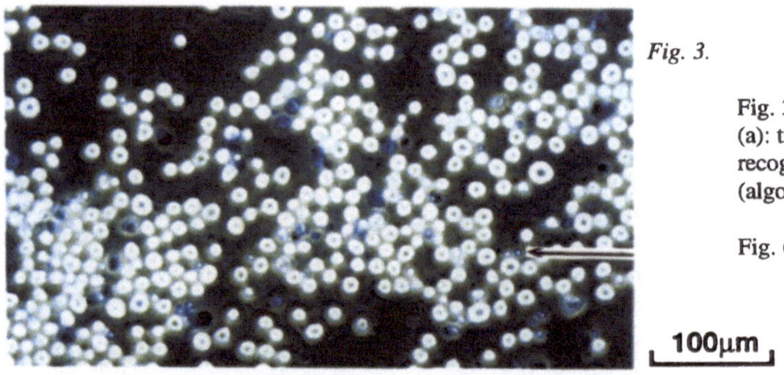

Fig. 3 (top). Viable cells and trypan blue stained dead cells. (a): trypan blue stained cells; (b): processed image Orange: recognized as a viable cell, blue: recognized as a dead cell (algorithm II and III).

Fig. 6. (left) Cell image with debris particles (arrow).

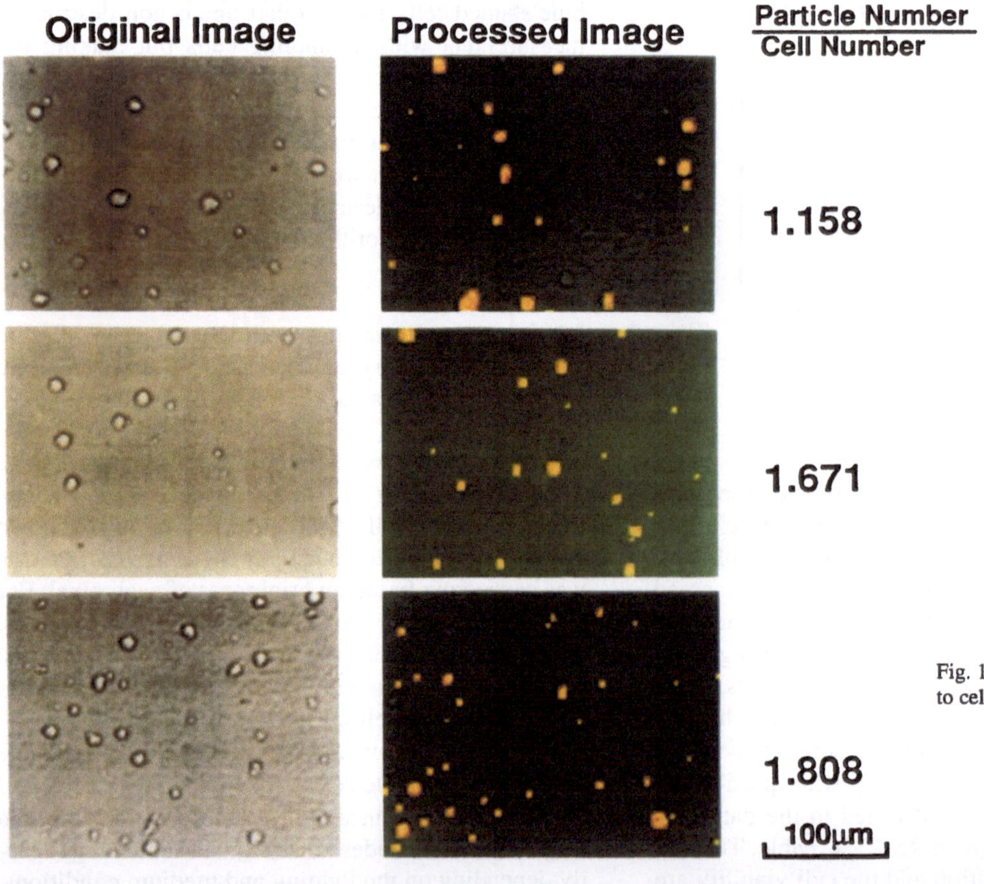

Fig. 11. Debris particle number to cell number.

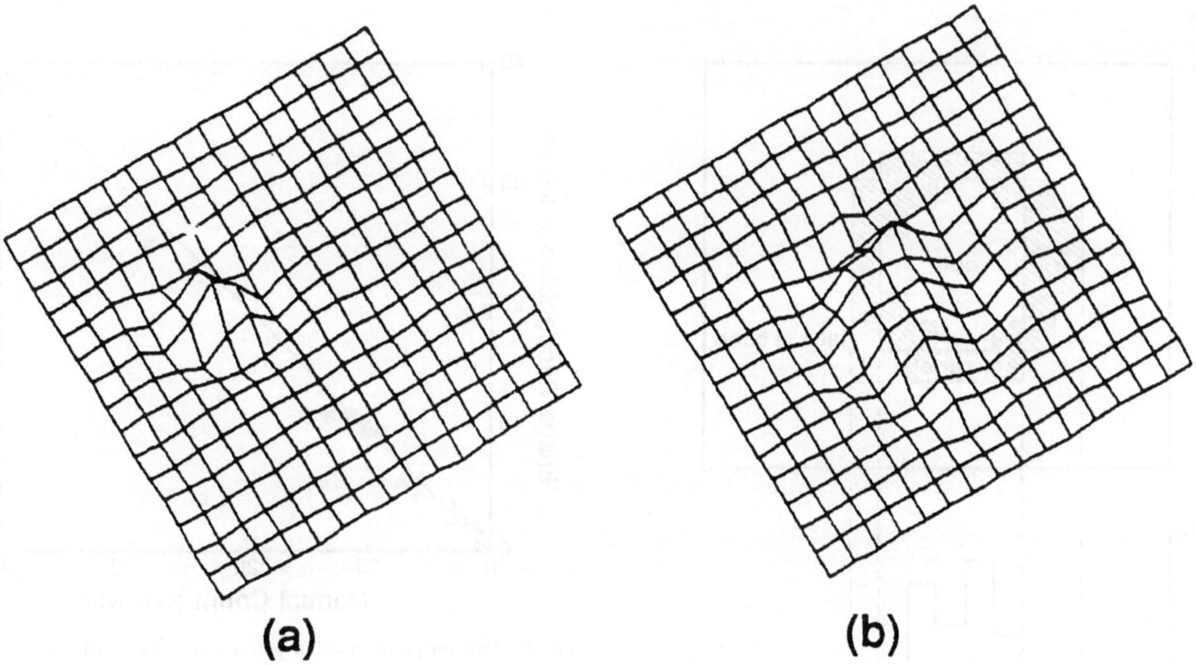

Fig. 7. Brightness distribution of viable cell and debris particle. (a): debris particle; (b): viable cell. 1 mesh equals 2 μm.

Therefore, the simple binarization method has serious limitations in the detection of cell viability.

II Outline recognition method for viable cells with trypan blue staining (destructive)

A new algorithm for accurate detection of viable cells was developed. This method was programmed based on the specific characteristics of the gray scale image, in which the viable cell image is surrounded by an outline having lower brightness, while the brightness of the center of the viable cell image is higher than the background. In other words, there is a steep brightness contrast between the center of a viable cell and its outline. A density difference between medium and cytoplasm caused light refraction, which causes the aforementioned brightness contrast. At first, the highest brightness peak is detected. Then the detected area is expanded until a steep brightness contrast is detected. This recognized contrast point is expanded along neighboring gaps until it forms a cirle. Thus viable cells are recognized by size and number.

III Dead cell recognition with trypan blue staining (destructive)

According to Fig. 4, dead cells include a set of lower brightness pixels. Conversely, the dark part of the outline surrounding the viable cell is very narrow. These specific characteristics of two-dimensional brightness distribution allow the accurate image recognition for trypan blue stained dead cells.

IV Outline recognition method for total cells (non-destructive)

In this method, sample was transferred to the flow chamber without any dilution nor staining. Although this method uses the same algorithm as the algorithm II, it cannot distinguish viable and dead cells. Therefore, measured cell number represents total cell number. Accordingly, viability could not be measured automatically by this algorithm alone.

V Cell debris particle measurement for cell viability (non-destructive)

In order to non-destructively measure cell viability, cell size distribution analysis was considered at first. Viable cells are often reported to be larger than dead cells (Broise *et al.*, 1991; Frame *et al.*, 1990; Rudt *et al.*, 1992; Sen *et al.*, 1989), and this might be used for distinguishing viable and dead cells. However, the ranges of the viable cell size and the dead cell size overlap. Moreover, the dead cell sometimes swells due to bleb formation and osmotic pressure (Resau *et al.*, 1988). Therefore, a simple cell size histogram cannot be used for measuring cell viability.

On the other hand, cell debris particles are often observed in culture broth, especially in high density

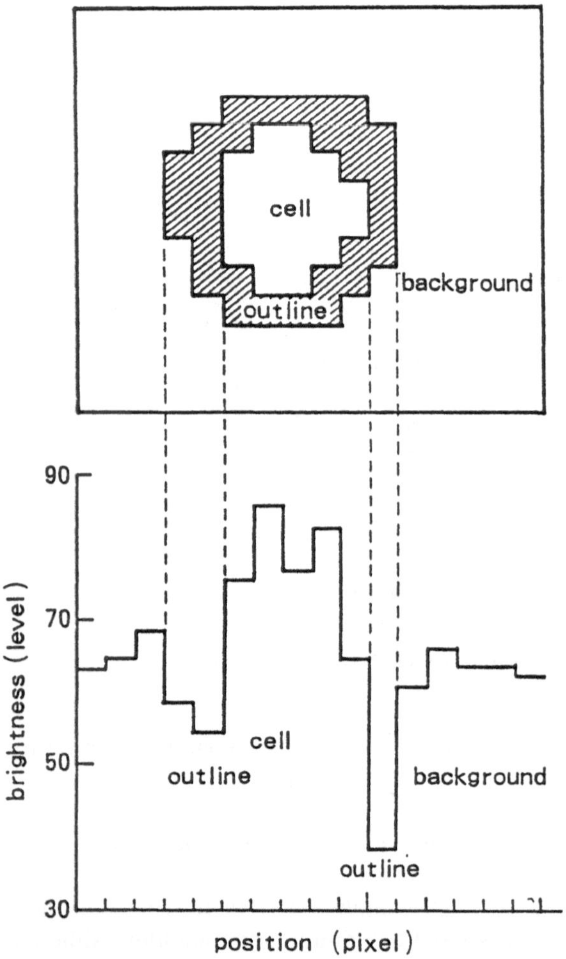

Fig. 5. Brightness of viable cell outline.

cultures (Goebel *et al.*, 1990; Rudt *et al.*, 1992). The aurthors focused on the existence of cell debris particles in the culture broth shown in Fig. 6 as a cell viability indicator. This allows the non-destructive measurement of cell viability. The brightness distributions of a cell debris particle image, compared with a viable cell image are illustrated in Fig. 7. Both the viable cell and the cell debris particle images have similar brightness distributions, but the sizes are different. Therefore, the algorithm for recognizing viable cells was modified and applied to the cell debris particle measurement.

Cell line

The mouse-mouse hybridoma, STK-1 was used in this study. The culture experiment was carried out using batch mode and semi-continuous mode with a periodic medium change using a 500 ml spinner flask (Corning,

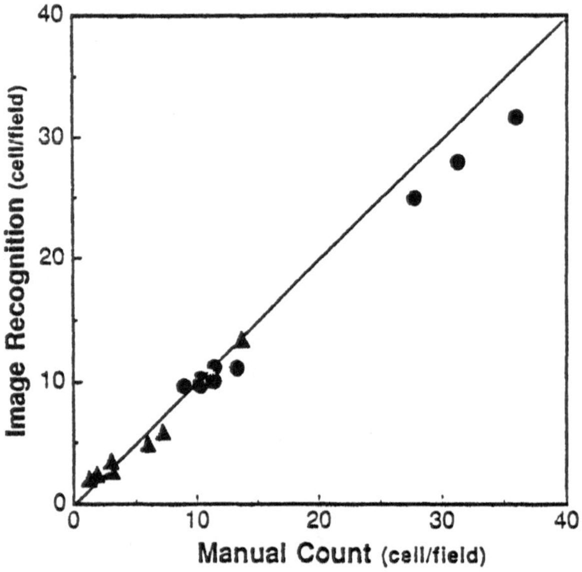

Fig. 8. Cell image recognition performance. (●): viable cells by algorithm II; (▲): trypan blue stained dead cells by algorithm III.

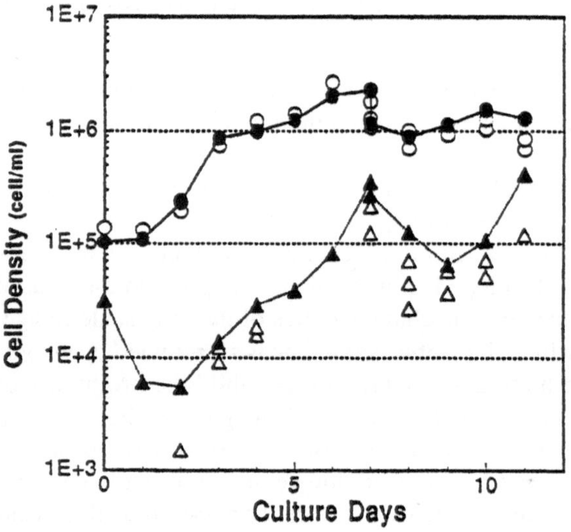

Fig. 9. Viable and dead cell measurement during batch culture. (—●—): viable cells by algorithm II; (○): viable cells manual measurement; (—▲—): dead cells by algorithm III; (△): dead cells manual measurement.

NY) filled with 500 ml medium, at 80 rpm and 37 °C. In the continuous culture, the whole amount of the culture medium was exchanged with new medium every four days. The oxygen was supplied into the vessel by surface aeration. The growth medium was Dulbecco MEM medium (Flow Laboratory) supplemented with 10% of new born calf serum.

For the manual measurement and trypan blue staining mesurement (algorithm I, II, and III), 0.3 to 0.5 ml

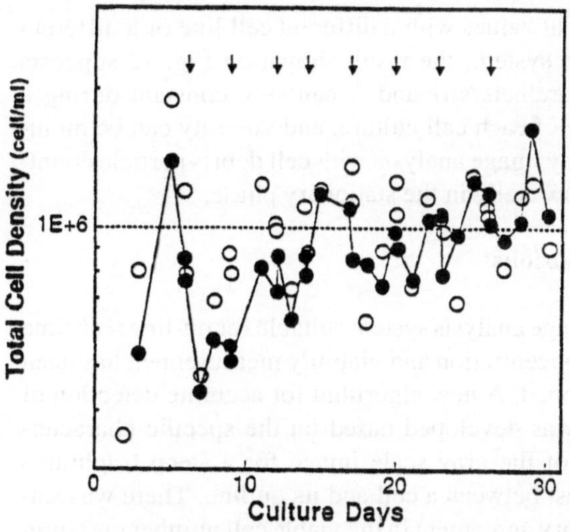

Fig. 10. Continuous culture result. (—•—): total cells by algorithm IV; (○): total cells manual measurement; (↓): medium exchange.

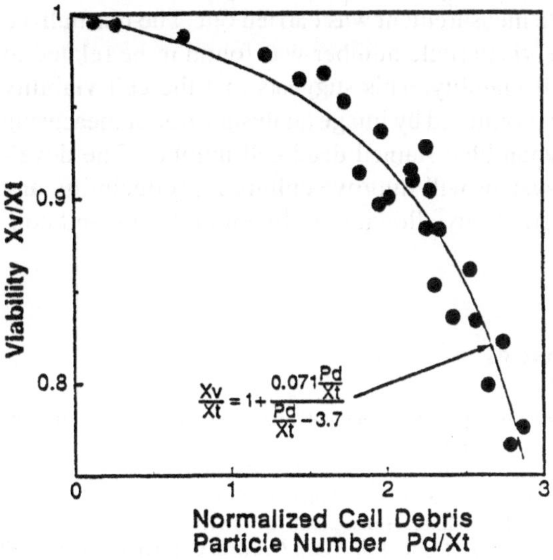

Fig. 12. Cell debris particle as viability indicator. (•): measured value; (—): estimating calculation.

of the culture broth was sampled manually and diluted using a phosphate buffer solution. The sample was stained with trypan blue, then measured by hemacytometer cell number counting and automatic image analysis. For the non-destructive measurement (algorithm IV), 1 ml of the culture broth was sampled off-line and injected through tubing into the flow chamber on the automatic XY stage of the microscope without any dilution nor staining.

Results and discussion

Image analysis for every algorithm took 3 to 5 seconds. However, cells needed to be settled down about 1 minute for a clear image. Thus maximum measuring cycle is expected to be 1 minute. This measuring cycle is short enough comparing with cell growth rate. Although the sampling was performed off-line in this experiment, on-line measurement will be feasible by connecting the flow chamber with a bioreactor and supplying culture liquid by peristaltic pump.

The result of the manual measurement and the image analysis of the destructive measurement (algorithm II and III) is shown in Fig. 8. Both the viable and dead cell measurements were successful. Another batch culture experiment was carried out for 11 days. The measurement results of the viable and dead cells are shown in Fig. 9. Good agreement between manual and image analysis (algorithm II and III) was found. This demonstrates the effectiveness of the image analysis method. However, a small error was observed in the results after 7 days in Fig. 10. This suggests that some particles which can be easily misunderstood to be deformed dead cells are present during the stage of cell viability decrease.

The continuous culture experiment was carried out changing the medium for one month. In this continuous culture, sample was transferred to the flow chamber without any dilution nor staining for simulating an on-line measurement, and total cell number was measured by algorithm IV. The result is shown in Fig. 10. There was satisfactory agreement for the total cell number measurement between manual and image analysis during the culture process. However, viability could not be measured automatically by algorithm VI alone.

In order to non-destructively measure the viability, the cell debris particles were investigated during the culture process. The results are illustrated in Fig. 11. The recognized image includes the small particles as well as cells. During the phase of necrosis, the cell undergoes degradation by autolysis and denaturation. Then, the membrane debris from fragmentation and distortion of organelles begins to be converted to large myelin figures, which occupy large areas of the cytoplasm and represent formation of bilayer configurations from the altered and hydrated lipid derivatives (Resau *et al.*, 1988). These bilayers may configure into small particles that will last after complete cell membrane destruction, and can be measured as a parameter of cell death and following cell disruption.

Figure 12 shows the measured relationship between normalized cell debris particle number, Pd/Xt, and manually measured cell viability, Xv/Xt, in the aforementioned continuous culture results.

Suppose cell and debris particle lysis to be enzymatic reactions by hydrolytic enzymes released from lysosome, then viable cell density, Xv, dead cell density, Xd, and cell debris particle density, Pd, can be described as the following Michaelis-Menten kinetic equations.

[*viable cells*]

$$\frac{dXv}{dt} = (\mu - \delta)Xv \tag{1}$$

[*dead cells*]

$$\frac{dXd}{dt} = \delta Xv - \frac{VcXd}{Xd + Kc} \tag{2}$$

[*cell debris*]

$$\frac{dPd}{dt} = \frac{VcXd}{Xd + Kc} - \frac{VpPd}{Pd + Kp} \tag{3}$$

where, μ: cell growth rate, δ: cell death rate, Vc: maximum rate of dead cell lysis, Vp: maximum rate of cell debris lysis, Kc, Kp: Michaelis constant.

Second term of equation (2), VcXd/(Xd+Kc), represents a lysis of dead cells in a form of Michaelis-Menten kinetics with substrate concentration Xd, maximum rate Vc, and Michaelis constant Kc. Similarly, second term of equation (3), VpPd/(Pd+Kp), represents a lysis of cell debris.

During the stationary phase, cell debris particle density, Pd, and total cell density, Xt, will change slowly. Therefore, by assuming dPd/dt to be zero, and Xt to be constant, equation (3) enables the following cell viability, Xv/Xt, estimation (algorithm V):

$$\frac{Xv}{Xt} = 1 - \frac{Xd}{Xt} = 1 - \frac{\alpha \frac{Pd}{Xt}}{\frac{Pd}{Xt} + \beta} \tag{4}$$

where

$$\alpha = \frac{Kc}{({}^{Vc}/{}_{Vp} - 1)Xt}$$

and

$$\beta = \frac{Kp}{(1 - {}^{Vp}/{}_{Vc})Xt}$$

From regression of the measured data shown on Fig. 12, the parameters α and β for this experiment was determined to be –0.071 and –3.7, respectively.

As shown on Fig. 12, this estimating equation represents the measured values very well for this experiment. Although the parameters α and β may have different values with a different cell line or a different culture system, the result shown on Fig. 12 suggests that parameters α and β can stay constant during a course of each cell culture, and viability can be monitored by image analysis with cell debris particle counting, especially in the stationary phase.

Conclusions

An image analysis system suitable for on-line real-time cell concentration and viability measurement has been developed. A new algorithm for accurate detection of cells was developed based on the specific characteristics of the gray scale image for a steep brightness contrast between a cell and its outline. There was satisfactory agreement in the viable cell number measurement between manual and image analysis during the culture process. A destructive measurement using trypan blue staining and a non-destructive measurement without staining cells but making use of cell debris particle measurement was carried out. The normalized cell debris particle number was found to be related to the cell viability. This suggests that the cell viability can be monitored by image analysis without measuring the trypan blue stained dead cell number. The developed system will improve culture reproducibility and productivity by allowing on-line monitoring and control.

References

Bradbury S (1979) Microscopical Image Analysis : Problems and Approaches. J. Microsc. 115: 137–150.

Broise Ddl, Noiseux M, Lemieux R & Massie B (1991) Long-Term Perfusion Culture of Hybridoma: A "Grow or Die" Cell Cycle System. Biotechnol. Bioeng. 38: 781–787.

Brooks RF & Shields R (1985) Cell Growth, Cell Division and Cell Size Homeostasis in Swiss 3T3 Cells. Exp. Cell Res. 156: 1–6.

Chuck AS & Palsson BO (1992) Population Balance between Producing and Nonproducing Hybridoma Clones Is very Sensitive to Serum Level, State of Inoculum, and Medium Composition. Biotechnol. Bioeng. 39: 354–360.

Frame KK & Hu W-S (1990) Cell Volume Measurement as an Estimation of Mammalian Cell Biomass. Biotechnol. Bioeng. 36: 191–197.

Goebel NK, Kuehn R & Flickinger MC (1990) Methods for Determination of Growth-Rate-Dependent Changes in Hybridoma Volume, Shape and Surface Structure during Continuous Recycle. Cytotechnol. 4: 45–57.

Livne A, Grinstein S & Rothstein A (1987) Volume-Regulating Behavior of Human Platelets. J. Cell. Physiol. 131: 354–363.

Miller SJO, Henrotte M & Miller AOA (1986) Growth of Animal Cells on Microbeads. I. *In situ* Estimation of Numbers. Biotechnol. Bioeng. 28: 1466–1473.

Pons M-N, Wagner A, Vivier H & Marc A (1992) Application of Quantitative Image Analysis to a Mammalian Cell Line Grown on Microcarriers. Biotechnol. Bioeng. 40: 187–193.

Resau JH & Trup BF (1988) Cell Injury, Differentiation, and Regeneration in Explant, Organ, and Cell Culture Models. In: Advances in Cell Culture. Vol. 6 (pp. 261–289) Academic Press, Inc., London.

Rudt S, Blunk T & Müller RH (1992) Quantification of Cell Subpopulations, Fractions of Dead Cells and Debris in Cell Suspensions by Laser Difractometry. Pharm. Ind. 54: 966–969.

Sen S, Srienc F & Hu W-S (1989) Distinct Volume Distribution of Viable and Non-Viable Hybridoma Cells: A Flow Cytometric Study. Cytotechnol. 2: 85–94.

Shields R, Brooks RF, Riddle PN, Capellaro DF & Delia D (1978) Cell Size, Cell Cycle and Transition Probability in Mouse Fibroblasts. Cell 15: 469–474.

Tucker KG, Chalder S, Al-Rubeai M, Thomas CR (1994) Measurement of Hybridoma Cell Number, Viability, and Morphology Using Fully Automated Image Analysis. Enzyme Microb. Technol. 16: 29–35.

Wheatley DN, Inglis MS & Foster MA (1987) Hydration, Volume Changes and Nuclear Magnetic Resonance Proton Relaxation Times of HeLa S-3 Cells in M-Phase and the Subsequent Cell Cycle. J. Cell Sci. 88: 13–23.

Address for offprints: S. Murakami, Kasado Works, Hitachi Ltd., 794 Higashitoyoi, Kudamatsu, Yamaguchi 744, Japan.

Cytotechnology **15**: 291–299, 1994.

Software sensors for the monitoring of perfusion cultures: Evaluation of the hybridoma density and the medium composition from glucose concentration measurements

François Pelletier, Christian Fonteix, Aurélio Lourenço da Silva, Annie Marc and Jean-Marc Engasser
Institut National Polytechnique de Lorraine, Laboratoire des Sciences du Génie Chimique, CNRS, 2, avenue de la Forêt de Haye, BP 172, F-54500 Vandoeuvre-les-Nancy Cedex, France

Key words: Estimator, Extended Kalman Filter, hybridoma, kinetic model, perfusion culture.

Abstract

New software sensors based on the Extended Kalman Filter technique have been developed for the monitoring of animal cell perfusion cultures. They use a kinetic model describing the growth, death and metabolism of hybridoma cells as a function of the medium composition. The model was initially validated on a batch culture and found to correctly predict the continuous perfusion culture kinetics, except for the production of ammonia and lactate. Using the measurement of a single component in the culture medium, in this case glucose, the Extended Kalman Filter provides an excellent evaluation of the time variation of the concentrations of living and dead cells, of glutamine and antibodies, during the whole perfusion culture for a retained cell density rising from 1 to 11 × 10^6 cells.ml^{-1} inside the reactor.

Nomenclature

Kinetic model

A, death rate constant (h^{-1}); D, dilution rate (h^{-1}); Glc, glucose concentration (mM); Gln, glutamine concentration (mM); k_b, specific death rate (h^{-1}); k_l, specific lysis rate (h^{-1}); K_{Gln}, glutamine saturation constant (mM); K_{Lac}, lactate inhibition constant (mM); K_{NH4}^{+}, ammonia inhibition constant (mM); Lac, lactate concentration (mM); Mab, monoclonal antibodies concentration (mg.l^{-1}); NH_4^+, ammonia concentration (mM); X, living cell density (cells.ml^{-1}); X_b, dead (blue) cell density (cells.ml^{-1}); Y, yield coefficients (mmol. 10^{-9} cells or mg. 10^{-9} cells).

Greek symbols

α, retention coefficient of antibodies; κ_{Gln}, glutamine kinetic constant for death rate expression (mM); κ_{Lac}, lactate kinetic constant for death rate expression (mM^{-1}); $\kappa_{\mathrm{NH_4^+}}$, ammonia kinetic constant for death rate expression (mM^{-1}); ν, specific consumption rates (mmol.h^{-1}. 10^{-9} cells); μ, specific growth rate (h^{-1}); $\mu_{\max}$, maximum specific growth rate (h^{-1}); π, specific production rates (mmol.h^{-1}. 10^{-9} cells or mg.h^{-1}. 10^{-9} cells).

Extended Kalman Filter

f, vector of system dynamics; **h**, measurement equation; K, Kalman gain matrix; t, time; **v**, measurement noise vector; **w**, system noise vector; **x**, state vector; **y**, measurement or output vector.

Superscripts and subscripts

â, estimation; ~, prediction; in, feeding.

Introduction

The efficient operation of high cell density perfusion (Graf & Schügerl, 1991; Rolf *et al.*, 1992; de la Broise *et al.*, 1992) or fed-batch (Omasa *et al.*, 1992) cultures requires an optimal feeding of nutrients during the phases of cellular growth and protein production. This necessitates a monitoring of the concentration of viable cells, of their metabolic activities and of the medium composition.

Unfortunately reliable physical sensors for on-line monitoring of cytoreactors are still rare. A feasible alternative is the use of indirect or software sensors which have been the subject of attention for many years (Bastin & Dochain, 1990; Stephanopoulos & San, 1984; Van der Heijden *et al.*, 1989) and are increasingly utilized in the fermentation industry. They consist of algorithms which, through the use of simulation models of the process, provide an estimation of the medium composition from a limited number of on-line or off-line analyses.

A new estimator for mammalian cell cultures, developed as a central part of an advanced control procedure, is presented. It entails a kinetic model initially established on a batch culture. The properties of the estimator are illustrated on a continuous perfusion culture yielding a high hybridoma cell density.

$$\mu = \mu_{max}\cdot\frac{Gln}{Gln + K_{Gln}}\cdot\frac{K_{Lac}}{Lac + K_{Lac}}\cdot\frac{K_{NH_4^+}}{NH_4^+ + K_{NH_4^+}}$$

$$k_b = A.\left(\kappa_{NH_4^+}.NH_4^+ + \kappa_{Lac}.Lac + \frac{\kappa_{Gln}}{Gln + \kappa_{Gln}}\right)$$

$$v_{Glc} = Y_{Glc}.\mu$$

$$v_{Gln} = Y_{Gln}.\mu$$

$$\pi_{NH_4^+} = Y_{NH_4^+}.\mu$$

$$\pi_{Lac} = Y_{Lac}.\mu$$

$$\pi_{Mab} = Y_{Mab}.\mu$$

Fig. 1. Specific rate expressions used in the kinetic model for cell growth and death, glucose and glutamine consumption, ammonia, lactate and antibodies production.

Living cells: $$\frac{dX}{dt} = (\mu - k_b - k_l).X$$

Dead cells: $$\frac{dX_b}{dt} = k_b.X$$

Glucose: $$\frac{dGlc}{dt} = -v_{Glc}.X + D.(Glc_{in} - Glc)$$

Glutamine: $$\frac{dGln}{dt} = -v_{Gln}.X + D.(Gln_{in} - Gln)$$

Lactate: $$\frac{dLac}{dt} = \pi_{Lac}.X + D.(Lac_{in} - Lac)$$

Ammonia: $$\frac{dNH_4^+}{dt} = \pi_{NH_4^+}.X + D.(NH_{4\,in}^+ - NH_4^+)$$

Antibodies: $$\frac{dMab}{dt} = \pi_{Mab}.X - D.(1-\alpha).Mab$$

Fig. 2. Mass balance equations for viable and dead cells, glucose, glutamine, ammonia, lactate and antibodies for a continuous perfusion culture. In the case of a batch culture the dilution rate is set to zero.

Materials and methods

Cell cultures

Cell line and culture medium

The 6H2 murine cell line provided by Sorin Biomedica (Sallugia, Italy) results from the fusion of mouse BALB/c spleen cells and Sp2.0 myeloma cells. The monoclonal antibody produced is an IgG2a directed against a high molecular weight melanoma-associated antigen. Cells have been routinely cultivated in T-flasks for 6 days, before transferred inside the bioreactor at a concentration of 0.15 to 0.2 × 10^6 cells.ml^{-1}.

Batch culture

The batch culture was performed in a 2 liters reactor (Inceltech-SGI, France) with 1.2 liter working volume. The DMEM/HamF12 (3:1) basal medium was supplemented with 15.6 mM glucose, 4 mM glutamine, 5% (v/v) Fetal calf serum, 2% (v/v) essential amino acids, 1% (v/v) non essential amino acids and antibiotics (100 UI.ml^{-1} penicillin, 100 μg.ml^{-1} streptomycin). The temperature and agitation levels were controlled at respectively 37 °C and 50 rpm. The dissolved oxygen concentration was regulated around the set point of

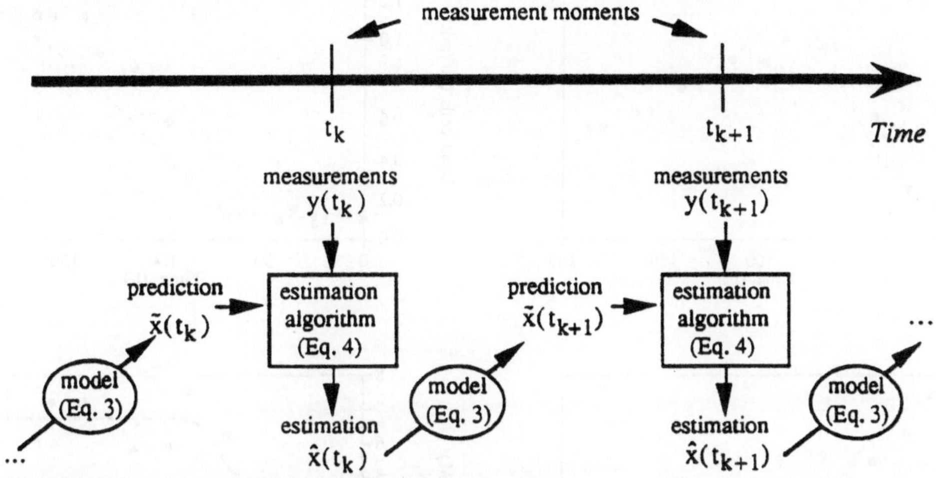

Fig. 3. Schematic illustration of the Extended Kalman Filter procedure which uses a kinetic model to continuously predict the state variables and periodic measurements to correct the model calculations.

Table 1. Parameters used for the kinetic model.

Parameter	Value	Unit
μ_{max}	0.065	h^{-1}
K_{gln}	0.3	mM
K_{Lac}	55	mM
$K_{NH_4^+}$	5	mM
A	0.006	h^{-1}
κ_{Gln}	0.1	mM
κ_{Lac}	0.0.14	mM^{-1}
$\kappa_{NH_4^+}$	0.05	mM^{-1}
Y_{Glc}	4.9	mmol.10^{-9} cells
Y_{Lac}	3.5	mmol.10^{-9} cells
Y_{Gln}	1.9	mmol.10^{-9} cells
$Y_{NH_4^+}$	1.9	mmol.10^{-9} cells
Y_{Mab}	18	mg.10^{-9} cells
k_l	0.0095	h^{-1}

40% by direct bottom sparging with pure oxygen and the pH was maintained at 7±0.1 by either CO_2 injection into the headspace or delivery of 0.2 N NaOH solution in the culture medium.

Perfusion culture

The perfused cell culture was performed in a 2 liters working volume Cytoflow® bioreactor (Inceltech-SGI, France) with two different dilution rates of 0.04 h^{-1} and 0.08 h^{-1}. The cells have been initially seeded in a DMEM/HamF12 (3:1) basal medium supplemented with 35 mM glucose, 4.5 mM glutamine, 5% (v/v) Fetal calf serum, 2% (v/v) essential amino acids, 1% (v/v) non essential amino acids and antibiotics (100 UI.ml^{-1} penicillin, 100 μg.ml^{-1} streptomycin). The composition of the feed medium was previously determined to yield high densities with the 6H2 hybridoma cell line (Pinton, 1991). It is the same as in the batch culture except for the glucose concentration which was fixed at 18 mM. The operating conditions, pH, pO_2, temperature and agitation were controlled in the same way as for the batch culture.

Analyses

In order to avoid some interpretation difficulties due to insufficient experimental points (Phillips *et al.*, 1991), samples were taken twice a day from the bioreactor for metabolic assays and cell numeration. The viable and dead cells densities were determined by trypan blue dye exclusion using a hemocytometer. The stained cells were treated as dead cells. The samples were centrifuged and the supernatants aliquoted and frozen for subsequent analyses. Glucose, lactate and glutamine concentrations were determined by enzymatic methods, while NH_4^+ ions concentrations were analyzed with a selective electrode (Orion). The monoclonal antibody concentration was measured by the ELISA technique as previously described (Pinton, 1991). All measurements have been realized threefold for each culture sample.

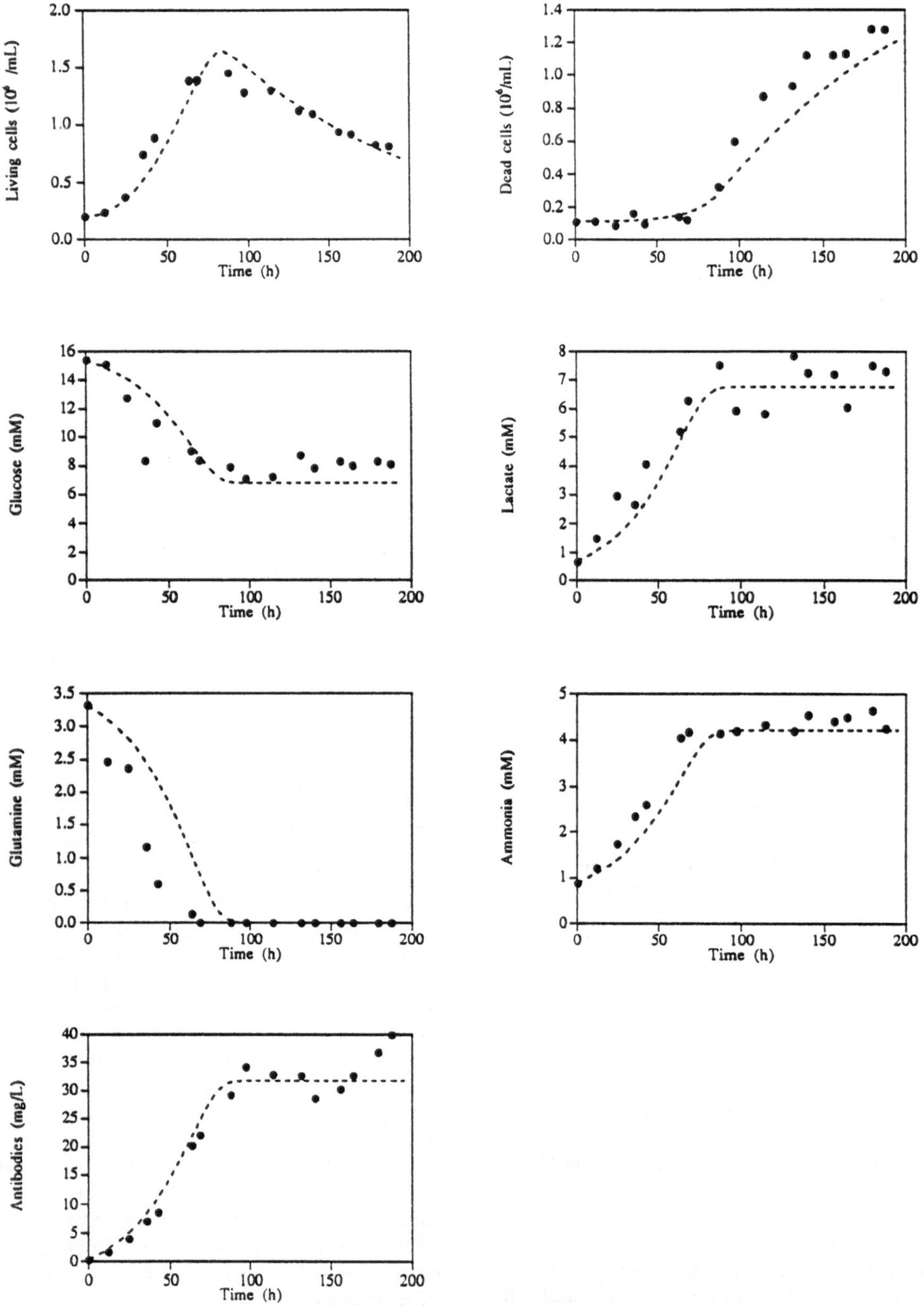

Fig. 4. Kinetics of a batch culture of 6H2 hybridoma cells under controlled pH and dissolved oxygen: (•) represent the experimental measurements and the dotted curve (- - -) the concentrations predicted by the kinetic model.

Kinetic model

The kinetic model for the hybridoma cultures used in this study has been contructed according to a previously proposed methodology (Goergen *et al.*, 1994). The selected rate expressions for cellular growth and death, nutrients consumption, metabolites and protein production which have been choosen are given in Figure 1.

Cellular growth is characterised by a maximum specific growth rate, μ_{max}. During cultures the specific growth rate, μ, usually decreases because either depletion of essential nutrients or accumulation of inhibitory metabolites. In the present case, as glucose was never limiting, the specific growth rate is described as a function of the concentration of glutamine, lactate and ammonia in the medium. Other amino acids were not found limiting in the investigated cultures (Pinton, 1991).

In batch culture, hybridoma lysis is usually negligible and the cellular death rate is given by the rate of appearance of cells counted as blue by the trypan blue technique. The kinetics expression for the specific rate of cellular death, k_b, describes the possible increase of death rate with limitation in glutamine or accumulation in lactate and ammonia.

For many investigated cell lines, the specific rates of nutrient uptake, ν, and metabolite or protein production, π, have been found growth associated and are thus described as proportional to the specific growth rate (Goergen *et al.*, 1992, Hiller *et al.*, 1993).

The rate expressions of Figure 1 have been used for the modelling of both the batch and continuous perfusion cultures. In the case of long term continuous cultures, lysis has been shown to represent a significant additional process for cell death (Goergen *et al.*, 1993). Thus, for the perfusion culture, a constant specific rate of cell lysis, k_l, was added to the previously introduced death constant, k_b, to express the global rate of cell death.

Mass balance equations

The set of differential equations describing the time variation of the concentrations of the different medium components for the perfusion culture is given in Fig. 2. The perfusion reactor, which consists of a stirred vessel with a microfiltration membrane on an external loop with a small recirculating volume (Pinton, 1991) can be considered as a well stirred homogeneous system. The derived equations also simulate the kinetics of batch cultures when the dilution rate is set to zero. For the perfusion culture, cells are totally retained by the microfiltration membrane. As previously reported for the filtration of antibodies (Piret & Cooney, 1990; Schmid *et al.*, 1992), a partial retention of the antibodies is integrated in the mass balance equations, with a retention coefficient, α, proportional to the cellular density. The degradation of glutamine is neglected, considering the relatively low level inside the reactor. The release of ammonium from dead cells was also considered as negligible.

Software sensors

A software sensor is an algorithm for the on-line estimation of concentrations which are not measurable in real time, based on a simulation model of the process and on one or several accessible measurements. The Extended Kalman Filter (EKF) used in this study is an extension of the Linear Kalman Filter (Kalman, 1960; Kalman & Bucy, 1961) to non-linear systems. Its application to the monitoring of microbial fermentation processes has been extensively described (Albiol *et al.*, 1993; Dubach & Märkl, 1992; Valero *et al.*, 1990). In this study, the EKF is used in its version with a continuous model for the process simulation and a discrete model for measurements (Miller & Leskiw, 1987). The general methodology is briefly outlined in the following.

The dynamic of the system, i.e. the time variation of the medium components concentrations, is represented by the general equation:

$$\frac{d\mathbf{x}}{dt} = \mathbf{f}(\mathbf{x}(t), t) + \mathbf{w}(t) \qquad (1)$$

where $\mathbf{x}$ is the state vector (the concentrations of the medium components), $\mathbf{f}$ is a vector of non-linear functions of $\mathbf{x}$ and time, which corresponds to the different rate expressions of formation or disappearance of the components, and $\mathbf{w}$ is the vector of the system random disturbances, which represents the degree of uncertainty of the kinetic model (process noise).

In this work, a single measurement, the concentration of glucose, was used for the Kalman filter. At a given time t_k, this measurement, $\mathbf{y}$, is related to the state vector by the following model:

$$\mathbf{y}(t_k) = \mathbf{h}(\mathbf{x}(t_k), t_k) + \mathbf{v}(t_k), \; k = 1, 2, \ldots \qquad (2)$$

where $\mathbf{h}$ is a vector of nonlinear functions of $\mathbf{x}$ and t, and $\mathbf{v}$ is the random error vector accounting for the uncertainties in the measurements. In our case, the vectors **y**, **h** and **v** are reduced to scalars (one-dimension vector) because only one variable is measured.

As schematically described in Fig. 3, the EKF proceeds in two steps: a prediction step, which take place between measurements times, and a correction step which only occurs at the times of measurement. The predicted state vector, $\tilde{x}(t_k)$, is given by integrating the system model between the previous and the new measurement times t_{k-1} and t_k, according to

$$\frac{d\tilde{x}}{dt} = f(\tilde{x}(t), t) \qquad (3)$$

The predicted state vector is then corrected by a term proportional to the difference between the *measured and predicted glucose concentration:

$$\tilde{x}(t_k) = \tilde{x}(t_k) + K_k(y(t_k) - h(\tilde{x}(t_k), t_k)),\ k = 1, 2, \ldots \qquad (4)$$

where the matrix K_k is the Kalman gain which is optimized at each measurement time.

The algorithm was implemented on an Apollo 9000 Model 720 GRX Hewlett-Packard workstation using C programming language. Glucose concentration in the culture medium was measured off-line at 10 hours intervals and the results transferred to the computer.

Results and discussion

Kinetic model development in batch cultures

For the 6H2 hybridoma cells cultures in a batch reactor, the measured concentrations of glucose, glutamine, living and dead cells, lactate, ammonia and monoclonal antibodies are presented in Fig. 4. A maximal cell density of 1.5×10^6 cells.ml^{-1} was reached after 80 hours of culture, corresponding to a total depletion of glutamine. Ammonia and lactate, which reached 8 mM and 5 mM concentrations respectively, may also contribute to the inhibition of growth rate and stimulation of cellular death.

The rate expression and mass balance equations described in Figs. 1 and 2 were used to model the hybridoma kinetics in the batch culture. Some of the model parameters were estimated directly from the experimental data (Goergen *et al.*, 1994). Others were evaluated with a genetic algorithm to obtain the best fit between the model and the experimental data (Bicking *et al.*, 1994). With these values of parameters (Table 1), except for some discrepancies for the utilisation of glutamine, the model yielded a satisfactory simulation of the hybridoma growth, death and metabolism.

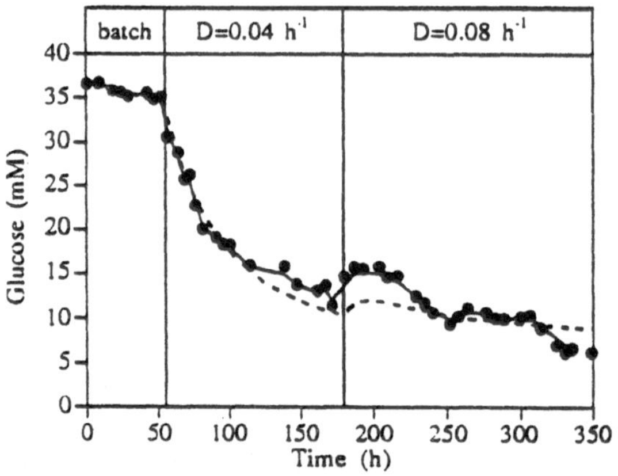

Fig. 5. Time variation of the glucose concentration in a continuous perfusion culture of 6H2 hybridoma cells under controlled pH and dissolved oxygen: (•) represent the experimental measurements, the dotted curve (- - -) the concentrations predicted by the kinetic model and the continuous curve (——) the concentrations estimated by the Extended Kalman Filter.

Medium composition estimation during a perfused culture

The hybridoma were cultured in a continuous perfusion bioreactor in which the medium is recirculated through an external microfiltration membrane to recycle the cells. The measured concentrations of nutrients, living and dead cells, metabolites and proteins inside the reactor during the 350 hours of culture are shown in Figure 6. After an initial batch phase, the perfusion culture was first operated for 130 hours at a dilution rate of 0.04 h^{-1}. During this period, the cell density increased to 5×10^6 cells.ml^{-1} and the antibody concentration to 70 mg.l^{-1}. In a latter phase when the dilution rate was set at 0.08 h^{-1}, the cell and antibody concentrations further increased to 11×10^6 cells.ml^{-1} and 120 mg.l^{-1}. This viable cell density is in the range of previously reported perfusion cultures (Al-Rubeai *et al.*, 1992; Büntemeyer *et al.*, 1992; Goebel *et al.*, 1990; Shintani *et al.*, 1991; Takazawa & Tokashiki, 1989).

The hybridoma perfusion culture was then used to test the capacity of the Kalman filter estimator to predict the medium composition with the model previously validated on the batch culture. In this case, a single component was measured, namely glucose.

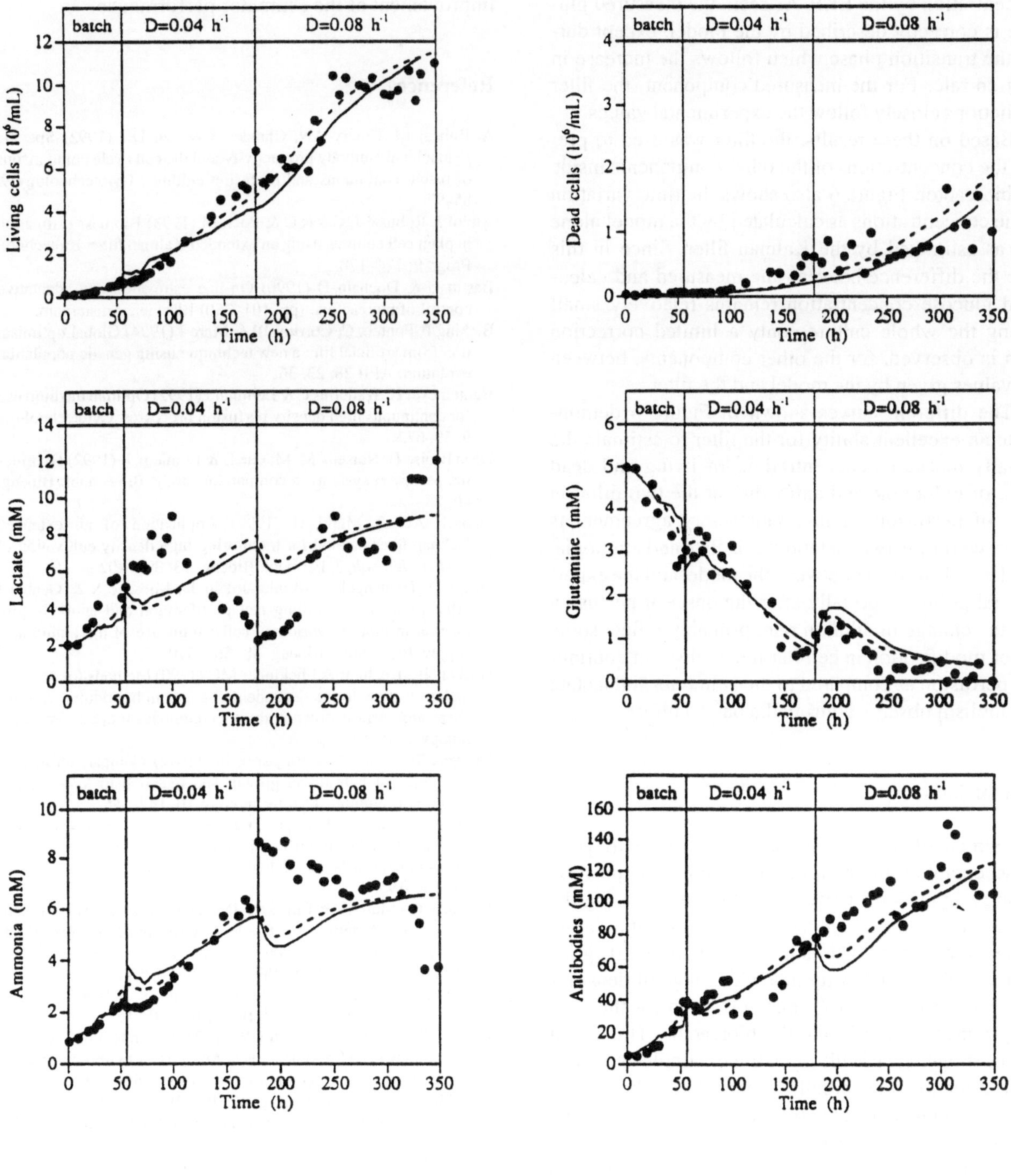

Fig. 6. Time variation of the concentrations of living cells, dead cells, lactate, glutamine, ammonia and antibodies inside the bioreactor during a continuous perfusion culture under controlled pH and dissolved oxygen: (•) represent the experimental measurements, the dotted curve (- - -) the concentrations predicted by the kinetic model and the continuous curve (——) the concentrations estimated by the Extended Kalman Filter.

Figure 5 shows, in addition to the experimental points, the glucose concentration predicted by the model as well as the continuous estimation of the glucose concentration by the EKF. As seen, the measured glucose is correctly described by the model, except during the transition phase which follows the increase in dilution rate. For the measured component, the filter predictions closely follow the experimental values.

Based on these results, the filter was used to predict the concentrations of the other components inside the bioreactor. Figure 6 also shows the time variation of the concentrations as calculated by the model alone and as estimated by the Kalman filter. Since in this case the difference between the measured and calculated glucose concentration remains relatively small during the whole culture, only a limited correction term is observed, for the other components, between the values given by the model and the filter.

The different curves shown in Figure 6 demonstrate an excellent ability for the filter to estimate the actually measured concentrations of living and dead cells, of glutamine and antibodies at the two dilution rates of perfusion. A less satisfactory agreement is observed for the two metabolites lactate and ammonia. The large differences between the model and the experimental points, especially after the onset of perfusion and the change in dilution rate, probably reflect some major modification in cellular metabolism in continuous perfusion as compared to the ammonia and lactate metabolism observed during the batch culture.

Conclusions

A Kalman filter type estimator was found to yield an acceptable estimation of the medium composition during a hybridoma perfusion culture when using the measurement of a single medium component in outlet stream, the glucose. The estimation is particularly good for the viable cell density in a wide range of densities between 1 and 11×10^6 cells.m l^{-1} and for the antibody concentration inside the bioreactor. This result has been achieved with a relatively simple model of hybridoma kinetics initially validated on a batch culture and found applicable for the continuous perfusion culture in a wide range of cell densities.

Significant discrepancies, however, remain between the predicted and measured concentrations of lactate and ammonia. This is probably indicative of important changes in cellular metabolism when cells are cultured in a continuous mode for extended periods. Thus, a better quantitative description of the production of lactate and ammonia by hybridoma in continuous perfusion cultures has to be achieved for a further improvement of the estimator performance.

References

Al-Rubeai M, Emery AN, Chalder S & Jan DC (1992) Specific monoclonal antibody productivity and the cell cycle comparisons of batch, continuous and perfusion cultures. Cytotechnology 9: 85–97.

Albiol J, Robusté J, Casas C & Poch M (1993) Biomass estimation in plant cell cultures using an extended Kalman filter. Biotechnol. Progr. 9: 174–178.

Bastin G & Dochain D (1990) On-line estimation and adaptative control of bioreactors. (pp. 101–250) Elsevier, Amsterdam.

Bicking F, Fonteix C, Corriou JP & Marc I (1994) Global optimisation from artificial life: a new technique using genetic population evolution. APII 28: 23–36.

Büntemeyer H, Wallerius C & Lehman J (1992) Optimal medium use for continuous high density perfusion processes. Cytotechnology 9: 59–67.

De la Broise D, Noiseux M, Massie B & Lemieux R (1992) Hybridoma perfusion systems: a comparison study. Biotechnol. Bioeng. 40: 25–32.

Dubach AC & Märkl H (1992) Application of an extended Kalman filter method for monitoring high density cultivation of *Escherichia coli*. J. Ferment. Bioeng. 73: 396–402.

Duval D, Demangel C, Munier-Jolain K, Miossec S & Geahel I (1991) Factors controlling cell proliferation and antibody production in mouse hybridoma cells: influence of the amino acid supply. Biotechnol. Bioeng. 38: 561–570.

Goebel NK, Kuehn R & Flickinger MC (1990) Methods for determination of growth-rate-dependent changes in hybridoma volume, shape and surface structure during continuous recycle. Cytotechnology 4: 45–57.

Goergen JL, Marc A & Engasser JM (1992) Comparison of specific rates of hybridoma growth and metabolism in batch and continuous cultures. Cytotechnology 10: 147–155.

Goergen JL, Marc A & Engasser JM (1993) Determination of cell lysis and death kinetics in continuous hybridoma cultures from the measurement of lactate dehydrogenase release. Cytotechnology 11: 189–195.

Goergen JL, Marc A & Engasser JM (1994) Biochemistry of cells in culture: Modeling. In: Doyle A, Griffiths JB & Newell DG (eds.) Cell and Tissue Culture: Laboratory procedures. Module 8B:2 (pp. 2.1-2.19) John Wiley and sons, Chichester.

Graf H and Schügerl K (1991) Some aspects of hybridoma cell cultivation. Appl. Microbiol. Biotechnol. 35: 165–175.

Hiller GW, Clak DS & Blanch HW (1993) Cell retention – Chemostat studies of hybridoma cells – Analysis of hybridoma growth and metabolism in continuous suspension culture on serum-free medium. Biotechnol. Bioeng. 42: 185–195.

Kalman RE (1960) A new approach to linear filtering and prediction problems. Trans. ASME: J. Basic Eng. 82: 33–45.

Kalman RE & Bucy RS (1961) New results in linear filtering and prediction theory. Trans. ASMEL: J. Basic Eng. 83: 95–108.

Miller KS & Leskiw DM (1987) An introduction to Kalman filtering with applications. (pp. 40–43) Robert E. Krieger Publishing Company, Malabar (Florida).

Omasa T, Ishimoto M, Higashiyama KI, Shioya S & Suga KI (1992) The enhancement of specific antibody production rate in glucose and glutamine controlled fed-batch culture. Cytotechnology 8: 75–84.

Phillips PJ, Marquis CP, Barford JP & Harbour C (1991) An analysis of some batch and continuous kinetic data of specific monoclonal antibody production from hybridomas. Cytotechnology 6: 189–195.

Pinton H (1991) Cultures perfusées d'hybridomes en réacteur agité avec recyclage externe des cellules: mise au point, cinétiques et performances. Ph. D. Thesis, Institut National Polytechnique de Lorraine, France.

Piret JM & Cooney CL (1990) Mammalian cell and protein distributions in ultrafiltration hollow fiber bioreactors. Biotechnol. Bioeng. 36: 902–910.

Rolf G, Walz WF, Noé W & Konrad A (1992) Safety and economic aspects of continuous mammalian cell culture. J. of Biotechnol. 22: 51–68.

Schmid G, Wilke CR & Blanch HW (1992) Continuous hybridoma suspension cultures with and without cell retention: kinetics of growth metabolism and product formation. J. Biotechnol. 22: 31–40.

Shitani Y, Kohno YI, Sawada H & Hitano K (1991) Comparison of culture methods for human-human hybridomas secreting anti-HBsAg human monoclonal antibodies. Cytotechnology 6: 197–208.

Stephanopoulos G & San KY (1984) Studies on on-line bioreactor identification. Biotechnol. Bioeng. 26: 1176–1188.

Takazawa Y & Tokashiki M (1989) High cell density perfusion culture of mouse-human hybridomas, Appl. Microbiol. Biotechnol. 32: 280–284.

Valero, F, Lafuente J, Poch M & Solà C (1990) Biomass estimation using on-line glucose monitoring by flow injection analaysis. Appl. Biochem. Biotechnol. 24: 591–601.

Van der Heijden RTJM, Hellinga C, Luyben KCAM & Honderd G (1989) State estimators (observers) for the on-line estimation of non-measurable process variables. Trends Biotechnol. 7: 205–209.

Address for offprints: F. Pelletier, Institut National Polytechnique de Lorraine, Laboratoire des Sciences du Génie Chimique, CNRS, 2, avenue de la Forêt de Haye, BP 172, F-54500 Vandoeuvre-lès-Nancy Cedex, France.

Cytotechnology **15:** 301–309, 1994.

Design of a bubble-swarm bioreactor for animal cell culture

F. Gudermann, D. Lütkemeyer and J. Lehmann
Institute of Cell Culture Technology, University of Bielefeld, P.O. Box 100131, 33501 Bielefeld, Germany

Key words: Aeration, stirred bioreactor, bubble-swarm, hybridoma, oxygen transfer

Abstract

A stationary bubble-swarm has been used to aerate a mammalian cell culture bioreactor with an extremely low gas flow rate. Prolonging the residence time of the gas bubbles within the medium improved the efficiency of the gas transfer into the liquid phase and suppressed foam formation. An appropriate field of speed gradients prevented the bubbles from rising to the surface. This aeration method achieves an almost 90% transfer of oxygen supplied by the bubbles. Consequently, it is able to supply cells with oxygen even at high cell densities, while sparging with a gas flow of only $0.22 \cdot 10^{-3}$ – $1.45 \cdot 10^{-3}$ vvm (30–200 ml/h).

The reactor design, the oxygen transfer rates and the high efficiency of the system are presented. Two repeated batch cultures of a rat-mouse hybridoma cell line are compared with a surface-aerated spinner culture. The used cell culture medium was serum-free, either with or without BSA and did not contain surfactants or other cell protecting agents. One batch is discussed in detail for oxygen supply, amino acid consumption and specific antibody production.

Introduction

The aeration of mammalian cell cultures by conventional direct gas sparging presents several problems. First there is the low efficiency of existing aeration units, which require high gas volume flow rates in order to transfer the appropriate quantity of oxygen into the culture broth. This leads to a high gas flow rate through the liquid surface and thus to foam formation. So the use of pure oxygen is frequently necessary to aerate high density cell cultures sufficiently. Increasing the efficiency of the aeration unit will reduce the net gas flows, therefore costs can be lowered.

Another serious problem is excessive foam formation, resulting in damage to the cultured cells. The negative effects of an increasing gas flow rate on animal cell cultures were first discussed by Handa *et al.*, 1985. The degree of cell damage has been linked to bubble size by Oh *et al.*, 1992 and Jöbes *et al.*, 1991. Zhang *et al.*, 1992, localized the lethal region in a bubble column reactor. They showed like Handa-Corrigan *et al.*, 1989, that cell damage occurs mainly at the liquid surface and in the foam layer. The main mechanisms causing this damage are the rapid bubble bursts with relatively high energy release, the shearing of cells in draining liquid films between bubbles in the foam and the physical loss of cells.

Surface active substances used for foam regulation and cell protection could have the disadvantage of a reduction in mass transfer (Prins and van 't Riet, 1987). Not only the transfer of gases in and out of the culture broth may be adversely affected but also the uptake of water soluble nutrients from the medium by the cells. Zhang *et al.*, 1992 stated that adsorption of antifoam agent, e.g. a hydrophobic water-insoluble silicone emulsion, occurs at the hydrophobic sites of cell surfaces and may cause the formation of a layer on the cell membrane. At sufficiently high concentrations this layer could resist the transport of water soluble nutrients into the cell.

Another important point is the product separation from the culture broth at the end or during the cultivation. To simplify the down-stream processing a defined and minimized medium composition is desirable (Jäger *et al.*, 1991). The absence of FCS, antifoam or Pluronic F-68 simplifies the separation procedure and therefore reduces costs.

One can summarize that neither bubble formation nor bubble rise but bursting of bubbles at the surface is the major mechanism of cell damage in a bubble aerated bioreactor. Surfactants should not be used, so an alternative foam preventing concept has to be designed with regard to the special demands of animal cells.

The basic idea of this concept is the improvement of the oxygen transfer due to prolongation of the residence time of gas bubbles within the medium. This would lead to a drastic decrease of the gas flow rate into the liquid and a better utilization of the bubbles' oxygen content. Hence, if the overwhelming amount of oxygen, brought as gas bubbles into the medium, can be dissolved, there will be only a neglectable foam formation.

An appropriate, downward directed liquid stream had to be generated e.g. in the central part of the culture vessel, to prevent the bubbles from rising to the liquid surface. The velocity field had to prevent bubbles within a certain range of diameter from reaching the surface, so a bubble-swarm will be formed at a defined region in the bioreactor. Therefore, the downward directed liquid velocity has to be high – up to 25 cm/s to prevent the largest bubbles from rising – in the upper and as low as possible in the lower part of the stream field – to keep small bubbles. The high liquid velocity probably affects growth and production rates of a cell culture adversely due to hydrodynamical forces. As it is well known that a culture medium with low protein content increases shear sensitivity of animal cells (Ozturk *et al.*, 1991; Hülscher *et al.*, 1988) two cultivations of a hybridoma cell line were carried out with different medium formulations. A serum- and BSA-free medium was first used and a BSA (bovine serum albumin) containing serumfree medium afterwards. The results of these two cultivations in the bubble-swarm aerated bioreactor were compared to those, obtained with a surface aerated spinner culture, which guarantees almost ideal cultivation conditions, especially without critical shear forces. Also the capacity of the bubble-swarm aeration and the efficiency of the system are discussed and compared with conventional bubble columns.

Materials and methods

Cells

The cell line used in this study is a rat-mouse hybridoma (HB 58, ATCC). This cell line secretes a rat monoclonal antibody IgG_1 type specific for mouse antibody κ light chains. The cells were precultivated in T-flasks, in spinner (Techne, Cambridge, UK) and in a membrane aerated Super Spinner with a working volume of 1 liter (Lehmann *et al.*, 1992; Heidemann *et al.*, 1994) stirred at 38 rpm in a standard CO_2 incubator (Heraeus, Hanau, FRG) at 5% CO_2 and 37 °C.

Medium

Two medium formulations were used: Medium A is a BSA- and serum-free medium (Jäger *et al.*, 1988). It consists of a 1:1 mixture of DMEM and Hams's F12 (Gibco, Eggenstein, FRG) and is supplemented with 50 μM ethanolamine, 2 mM sodium pyruvate, 10 mg/l human transferrin (Fe saturated) and 10 mg/l bovine insulin. Except for an addition of 500 mg/l BSA medium B is of the same formulation as medium A.

Analytical method

Two samples per day were taken from the bioreactor vessel and analyzed as follows: the cell number was counted microscopically by trypan blue exclusion. Glucose and lactate were analyzed using the YSI 2700 analyzer (Yellow Springs Instruments, Ohio, USA).

Ammonium was determined using an ammonia selective electrode with microprocessor pH/Ion meter pMX 2000 (WTW, Weilheim, FRG). The free amino acids contained in the sample were analyzed by an automated reversed phase HPLC system (Kontron, Neufahrn, FRG) with pre-column derivatization using the OPA method (Büntemeyer *et al.*, 1991). Antibody concentration in the supernatant was analyzed using a kinetic sandwich ELISA method and the EL 311 autoreader (Tecnomara, Fernwald, FRG).

Bioreactor concept

The used bioreactor vessel is a Biostat MD glass fermenter (B. Braun Biotech International GmbH, Melsungen, FRG) with a total volume of 3 liters that was modified as shown in Fig. 1.

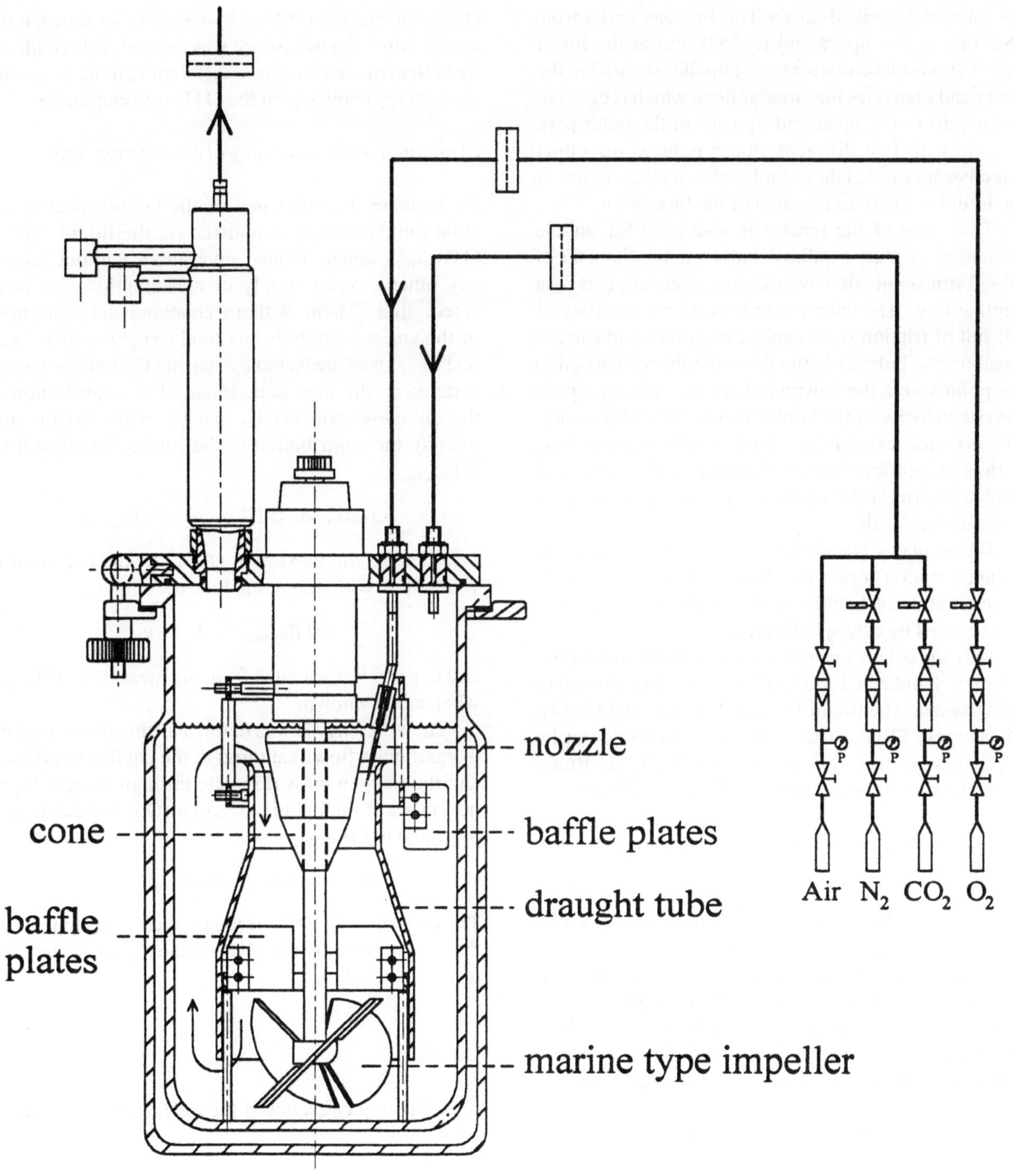

Fig. 1. Schematic diagram of the Bubble-Swarm Bioreactor. Oxygen is sparged through the nozzle into the draught tube. The shape of the draught tube and the cone mounted on the stirrer shaft create a velocity field that prevents the bubble-swarm from rising to the surface.

In the center of the aeration unit the draught tube with its special shape is installed to build up the liquid stream described earlier. The diameter varies from 55.5 mm at the upper end to 85.0 mm at the lower end. A three-blade marine type impeller situated in the lower end generates the circular flow, which is directed downward in the inner and upward in the outer part. A cone with two different slopes reduces the upper effective aperture of the draught tube and thus increases the liquid velocity in this area of the bioreactor.

This zone of the reactor is also used for bubble formation. A glass capillary with an outer diameter of 30–50 μm points downwards and is surrounded by a laminar flow. The bubbles are torn off the capillary as a result of friction force against buoyancy and surface tension. The bubbles in the draught tube are carried to the point where the downward stream velocity equals the rise velocity of the bubble caused by its buoyancy. The distance traveled by a bubble in the draught tube is thus proportional to the diameter of the bubble. A bubble-swarm builds up while sparging air or oxygen through the nozzle.

Baffle plates are placed at two different zones to reduce rotation inside the draught tube. This avoids concentration of bubbles in the center of the stream-field caused by centripetal force.

pO_2 and pH are measured with conventional probes (Ingold, Steinbach, FRG) and controlled by a standard DCU system (B. Braun Biotech International GmbH, Melsungen, FRG). Data acquisition is carried out by a process control system (Micro MFCS, B. Braun Biotech International GmbH, Melsungen, FRG).

Measurement of the volume flow rate through the nozzle

To determine the volume flow rate through the nozzle tip it was brought into air saturated water. Air was sparged through the capillary at a certain pressure and the bubbles built by the nozzle were caught up by a water filled glass cylinder placed above the capillary. The oxygen mass flow rate was calculated from the time needed to displace a defined volume of water.

Measurement of the oxygen transfer rate OTR

The dynamic method was used to evaluate the oxygen transfer rates. The bioreactor contained 2.3 liters of sterile basal medium at a temperature of 37 °C. The pO_2 was continuously measured and the data was automatically recorded in one minute intervals by the Micro-MFCS. First, the oxygen was displaced using nitrogen. The dynamical measurement with the calibrated electrode was then started by sparging air through the nozzle while the headspace was aerated with 10 l/h air. The effect of air sparging at different volume flow rates through the capillary on the OTR was examined.

Evaluation of the efficiency of an aeration unit

To evaluate the efficiency of the bubble-swarm aeration unit, the oxygen transfer via the liquid surface $OTR_{surface}$ under defined conditions without having any other oxygen supply or consumption was measured, first. Then, without changing the conditions of the surface aeration, the total oxygen transfer rate (OTR_{total}) was measured by starting the bubble-swarm aeration in the next experiment. The contribution of the bubble-swarm OTR_{bs} can be evaluated by subtracting the contribution of the surface aeration from OTR_{total}

$$OTR_{bs} = OTR_{total} - OTR_{surface}$$

One obtains the maximum OTR_{bs}, if the calculations were made with dO_2 equal to zero. In this case

$$OTR_{bsmax} = k_L a \cdot c_L$$

and $k_L a$ can be evaluated from the measured OTR_{bsmax} when $c_L{}^*$ is known.

The efficiency E was defined as the quotient of the oxygen mass flow rate through the capillary $m(O_{2cap})$ and the oxygen mass flow rate through the gas-liquid interfaces of the bubble-swarm $m(O_{2bs})$ which is proportional to OTR_{bsmax}.

$$E = m(O_{2cap})/m(O_{2bs}) \cdot 100$$

The efficiency is 100% if the oxygen sparged through the nozzle is totally dissolved in the liquid. In this case $m(O_{2cap}) = m(O_{2bs})$.

Results

Capacity and efficiency of the bubble-swarm aeration concept

It was possible to create a stationary bubble-swarm (Fig. 2) using the bioreactor geometry shown in Fig. 1. The contribution of the bubble-swarm to the oxygen transfer into the medium was examined and compared to a conventional bubble column. To evaluate

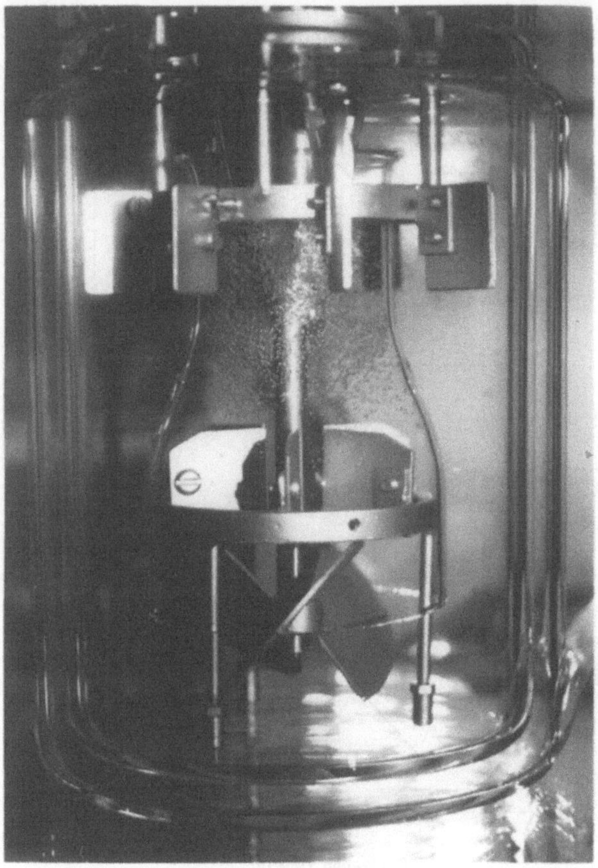

Fig. 2. Bubble-swarm formation inside the draught tube. Basal medium with 10 mg/l bovine insulin but without phenolred was used. The stirrer speed was set to 60 rpm.

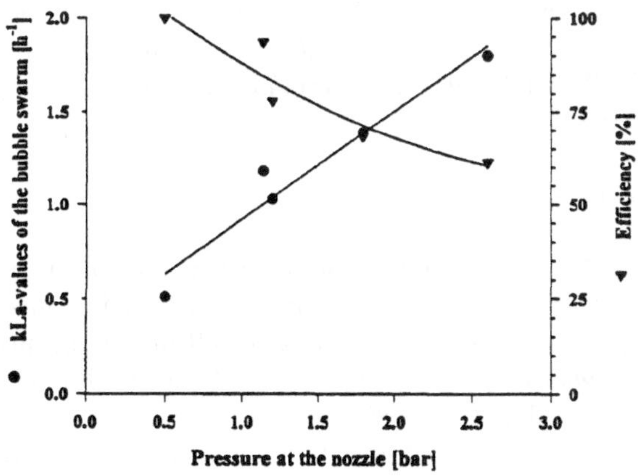

Fig. 3. The plot of the k_La-values of the bubble-swarm and the efficiency of aeration versus the pressure at the nozzle shows that increasing pressure results in increasing k_La-values.

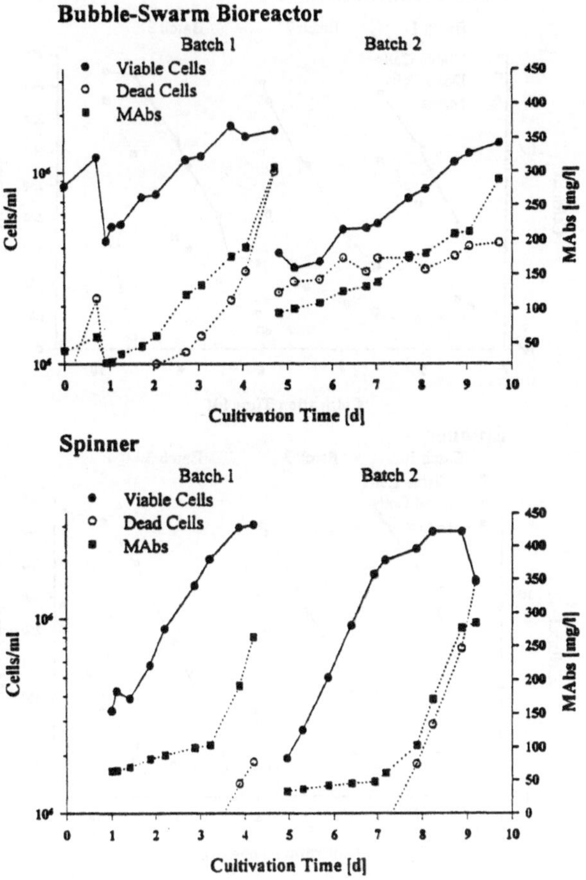

Fig. 4. Time courses of repeated batch hybridoma culivations in the Bubble-Swarm Bioreactor and a surface aerated spinner culture.

the efficiency of the system the maximum OTR of the bubble-swarm was measured by sparging air through the nozzle into sterile basal medium and aerating the headspace with 10 l/h air. The pressure at the capillary was varied from 0.5 to 2.6 bar resulting in volume flow rates through the nozzle of $0.23 \cdot 10^{-3}$ vvm to $1.26 \cdot 10^{-3}$ vvm (32 ml/h-174 ml/h), respectively while headspace aeration remained constant. The OTR_{bsmax} rises with increasing pressure from 110 μmol $O_2/(l \cdot h)$ to 388 μmol $O_2/(l \cdot h)$ (Fig. 3). The corresponding k_La-values of the bubble-swarm aeration were 0.5 h^{-1} up to 1.8 h^{-1}. Evaluating the quotient of the mass flow rate through the capillary and the mass flow rate through the gas-liquid interfaces of the bubble-swarm the efficiency can be represented. At low pressure the efficiency was greater than 90%. An increase in pressure results in decreasing efficiency, but at 2.6 bar still 61.4% of the oxygen carried by the air bubbles was transferred into the medium.

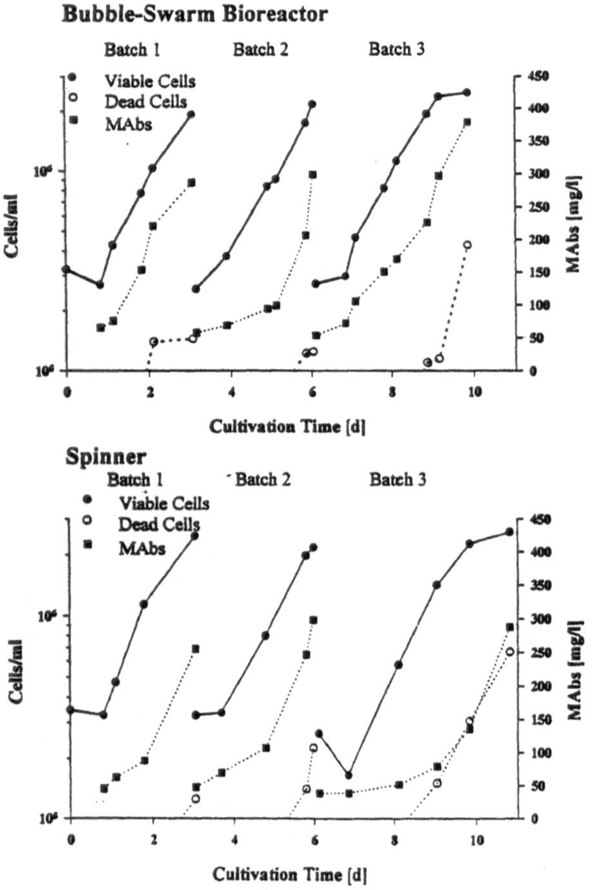

Fig. 5. Time courses of three repeated batch hybridoma cultivations in the Bubble-Swarm Bioreactor and the corresponding surface aerated spinner batch cultivations.

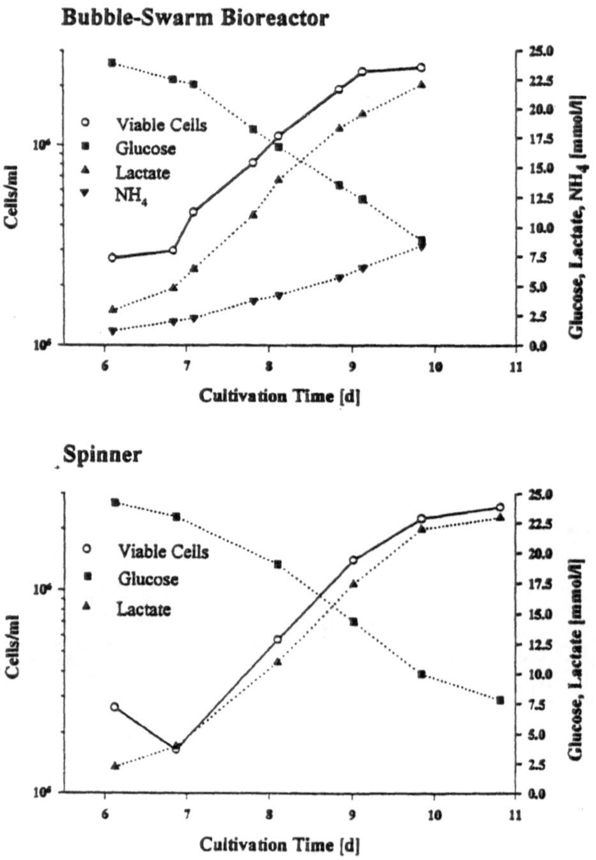

Fig. 6. Detailed consideration of the third batch of the cultivation in the Bubble Swarm Bioreactor and the corresponding control batch shown in Fig. 5. Glucose consumption and lactate production are of the same order in the Bubble-Swarm Bioreactor and the spinner culture.

Zhang *et al.*, 1992 determined the oxygen transfer rates in a bubble aerated animal cell bioreactor for different medium formulations and three different spargers. They sparged air with flow rates of $66.7 \cdot 10^{-3}$ vvm to $400.0 \cdot 10^{-3}$ vvm and measured k_La values ranging from about 2 h^{-1} (ring sparger, lowest air flow) to 75 h^{-1} (G-4 sinter sparger, highest air flow) in basal medium stirred with 60 rpm at 36.5 °C. The efficiency of these systems is less than 7% with the exception of the G-4 sinter sparger used with a volume flow rate of $66.7 \cdot 10^{-3}$ vvm. This configuration reached almost 15% of efficiency. The efficiency of the ring sparger, however, never exceeded 1%.

The oxygen consumption of different hybridoma cell lines varies between 130–330 μmol $O_2/(10^9$ cells·h) (Spier *et al.*, 1984). Measurement of the oxygen uptake rate of the utilized hybridoma yielded 216 μmol $O_2/(10^9$ cells·h). Consequently, a cell density of $1.8 \cdot 10^6$ cells/ml could be supplied by air sparging. Utilizing pure oxygen for a cell density of up to $8.6 \cdot 10^6$ cells/ml the oxygen supply can satisfy the oxygen demand.

Cultivation 1

To test the aeration unit and its effects on cell growth two batch cultivations were made. The BSA-free medium A which did not contain any cell protecting proteins was used first. The Bubble-Swarm Bioreactor was inoculated with 800 ml cell suspension from a membrane aerated Super-Spinner and an additional 400 ml of fresh medium. The cell density was $8.4 \cdot 10^5$ cells/ml. The culture was surface aerated and stirred at 60 rpm. After 18.5 hours 1100 ml of fresh medium was added. Sparger aeration was then started with the

Bubble-Swarm Bioreactor containing a total volume of 2.3 liters. This allowed the circulation of the fluid through the draught tube and the formation of the bubble-swarm.

Parallel to this a surface aerated spinner culture was inoculated with cells of the same Super-Spinner and the same medium that was used to inoculate the Bubble-Swarm Bioreactor. This spinner culture was used as reference for cell growth, antibody production and other metabolic parameters to those of the bubble-swarm aerated bioreactor culture.

The time courses of the first two batches in the Bubble-Swarm Bioreactor and the corresponding spinner culture are shown in Fig. 4. Cell growth was reduced in the Bubble-Swarm Bioreactor. The maximum specific growth rate μ_{max} was 0.75 d^{-1}, whereas 1.07 d^{-1} was observed in the spinner culture. Thus, one batch in the Bubble-Swarm Bioreactor took approximately one day longer than in the spinner culture. At the end of the batch the cell density in the spinner culture ($\sim 3 \cdot 10^6$ cells/ml) was two times higher than in the Bubble-Swarm Bioreactor. Also the viability was higher in the spinner culture. Only at the end of the logarithmic growth phase did the number of dead cells exceed $2 \cdot 10^5$ cells/ml, whereas this value was already reached in the logarithmic growth phase of the Bubble-Swarm Bioreactor culture. In spite of different antibody production rates of 108.8 mg/(10^9cells · d) for the spinner culture and 71.3 mg/(10^9 cells · d) for the Bubble-Swarm Bioreactor culture, the antibody concentration at the end of the batch was about 300 mg/l in both systems, due to different batch duration.

Cultivation 2

To examine the influence of higher protein concentration on cell viability medium B (containing 500 mg/l BSA) was utilized for the second cultivation in the Bubble-Swarm Bioreactor and the spinner culture.

Figure 5 compares three repeated batches of the Bubble-Swarm Bioreactor and three batches of the surface aerated spinner. Batch durations, final cell densities, viability and final antibody concentration were approximately the same in both systems.

A detailed consideration of the third batch of the Bubble-Swarm Bioreactor cultivation and the corresponding spinner batch is plotted in Fig. 6. Glucose consumption and lactate production in the Bubble-Swarm Bioreactor were of the same order as those in the spinner culture (Table 1).

The electrode readings of the controlled oxygen- and pH-values during the third batch in the Bubble-Swarm Bioreactor are plotted in Fig. 7. In the first phase mixed gas (air, nitrogen and CO,) was sparged through the nozzle with constant pressure and was also used for headspace aeration $72.5 \cdot 10^{-3}$ vvm (10 l/h). At day 8 of the cultivation pure oxygen was sparged through the nozzle to supply the cell culture. pH control, especially CO_2 removement was performed via the headspace (20 l/h air). From this point on oxygen supply and CO_2 exchange were separated. Unfortunately, it was not possible to control the pressure at the nozzle dependent on the actual pO_2-level with the standard DCU system. A certain pressure was chosen and kept constant for hours, so drops of the pO_2-level were caused by an altered oxygen demand of the culture due to cell growth. At the end of the batch it was difficult to choose a suitable pressure to supply the cell culture with the result of an alterating pO_2-level.

Discussion

The bubble-swarm aeration method was able to supply the cell culture with oxygen, but CO_2, had to be removed via the surface, because only a small percentage of the gas sparged into the medium reached the surface. The Bubble-Swarm Bioreactor cultivations showed that it was possible to control the pH-level up to a cell density of $2.5 \cdot 10^6$ cells/ml.

In contrast to the calculations above only $2 \cdot 10^6$ cells/ml can be supplied with oxygen at a pressure of 2.6 bar applied to a capillary that has already been used for three weeks. The reason for the reduced gas flow rates through the nozzle after a certain period of time might be attributed to attached cells at the capillary surface. At the end of the cultivation the used nozzle was overgrown with cells. Especially the fine tip of the capillary and parts of the aperture were coated.

During the OTR measurements and the two Bubble-Swarm Bioreactor cultivations significant foam formation did not occur. While sparging with $1.50 \cdot 10^{-3}$ vvm (207 ml/h) foam formation reached its maximum, but even then only a monolayer of small bubbles coated the surface. This shows that the addition of foam regulating surfactants was not necessary when using the bubble-swarm aeration unit.

Nevertheless, BSA has to be added to the serum-free medium to protect the cells against shear forces. Cell damage occurs in the Bubble-Swarm Bioreactor only while using medium A with a very low protein

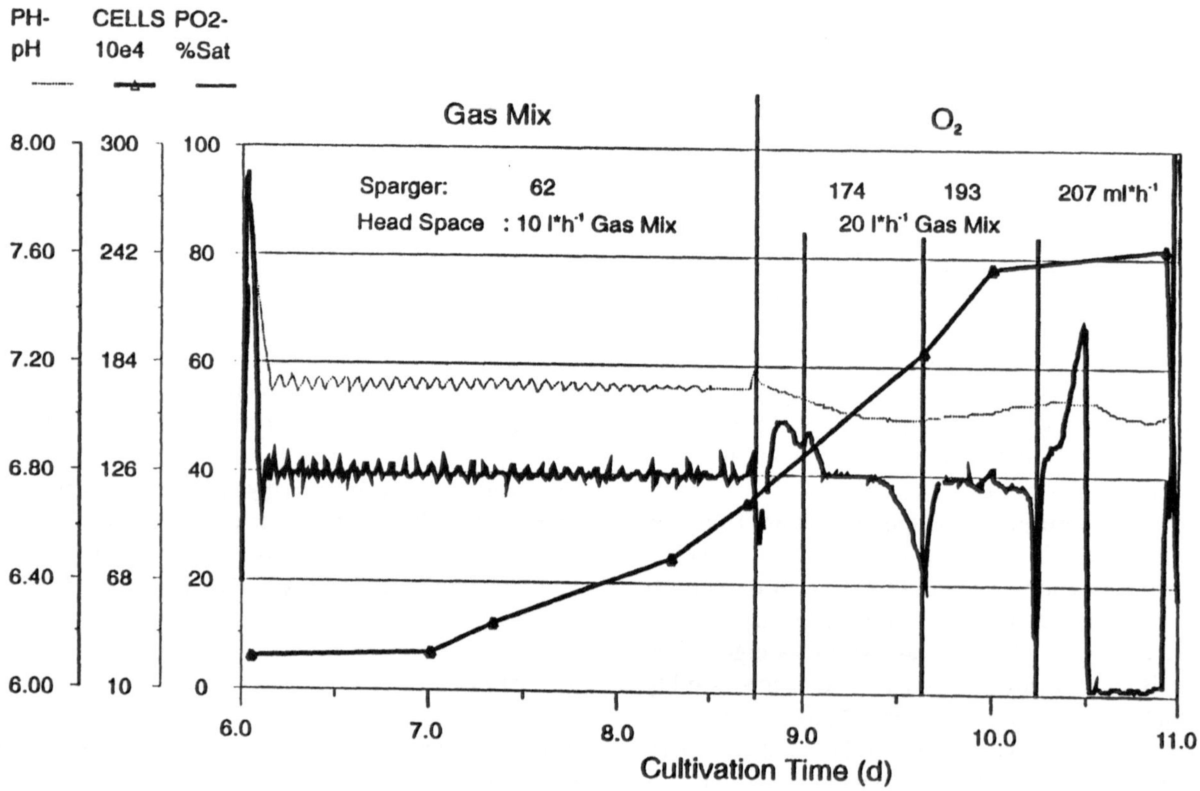

Fig. 7. Time plot of the cell density and controlled pO_2- and pH-values during the third batch of the Bubble-Swarm Bioreactor cultivation of Fig. 5.

Table 1. Comparison of specific growth rate, antibody production and metabolic rates in the third batch of the cultivation in the Bubble-Swarm Bioreactor and the corresponding spinner control batch shown in Fig. 5.

	μ_{max} $[d^{-1}]$	q_{MAbs} $[mg/(10^9 \cdot cell \cdot d)]$	$q_{Glucose}$ $[mmol/(10^9 \cdot cells \cdot d)]$	$q_{Lactate}$ $[mmol/(10^9 \cdot cells \cdot d)]$	$q_{Ammonia}$ $[mmol/(10^9 \cdot cells \cdot d)]$
Bubble-Swarm Bioreactor	0.87	139.4	5.78	7.16	2.69
Spinner culture	0.90	113.3	5.78	8.11	

content (cultivation 1). The fact that the addition of BSA did not affect the foam formation in the Bubble-Swarm Bioreactor cultivation, especially a foam reduction was not observed, showed that the higher growth rate of cultivation 2 cannot be related to an altered foam formation. This indicates that shear forces e.g. in the zone of high velocity which is located in the upper part of the draught tube were the reason of the reduced cell growth in cultivation 1.

Small bubbles with low rise velocities are the assumption of the formation of a bubble-swarm. The diameter of the bubbles build by a nozzle is proportional to the radius of its orifice. So capillaries with small outer diameters are required. A nozzle with a radius of 5 m built the smallest bubbles but showed a high tendency of clogging. In order to prevent clogging the orifice diameter in the final aeration unit is set to 30–50 μm. This also yields a suitable bubble size and gas flow rate. Both cultivations and the necessary calibrations were made with the same nozzle, sparging continuously for more than three weeks without having to be cleaned.

Conclusion

The aeration of a bioreactor with a stationary bubble-swarm satisfied the special demands of mammalian cell cultures. A rat-mouse hybridoma cell line was successfully cultivated in serum-free medium. The comparison of the Bubble-Swarm Bioreactor cultivated cells with those, grown in a spinner bottle showed no changes of physiological parameters when BSA containing serum-free medium was used. The high efficiency of the system allowed sparging with low gas flow rates, thus only a small amount of oxygen was needed to supply the cell culture. It was also possible to reduce the gas flow rate from the liquid into the headspace, because bubble disengagement at the surface was minimized, due to high efficiency. This led to a suppression of foam formation without using additional surface active substances for foam regulation.

Further work will concentrate on an alternative nozzle material that prevents cell attachment, optimization of hydrodynamical parameters (impeller or draught tube geometry) and the development of a pressure control system.

References

Büntemeyer H, Lütkemeyer D and Lehmann J (1991) Optimization of serum-free fermentation processes for antibody production. Cytotechnology 5: 57–67.

Handa A, Emery AN and Spier RE (1985) On the evaluation of gas-liquid interfacial effects on hybridoma viability in bubble column bioreactors. Develop. Biol. Standard. 66: 241–253.

Handa-Corrigan A, Emery AN and Spier RE (1989) Effect of gas-liquid interfaces on the growth of suspended mammalian cells: mechanisms of cell damage by bubbles. Enzyme Microb. Technol. 11: 230–235.

Heidemann R, Riese U, Lütkemeyer D, Büntemeyer H and Lehmann J (1994) The Super-Spinner: a low cost animal cell culture bioreactor for the CO_2, incubator. Cytotechnology 14: 1–9.

Hülscher M and Onken U (1988) Influence of bovine serum albumin on the growth of hybridoma cells in airlift loop reactors using serum-free medium. Biotechnology Letters 10: 689–694.

Jäger V, Lehmann J and Friedl P (1988) Serum-free growth medium for the cultivation of a wide spectrum of mammalian cells in stirred bioreactors. Cytotechnology 1: 319–329.

Jäger V (1991) Serum-free media suitable for upstream and downstream processing. In: Spier RE, Griffiths JB, Meignier B (eds.) Production of Biologicals from Animal Cells in Culture. Research, Development and Achievements. Butterworths, Kent, UK.

Jöbes I, Martens D and Tramper J (1991) Lethal events during gas sparging in animal cell culture. Biotechnol. Bioeng. 37: 484–490.

Lehmann J, Heidemann R, Riese U, Lütkemeyer D and Büntemeyer H (1992) Der Superspinner. Ein 'Brutschrank-Fermenter' für die Massenkultur tierischer Zellen. Bioengineering 5+6: 36–38.

Oh SKW, Nienow AW, Al-Rubeai M and Emery AN (1992) Further studies of the culture of mouse hybridomas in an agitated bioreactor with and without continuous sparging. J. Biotechnol. 22: 245–270.

Ozturk SS and Palsson BO (1991) Examination of serum and bovine serum albumin as shear protective agents in agitated cultures of hybridoma cells. J. Biotechnol. 18: 13–28.

Prins A and van 't Riet K (1987) Proteins and surface effects in fermentation: foam, antifoam and mass transfer. Tibtech 5: 296–301.

Spier RE, Griffiths JB (1984) An examination of the data and concepts germane to the oxygenation of cultured animal cells. Dev. Biol. Standard. 55: 81–92.

Zhang S, Handa-Corrigan A and Spier RE (1992) Foaming and media surfactant effects on the cultivation of animal cells in stirred and sparged bioreactors. J. Biotechnol. 25: 289–306.

Zhang S, Handa-Corrigan A and Spier RE (1992) Oxygen transfer properties of bubbles in animal cell culture. Biotechnol. Bioeng. 40: 252–259.

Address for offprints: F. Gudermann, University of Bielefeld, Technical Faculty, Institute of Cell Culture Technology, P.O. Box 10 01 31, 33501 Bielefeld, Germany.

Cytotechnology **15:** 311–320, 1994.

Cells and bubbles in sparged bioreactors

Jeffrey J. Chalmers
Department of Chemical Engineering, Ohio State University, 140 W 19th Ave., Columbus, OH 43210 U.S.A.

Key words: Animal cells, bioreactors, bubbles, damage, rupture

Abstract

Ever since animal cells have been grown *in-vitro*, various techniques have been used to supply the cells with oxygen. The most simple and commonly used 'large-scale' technique to provide oxygen is through the introduction of gas bubbles. However, almost since the beginning of *in-vitro* cell culture, empirical observations have indicated that bubbles can be detrimental to the cells. This review will discuss the background of the problem, review the relevant research on the topic, attempt to provide a coherent summary of what we know from all of this research, and finally outline what still needs to be investigated. Specific topics to be covered include: experimental correlations of cell damage with bubbles, cell attachment to bubbles, the hydrodynamics of bubble rupture, bioreactor studies, visualization studies, and computer simulations and qualification of cell death as a result of bubble rupture.

Introduction

During the last decade, the number of different types of proteins produced from animal and insect cells has greatly increased. While this growth has been rapid, considerable confusion exists with respect to whether these types of cells are so 'shear sensitive' that they can not be grown in bioreactors. Without the introduction of protective additives this 'shear sensitivity' can be so severe that substantial cell death occurs and negative growth rates in bioreactors can be observed.

Animal cells, derived from tissue, have been cultured in suspension culture for over 35 years (Swim and Parker, 1960; Runyan and Gyer, 1963). These first researchers recognized that the introduction of bubbles into cell culture resulted in cell damage. These researchers also recognized that the introduction of 'protective additives' such as Pluronic F-68 could prevent cell damage in agitated and bubble aerated bioreactors. Kilburn and Webb, in 1968, suggested that this protection was the result of Pluronic F-68 creating a 'highly condensed interfacial structure of adsorbed molecules' at the cell-medium interface.

While this early work suggested that 'shear sensitivity' was the result of cells interacting with bubbles, a large amount of confusion exists to this day as to the actual mechanism by which suspended cells are damaged. All of the suggested causes of suspended cell damage can be classified into five possible mechanisms: one that involves purely hydrodynamic forces acting on the cell due to the mixing in the vessel, and the other four that involve interactions with bubbles. These cell-bubble damage mechanisms are: (1) cell interactions with bubble generation at the sparger, (2) cell interactions with rising bubbles, (3) cell interactions with bubbles coalescing and breaking up, (4) and cell interactions with bubbles at the air-medium interface.

This review is limited to cell interactions with bubbles and will discuss the results of research from several laboratories which used several different techniques to study cell damage. Taken together, the results firmly indicate the primary mechanisms of suspended cell damage. The reviewed research will be grouped into: bioreactor studies, correlational studies, regions for cell damage as a result of aeration, visualization studies, and studies on the rupture of bubbles at gas-liquid interfaces with respect to cells.

Bioreactor studies

In the early 1980's the amount of research and the number of publications on the large scale cultivation of suspended cells increased greatly. However, much of this early work centered more on the rpm of the impeller with respect to the sensitivity of cells to damage and not on sparging of the cell suspension with gas (Lee *et al.*, 1988; Backer *et al.*, 1988; Dodge and Hu, 1986; Tramper *et al.*, 1986). This was probably the result of a number of observations and a lack of understanding of the hydrodynamics in agitated bioreactors. One such observation was that anchorage-dependent cells can be damaged by low level shear generated in the impeller regions of bioreactors (Sinskey *et al.*, 1981; Hu *et al.*, 1985; Croughan *et al.*, 1987; Cherry and Papoutsakis 1986).

In contrast, two research groups demonstrated that suspended cells can withstand much higher impeller rpm's, with the associated higher hydrodynamic forces, than is commonly used. Kunas and Papoutsakis (1990a) and Oh *et al.* (1989) reported that hybridoma cells can withstand impeller rpm's from 100 to 450 in one to two liter bioreactors with no detectable detrimental effect on the cells. Typical rpm's for these systems are on the order of 50 to 100. Both authors reported that only when air was sparged into the vessel did cell damage occur.

To further determine the effect of bubbles on cells, Kunas and Papoutsakis (1990b) developed a bioreactor that removed the air-liquid interface from the top of the vessel. This system completely prevented the formation of a central vortex around the impeller shaft. This removal of the central vortex correspondingly prevented bubble entrainment except bubbles of the 50 to 400 μm size. These small bubbles followed the flowing liquid and did not rise to the top and rupture. In this mode, the bioreactor could be operated at impeller rates up to 700 rpm without cell damage. These results lead Kunas and Papoutsakis (1990a) to state, 'Only when entrained bubbles interact with a freely moving gas-liquid interface, such as what exists between the culture medium and the gas headspace, does significant cell damage occur.'

These results firmly indicate that suspended cell damage, from purely hydrodynamic effects (mixing of the impeller), does not occur under 'normal' agitated, bubble aerated, bioreactor operations. However, it does not rule out mixing in the impeller when gas interfaces are present.

Correlational studies

In 1988, Tramper *et al.* published a paper in which they argued, from a mathematical approach based on experimental evidence, that cell damage can be correlated with a 'hypothetical killing volume' associated with each bubble in a bioreactor. This volume can be determined from the following equation:

$$K_d = 24FX/(\Pi^2 d_b^3 D^2 H) \qquad (1)$$

where

X = *hypothetical killing volume*
k_d = first order death rate constant
F = air flow rate into vessel
d_b = air bubble diameter
D = column diameter
H = height of column.

This relationship was verified in a variety of experimental conditions: different height to diameter ratios, bubble diameters, and cell types (Jobses *et al.*, 1991; Martens *et al.*, 1992). An important implication of this relationship is that this 'hypothetical killing volume' is independent of the height and diameter of the bubble column and it is also independent of the number of air bubbles; i.e., it only depends on the size and/or volume of the bubble. Of even greater significance of this work is the correlation of cell damage with a specific volume asociated with each bubble.

While these authors do not suggest what this 'hypothetical killing volume' associated with each bubble is or where on the bubble or in the bubble's path it is located, it has important implications in the relationship between bubbles and cell damage.

A second correlational model was proposed by Wang *et al.* (1994). In an argument similar to Tramper *et al.* (1988) it is suggested that the local, specific cell death rate is related to the local specific bubble interfacial area and the local bubble breakup/bursting frequency:

$$p = (k_2 s/K)a \qquad (2)$$

where

p = local specific cell death rate (h^{-1})

k_2 = intrinsic cell inactivation rate constant (cells m^{-3} h^{-1})

K = Michaelis-Menton saturation constant for cells adsorbed into an 'inactivation zone' around a bubble

a = local specific bubble interfacial area (bubble surface area/medium volume, m^{-1})

Arguing that this bubble breakup/bursting can take place at several locations in the bioreactor and averaging over the entire bioreactor volume, the model predicts that the specific cell death rate is linearly proportional to the specific interfacial area (to a first order approximation). Since the model averages over all of the medium in the bioreactor, the model does not distinguish between different regions, for example: bubble formation at the sparger, bubble bursting at the free surface, or bubble interactions in the impeller region. Analyzing the experimental work of Oh *et al.* (1989), Wang *et al.* (1994) demonstrated that this linear specific death rate, with respect to gas interfacial area, is confirmed. Like the model of Tramper *et al.* (1988), this model, without a mechanistic basis, demonstrates the importance of gas interfacial area with respect to cell damage. However, it is more general in that it applies for systems in which air is not sparged since the air flow rate into the vessel is not explicitly accounted for.

Regions for cell damage as a result of aeration

The possible cell damage that can result from gas sparging can be classified into three distinct regions in a bioreactor: the bubble generation region, the bubble rising region, and the bubble disengagement region. In the region of bubble rise, bubbles coalescence or breakup may result in cell damage. In addition, if the bioreactor is agitated and the impellers are above the bubble generation site, it has been suggested by Wang *et al.* (1994) that it is also possible that cell damage can take place in the region where bubbles coalesce and break up in response to impeller generated fluid shear.

Bubble rising

Since the hypothetical killing volume correlation demonstrates that increasing the height to diameter ratio for a given aeration rate decreased cell damage, the distance that a bubble rises does not seem to affect cell viability. This conclusion is confirmed by Orton and Wang (1991) when they presented photographic evidence, using cell viability dyes, indicating that cell damage takes place at the gas-medium interface in a sparged bubble column without an impeller.

Bubble generation

Very few studies have been conducted on the effect of different spargers on cell viability. One such study is that by Murhammer and Goochee (1990) comparing two different airlift bioreactors. These two airlift systems, a Ventrex Laboratories (Portland, ME) Cellift and a Kontes (Vineland, NJ) autoclavable airlift, have similar dimensions and were operated under similar conditions. However, while cells grew well in the Cellift with Pluronic F-68 present (when compared to spinner flask culture) they would not grow or grew poorly in the Kontes system whether or not Pluronic F-68 was present.

The only apparent difference between the two systems is the pressure drop of the air through the sparger at the base of the two vessels. If a pressure drop of 3.0 psi or greater is present, no cell growth is observed with or without Pluronic F-68, yet if the pressure drop is lowered below 2.5 psi the cells begin to grow. In the Ventrex system the pressure drop is never greater than 1.0 psi. Murhammer and Goochee suggest that the increased air velocity leaving the gas orifice, indicated by the higher pressure drop, leads to higher levels of fluid turbulence and, consequently, higher cell damage. Further work is needed to verify this hypothesis.

Bubble disengagement region

When a gas bubble reaches the top af a vessel, several fates await it. If no other bubbles are present at the gas-liquid interface, it will partially rise out of the liquid, forming a thin film separating the gas above the bubble from the gas within. Depending on the compounds in the liquid this film will either rapidly drain and the bubble will rupture, or the film will stabilize and the bubble will remain at the interface for a relatively long period of time. If this time is sufficiently long, other bubbles will join it and a foam layer will form. If a foam layer is present at the top of the vessel, a bubble larger than those in the foam will slowly rise through the foam, potentially coalescing with other bubbles in the foam to form bigger bubbles.

Visualization studies

With respect to bubbles at the interface, a number of fates await bubble-bound cells. Using a microscope-video system, Handa *et al.* (1987) observed that cells oscillated near rapidly rupturing bubbles, and suggested that this oscillation damaged cells. It was also proposed that cells could be damaged as a result of physical shearing as the liquid drains around the bubbles. Finally, it was believed that there is actual physical loss of cells in the bubble foam.

While Handa *et al.* (1987) pioneered the use of visual techniques to develop a mechanistic understanding of cell damage as a result of bubbles, the system was limited in both depth of field, field of view and filming speed. Bavarian *et al.* (1991), Chalmers and Bavarian (1991), and Garcia-Briones and Chalmers (1992) used a system that did not have these limitations, and were able to observe that insect cells, suspended in medium without protective additives such as Pluronic F-68, attached to bubbles.

In Fig. 1 several photographs, taken from a video tape of these cell-bubble observations are presented. In each photograph, the micro-bubbles (10 to 100 microns in diameter) were created through electrolysis and are rising from bottom to top. These bubbles are black while the cells (TN-368 or SF-9 insect cells) are a lighter gray level. In 1a through 1c individual cell-bubble interactions are presented while in 1d and 1e clumps of cells interacting with several bubbles is shown.

Fig. 2 depicts cells interactions, or the lack thereof, with a bubble film, at an air-medium interface. In 2a the film of a 3.5 mm bubble in TNM-FH insect cell medium (10% serum without protective additives) is shown just after the bubble came to rest at the interface. However, in 2b a bubble is shown which has no cells attached to the film. The only difference between it and the bubble shown in 2a is the presence of 0.2% Pluronic F-68.

In Fig. 3 cells attached to the air-medium interface, without bubbles present (3a), and cells trapped in the foam layer at the top of the bubble column are presented (3b).

All of these results indicate that cells can be, depending on the medium used, closely associated with the gas-liquid interfaces. In light of the studies discussed previously, it is this close association that is detrimental to cells.

Studies on the rupture of bubbles at gas-liquid interface with respect to cells

The dynamics of the rupture of a bubble at a gas-liquid interface is complex and the equations that define the process cannot be solved analytically. Consequently, it is not possible to obtain exact solutions of the velocity and the resultant hydrodynamic forces associated with the process. This rupture process can be broken down into two steps: the rupture and retraction of the bubble film and the bubble cavity collapse.

The actual mechanisms of these two processes have been discussed in several articles (MacIntyre, 1972; Chalmers and Bavarian, 1991; Garcia-Briones *et al.* 1994) and will only briefly be reviewed here. When the bubble film breaks, the liquid in the film rapidly retracts forming a toroidal ring. Once this ring reaches the edge of the cavity, liquid flows down into the cavity until it reaches the bottom. At the bottom, a point of stagnation is obtained and two jets result: an upward jet that can achieve significant distances above the gas-liquid interface, and a lower jet which flows into the liquid beneath the cavity.

Computer simulations of bubble rupture

To determine if the hydrodynamic forces associated with the rupture of a gas bubble are sufficient to damage cells, Boulton-Stone and Blake (1993) and Garcia-Briones *et al.* (1994), conducted computer simulations of the process of bubble rupture.

Boulton-Stone and Blake modeled the bubble rupturing process numerically using a boundary integral method. With these solutions the predicted fluid velocities and shapes of the dynamic gas-liquid interfaces were compared to experimental observations and values in the literature. Both the pressure distribution and the energy dissipation rates as a bubble ruptures were also determined.

Garcia-Briones *et al.* (1994) used the computer program Flow 3-D (Flowscience, Los Alamos, NM) on a Cray Y-MP computer to determine the hydrodynamics associated with a bubble rupture. This code produces a transient solution of the Navier-Stokes equations.

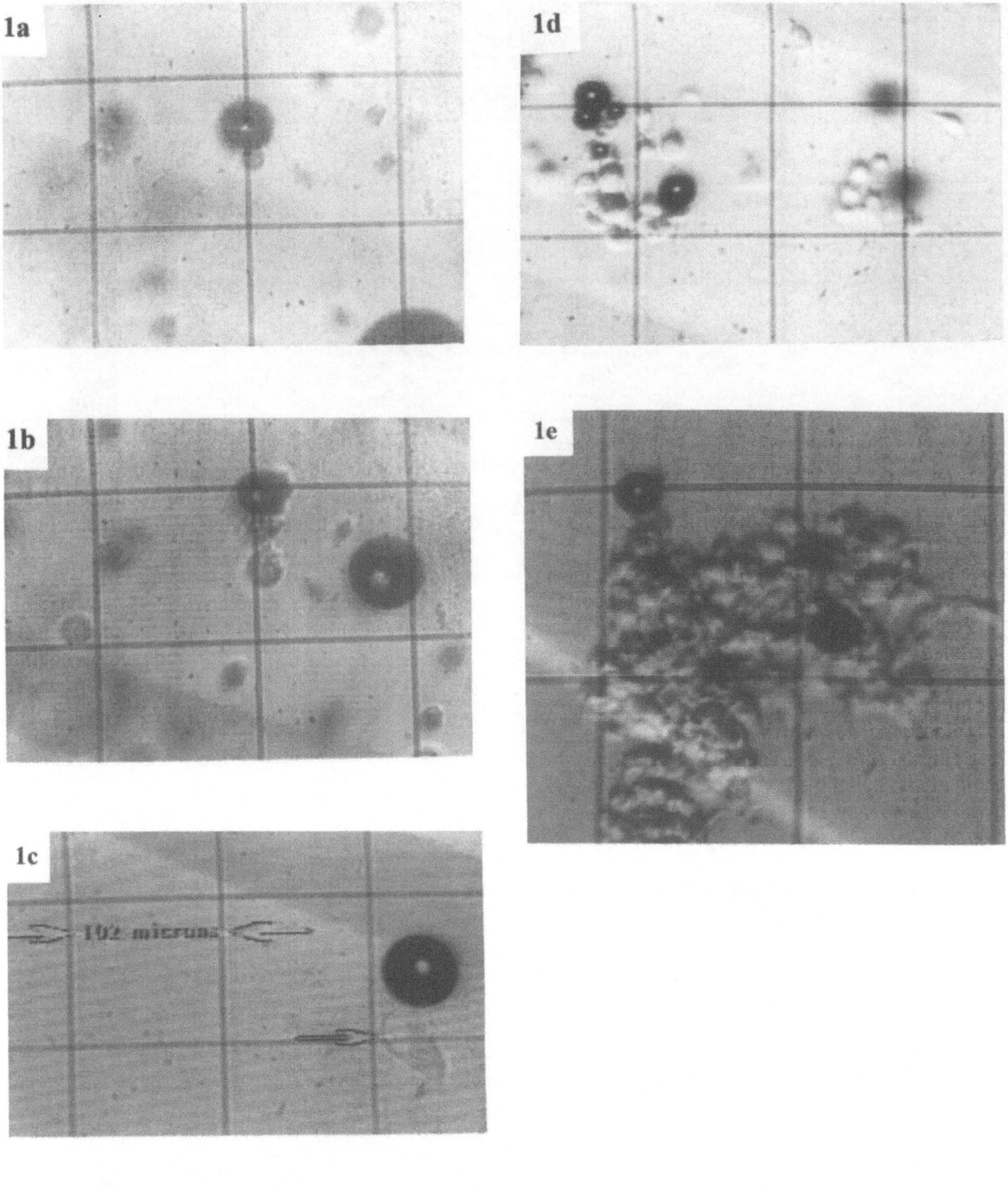

Fig. 1. Photographs of individual and clumps of cells attached to rising bubbles. (Reprinted with permission from Biotechnology Progress 7: 140–150 1991).

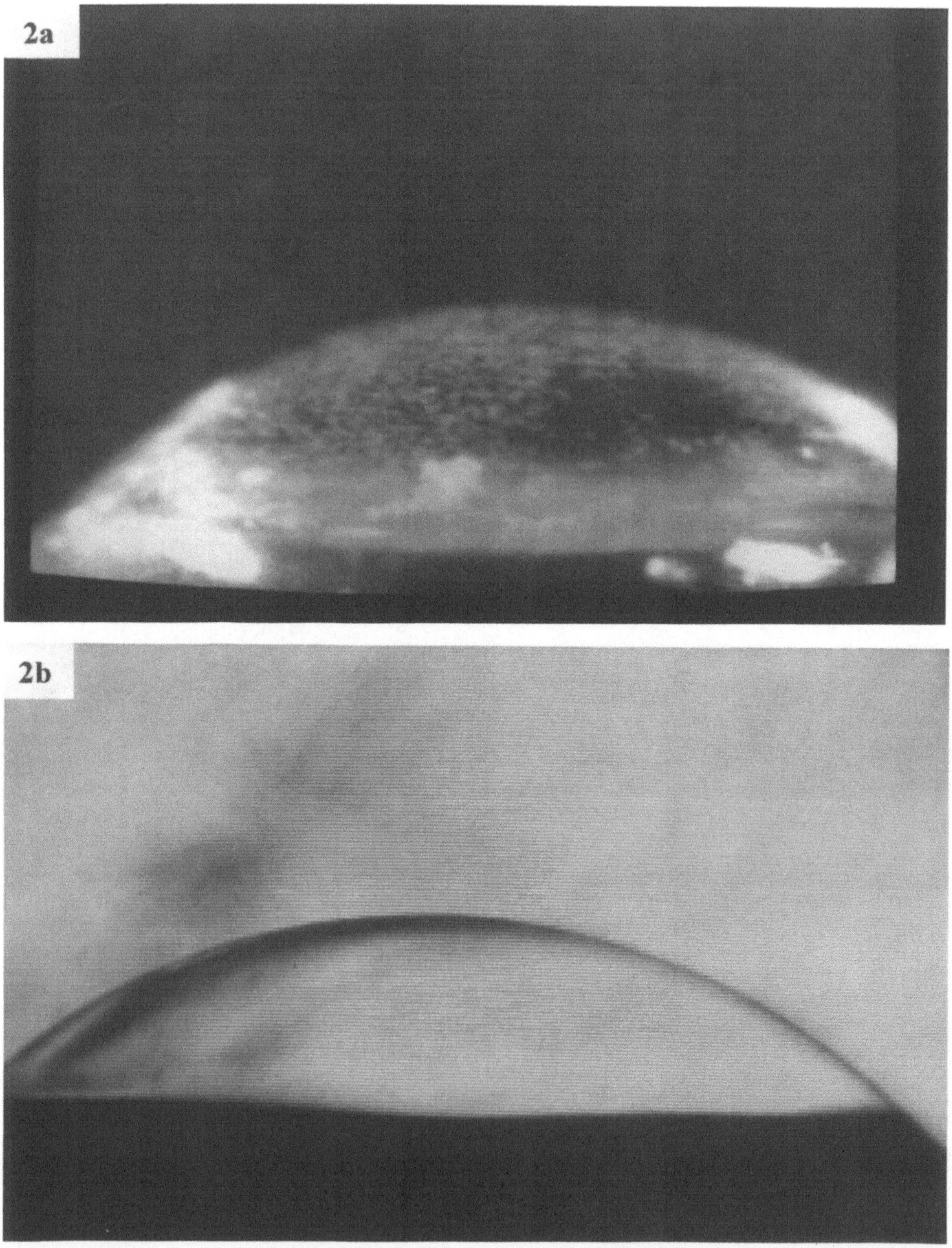

Fig. 2. Photographs of a bubble resting at the air-medium interface. Cells appear as small white spots on the bubble film. 2a is immediately after the bubble came to rest on the interface. 2b is a bubble in identical conditions except for the presence of 0.2% Pluronic F-68 in the medium. (Reprinted with permission from Ann. N.Y. Acad. Sci. 665: 219–229).

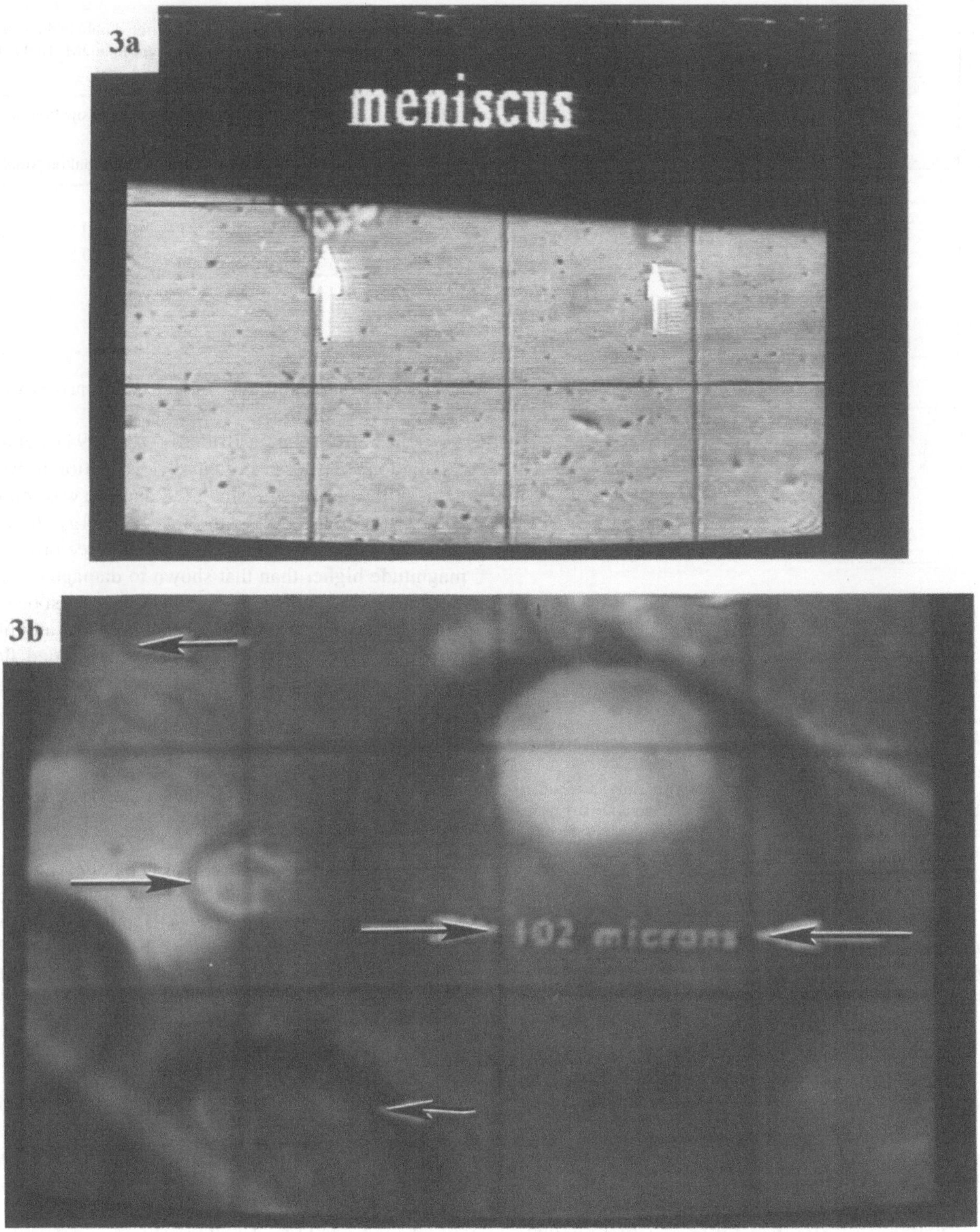

Fig. 3. Photographs of video images of cells attached to the air-medium interface and in the foam layer. (Reprinted with permission from Biotechnol. Progr. 7: 140–150 1991).

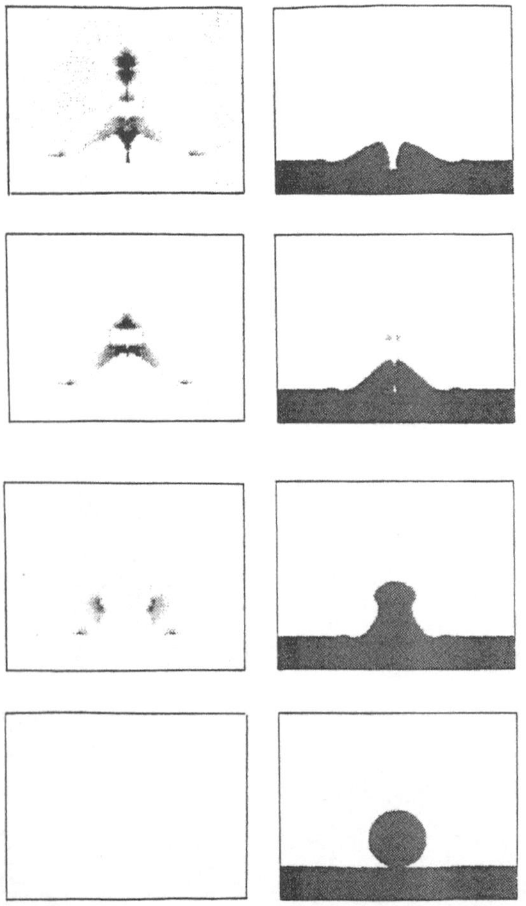

Fig. 4. Fluid fraction (left) and dissipation function distribution (right) for a 0.77 mm diameter bubble. The four successive pairs correspond to 0.0, 3×10^{-4}, 4.3×10^{-4}, and 5.5×10^{-4} seconds after bubble rupture. (Reprinted with permission from Chem. Eng. Sci. 49: 2301-2320).

Fig. 4 presents the gas-liquid interface of a rupturing bubble at specific time intervals. With the solution obtained from this code, the transient rate of energy dissipation per unit volume in the regions close to the bubble interface during the rupturing process was calculated. These rates were then compared to the transient rates of energy dissipation needed to damage cells from other experimental studies of cell damage in well-defined flows. Fig. 4 also presents the regions of highest energy dissipation for the given time intervals presented. Comparing these energy dissipations rates, both groups (Boulton-Stone and Blake (1993) and Garcia-Briones *et al.* (1994)) reported that the rate of energy dissipation decreases rapidly as bubble size increases. The actual values of these rates for several bubble diameters agreed between the two groups within an order of magnitude. Table 1 summarizes these values.

Table 1. Rates of energy dissipation for differed bubble diameters (from Garcia-Briones *et al.* 1994, Boulton-Stone JM, Blake JR 1993)

Bubble diameter (mm)	Total elapsed time (sec)	Maximum energy dissipation rate $Jm^{-3}\ s^{-1}$ Garcia-Briones	Boulton-Stone
0.77	5.5×10^{-4}	9.52×10^{7}	-
1.77	2.0×10^{-3}	1.66×10^{7}	4.0×10^{8}
6.32	1.0×10^{-2}	9.4×10^{4}	8.0×10^{3}

Table 2, from Garcia-Briones *et al.* (1994) includes calculations of the rates of energy dissipation in well-defined flow devices in which cell damage was reported. As can be observed, the rate of energy dissipation of a 0.77 mm bubble rupturing is three orders of magnitude higher than that shown to damage cells in well-defined flow devices. However, care must be taken in comparing the energy dissipation when a bubble ruptures versus the studies in the well-defined flow devices. The time scale during a bubble rupture is on the order of milliseconds while that for all of the studies, except Augenstein *et al.* (1971) are on the order of seconds.

Bubble rupture experiments with cells

Based upon the observation that cells attach to bubbles and the bubble film, and the computer simulations of Garcia-Briones *et al.* (1994), Trinh *et al.* (1994) quantified the number of insect cells killed per bubble rupture in medium with and without Pluronic F-68. On average, a single 3.5 mm bubble rupture will kill 1050 insect cells suspended in normal growth medium without Pluronic F-68. However, when Pluronic F-68 is present, no cells are killed.

To further determine the damage mechanism, Trinh *et al.* (1994) captured the upward jet that results from the bubble rupture. When no Pluronic F-68 was present, the concentration of cells in this upward jet was twice (on average) that in the bulk liquid surrounding the bubble. In addition, effectively all the cells in this upward jet were killed in the bubble rupture process. In related experiments, on average, 292 cells were present on the 3.5 mm bubble film (when no protective additive was present).

Table 2. Rates of energy dissipation in which cell damage was reported in well-defined flow devices (From Garcia-Briones *et al.* 1994)

Cell type	Instrument	Reference	Rate of cell damage (% min^{-1})	Dissipation function ($Jm^{-3}\ s^{-1}$)
insect	cone and plate	Goldblum *et al.*	33.5	3.15×10^4
hybridoma	concentric cylinder	Schurch *et al.*	3.4	2.20×10^4
hybridoma	double cup and bob	Smith *et al.*	(1)	5.81×10^2
mammalian	capillary	Augenstein *et al.*	16900	4.80×10^7

(1) At 15 hrs. cell viability was 73% (78% at time = 0) as compared with 85% for a control.

Since both the photographic evidence of MacIntyre (1968, 1972) and the computer simulations of Garcia-Briones *et al.* (1994) indicate that the fluid in the upward jet originated from a thin film surrounding the bubble, Trinh *et al.* suggested that the 'hypothetical killing volume' suggested by Tramper *et al.* (1988) is in fact the thin layer of fluid surrounding the bubble cavity before it ruptures. The observations that the concentration of cells in the upward jet is twice as high as that in the bulk can be explained if the concentration of cells absorbed onto the bubble cavity interface before the bubble ruptured is twice the concentration of the bulk. This explanation is also consistent with the observation that Pluronic F-68 prevents cells from adhering to the gas-liquid interface and that there are no cells in the upward jet. Since the liquid in the upward and lower jets experience the highest hydrodynamic forces, and if Pluronic F-68 prevents cells from being contained in the jets, it is consistent that very little cell death is observed with Pluronic F-68 present when a bubble ruptures.

Conclusion

On the basis of the results discussed above, the following conclusions can be made. First, rupturing bubbles at gas-liquid interfaces with attached cells results in cell death. Second, additives that prevent cell attachment to gas-liquid interfaces protect cells from damage. Third, a mechanistic understanding of the actions of these additives is needed.

A further point is that only one study of sparging on microcarrier cultures has been published (Aunins, 1986). Further studies are needed to determine if the observations reported above for suspended cells are also true for cells attached to microcaniers.

Acknowledgments

The author wishes to thank the National Science Foundation for financial assistance and all the graduate students who have worked with him.

References

Augenstein DC, Sinskey AJ, Wang DIC (1971) Effect of shear on the death of two strains of mammalian tissue cells. Biotechnol. Bioeng. 13: 409–418.

Aunins JG, Croughan MS, Wang DIC (1986) Engineering developments in homogeneous culture of animal cells: Oxygenation of reactors and scaleup. Biotechnol. Bioeng. 17: 399–723.

Backer MP, Metzger LS, Slaber PL, Nevitt KL, Boder GB (1988) Large-scale production of nonoclonal antibodies in suspension culture. Biotechnol. Bioeng. 32: 993–1000.

Bavarian F, Fan LS, Chalmers JJ (1991) Microscopic visualization of insect cell-bubble interactions. I: Rising bubbles, air-medium interface, and the foam layer. Biotechnol. Prog. 7: 140–150.

Boulton-Stone JM, Blake JR (1993) Gas-bubbles bursting at a free surface. J. Fluid Mech. 154: 437–466.

Cherry RS, Papoutsakis ET (1986) Hydrodynamic effects on cells in agitated tissue culture reactors. Bioproc. Eng. 1: 29–41.

Chalmers JJ, Bavarian F (2991) Microscopic visualization of insect cell-bubble interactions. II: The bubble film and bubble rupture. Biotechnol. Prog. 7: 151–158.

Croughan MS, Hamel JF, Wang DIC (1987) Hydrodynamic effects in animal cells grown in microcarrier cultures. Biotechnol. Bioeng. 29: 130–141.

Dodge TC, Hu WS (1986) Growth of hybridoma cells under different agiatation conditions. Biotechnol. Letters 8: 683–686.

Garcia-Briones MA, Brodkey RS, Chalmers JJ (1994) Computer simulations of the rupture of a gas bubble at a gas-liquid interface and its implications in animal cell damage. Chem. Eng. Sci. 49: 2301–2320.

Garcia-Briones MA, Chalmers JJ (1992) Cell-bubble interactions: Mechanims of suspended cell damage. Ann. N.Y, Acad. Sci. 665: 219–229.

Goldblum S, Bae Y, Hink WF, Chalmers JJ (1990) Protective effect of methylcellulose and other polymers on insect cells subjected to laminar shear stress. Biotechnol. Prog. 6: 383–390.

Handa-Corrigan A, Emary AN, Spier RE (1989) Effect of gas-liquid interfaces on the growth of suspended mammalian cells: mechanisms of cell damage by bubbles. Enzyme Microb. Technol. 11: 230–235.

Handa A, Emary AN, Spier RE (1987) On the evaluation of gas-liquid interfacial effects on hydridoma viability in bubble column bioreactors. Dev. Biol. Stand. 66: 241–253.

Hu WS, Meier J, Wang DIC (1985) A mechanistic analysis of the inoculum requirement for the cultivation of mammalian cells on microcarriers. Biotechnol. Bioeng. 27: 585–595.

Jobses I, Martens D, Tramper J (1991) Lethal events during gas sparging in animal cell culture. Biotech. Bioeng. 37: 484–490.

Kilburn DG, Webb FC (1968) The cultivation of animal cells at controlled dissolved oxygen partial pressure. Biotechnol. Bioeng. 10: 801–814.

Kunas KT, Papoutsakis ET (1990a) Damage mechanisms of suspended animal cells in agitated bioreactors with and without bubble entrainment. Biotechnol. Bioeng. 36: 476–483.

Kunas KT, Papoutsakis ET (1990b) The protective effect of serum against hydrodynamic damage of hydridoma cells in agitated and surface-areated bioreactors. J. Biotechnol. 15: 57–70.

Lee GM, Huard TK, Kaminski MS, Palsson BO (1988) Effect of mechanical agitation on hydridoma cell growth. Biotechnol. Letters 10: 625–628.

MacIntyre F (1972) Flow patterns in breaking bubbles. J Geophys. Res. 77: 5211–5228.

MacIntyre F (1968) Bubbles: a boundary-layer 'microtome' for micron-thick samples of a liquid surface. J. Phys. Chem. 72: 589–592.

Martens DE, de Gooijer CD, Beuvery EC, Tramper J (1992) Effect of serum concentration on hybridoma viable cell density and production of monoclonal antibodies in CSTRs and on shear sensitivity in air-lift loop reactors. Biotechnol. Bioeng. 39: 891–897.

Murhammer DW, Goochee CF (1990) Sparged animal cell bioreactors: mechanism of cell damage and Pluronic F-68 protection. Biotechnol. Prog. 6: 391–397.

Oh SKW, Nienow AW, Al-Rubeai M, Emary AN (1989) The effect of agiatation intensity with and without continuous sparging on the growth and antibody production of hybridoma cells. J. Biotechnol. 12: 45–62.

Orton D, Wang DIC (1991) Fluorescent Visualization of Cell Death in Bubble Areated Bioreactors. Cell Culture Engineering III, Engineering Foundation, Feb. 2-7,

Ruyan WS, Gyer RP (1963) Growth of L cell suspensions on a Warburg apparatus. Proc. Soc. Bio. Med. 103: 252–254.

Schurch U, Kramer H, Einsle A, Widmer F, Eppenberger HM (1988) Experimental evaluation of laminar shear stress on the behaviour of hybridoma mass cell cultures producing monoclonal antibodies against mitochondrial creatine kinase. J. Biotechnol. 7: 179–184.

Sinskey AJ, Fleischaker RJ, Tyo MA, Giard DJ, Wang DIC (1981) Production of cell derived products: virus and interferon. Ann. N.Y. Acad. Sci. 369: 47–59.

Smith CG, Greenfield PF, Randerson DH (1987) A technique for determining the shear sensitivity of mammalian cells in suspension culture. Biotechnol. Tech. 1: 39–44.

Swim HE, Parker RF (1960) Effect of Pluronic F-68 on growth of fibroblasts in suspension on rotary shakers. Proc. Soc. Biol. Med. 103: 252–254.

Tramper J, Smit JD, Straatman J, Valk JM (1988) Bubble-column design for growth of fragile insect cells. Bioprocess Engin. 3: 37–41.

Tramper J, Williams JB, Joustra D (1986) Shear sensitivity of insect cells in suspension. Enzyme Microb. Technol. 8: 33–36.

Trinh K, Garcia-Briones MA, Hink FH, Chalmers JJ (1994) Quantification of damage to suspended insect cells as a result of bubble rupture. Biotechnol. Bioeng. 43: 37–45.

Wang NS, Yang JD, Calabrese RV, Chang KC (1994) Unified modeling framework of cell death due to bubbles in agitated and sparged bioreactors. J. Biotechnol. 33: 107–122.

Address for offprints: J.J. Chalmers, Department of Chemical Engineering, Ohio State University, 140 W 19th Ave., Columbus, OH 43210 U.S.A.

Cytotechnology **15:** 321–328, 1994.

Quantitative investigations of cell-bubble interactions using a foam fractionation technique

W. S. Tan, G. C. Dai & Y. L. Chen
Laboratory of Cell Culture Technology, Research Institute of Biochemical Engineering, East China University of Science & Technology, 130 Meilong Road, Shanghai 200237, P.R. China

Key words: Adsorption, cell-bubble interactions, foam fractionation, hybridoma cells, Pluronic F68, serum

Abstract

Previous work by the authors and others has shown that suspended animal cell damage in bioreactors is caused by cell-bubble interactions, regardless whether the bubbles are from bubble entrainment or direct gas sparging. As approach to measure the adsorptivity of animal cells to bubbles, a modified batch foam fractionation technique has been developed in this work and proven to be applicable. By using this technique, the number of cells adsorbed per unit bubble surface area and the adsorption coefficients have been measured to quantify hybridoma cell-bubble interactions, and the preventive effects of serum and Pluronic F68 on these interactions. It was demonstrated quantitatively that the hybridoma cells adhere to bubbles spontaneously and significant numbers exist in the foam, and that both the serum and Pluronic F68 provide strong prevention to these cell-bubble interactions. The results obtained provide criteria for bioreactor operation and medium formulation to prevent cell-bubble interactions and cell damage in the culture processes.

Abbreviations: NBCS – new born calf serum; SFM – serum-free medium.

Introduction

With the development of cell fusion and recombinant DNA technologies, large quantities of valuable biologicals can be produced by animal cell culture technology. A prominent example of such an animal cell is the hybridoma cell, used for *in vitro*, large scale production of monoclonal antibodies. In large scale suspension cultures of animal cells in bioreactors, the agitation and/or sparging are the most practical ways of supplying sufficient oxygen and mixing for cell growth. However, significant cell damage or injury has been observed to result from excessive agitation and gas sparging. There have been many studies on animal cell damage or injury in bioreactors (Handa-Corrigan *et al.*, 1987, 1989; Tramper *et al.*, 1986, 1988; Jobses *et al.*, 1991; Murhammer and Goochee, 1988, 1990b; Bavarian *et al.*, 1991; Chalmers and Bavarian, 1991; Kunas and Papoutsakis, 1990b; Gardner *et al.*, 1990; Michaels *et al.*, 1992; Oh *et al.*, 1989, 1992; Cherry and Hulle, 1992; Tan *et al.*, 1993). Because the agitation intensity of most bioreactors for animal cell suspended cultures rarely exceeds 300–500 rpm, independent of impeller size and design. The Kolmogorov eddy sizes produced in such bioreactors are usually much larger than the cell size. Therefore, it can be concluded from the previous investigations that the cell damage either in sparged (bubble column, airlift) or stirred bioreactors is most likely the result of cell-bubble interactions.

There are three possible regions of cell-bubble interactions: the bubble formation and injection region at the sparger, the bulk medium in which the bubble rises, and the bubble disengagement region at the air-medium interface (Tramper *et al.*, 1986, 1988). Recently, the work carried out by Handa-Corrigan *et al.* (1989), Kunas and Papoutsakis (1990b) and Jobses *et al.* (1991) demonstrated that cell damage takes place at the region of bubble disengagement at the medium-air interface. At the same time, three possible mechanisms for this cell death were proposed:

(1) oscillatory disturbances caused by rapidly bursting bubbles, (2) physical shearing of the cells resulted from the liquid film draining around the bubbles, and (3) physical loss of cells in the foam. By using a microscopic, high-speed video-system, the adsorption of cells to the bubbles and the transport of these cells into the foam layer have been observed (Bavarian *et al.*, 1991). Based on experimental results, Tan *et al.* (1993) inferred that only the cells adsorbed on the bubbles can probably be damaged in a similar manner to those mentioned above. To demonstrate that the bubble rupturing process kills cells, Garcia-Briones and Chalmers (1992) collected a sample from the upward jet, which resulted from a bubble rupturing at an air-medium interface. This sample contained a large number of cells and approximately 90% of the cells were dead. Using a more advanced microscopic system than Bavarian *et al.* (1991), Garcia-Briones and Chalmers (1992) were able to visualize a large area of the bubble film of a single bubble at the gas-liquid interface and reported that a large number of cells (>300) could be observed on the film. Recently, Trinh *et al.* (1994) found that approximately 1050 cells were killed per single, 3.5–mm bubble rupture and approximately the same number of dead cells were present in the upward jet. From this point of view, the adsorption of cells to bubble surfaces is the step controlling the cell damage and losses in bioreactors and plays an important role in the successful cultivation of animal cells in stirred and sparged bioreactors. To our knowledge, no general and reliable technique is yet available in the literature with respect to quantification of the adsorption of cells on bubbles and cell enrichment in the foam.

The protective effects of additives including serum, Pluronic F68, polyethylene glycol (PEG), bovine serum albumin and several protein mixtures on animal cells have been examined by various investigators (Handa-Corrigan *et al.*, 1989. Kunas and Papoutsakis, 1989, 1990a; Michaels *et al.*, 1991a, 1991b; Murhammer and Goochee, 1988, 1990a,. 1990b; Ozturk and Palsson, 1991; Ramirez and Mutharasan, 1990, 1992; Smith and Greenfield, 1992). It was proposed that these additives protect cells from damage by biological effects on cells to make them more shear resistant and/or by physical ones to change the shear forces produced. Recently, Garcia-Briones and Chalmers (1992), Zhang *et al.* (1992) and Tan *et al.* (1993) suggested that Pluronic F68, due to its surface active properties, changes the physicochemical characteristics of the foam formed and prevents cells from adhesion to bubbles.

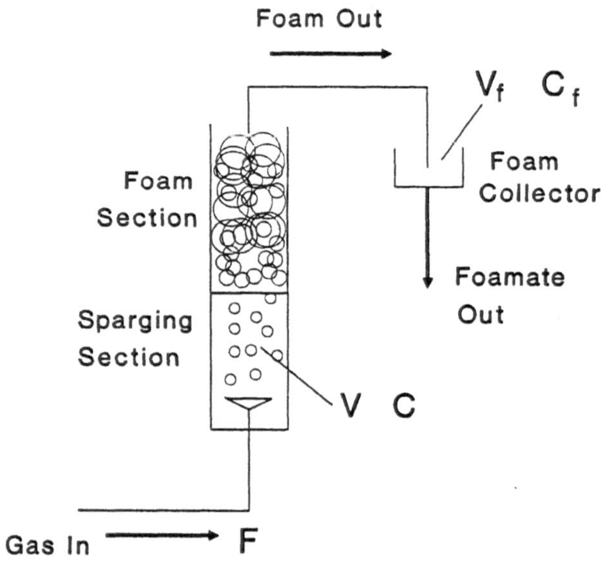

Fig. 1. Diagram of the modified batch foam fractionation.

In this work, a modified batch foam fractionation technique has been developed. By using this technique, the number of cells adsorbed on bubble surfaces and the adsorption coefficients have been measured to quantify the hybridoma cell-bubble interactions and the prevention provided by serum and Pluronic F68. Based on the results obtained, the mechanism of cell death and the protection by additives will be discussed.

Description of foam fractionation technique

In order to quantify the hybridoma cell adsorption and enrichment in the foam layer, a modified batch foam fractionation technique has been developed, diagramed in Fig. 1. Gas is sparged into the lower end of the apparatus at a constant flow rate F (ml min^{-1}). Rising through the cell suspension, the bubbles in which the cells are trapped enter the foam. The foam rises through the foam section and is collected as it exits the top. With an attempt to find correlations between the measurable data and the number of cells adsorbed on bubble surfaces, a model is derived on the basis of the following assumptions.

(1) The foam does not rupture and drain back into the bulk suspension before exiting.

(2) Once adsorbed on bubbles, the cells do not leave the bubble surfaces as the foam rises until the foam exists.

(3) The changes of the physical properties, such as surface tension, of the cell suspension during operation are negligible.

(4) No cell growth and metabolism occur over the short duration of the foam fractionation.

(5) As the bubbles just depart from the cell suspension and enter the foam phase, the adsorption of cells on these bubbles reaches equilibrium with the viable cell density in the bulk solution, following the equation

$$\Gamma_c = f(C) \quad (1)$$

with Γ_c being the number of cells adsorbed per unit bubble surface area (cell cm^{-2}), C the viable cell density in the bulk cell suspension (cell ml^{-1}).

The number of cells adsorbed per unit surface area Γ_c is considered to be proportional to C. Thus, equation (1) becomes:

$$\Gamma_c = \alpha C \quad (2)$$

where α is the adsorption coefficient (cm). This proportionality follows from the analogy of the adsorption characteristics of surface-active agents to interface in dilute solutions. The assumption of constant physical properties of the cell suspension during foam fractionation is implicit in equation (2).

Obviously, the batch foam fractionation is a time-varying process. As shown in Fig. 1, the volume and viable cell density for a given cell suspension are V (ml) and C (cell ml^{-1}) at time t, respectively. After a interval of time dt, the cell suspension volume decreases to V – dV, producing foam liquid with volume of dV_f which equals –dV. In this foam liquid the density of cells trapped by bubbles is C_f. Meanwhile, the viable cell density in the bulk cell suspension becomes C – dC. Thus, the mass balance for the viable cells over dt is

$$V\,C = dV_f\,C_f + (V - dV)(C - dC) \quad (3)$$

where $dV_f = -dV$. With higher derivatives being neglected, the above equation becomes

$$\frac{dV}{V} = \frac{dC}{C_f - C} \quad (4)$$

Equation (4) can also be written in an integrated form as follows

$$\ln\frac{V_0}{V} = \int_C^{C_0} \frac{dC}{C_f - C} \quad (5)$$

where the initial condition is taken as $V = V_0$ and $C = C_0$ at $t = 0$. To integrate the above equation, the correlation between C_f and C should be known.

Over the short duration of time dt, the number of cells trapped by bubbles from the bulk suspension into the foam phase equals:

$$dN = \Gamma_c\,dA = \alpha\,C\,dA \quad (6)$$

where dA is the surface area of bubbles produced during dt, cm^2. As defined, C_f is

$$C_f = \frac{dN}{dV_f} = \frac{\alpha\,C\,dA}{dV_f} = \frac{\alpha}{\frac{dV_f}{dA}}C = \frac{\alpha}{\delta}C \quad (7)$$

where $\delta = \frac{dV_f}{dA}$, cm, is taken as the equivalent thickness of the liquid film of the foam. It should be noted that δ is not the actual thickness of the liquid film, due to liquid drainage and bubble coalescence with the foam rising in the foam section. It is defined as the ratio of foam liquid volume to bubble surface area produced over dt. Obviously, δ will be dependent on both the physical nature of the liquid and the gas flow rate. For a given cell suspension and constant gas flow rate, both α and δ are constants, meaning that C_f is proportional to C. Thus, equation (7) can be written as

$$C_f = \beta\,C \quad (8)$$

where β is the proportionality coefficient.

Substituting equation (8) into (5) and integrating yields

$$\ln\frac{V_0}{V} = \frac{1}{\beta - 1}\ln\frac{C_0}{C} \quad (9)$$

or

$$\beta = \frac{\ln\frac{C_0}{C}}{\ln\frac{V_0}{V}} + 1 \quad (10)$$

It can be seen that, by only measuring the volumes of cell suspension and viable cell densities at the beginning and end of the foam fractionation process, β can be estimated easily from equation (9) or (10).

From the beginning (t=0) to the end (t=t) of the foam fractionation, the volume of foam liquid produced equals to

$$V_f = \int_{V_0}^{V} -dV = V_0 - V \quad (11)$$

and the surface area of bubbles sparged can be calculated approximately as

$$A = \int_0^t \frac{F}{\frac{\pi}{6} d_b^3} \cdot \pi d_b^2 dt = \frac{6 F t}{d_b} \quad (12)$$

where d_b is the average diameter of bubbles, cm.

Combining equations (11) and (12), δ can be estimated from the following equation:

$$\delta = \frac{V_f}{A} = \frac{(V_0 - V)d_b}{6 F t} \quad (13)$$

Thus, the adsorption coefficient α is

$$\alpha = \beta\delta \quad (14)$$

By measuring V, V_0, C and C_0 at certain experimental conditions (F, d_b and t) and combining equations (10), (13) and (14), α and $\Gamma_c = \alpha C_0$, which represent the cell adsorptivity on bubbles and enrichment in foam for a given cell suspension, can be determined.

Materials and methods

Cell culture

The cell line used in this study was 2F7 mouse-mouse hybridoma cells (Shanghai Institute of Cancer SIC, Shanghai, P.R. China) producing an IgG_{2a} antibody against the human small cell lung cancer. The hybridoma cells were grown in a 1.5-liter CelliGen bioreactor system (New Brunswick Scientific Co., Edison, NJ, USA) with a working volume of 1.3 liter in semicontinuous mode at a dilution rate of 0.55 day^{-1}. The serum-free medium (SFM), used for the bioreactor culture and developed by SIC, was a mixture of DMEM/F12 (1:1) (D8900, Sigma Chemical Co., St. Louis, MO, USA) and some additives, and supplemented with 50 units ml^{-1} of each penicillin and streptomycin. The bioreactor was kept at an agitation speed of 60 rpm, temperature 36.8 °C, pH 7.25 and dissolved oxygen (DO) 60% air saturation, and employed surface aeration. The pH was automatically controlled at the set point by adding 5.6% sodium bicarbonate or using the CO_2 content of the gas phase.

Foam fractionation experiments

The batch foam fractionation experiments were carried out in a 100-mL graduated measuring cylinder and performed on 100ml samples from the bioreactor culture. Gas was fed into the bottom of the cylinder at a constant flow rate. The sparger had ten nozzles with diameter of 0.25mm, generating bubbles ranged in size from 0.22 to 0.26cm in diameter. Foam rose through the cylinder and was collected as it exited the top. The height of the foam section inside the cylinder was 7cm at the beginning of experiments. To ensure a smooth flow of the foam out of the cylinder, the cylinder was held tipped up at a 45° angle.

To examine the effects of serum and Pluronic F68 on cell-bubble interactions, new born calf serum (NBCS) (East China University of Science & Technology, Shanghai 200237, P.R. China) and Pluronic F68 (BASF, Ludwigshafen, Germany) were directly added to the samples at the concentrations required just prior to fractionation.

Measurements

At the beginning and end of the experiments, samples (about 1 ml) were drawn to count the viable cell densities in the cell suspension. Duplicate counts were performed on each sample by the trypan blue exclusion method in a hemacytometer.

Some considerations for experiments

The gas flow rate selected should ensure that the foam with the captured cells flows out the cylinder smoothly, not rupturing and draining back into the bulk solution. The appropriate range of the gas flow rates should be determined experimentally.

Actually, the enrichment of surface active materials, such as proteins, from the cell suspension to the foam is inevitable, because of hydrophobic interactions. Thus, as a result of this fractionation, the physical properties of the bulk suspension, such as foam-forming ability, foam stability and surface tension, will change, causing the adsorption coefficient of cells on bubble surfaces to deviate from constant during foam fractionation and the true value. In order to make α applicable and reliable, the volume of the foam liquid produced should be small during the experiments. In this work, $V_0 - V$ was controlled in the range of 10—20 ml.

Results and discussion

Determination of the appropriate gas flow rates

Figure 2 shows the effects of gas flow rate on the experimental results. It can be seen that the liquid film

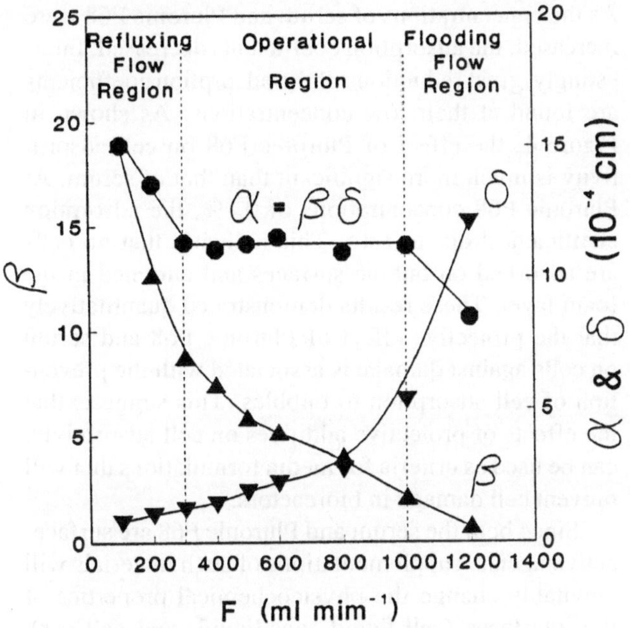

Fig. 2. Selection of the appropriate gas flow rates in foam fractionation experiments. Viable cell density: 13.3×10^5 cell ml^{-1}.

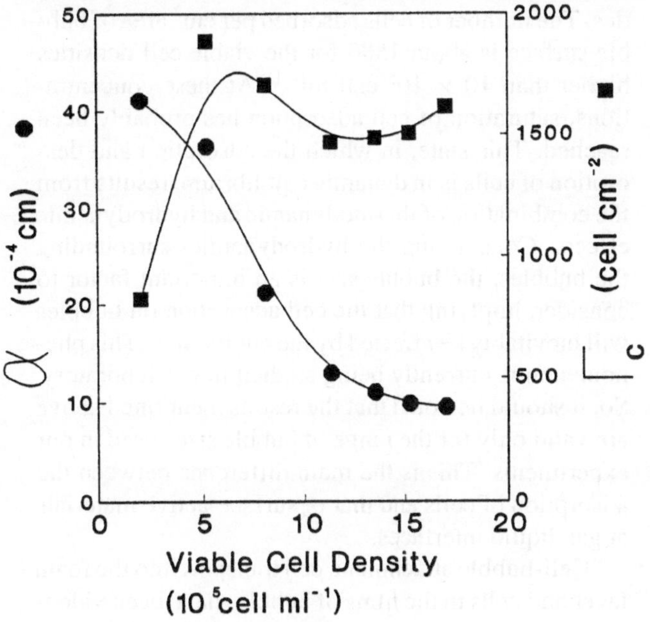

Fig. 3. Adsorptivity of the hybridoma cells cultured in serum-free medium (SFM). Bubble size: 0.22—0.26 cm.

thickness of the foam δ increases, but β decreases, with the gas flow rate increasing. This is attributed to a reduction of liquid drainage as the foam rises faster. This means that more liquid will be carried by the foam at higher gas flow rates. It is interesting that, although both δ and β are dependent on gas flow rates, the adsorption coefficient α is almost a constant in the range of gas flow rates from 300 to 1000 ml min^{-1}, independent of F. This constant is close to the true adsorption coefficient for a given sample. For the example shown in Figure 2, the viable cell density in this cell suspension was 13.3×10^5 cell ml^{-1}, and the adsorption coefficient α and the number of cells adsorbed per unit bubble surface Γ_c were estimated to be about 11.4×10^{-4} cm and 1516 cell cm^{-2}, respectively.

In the region of Figure 2 corresponding to low gas flow rates, the foams produced are not able to exit completely. Some bubbles rupture and drain back into the bulk solution. Since the cells trapped in these bubbles remain in the foam section, the measured adsorption coefficients are biased high. This region is called the *Refluxing Flow Region*. On the other hand, in the range of high gas flow rates, the foams carrying large amounts of liquid out of the cylinder result in α being biased low. This region is called the *Flooding Flow Region*. Obviously, to obtain true and reliable results, the gas flow rates should be selected from the region between the above two. In this region, called the *Operational Region*, the adsorption coefficients are independent of gas flow rates. For the 100-mL graduated measuring cylinder with inner diameter of 2.8-cm used in this work, the range of appropriate gas flow rates is 300–1000 ml min^{-1}.

Also by foam fractionation experiments, Cherry and Hulle (1992) used α as a characteristic paramter of the adsorption of insect cells (Sf9) to the foam layer. Their α was the same parameter as β used in this study. Having found an average α of 0.6, they concluded that insect cells were not enriched in the foam layer. Actually, the results in Figure 2 show that β is strongly affected by the gas flow rates, implying that β alone can not be used as a measure of cell adsorptivity and enrichment in the foam layer.

Hybridoma cell adsorptivity on bubble surfaces

The adsorption coefficients and the numbers of hybridoma cells adsorbed per unit bubble surface, corresponding to different densities of the cells cultured in serum-free medium (SFM), are shown in Figure 3. It was found that the adsorption coefficients decrease as viable cell density increases. In contrast, the number of the cells adsorbed does not change at high cell densities. The number of cells adsorbed per cm^2 area of bubble surface is about 1500 for the viable cell densities higher than 10×10^5 cell ml^{-1}. At these concentrations, saturation of cell adsorption has probably been reached. This state, in which the adsorption and desorption of cells is in dynamic equilibrium, results from the combination of thermodynamic and hydrodynamic effects. Considering the hydrodynamics surrounding the bubbles, the bubble size is an important factor to consider, implying that the cell adsorption on bubbles will inevitably be affected by the bubble size. This phenomenon is currently being studied in our laboratory. So, it should be noted that the results mentioned above are valid only for the range of bubble sizes used in our experiments. This is the main difference between the adsorption of cells and that of surface-active materials at gas-liquid interfaces.

Cell-bubble attachment, cell transport into the foam layer and cells in the films of bubbles have been videotaped (Bavarian *et al.*, 1991; Chalmers and Bavarian, 1991; Garcia-Briones and Chalmers, 1992), and now the results obtained from the foam fractionation experiments show quantitatively that significant numebrs of hybridoma cells do exist in the foam. Although cell damage does not take place in the process of cell adsorption, only those cells adsorbed on bubbles and existing in the foam layer can be destroyed by the shear stresses produced by bubble rupture, as Handa-Corrigan *et al.* (1989), Chalmers and Bavarian (1991), and Cherry and Hulle (1992) proposed. In other words, for cell death in sparged bioreactors, the event that occurs first is the cell adsorption to bubbles, meaning that the cell death rate in the bioreactors will be dependent on the cell adsorptivity and the bubble surface area produced. From this point of view, if cell adhesion to bubbles is prevented, and/or the number of cells captured by bubbles is decreased, by either reducing the cell adsorptivity or decreasing the bubble surface area produced, cell damage in bioreactors will be suppressed in efficiency or eliminated altogether.

Effects of serum and Pluronic F68 on cell-bubble adhesion

After the cell suspensions were taken from the bioreactor culture, the NBCS and Pluronic F68 were added to samples just prior to sparging. The results are shown in Figure 4. It is found that both serum and Pluronic F68 provide strong preventive effects to the cell adsorption on bubbles, and their effects are strongly dependent on their concentrations in the bulk cell suspensions. As the concentrations of serum and Pluronic F68 were increased, the adsorption coefficients decreased. Interestingly, great reductions of the adsorption coefficients are found at their low concentrations. As shown in Figure 4, the effect of Pluronic F68 on cell adsorptivity is much more significant than that of serum. At Pluronic F68 concentrations of 0.1%, the adsorption coefficient drops to zero. This indicates that no cells are adsorbed on bubble surfaces and enriched in the foam layer. These results demonstrated quantitatively that the protective effect of Pluronic F68 and serum on cells against damage is associated with the prevention of cell adsorption to bubbles. This suggests that the effects of protective additives on cell adsorptivity can be used as criteria for media formulations that will prevent cell damage in bioreactors.

Since both the serum and Pluronic F68 are surface-active agents, supplementations of such materials will inevitably change the physicochemical properties of the interfaces (cell-liquid, gas-liquid, and cell-gas), due to the hydrophobic interactions. Garcia-Briones and Chalmers (1992), Zhang *et al.* (1992) and Tan *et al.* (1993) suggested that the protection of Pluronic F68 was attributed to these changes, such as decreasing the surface tension of the medium. The effects of those additives on the behavior of interfacial phenomena in animal cell culture bioreactors should be investigated further.

Conclusions

The modified batch foam fractionation technique has been developed in this work and proven to be suitable for measuring the adsorptivity of hybridoma cells on bubbles. By measuring, it has been demonstrated quantitatively that the hybridoma cells adhere to bubbles spontaneously and significant numbers of cells exist in the foam. These results suggest that, by either reducing the cell adsorptivity or decreasing the bubble surface area produced, cell damage in bioreactors will be suppressed and prevented. As protective additives, the effect of serum and Pluronic F68 on cells has been found to be the result of preventing cells from adhesion to bubbles. In future, the effects of hydrodynamics surrounding the bubbles on cell-bubble adhesion, and the effects of protective additives on interfacial phenomena in animal cell bioreactors should be studied.

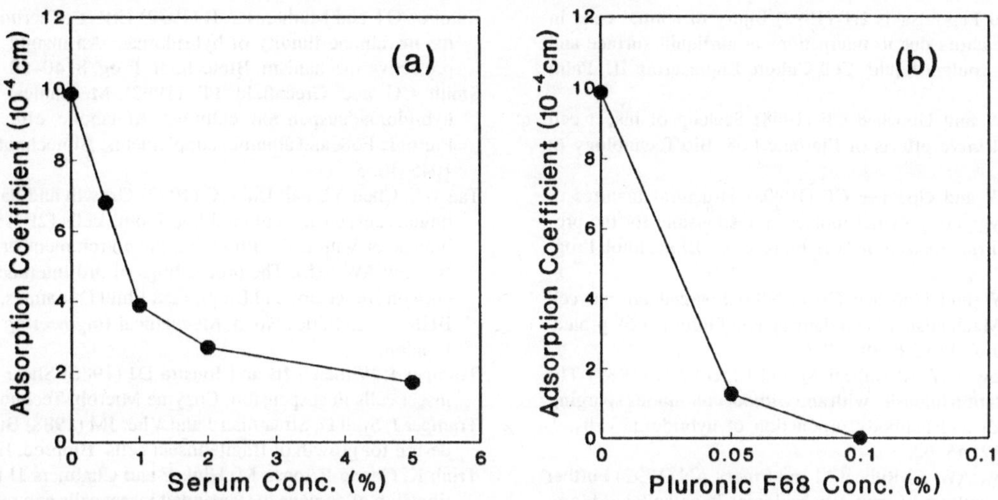

Fig. 4. Effects of serum (a) and Pluronic F68 (b) on cell-bubble adhesion. The hybridoma cells were cultured in SFM. Viable cell density: 15.3 $\times 10^5$ cell ml^{-1}.

Notes

1. A = surface area of the bubbles, cm^2
2. C = viable cell density in bulk cell suspension, cell ml^{-1}
3. C_f = density of cells trapped by bubbles in foam liquid, cell m^{-1}
4. C_0 = initial viable cell density in bulk cell suspension, cell ml^{-1}
5. d_b = average diameter of the bubbles, cm
6. F = gas flow rate, ml min^{-1}
7. N = number of cells captured by bubbles from bulk cell suspension to foam phase
8. t = time, min
9. V = volume of the bulk cell suspension, ml
10. V_f = volume of the foam liquid, ml
11. V_0 = initial volume of the bulk cell suspension, ml
12. α = adsorption coefficient, cm
13. $\beta = \frac{C_f}{C}$
14. $\delta = \frac{dV_f}{dA}$, equivalent thickness of the liquid film of the foam, cm
15. Γ_c = number of cells adsorbed per unit bubble surface area, cell cm^{-2}

References

Bavarian F, Fan LS and Chalmers JJ (1991) Microscopic visualization of insect cell-bubble interactions. I: Rising bubbles, air-medium interface, and the foam layer. Biotechnol. Prog. 7: 140–150.

Chalmers JJ and Bavarian F (1991) Microscopic visualization of insect cell-bubble interactions. II: The bubble film and bubble rupture. Biotechnol. Prog. 7: 151–158.

Cherry RS and Hulle CT (1992) Cell death in the thin film of bursting bubbles. Biotechnol. Prog. 8: 11–18.

Garcia-Briones M and Chalmers JJ (1992) Cell-bubble interactions: mechanisms of suspended cell damage. Ann. NY Acad. Sci. 665: 219–229.

Gardner AR, Gainer JL and Kirwan DJ (1990) Effects of stirring and sparging on cultured hybridoma cells. Biotechnol. Bioeng. 35: 940–947.

Handa-Corrigan A, Emery AN and Spier RE (1987) On the evaluation of gas-liquid interfacial effects on hybridoma viability in bubble column bioreactors. Develop. Biol. Standard. 66: 241–253.

Handa-Corrigan A, Emery AN and Spier RE (1989) Effects of gas-liquid interfaces on the growth of suspended mammalian cells: Mechanisms of cell damage by bubbles. Enzyme Microb. Technol. 11: 230–235.

Jobses I, Martens D and Tramper J (1991) Lethal events during gas sparging in animal cell culture. Biotechnol. Bioeng. 37: 484–490.

Kunas KT and Papoutsakis ET (1989) Increasing serum concentrations decrease cell death and allow growth of hybridoma cells at higher agitation rates. Biotechnol. Lett. 11: 525–530.

Kunas KT and Papoutsakis ET (1990a) The protective effect of serum against hydrodynamic damage of hybridoma cells in agitated and surface-aerated bioreactors. J. Biotechnol. 15: 57–70.

Kunas KT and Papoutsakis ET (1990b) Damage mechanisms of suspended animal cells in agitated bioreactors with and without bubble entrainment. Biotechnol. Bioeng. 36: 476–483.

Michaels JD, Petersen JF, McIntire LV and Papoutsakis ET (1991a) Protection mechanisms of freely suspended animal cells (CRL8018) from fluid mechanical injury. Viscometric and bioreactor studies using serum, Pluronic F68 and polyethylene glycol. Biotechnol. Bioeng. 38: 169–180.

Michaels JD and Papoutsakis ET (1991b) Polyvinyl alcohol and polyethylene glycol as protectants against fluid-mechanical injury of freely–suspended animal cells (CRL8018). J. Biotechnol. 19: 241–258.

Michaels JD and Papoutsakis ET (1992) Injury of animal cells in agitated bioreactors due to interactions at air/liquid surface and in the bulk, turbulent liquid. Cell Culture Engineering III, Palm Coast, FL.

Murhammer DW and Goochee CF (1988) Scaleup of insect cell cultures: Protective effects of Pluronic F68. Bio/Technology 6: 1411–1418.

Murhammer DW and Goochee CF (1990a) Structural features of nonionic polyglycol polymer molecules responsible for the protective effect in sparged animal cell bioreactors. Biotechnol. Prog. 6: 142–148.

Murhammer DW and Goochee CF (1990b) Sparged animal cell bioreactors: Mechanism of cell damage and Pluronic F68 protection. Biotechnol. Prog. 6: 391–397.

Oh SKW, Nienow AW, Al-Rubeai M and Emery AN (1989) The effects of agitation intensity with and without continuous sparging on the growth and antibody production of hybridoma cells. J. Biotechnol. 12: 45–62.

Oh SKW, Nienow AW, Al-Rubeai M and Emery AN (1992) Further studies of the culture of mouse hybridomas in an agitated bioreactor with and without continuous sparging. J. Biotechnol. 22: 245–270.

Ozturk SS and Palsson BO (1991) Examination of serum and bovine serum albumin as shear protective agents in agitated cultures of hybridoma cells. J. Biotechnol. 18: 13–28.

Ramirez OT and Mutharasan R (1990) The role of the plasma membrane fluidity on the shear sensitivity of hybridomas grown under hydrodynamic stress. Biotechnol. Bioeng. 36: 911–920.

Ramirez OT and Mutharasan R (1992) Effect of serum on the plasma membrane fluidity of hybridomas: An insight into its shear protective mechanism. Biotechnol. Prog. 8: 40–50.

Smith CG and Greenfield PF (1992) Mechanical agitation of hybridoma suspension cultures: Metabolic effects of serum, Pluronic F68, and albumin supplements. Biotechnol. Bioeng. 40: 1045–1055.

Tan WS, Chen YL and Dai GC (1993) Growth and damage of continuous suspension cultured hybridoma cells (2F7) in an agitated bioreactor with and without bubble entrainment or sparging. In: Nienow AW (ed.). The proceedings of 3rd International Conference on Bioreactor and Bioprocess Fluid Dynamics. (p. 153–161) BHR Group Series No. 5, Mechanical Engineering Publications, London.

Tramper J, Williams JB and Joustra DJ (1986) Shear sensitivity of insect cells in suspension. Enzyme Microb. Technol. 8: 33–36.

Tramper J, Smit D, Straatman J and Vlak JM (1988) Bubble-column design for growth of fragile insect cells. Bioproc. Eng. 3: 37–41.

Trinh K, Garcia-Briones M, Hink F and Chalmers JJ (1994) Quantification of damage to suspended insect cells as a result of bubble rupture. Biotechnol. Bioeng. 43: 37–45.

Zhang S, Handa-Corrigan A and Spier RE (1992) Foaming and media surfactant effects on the cultivation of animal cells in stirred and sparged bioreactors. J. Biotechnol. 25: 289–306.

Address for correspondence: W.S. Tan, Laboratory of Cell Culture Technology, Research Institute of Biochemical Engineering, East China University of Science & Technology, 130 Meilong Road, Shanghai 200237, P.R. China.

Cytotechnology **15:** 329–335, 1994.

Prediction of mechanical damage to animal cells in turbulence

C.R. Thomas, M. Al-Rubeai and Z. Zhang
BBSRC Centre for Biochemical Engineering, School of Chemical Engineering, University of Birmingham, Edgbaston, Birmingham B15 2TT, UK

Key words: Animal cells, cell strength, disruption, micromanipulation, modelling, turbulent flows

Abstract

In previous work a model was proposed for estimation of disruption of animal cells in turbulent capillary flows using information about the hydrodynamics, and cell mechanical properties determined by micromanipulation. The model assumed that the capillary flow consists of a laminar sublayer and a homogeneous turbulent region, and within the latter eddies of sizes similar to or smaller than the cells interact with those cells, causing local surface deformations. The proposed mechanism of cell damage was that such deformations result in an increase in membrane tension and surface energy, and that a cell disrupts when its bursting membrane tension and bursting surface energy are exceeded. The surface energy of the cells was estimated from the kinetic energy of appropriate sized eddies. To test the model, cells were disrupted in turbulent flows in capillaries at mean energy dissipation rates ranging from 800 to 2×10^4 Wkg^{-1}. The model assumed that the specific lysis rate is almost independent of the number of passes, which was verified by the experimental data. The implication was that despite the damage the cell mechanical properties did not change markedly during multiple recirculations through the capillaries. On average the model underestimated the cell disruption by about 15%. Although the model gave reasonably good predictions, it lacks proper explanation of the independence of the specific lysis rate on the number of passes. In this paper it is shown that this problem can be resolved in principle by consideration of the localisation of the energy dissipation in turbulent capillary flows. The necessity of further modelling of cell-turbulence interactions is demonstrated.

Introduction

Animal cells can be disrupted in sparged bioreactors, which limits the design and operation of large scale suspension cultures. In order to minimise the disruption, it is essential to understand the mechanisms by which this takes place. Ideally, a mechanistic model is needed to describe cell-hydrodynamic interactions so that cell disruption can be predicted using information about the hydrodynamics and about cell mechanical properties, such as cell bursting membrane tension, elastic area compressibility modulus and cell diameter. The latter parameters can be determined by micromanipulation (Zhang *et al.*, 1991; 1992), but sparged bioreactor hydrodynamics are complex and have not yet been well characterised. For the development of appropriate models to describe cell-hydrodynamic interactions, the logical procedure is therefore to begin by investigating cell behaviour in relatively simple flow fields.

Cell disruption in laminar flows has been investigated by Born *et al.* (1992). It was assumed that cells were deformed into ellipsoids in the laminar flow. Cell deformation was assumed to cause an increase in cell membrane tension. Each cell has its intrinsic bursting membrane tension. If the cell bursting membrane tension is exceeded the cell will be disrupted. Using information about cell mechanical properties measured by micromanipulation, a model was developed to estimate the cell disruption in laminar flows. To test the model, animal cells were exposed to shear stresses ranging from 124 Nm^2 to 577 Nm^2 for up to 3 mins in a cone and plate viscometer. The model could predict successfully losses of viable cells due to laminar shear stress, within a maximum error of less than 30%.

After such success, a model to estimate cell disruption in turbulent capillary flows was also proposed (Zhang *et al.*, 1993). Again the model consisted of two aspects: modelling of the hydrodynamics and a description of the cell-turbulence interactions. The flow field in a capillary was classified as a laminar sublayer close to the wall and a homogeneous turbulent region elsewhere. The turbulent flow field consists of eddies of different sizes. When a cell is exposed to a large eddy it will be entrained, but for eddies approaching or smaller than the cell size, there can be energy exchanges between the cell and eddies. It was assumed that the energy transferred to a cell will be converted to cell surface energy. A cell will deform locally, with indents and protuberances in the surface. These deformations will cause an increase in cell surface area and consequently an increase in membrane tension and surface energy. As with the laminar flow model, the cell disruption mechanism was presumed to be that a cell will be disrupted if its bursting membrane tension is exceeded. The energy carried by a cell was assumed to be equal to the kinetic energy of a liquid drop with the same size of the cell. By using these assumptions, a model was proposed to estimate the cell disruption per pass in turbulent capillaries. The model also assumed that the distribution of cell mechanical properties did not change significantly even during many passes. In order to check the model, cells in a holding flask were forced to recirculate through capillaries continuously for up to one and a half hours (up to 300 passes) and the actual cell losses were measured by counting the number of cells in the holding flask at different times. In all cases, the actual viable cell number showed almost a first-order decay with time, as also observed by other researchers (Augenstein *et al.*, 1971; McQueen *et al.*, 1987; McQueen and Bailey, 1989). Although the predictions of the model were reasonably consistent with experimental data, there remains a serious conceptual problem with the model. Why is the specific lysis rate almost independent of the number of passes when one might expect the weaker cells in the population to be disrupted first? Conceptually, if the flow field were homogeneous, cells would pick up the same amount of energy from the eddies during any pass. If this energy were large enough, any (weaker) cells that could be disrupted should be during their first pass through the capillaries, whilst strong cells should not be affected at all. Considering the residence time distribution of the cells in the presumably well-mixed holding flask, after four passes about 99% of cells should have passed through the capillary, and essentially all possible cell damage should have occurred. In fact, for some runs there were still cells being disrupted after 100 passes. In this paper, the aspect of the model concerning the hydrodynamics of turbulent capillary flows is reconsidered in order to answer this remaining question of why the specific lysis rate appears independent of the number of passes. Some further experimental data on cell disruption in a capillary and in a closed stirred tank are presented to support the theoretical developments.

The effect of the local energy dissipation rate on cell disruption

Normally, the flows in experimental capillaries are turbulent developing flows. The hydrodynamics for such developing flows are not well understood (Salami, 1986). In this paper the hydrodynamics of well developed turbulent pipe flows are assumed to be applicable, at least qualitatively, to developing turbulent flows in capillaries. For well developed flows, there are some data about the hydrodynamics, such as the distribution of the mean axial velocity and the distribution of the mean energy dissipation rate per unit mass (Laufer, 1954). Immediately adjacent to the inner wall of the tube there exists a laminar sublayer. The flow in a capillary tube consists of this laminar sublayer and turbulent flow elsewhere. The distribution of the mean axial velocity in the turbulent zone can be expressed by

$$u = u_{max}\left(1 - \frac{r}{R_t}\right)^{1/m} \tag{1}$$

where u_{max} is the fluid velocity at the centre line of the pipe, r is the radial distance from the centre line, R_t is the radius of the pipe, and m is a constant which depends on the Reynolds number. Normally m=7 for a wide range of Reynolds numbers (Benedict, 1980). There are also significant differences in the time-averaged energy dissipation rate per unit mass across turbulent region. Laufer (1954) has measured the radial distribution of the time-averaged energy dissipation rate per unit mass. The value of the energy dissipation rate per unit mass at the position $r/R_t=0.99$ is about 80 times that at $r/R_t=0$. It might be reasonable therefore to assume that cell disruption would be most likely to take place in a very small region close to the wall, whilst in other regions cells might survive because of lower local energy dissipation rates.

Previous work (Zhang *et al.*, 1993) proposed a cell disruption criterion:

$$\frac{3}{11}\rho\varepsilon^{2/3}R^{5/3} \geq T_b(1 + T_b/K) \qquad (2)$$

or

$$\varepsilon \geq \left[\frac{11T_b(1 + T_b/K)}{3\rho R^{5/3}}\right]^{3/2} \qquad (3)$$

where ρ is the fluid density of the medium, ε is the mean local energy dissipation rate per unit mass, R is the radius of the cell, T_b is the cell bursting membrane tension and K is the elastic area compressibility modulus. Equation (2) was obtained using the assumption that the time scales of turbulence are much less than cell disruption times. However, if the cells were purely elastic (as was otherwise assumed), they should be disrupted instantaneously when exposed to sufficiently high energy dissipation rates. Due to the randomness of turbulence, the local energy dissipation rate per unit mass will fluctuate with time. In that case, the instantaneous energy dissipation rate per unit mass ε_i should be used to replace ε to determine the cell disruption condition. Therefore

$$\varepsilon_i \geq \left[\frac{11T_b(1 + T_b/K)}{3\rho R^{5/3}}\right]^{3/2} \qquad (4)$$

The ratio between the maximum energy dissipation rate per unit mass and the mean depends on the properties of the turbulent flows. Meneveau and Sreenivasan (1991) found the ratio was about 60 in the atmospheric surface layer at a high Reynolds number of 7×10^6 and was about 20 in a laboratory boundary layer at a moderate Reynolds number of 32000. This ratio is not known in the developing capillary flow. However, it would be reasonable to assume ε_i is at least an order of magnitude greater than ε.

Suppose for all cells the inequality in equation (4) is satisfied in some zone in the capillary. If there is no radial migration of cells during flow through the capillary any cells entering this zone will be disrupted, provided the fluctuations in ε are rapid relative to the residence time in the capillary. By considering the mean axial fluid velocity distribution, the fraction of the total capillary volume within which cells are disrupted should be a good approximation for the cell specific lysis rate.

Previous work (Zhang *et al.*, 1993) also gave a relationship between the viable cell concentration C_i, and the number of passes N,

$$C_i = C_0 e^{-pN} \qquad (5)$$

where C_0 was the initial viable cell concentration in the flask and p was the specific lysis rate which was assumed to be constant. For a given cell culture, the higher the capillary mean energy dissipation rate per unit mass (and therefore the local mean and maximum instantaneous energy dissipation rates per unit mass), the bigger the region within which cells might be disrupted. Conversely, for a given capillary mean energy dissipation rate per unit mass, the weaker the cells (for example NS1 myelomas rather than TB/C3 hybridomas), the bigger the region where disruption will occur. In both cases the bigger the disruption region, the higher the specific cell lysis rate will be. With the simplifying assumptions outlined in Zhang *et al.* (1993) and given about the specific lysis rate should be independent of the number of passes. These qualitative conclusions are reasonably consistent with previous experimental data. For example, for the cell cultures used in the work of Zhang *et al.* (1993), using equation (4), the energy dissipation rate per unit mass ε_i which is required to disrupt about 95% of all cells will be of the order of 10^5 to 10^6 Wkg^{-1}. The overall mean energy dissipation rate per unit mass in the capillaries used in that study was of the order of 10^3 to 10^4 Wkg^{-1} (Zhang *et al.*, 1993). As discussed earlier, the local mean energy dissipation rate per unit mass near the capillary wall might be one order of magnitude greater than the overall mean. Furthermore, the instantaneous energy dissipation rate there might have values higher by at least another order of magnitude. Clearly the instantaneous energy dissipation rates per unit mass in a small zone near the capillary wall could be of the same order of magnitude as those needed to disrupt nearly all cells according to (4). It is possible that the reasonably good agreement found using the model of Zhang *et al.* (1993) was fortuitous, resulting from an underestimate of the energy dissipation rates and an overestimate of the capillary volume in which cells might be disrupted. Although the concept of a small disruption zone seems adequate from an order of magnitude argument, accurate prediction of cell disruption in capillaries will be impossible without better understanding of the hydrodynamics of developing flows.

As mentioned earlier, the concept of a small zone within which the cells can be disrupted, and a larger zone where cells are not affected is an approximation. Between these two zones, there could be a transition zone within which only a fraction of the cells can be disrupted because cells have a range of strengths. It should be noted that there is normally a wide distribution of cell mechanical properties (Zhang *et al.*, 1992). Theoretically the mean strength of the surviving cells should become greater as the weaker cells are destroyed, and the specific lysis rate should therefore decline with increasing number of passes. This decline might be obscured by cell counting errors and be less obvious because of backmixing in the holding flask. However, if the local energy dissipation rate per unit mass in the zone close to a capillary wall was only high enough to disrupt some of the cells entering that zone, there should be some strong cells which could survive an unlimited number of passes (assuming no other effects due to holding occurred). Unfortunately, this situation was not realised in the work of Zhang *et al.* (1993) because of the difficulty of maintaining turbulence in capillaries at lower levels of energy dissipation rate than those used.

To test the concepts outlined above, two types of experiments were undertaken. One was single-pass experiments. Cells held in a flask were pumped through a capillary and were collected in another flask. In such a system, all cells will be subjected to the same number of passes. The other experiments were conducted in a closed stirred tank where turbulence can be achieved at much lower energy dissipation rates per unit mass than is possible in capillaries.

Materials and methods

Cell culture

TB/C3 hybridoma cells were maintained in 50–200 ml Duran bottles on RPMI-1640 medium supplemented with 5% fetal calf serum. Once the viable cell concentration in a semicontinuous culture had reached 3 to 7 $\times$ 10^5 ml^{-1} (at various times after the previous dilution), cell suspension was transferred into the shear devices.

Flow apparatus

The flow apparatus consisted of a peristaltic pump, two holding flasks and a capillary test section connected with wide bore silicon tubing. The capillary tube had a diameter of 0.51 mm and a length of 10 mm. One flask had a top port which enabled fluid exit and the other had two ports which enabled sampling and fluid entry. The flow apparatus was at room temperature (24 °C). The cell suspension was pumped through the capillary in repeated single passes, being careful not to pump air from flask to flask. Flow rates of 0.95 mls^{-1} (Re=2380, overall mean ε=5160Wkg^{-1}, assuming a fluid density of 1000 kgm^{-3} and a viscosity of 10^{-3} Pas) and 1.18 mls^{-1} (Re=2940, overall mean ε=9330 Wkg^{-1}) were used.

Closed stirred tank

Cells were agitated for up to three hours in a 2L stirred fermentor (Infors, Switzerland) fitted with a single standard Rushton turbine. The fermentor diameter was 115 mm and its height was 224 mm. The diameter of the impeller was 57.5 mm. The free gas-liquid interface was eliminated by completely filling the vessel. The temperature of testing was controlled at 24 °C. Rotational speeds of 1000 rpm (Re=55100; tip speed 3.0 ms^{-1}) and 1500 rpm (Re=82650; tip speed 4.5 ms^{-1}) were used.

The concentration of cells in samples taken from the flow apparatus and the stirred vessel were measured by counting live and dead cells in a hemocytometer after staining by Trypan Blue. The mechanical properties of the viable cells in the holding flask for the repeated single pass experiment or in the stirred vessel were measured regularly by micromanipulation at room temperature, using the method of Zhang *et al.* (1991; 1992).

Results and discussion

Figure 1 shows that the fraction cells remaining versus the number of passes for a repeated single pass experiment. Since all cells were subjected to the same number of passes, it is expected some kind of decay in specific lysis rate might be detected. Clearly, the viable cell concentrations for the two runs do deviate from first-order decay; the slope of each curve declines with number of passes. Curve 1 can be fitted by the equation $\ln(C_i/C_0)=-0.25N^{0.56}$ with a mean deviation of 0.027. If Curve 1 were fitted by a straight line, the mean deviation would be 0.093, which is significantly greater than that due to cell counting errors. Similarly, Curve 2 can be fitted by $\ln(C_i/C_0)=-0.28N^{0.67}$ with the mean deviation of 0.027, whilst a linear correlation would not be acceptable. This is consistent with the concepts described earlier i.e. cell disruption in a small region of the capillary, and a bias towards dis-

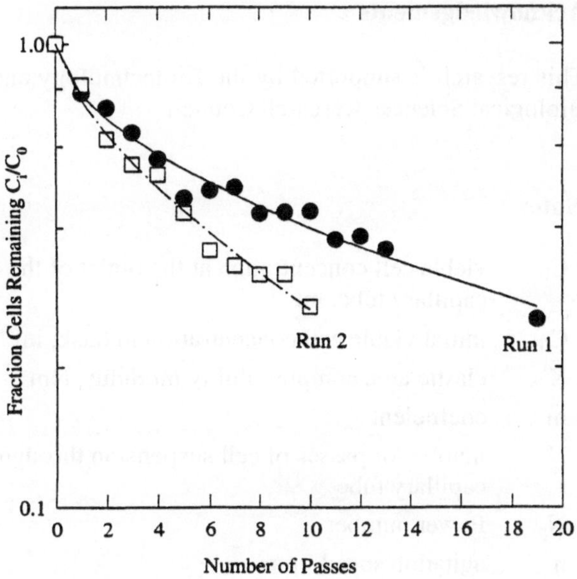

Fig. 1. Relationship between the viable cell concentration in a capillary tube and the number of passes. Run 1: ε_m=5160 Wkg^{-1}; Run 2: ε_m=9330 Wkg^{-1}.

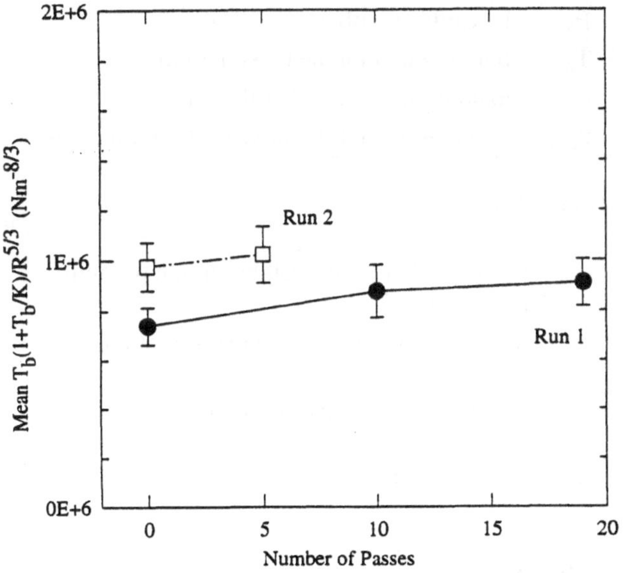

Fig. 2. Parameter from equation (4) representing cell strengths against number of passes in the capillary. The flow parameters of Run 1 and Run 2 are the same as in Figure 1. Error bars represent the standard error of the mean.

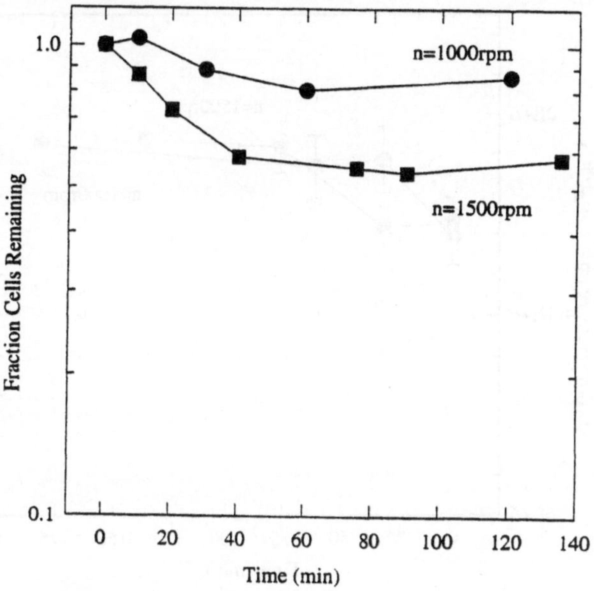

Fig. 3. Cell disruption in a closed stirred tank.

ruption of weaker cells due to some transition zone. Figure 2 gives the relationship between the mean value of the parameter as defined by the right-hand side of equation (4), which might be used to characterise cell strength, and the number of passes. It can be seen that the mean cell strength for each run increased slightly with the number of passes. The data presented in Fig. 1 and Fig. 2 are consistent with each other. Backmixing probably obscured this effect in the previous work of Zhang *et al.* (1993), and in that of other workers using recirculating flows through capillaries. In such systems, damaged cells are returned to a large reservoir of cells less exposed to the turbulence in the capillary, making it less easy to observe the damage, especially as cell counting errors are always significant.

Normally, turbulence can be maintained in stirred tanks at much lower mean energy dissipation rates per unit mass than in capillaries. However there is a wide distribution in the local energy dissipation rates, and the maximum mean local energy dissipation per unit mass in the impeller discharge zone could be about 10 to 100 times the mean for the whole vessel (Calabrese and Stoots, 1989; Wu and Patterson, 1989; Geisler, 1991). As shown by Fig. 3, there were significant losses in viable concentration during the first half hour at both agitation speeds. After that the viable cell concentrations did not change significantly. At 1000 rpm, the mean energy dissipation rate per unit mass was 7.2 Wkg^{-1}, assuming a power number N_p of 5.8 (Nienow and Miles, 1971), and about 20% of the cells were disrupted in total. At 1500 rpm the mean energy dissipation rate per unit mass was 24 Wkg^{-1} and about 40% of the cells was lost. As in Figure 2, the parameter reflecting the cell strength is plotted versus time in Fig.

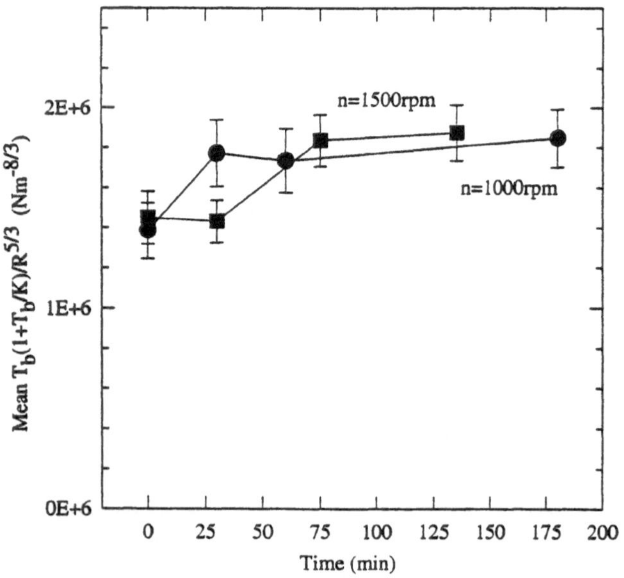

Fig. 4. Parameter from equation (4) representing cell strengths in the stirred vessel versus time. Error bars represent the standard error of the mean.

4. This demonstrates again that cells became stronger as cells were disrupted. In the late period of the experiment for both agitation speeds, both the viable cell concentration and the cell strength parameter values levelled off. This strongly suggests that in a closed stirred tank only the local energy dissipation rate per unit mass is critical to cell disruption and it was not high enough in these experiments to disrupt all the cells.

Conclusion

The localisation of the energy dissipation rate in turbulent capillaries and its possible effect on cell disruption has been considered to explain the relationship between the specific lysis rate and the number of passes. The experimental results from repeated single passes in the capillary and from cell disruption in the closed stirred tank suggest a bias toward disruption of weaker cells in turbulent flows. Such theoretical considerations and experimental observations might be used as a basis for further quantitative modelling of cell disruption in turbulent flows, although more detailed hydrodynamic information will certainly be needed if good predictions of disruption are to be achieved.

Acknowledgements

This research is supported by the Biotechnology and Biological Sciences Research Council.

Notes

C_i	viable cell concentration at the outlet of the capillary tube, ml^{-1}
C_0	initial viable cell concentration in flask, ml^{-1}
K	elastic area compressibility modulus, Nm^{-1}
m	coefficient
N	number of passes of cell suspension through capillary tube
N_p	Power number
n	agitation speed, rpm
p	specific cell lysis rate, $pass^{-1}$
r	distance from centre line of capillary tube, m
R	cell radius, m
R_t	radius of capillary tube, m
Re	Reynolds number
T_b	bursting membrane tension, Nm^{-1}
u	axial fluid velocity in tube, ms^{-1}
u_{max}	maximum axial liquid velocity in tube, ms^{-1}

Greek symbols

ε	mean local energy dissipation rate per unit mass, Wkg^{-1}
ε_i	instantaneous energy dissipation rate per unit mass, Wkg^{-1}
ε_m	mean overall energy dissipation rate per unit mass, Wkg^{-1}
ρ	density of fluid, kgm^{-3}

References

Augenstein DC, Sinksey AJ and Wang DIC (1971) Effect of shear on the death of two strains of mammalian tissue cells. Biotechnol. Bioeng. 13: 409–418.
Benedict RP (1980) Fundamentals of pipe flow. John Wiley and Sons, New York.
Born C, Zhang Z, Al-Rubeai M and Thomas CR (1992) Estimation of disruption of animal cells by laminar shear stress. Biotechnol. Bioeng. 40: 1004–1010.
Calabrese RV and Stoots CM (1989) Flow in the impeller region of a stirred tank. Chem. Eng. Progress 85: 43–50.

Geisler RK (1991) Fluidynamik und Leistungseintrag in turbulent geruhrten Suspensionen. PhD thesis, Technischen Universität München.

Laufer J (1954) The structure of turbulence in fully developed pipe flow. National Advisory Committee for Aeronautics Report 1174, 1–18.

McQueen A and Bailey JE (1989) Influence of serum level, cell line, flow type and viscosity on flow-induced lysis of suspended mammalian cells. Biotechnol. Letts. 11: 531–536.

McQueen A, Meilhoc E and Bailey JE (1987) Flow effects on the viability and lysis of suspended mammalian cells. Biotechnol. Letts. 9: 831–836.

Meneveau C and Sreenivasan KR (1991) The multifractal nature of turbulent energy dissipation. J. Fluid Mech. 224: 429–484.

Nienow AW and Miles D (1971) Impeller power numbers in closed vessels. Ind. Eng. Chem. Process Des. Develop. 10: 41–43.

Salami LA (1986) An investigation of turbulent developing flow at the entrance to a smooth pipe. Int. J. Heat and FLuid Flow 7: 247–257.

Wu H and Patterson GK (1989) Laser-Doppler measurements of turbulent-flow parameters in a closed mixer. Chem. Eng. Sci. 44: 2207–2221.

Zhang Z, Ferenczi MA, Lush AC and Thomas CR (1991) A novel micromanipulation technique for measuring the bursting strength of single mammalian cells. Appl. Microbiol. Biotechnol. 36: 208–210.

Zhang Z, Ferenczi MA and Thomas CR (1992) Micromanipulation technique with a theoretical cell model for determining mechanical properties of single mammalian cells. Chem. Eng. Sci. 47: 1347–1454.

Zhang Z, Al-Rubeai M and Thomas CR (1993) Estimation of disruption of animal cells by turbulent capillary flow. Biotechnol. Bioeng. 42: 987–993.

Address for correspondence: C.R. Thomas, BBSRC Centre for Biochemical Engineering, School of Chemical Engineering, University of Birmingham, Edgbaston, Birmingham B15 2TT, UK.

Cytotechnology **15:** 337–349, 1994.

Repeated-batch cultures of Baby Hamster Kidney cell aggregates in stirred vessels

J.L. Moreira[1], A.S. Feliciano[1], P.C. Santana[1], P.E. Cruz[1], J.G. Aunins[2] & M.J.T. Carrondo[1,3]
[1] *Instituto de Biologia Experimental e Tecnológica / Instituto de Tecnologia Quimíca e Biológica (IBET/ITQB), Apartado 12, 2780 Oeiras, Portugal;* [2] *Merck & Co., Rahway, 07065 NJ, U.S.A;* [3] *Laboratório de Engenharia Bioqúmica, FCT/UNL, 2825 Monte da Caparica, Portugal*

Key words: BHK, hydrodynamic size control, natural aggregates, cell retention, repeated-batch culture, stirred vessels.

Abstract

Natural aggregates of Baby Hamster Kidney cells were grown in stirred vessels operated as repeated-batch cultures during more than 600 hours. Different protocols were applied to passaging different fractions of the initial culture: single cells, large size distributed aggregates and large aggregates. When single cells or aggregates with the same size distribution found in culture are used as inoculum, it is possible to maintain semi-continuous cultures during more than 600 hours while keeping cell growth and viability. These results suggest that aggregate culture in large scale might be feasible, since a small scale culture can easily be used as inoculum for larger vessels without noticeable modification of the aggregate chacteristics. However, when only the large aggregates are used as inoculum, it was shown that much lower cell concentrations are obtained, cell viability in aggregates dropping to less than 60%. Under this 'selection' procedure, aggregates maintain a constant size, larger than under batch experiments, up to approximately 400 hours; after this time, aggregate size increases to almost twice the size expected from batch cultures.

Introduction

Baby Hamster Kidney (BHK) and Chinese Hamster Ovary (CHO) cells are very widely used in industry for the production of biologicals; since the 1960's, BHK have been accepted for the production of vaccines (Pay *et al.*, 1985; Radlett, 1987). Both BHK and CHO are anchorage-dependent cells often forming natural aggregates when grown in suspension in stirred vessels (Litwin, 1991; Moreira *et al.*, 1994a; Peshwa *et al.*, 1993; Renner *et al.*, 1993). Other anchorage-dependent cells, like Vero, 293 and Swine Testicular cells have also been reported as naturally aggregative in stirred vessels (Goetghebeur and Hu, 1991; Litwin, 1992; Peshwa *et al.*, 1993). In microcarrier cultures of diploid fibroblasts, the formation of aggregates is sometimes observed (Junker *et al.*, 1992), and the decrease of microcarrier concentration of BHK cell culture leads to the formation of multilayers, clumps and aggregates (Alves *et al.*, 1994; Ogata *et al.*, 1993). Small microspheres have been used to induce the formation of aggregates for CHO, Vero, 293 and Swine Testicular cells (Goetghebeur and Hu, 1991; Perusich *et al.*, 1991).

The cell culture method of choice is very important, since culture conditions influence protein glycosylation and thus its structure and activity (Goochee and Monica, 1990; Maiorella *et al.*, 1993). BHK cells grown in suspension and on microcarriers produce recombinant proteins whose glycosylation pattern can be profoundly affected by the culture method (Gawlitzek *et al.*, 1994). It was reported that BHK morphology (flat and elongated on the microcarrier surface, rounded in aggregates or single cells) affects the relation between cell growth and productivity (Racher *et al.*, 1994).

Aggregates can easily be retained inside reactors due to their large size (in the order of hundreds of microns) by sedimentation, while single cells require additional equipment, e.g. filters or centrifuges. To use aggregates in bioreactors, the control of their size is essential: if too large aggregates are produced, necrotic centres can be formed, reducing productivity and increasing downstream processing problems.

Aggregates have other potential uses, such as artificial organs for transplantation (Hirai *et al.*, 1991) or bioartificial organs (Nyberg *et al.*, 1993).

Aggregate characteristics are dependent upon the specific cell line used, medium composition and agitation rate (Tolbert *et al.*, 1980). Microsphere induced aggregates with sizes ranging between 90 and 500 μm were obtained (Goetghebeur and Hu, 1991), depending on the cell line (Vero, CHO, 293 or Swine Testicular cells). Medium composition has a large influence on aggregate characteristics: Vero cells were grown in several different serum free media with aggregate size and density being dependent on the media (Litwin, 1992). 293 cell aggregate formation, packing and viability are largely influenced by calcium concentrations (Peshwa *et al.*, 1993), whereas CHO cell aggregates could be eliminated by the use of a medium with reduced concentration of amino acids and inorganic salts (Boraston *et al.*, 1992). Serum-free medium is important for the use of animal cells in large-scale. In a previous work submitted for publication elsewhere (Moreira *et al.*, 1994b) the authors showed that serum-free medium can be successfully used to grow animal cell aggregates in stirred vessels; if cell growth is not affected, aggregate characteristics (size, compactness and adherent fraction) are generally maintained.

Hydrodynamics has been identified as one of the most important variables for aggregate size control in batch experiments (Moreira *et al.*, 1994a); BHK viable aggregates with sizes between 80 and 200 μm were obtained in the studied range of agitation rates (70 to 20 rpm). In order to use natural aggregates in bioreactors, longer bioreaction periods would normally be required. Furthermore, information relating to aggregate size stability and cell growth should be available to allow control strategies to be implemented. In previous work, repeated-batch cultures of BHK cells were maintained during more than 600 hours; after achieving maximum cell growth the suspension was transferred into the new vessel twice a week without the separation of cells and aggregates. Total cell concentration, growth and viability, and all aggregate characteristics were maintained, similarly to the batch cultures under the same conditions (Moreira *et al.*, 1994c). Nevertheless passaging of cells was only made without the separation of any fraction of the original culture. In this report we describe the results of the culture of BHK natural aggregates in stirred tanks operated during long periods of time as repeated-batch cultures, with retention of different fractions of the culture: single cells, normal size aggregates and large aggregates.

Materials and methods

Reagents

BHK NOVO cells genetically modified to secrete Alkaline Phosphatase (AP) were obtained from Dr. Hansjorg Hauser (GBF, Braunschweig, Germany). The cells were grown in Dulbecco's Modified Eagle's Medium (DMEM), supplemented with 5% fetal calf serum (FCS), 4 mM glutamine and 1 g/l glucose (all final concentrations and from Gibco, Glasgow, UK). Trypsinization was performed using a 0.2% (w/v) trypsin/EDTA solution (Gibco, Glasgow, UK). Cells for inoculum were grown in T-flasks and roller bottles. All cultures were performed at 37 °C in a 7% CO_2 atmosphere.

Bioreactor system and passaging procedures

Bioreactor studies were performed in 250 ml Wheaton Magna-Flex spinners (with a suspended ball impeller) (Techne, USA) at 45 rpm. Spinner culture medium was as described above, but a final glucose concentration of 5 g/l was used instead. Inoculation density was 2×10^5 cells/ml.

Twice a week, cells were taken from the spinner flask and 2×10^5 cells/ml reinoculated in a new spinner. Several experiments have been performed, as described in Fig. 1: I) after 5 minutes of sedimentation, suspended single cells and small aggregates were carefully removed, concentrated by centrifugation and inoculated in a new spinner; II) after 5 minutes of sedimentation, 200 ml of the culture were carefully removed, and the remaining fraction (mainly large aggregates) used as inoculum for the new spinner; III) similar to experiment II, but using 10 minutes of sedimentation (a larger aggregate size range was transferred into the new vessel). The same fraction of medium was exchanged during passaging in each experi-

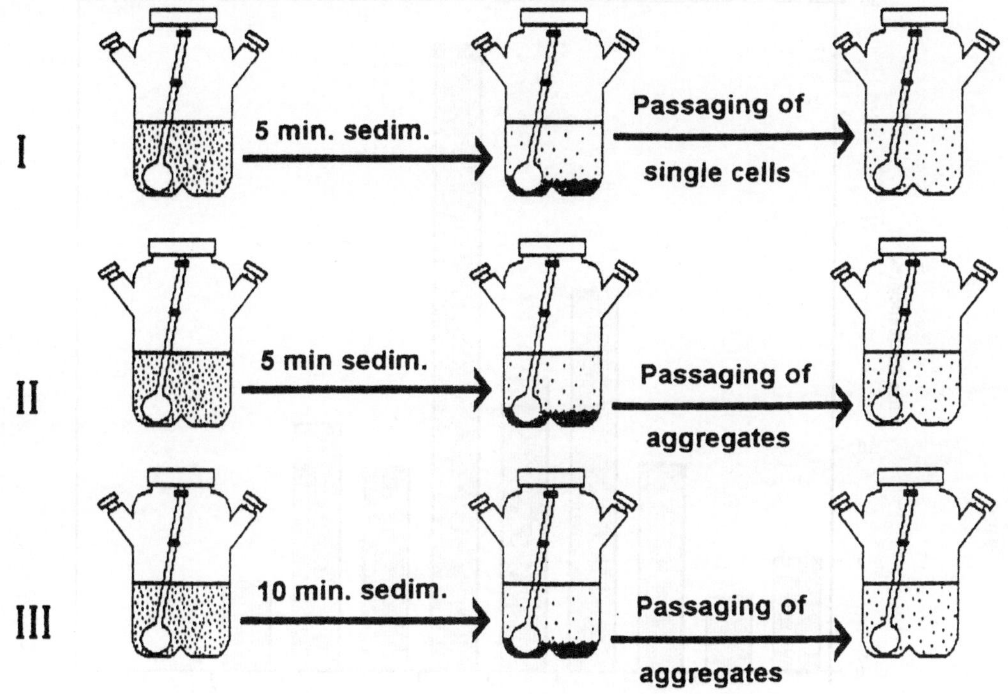

Fig. 1. Passaging procedures for the different repeated-batch cultures.

ment (approximately 96% of the culture volume). As a blank to prove that the culture size distribution is not affected by the centrifugation step, experiments were also performed without centrifugation (thus using different culture volumes during passaging), similar results being obtained.

Analytical

Sampling was performed as described elsewhere, optimising aggregate dissociation at negligible cell damage (Moreira *et al.*, 1994d). In short, homogeneous aliquots were removed from the culture, and 50 μl observed in an inverted microscope for aggregate size and number quantification (using a micropipette with an enlarged orifice tip and a 96 well plate); the measurement of at least 50 aggregates has been shown to be statistically significant. Total and viable cell concentrations were evaluated using trypan blue dye exclusion, after aggregate dissociation by passing a 1:1 dilution of homogeneously suspended cells in trypan blue (Gibco, Glasgow, UK) 50 times through a normal 200 μl micropipette tip (Gilson, Villiers, France). The comparison with nuclei counts has been previously presented (Moreira *et al.*, 1994d), with similar results being obtained. The single cell concentration was obtained after settling the aggregates inside the sampling tube (5 to 10 minutes inside a 14 × 45 mm sample tube leads to the preservation of single cells in suspension).

Results and discussion

Aggregate size distribution (result of the measurement of more than 100 aggregates) 72 hours after the initial inoculation is presented in Fig. 2. This size distribution is the initial one for all experiments, since the same initial culture was used to start the three different cultures.

The major fraction of aggregates has sizes ranging 100 to 160 μm, but small (40 to 80 μm) and very large aggregates (180 to 220 μm) could be found. After 5 minutes of sedimentation only the large aggregates were decanted and the separation between small aggregates and single cells is not complete. Thus, after passaging of single cells and small aggregates (experiment I) an increase in single cell fraction and a large decrease in aggregate size was obtained. In both experiments, where mainly aggregates were transferred into the new vessel (experiments II and III), a decrease in the fraction of single cells was observed. In experiment II only the larger aggregates were transferred into

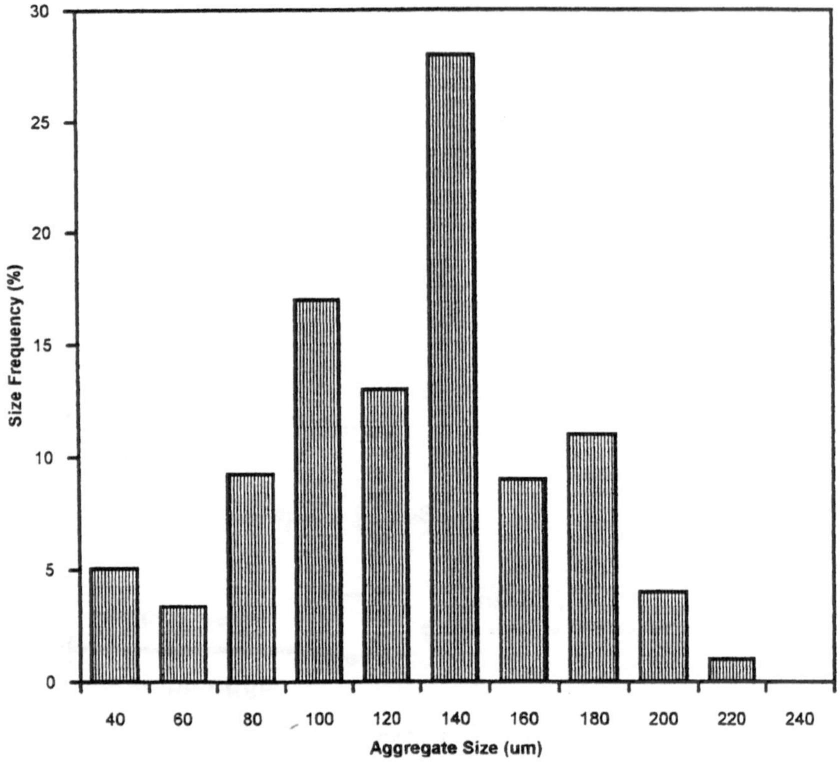

Fig. 2. Aggregate size distribution 72 hours after the initial inoculation.

the new vessel and thus the average aggregate diameter was normally larger than the value obtained in the other experiments, whereas in experiment III aggregates with size near the initial distribution were transferred into the new vessel, covering an intermediate situation between experiment I and II. Re-aggregation is a fast phenomenon, and soon after passaging the aggregate size distribution changed, making the determination of aggregate size distributions immediately after passaging difficult.

Viable cell concentrations are presented in Fig. 3. The number of cell doublings is presented in Fig. 4A, and the number of cells accumulated during the experiments per volume of feed (cell yield on medium) is shown in Fig. 4B. When single cells and large size distributed aggregates were transferred into the new vessel (experiments I and III) similar results were obtained (Fig. 3A and C), up to 2×10^6 cells/ml being achieved at the end of the growth phase; slightly larger cell concentrations were achieved when large size distributed aggregates were transferred into the new vessel (experiment III), thus justifying the differences found between those experiments presented in Fig. 4B. Also cell population doublings were similar for both experiments (Fig. 4A); these results are in close agreement with the data obtained in batch cultures (Moreira *et al.*, 1994a), in repeated-batch culture with no special cell separation before passaging (Moreira *et al.*, 1994c) and in attached static cultures (data not shown). Nevertheless when basically large aggregates were transferred into the new vessel (experiment II), cells grew to less than 1×10^6 cells/ml (Figs. 3B and 4B), clearly indicating a disadvantage of using large aggregates as inoculum. Cell doublings were also lower in this case than for the other experiments (Fig. 4A).

Using the sampling method described above, it is possible to obtain viability data for the single and aggregated cells. The average viability of each batch cycle for all experiments is presented in Fig. 5. No large differences in viability of single cells were observed between experiments (Fig. 5A); viabilities between 80 and 90% were achieved, similarly to batch cultures (Moreira *et al.*, 1994a). Comparable results were obtained for cells in aggregates when single cells and large size distributed aggregates were transferred into the new vessel (experiments I and III) (Fig. 5B); however, it is possible to observe that cell viability dropped to the 60 to 50% range in experiment II, where only

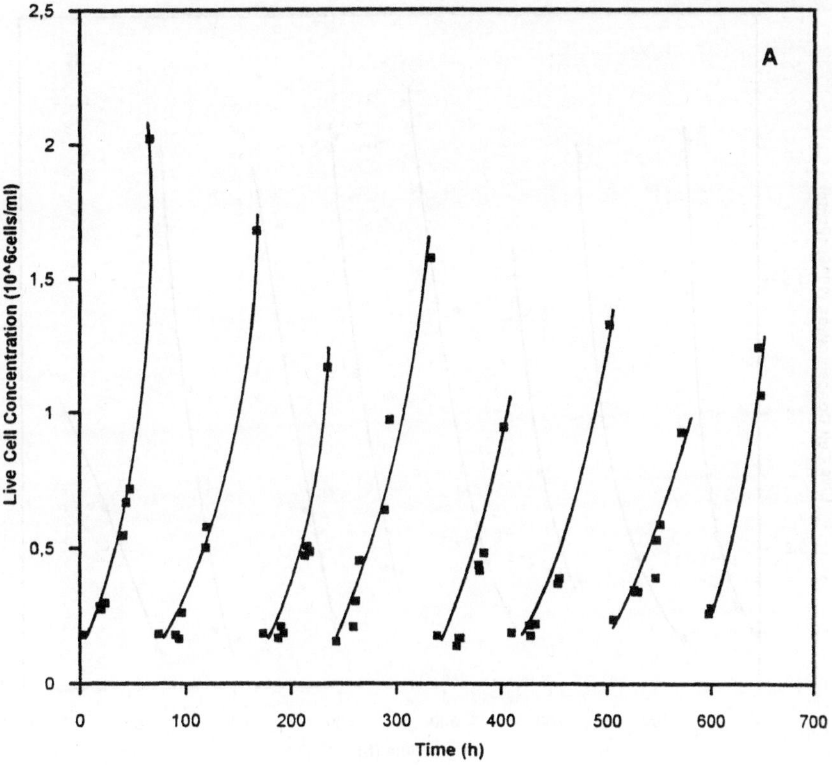
A
Live Cell Concentration (10^6cells/ml)
2,5
2
1,5
1
0,5
0
0
100
200
300
400
500
600
700
Time (h)

Fig. 3a.

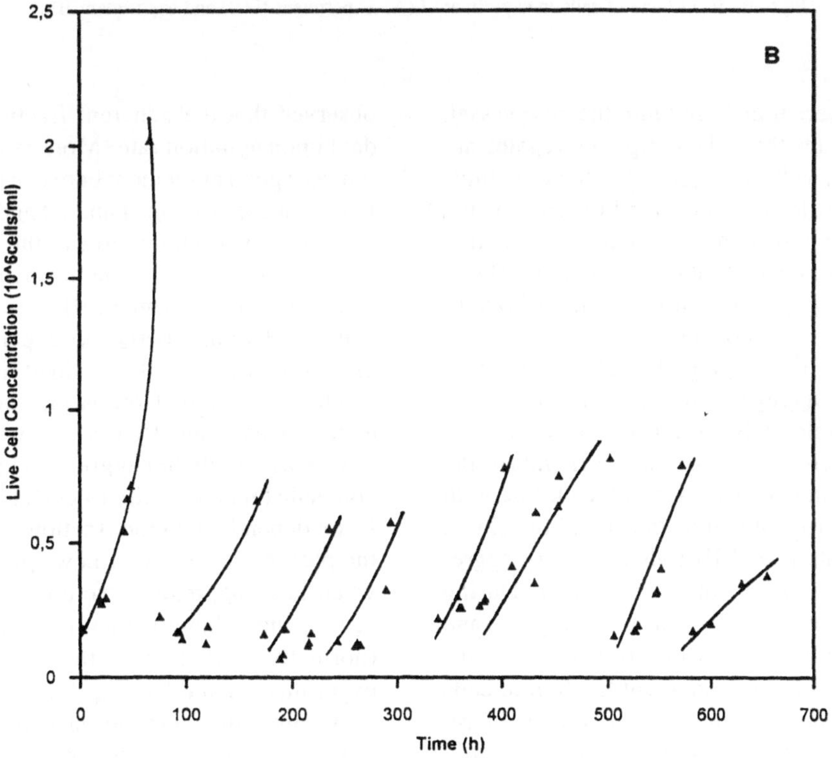
B
Live Cell Concentration (10^6cells/ml)
2,5
2
1,5
1
0,5
0
0
100
200
300
400
500
600
700
Time (h)

Fig. 3b.

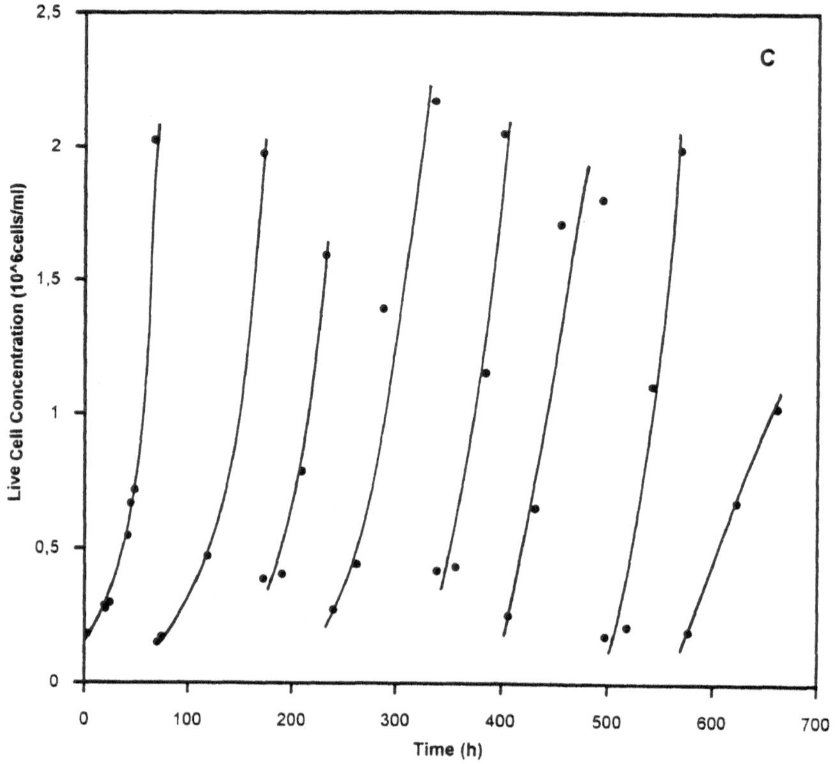

Fig. 3c.

Figs. 3a–3c Growth of cells in experiment I (A), experiment II (B) and experiment III (C).

large aggregates were transferred into the new vessel. This clearly indicates that when large aggregates are used as inoculum, cells in aggregates have a limited growth and start losing their viability whereas the single cells (resulting from the inoculation procedure, growth or erosion of the cells at the aggregate surface) maintain their viability. The cause for this behaviour of the large aggregates is not clear.

Immediately after passaging, the adherent fraction (fraction of cells in aggregates) dropped in experiment I (where mainly single cells were transferred into the new vessel) and increased in experiment II and III; the aim of these experiments was to start the culture with single cells in experiment I, and almost without single cells in experiments II and III (where mainly aggregates were transferred into the new vessel). During each batch cycle, cells grew both as aggregates and single cells; as can be seen in Fig. 6 there is no difference between the average adherent fraction of cells (the ratio between the concentration of cells in aggregates and total cells) in experiments I and III (approximately 60% in both cases). In previous work it was observed that the adherent fraction is largely dependent upon agitation rate (Moreira *et al.*, 1994a). There is a very good agreement between the data presented in Fig. 6 for experiments I and III and the data previously obtained for batch cultures at the same agitation rate. These results can be explained by rapid shedding from aggregates in experiment III to re-establish the single cell population, whereas in experiment I single cells grew similarly to cells obtained from T-flasks used in batch inoculation. Comparing the data obtained for experiments I and III (passaging of single cells and large size distributed aggregates) it is thus possible to conclude that the ability of cells to grow as aggregates is not dependent on the fraction of single cells used in the passage of cells to a new spinner flask. However, when large aggregates were transferred into the new vessel, the adherent fraction always remained larger (normally close to 80%) than the values obtained for experiments I and III (Fig. 6).

Control of aggregate size in bioreactors is very important. The diameter of the aggregates obtained in this study is presented in Fig. 7. For experiment I

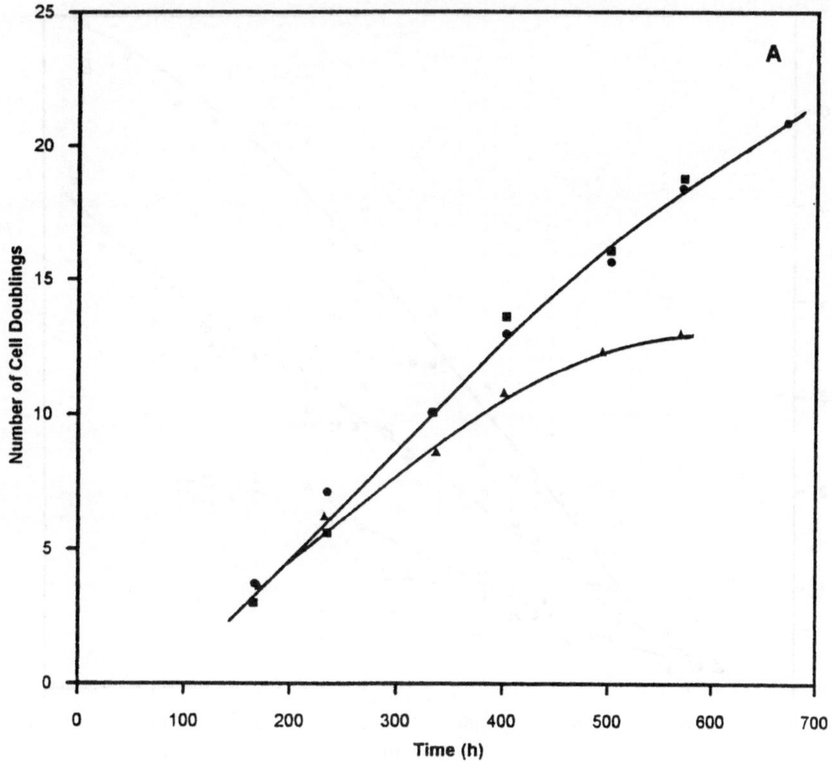

Fig. 4a.

and III the mean aggregate diameter was in the range of 140 μm with size distributions similar to the initial culture (presented in Fig. 2).

From our previous work in batch culture it was verified that hydrodynamics is one of the most important variables for size control (Moreira *et al.*, 1994a), and using the same vessel at 45 rpm, aggregates with similar diameters to those presented in Fig. 7 when single cells and large size distributed aggregates are transferred into the new vessel (experiments I and III) were obtained. Similar results were obtained in a BHK repeated-batch culture operated during more than 700 hours with passaging of the homogeneous suspension without cell or aggregate separation (Moreira *et al.*, 1994c).

For experiment II, where only large aggregates were transferred into the new vessel, during the initial 400 hours of operation, aggregates diameters remained in the range 160 to 170 μm (Fig. 7), thus larger than the aggregates in the other experiments. After that time, aggregates grew in size up to diameters almost twice as large as those obtained for the other experiments. After 500 hours this experiment was stopped, since after passaging of these large aggregates almost no cell growth was observed (data not shown).

As can be seen in Fig. 8, the concentration of aggregates was much reduced when large aggregates were transferred into the new vessel (experiment II). Since in experiment II smaller total cell concentrations, larger adherent fraction and larger aggregates were obtained (Figs. 3, 6 and 7), a reduced concentration of aggregates could be anticipated.

Cell density in aggregates (the number of cells present in one cm^3 of aggregate, obtained from cell concentration in aggregates and mean aggregate diameter) when aggregates are transferred into the new vessel (experiments II and III) is normally in the range 3 to 5 $\times$ 10^8 cells/ml aggregate, similarly to the data obtained in batch cultures (Moreira *et al.*, 1994a). Aggregates originated from single cells (experiment I) normally have a lower cell density than when aggregates are transferred into the new vessel (2 to 4 $\times$ 10^8 cells/ml aggregate). This might explain a slight increase in aggregate concentration found for experiment I (Fig. 8). This also suggests that when aggregates are transferred into the new vessel cell reorganization inside aggregates might occur, originating

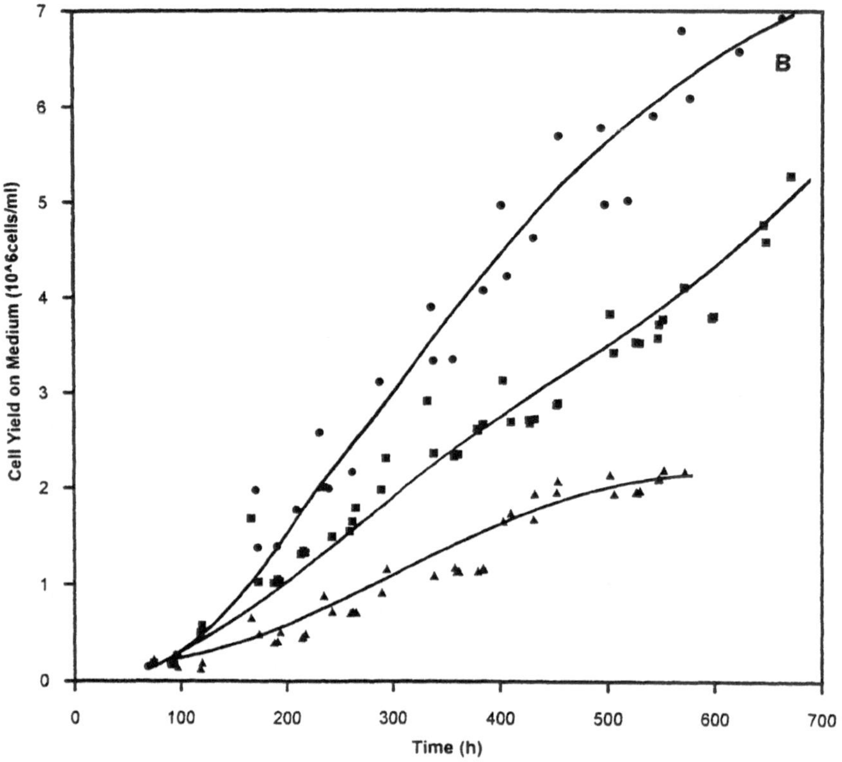

Fig. 4b.

Figs. 4a–4b Number of cell doublings (A) and cell yield on medium (B) (■ – experiment I: ▲ – experiment II; • – experiment III).

denser particles, due to aggregate breakage and re-aggregation. Nevertheless since both experiments II and III (passaging of aggregates) lead to aggregates with similar densities, the decrease in cell growth and viability observed when large aggregates are passed (experiment II) can not be associated with the formation of very dense particles with diffusional problems.

The decrease in cell growth and viability of cells in aggregates can not be explained by the higher aggregate diameter (depicted in Fig. 7), since during 400 hours aggregate size remained close to 170 μm (already with low cell growth and viability) and only then increased to 240 μm.

The formation of very large viable aggregates of different cell lines has also been reported: hepatocytes with 200 μm (Hirai *et al.*, 1991), BHK cells with 200 μm (Moreira *et al.*, 1994a), Vero cells with 250 μm (Goetghebeur *et al.*, 1991; Perusich *et al.*, 1991), CHO cells up to 360 μm and Swine Testicular cells up to 380 μm (Goetghebeur *et al.*, 1991); cell viability of those aggregates was always higher than 85%. Only at 600 μm were aggregates of 293 cells found to have necrotic centres with dead cells (Peshwa *et al.*, 1993). Although no difference was established between proliferating and quiescent cells.

Aggregates of tumour cells (spheroids) were reported with sizes above 800 μm, having a central zone with necrotic cells, an intermediate region with non proliferative but alive cells and an outside zone where cells grew easily. The thickness of the proliferative and quiescent rims of EMT6/R0 spheroids ranged between 50 to 200 μm and 200 to 600 μm, respectively (Mueller-Klieser, *et al.*, 1986; Landry and Freyer, 1984; Freyer and Sutherland, 1985). Those results are in good agreement with the high cell viability and growth observed in batch cultures with aggregates up to 200 μm (Moreira *et al.*, 1994a) and can not explain the decrease in cell viability and growth observed in experiment II (when large aggregates are transferred into the new vessel), since necrotic centres are not expected for aggregates of 170 μm.

We have previously shown that the size of aggregates grown in batch cultures is controlled by the

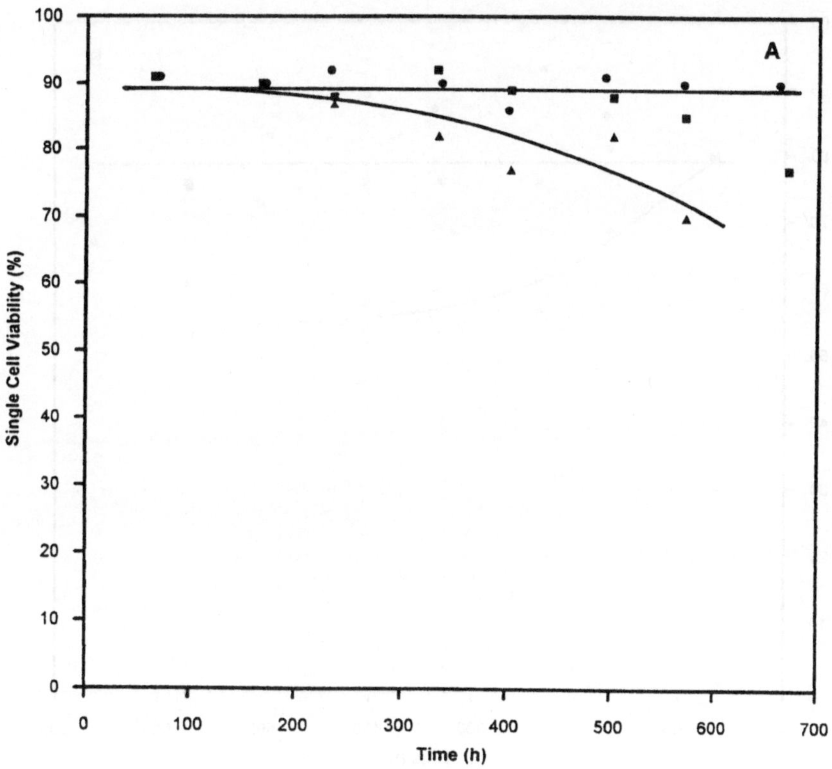

Fig. 5a.

hydrodynamic of the liquid, independently of vessel and impeller size and geometry (Moreira *et al.*, 1994e); aggregate size can be correlated with the Kolmogoroff's microscale theory (Kawasa and Moo-Young, 1990). Thus it is possible to assume that the 'critical eddies' for aggregate size control exercise a force on the aggregates that has the same order of magnitude of the adhesiveness forces between cells in the aggregates. The results of experiments I and III (where single cells and large size distributed aggregates were transferred into the new vessel) clearly indicate that the adhesiveness force between cells is constant throughout the experiments. When large size aggregates are transferred into the new vessel (experiment II), the observed increase in size immediately after the first passaging can be associated with an increase in the adhesiveness force between cells. In a recent report it was observed that CHO cell aggregates are formed around decaying and dead cells, with deoxyribonucleic acid (DNA) released from the cells mediating cell-cell adhesion (Renner *et al.*, 1993). Since a decrease in cell viability was observed immediately after the first passaging (Fig. 5B), the increase in free DNA can thus justify the increase in aggregate size. Nevertheless the reason why the increase to much larger sizes (from 170 to 240 μm) only takes place 400 hours after the beginning of the experiment is not clear.

Conclusions

It was observed that when single cells or large size distributed aggregates are used as inoculum for passaging, it is possible to maintain semicontinuous cultures during more than 600 hours. Cell growth, viability, adherent fraction, aggregate size and concentration are similar in both cases (experiments I and III), in agreement with the data obtained from batch cultures. From these experiments it is also possible to conclude that aggregated cells lead to new aggregates, similar to those obtained with single cells (either resulting from T-flasks or single cells from a stirred vessel). This also corroborates the conclusions previously obtained in batch culture that the adherent fraction and aggregate size are basically dependent on hydrodynamics (Moreira *et al.*, 1994a).

On the other hand, if only large aggregates are used as inoculum, it was shown that much lower cell con-

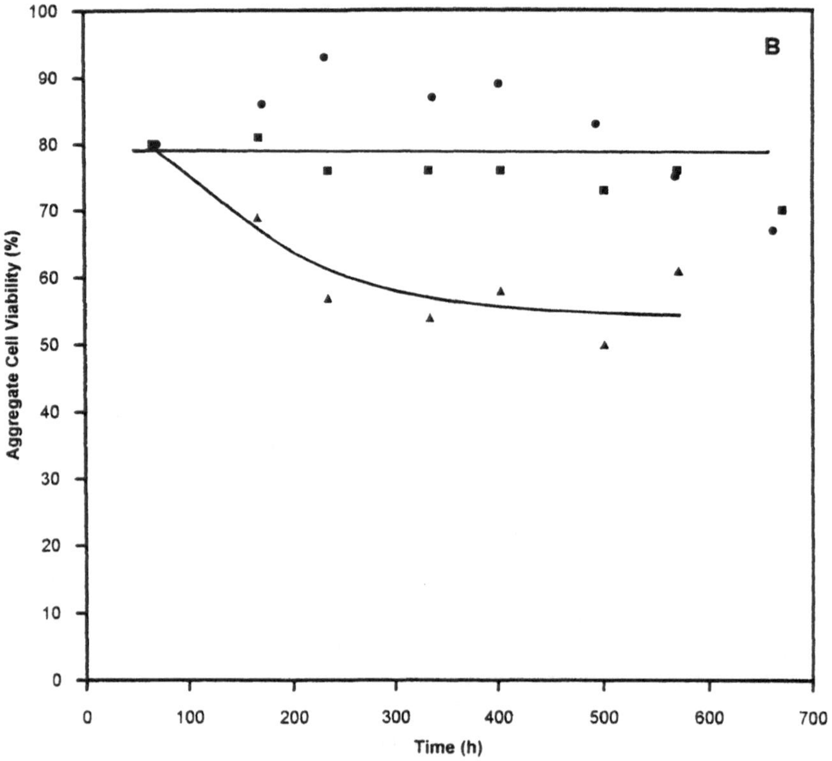

Fig. 5b.

Figs. 5a–5b Single cell (A) and aggregate cell (B) viabilities (■ – experiment I: ▲ – experiment II; • – experiment III).

centrations are obtained. Cells in these large aggregates become non viable whereas single cells remain viable (Fig. 5). Aggregates normally remain slightly larger than in other experiments (I and III), but their size remains constant during more than 400 hours (between 160 and 170 μm) and then increases to 240 μm. Since larger viable aggregates were obtained in batch cultures (Moreira *et al.*, 1994a), the existence of diffusional limitations does not explain the formation of necrotic centres inside aggregates in experiment II, as cell density is unchanged. The increase in size when large aggregates are transferred into the new vessel can be associated with an increase in cell adhesiveness strength in aggregates, probably due to the release of DNA of dead cells in those aggregates, thus associated with the decrease in aggregate viability experimentally observed.

These results also evidence the easy application of aggregates in scale up, since a small scale culture can easily be used as inoculum for larger vessels without the modification of aggregate characteristics. Since reinoculation of cells adhered to macroporous supports is a difficult task, this clearly constitutes an advantage for aggregates. Possible disadvantages like the reduction of cell growth or the formation of necrotic centres were not observed for batch cultures of aggregates with sizes up to 350 μm (Goetghebeur and Hu, 1991; Moreira *et al.*, 1994a). The use of 250 μm aggregates for vaccine production was also reported (Perusich *et al.*, 1991); a decrease in the kinetics of virus production was observed, but final titters similar to those obtained on microcarrier culture were obtained. From our previous knowledge (Moreira *et al.*, 1994e) it is also expected that if the hydrodynamic conditions of large volume vessels are controlled, aggregate characteristics (size, compactness, adherent fraction and concentration) can be reasonably maintained.

One of the advantages of the use of aggregates in stirred vessels is the easy cell retention by sedimentation. The optimal utilisation of decanters for a certain perfusion rate is a function of aggregate size. From this work it is possible to conclude that if the decanter is designed only for the retention of large aggregates it will not be possible to maintain high viability and

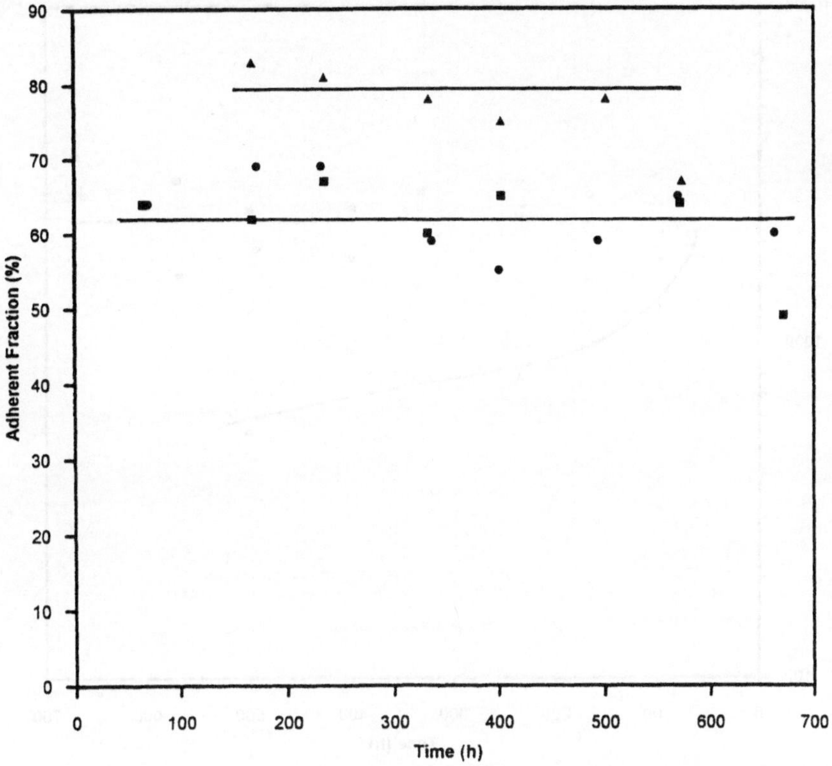

Fig. 6. Adherent fraction (■ – experiment I: ▲ – experiment II; • – experiment III).

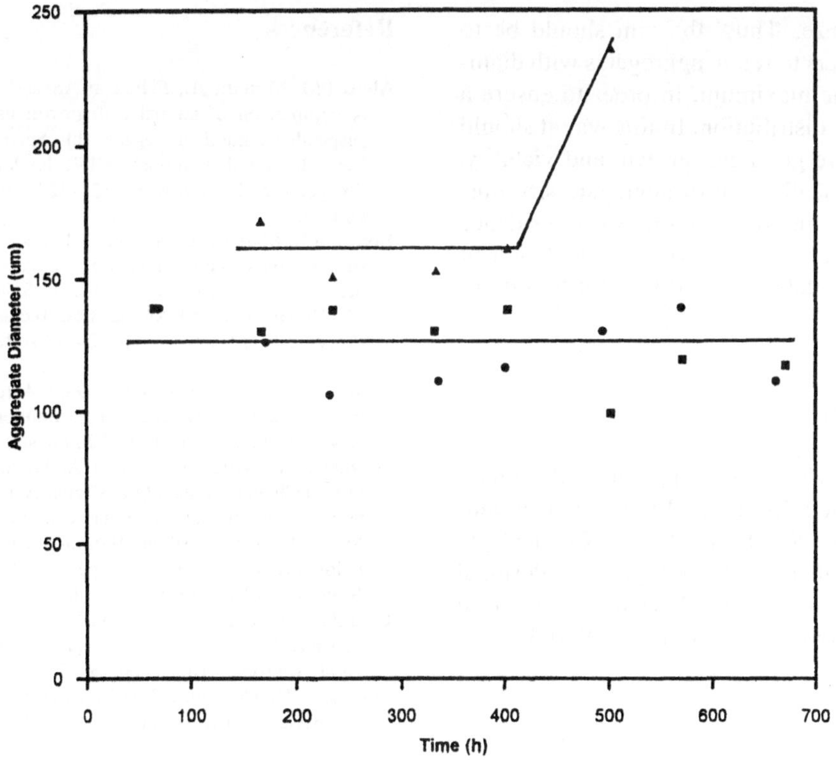

Fig. 7. Aggregate diameter (■ – experiment I: ▲ – experiment II; • – experiment III).

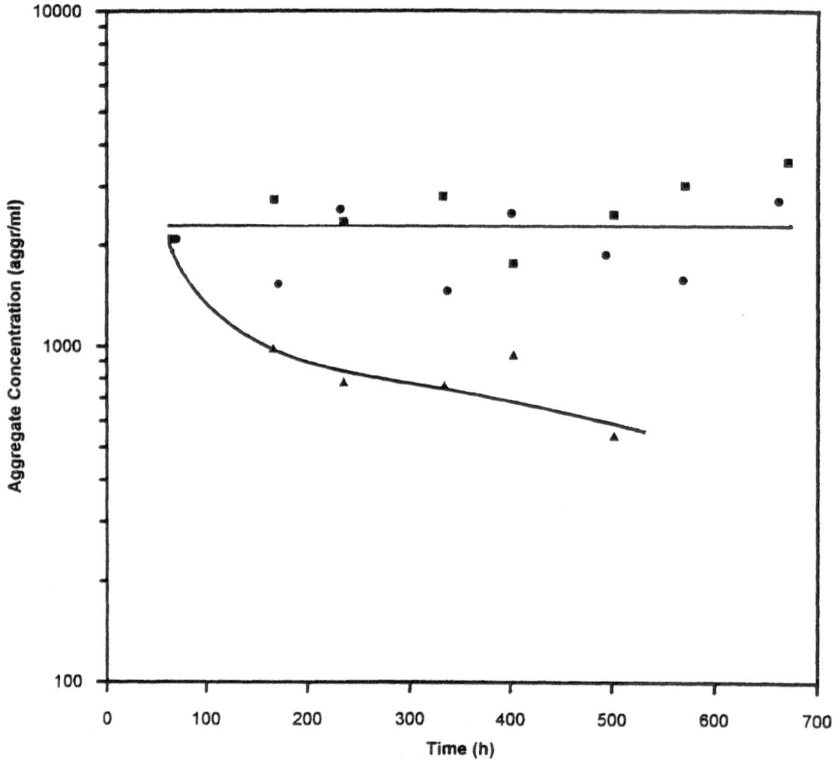

Fig. 8. Aggregate concentration (■ – experiment I: ▲ – experiment II; • – experiment III).

control aggregate size. Thus, the aim should be to design 'fast' decanters to retain aggregates with diameter smaller than the maximum, in order to ensure a large aggregate size distribution. In this way it should be possible to ensure good cell growth and viability, and to hydrodynamically control aggregate size during long periods of time; furthermore, since residence times in the decanter will be kept low, depletion of oxygen and other metabolites will be kept to a minimum.

Acknowledgements

The authors acknowledge and appreciate the financial support received from the European Community (BRIDGE PL 890035), Junta Nacional de Investigação Científica e Tecnológica – Portugal (PMCT/C/BIO/882/90 and BD/241/90-IF) and from Program PEDIP 2 under grant number 1495/F3.

References

Alves PM, Moreira JL, Feliciano AS and Carrondo MJT (1994) A comparison of animal cell growth using microcarriers and suspended natural aggregates. In: Spier RE, Griffiths JB and Berthold W (eds.) Animal Cell Technology: Products of Today, Prospects for Tomorrow (pp. 321–323). Butterworth-Heinemann, Oxford.

Boraston R, Marshall C, Norman P, Renner G and Warner J (1992) Elimination of cell aggregation in suspension cultures of chinese hamster ovary (CHO) cells. In: Spier RE, Griffiths JB and MacDonald C (eds.) Animal Cell Technology: Developments, Processes and Products. (pp. 424–426) Butterworth-Heinemann, Oxford.

Freyer JP and Sutherland RM (1985) A reduction in the *in situ* rates of oxygen and glucose consumption of cells in EMT6/R0 spheroids during growth. J. Cell. Physiol. 124: 516–524.

Gawlitzek M, Villers C, Verbert A, Wagner R and Conradt HS (1994) Changes in the glycosylation pattern of recombinant proteins effected by defined culture conditions of BHK-21 cells. In: Spier RE, Griffiths JB and Berthold W (eds.) Animal Cell Technoloy: Products of Today, Prospects for Tomorrow (pp. 649–651). Butterworth-Heinemann, Oxford.

Goetghebeur S and Hu W-S (1991) Cultivation of anchorage-dependent animal cells in microsphere-induced aggregate culture. Appl. Microbiol. Biotechnol. 34: 735–741.

Goochee CF and Monica T (1990) Environmental effects on protein glycosylation. Bio/Technol. 8: 421–427.

Hirai Y, Takebe K, Nakajima M, Takashina M and Iizuka M (1991). Extended expression of liver functions of hepatocytes in collagen-contained cell aggregates (cell packs). Cytotechnol. 6: 209–217.

Junker BH, Wu F, Wang S, Waterbury J, Hunt G, Hennessey J, Aunins JG, Lewis J, Silberklang M and Buckland BC (1992) Evaluation of a microcarrier process for large-scale cultivation of attenuated hepatitis A. Cytotechnol. 9: 173–187.

Kawase Y and Moo-Young M (1990) Mathematical models for design of bioreactors: applications of Kolmogoroff's theory of isotropic turbulence. Chem. Eng. J. 43: 19–41.

Landry J and Freyer LP (1984) Regulatory mechanisms in spheroidal aggregates of normal and cancerous cells. Recent Results Cancer Research. 95: 50–66.

Litwin J (1991) The growth of CHO and BHK cells in suspended aggregates in serum free medium. In: Spier RE, Griffiths JB and Meignier B (eds.) Production of Biologicals from Animal Cells in Culture. (pp. 429–433) Butterworth-Heinemann, Oxford.

Litwin (1992) The growth of Vero cells in suspension as cell-aggregates in serum-free media. Cytotechnol. 10: 169–174.

Maiorella BL, Winkelhake J, Young J, Moyer B, Bauer R, Hora M, Andya J, Thomson J, Patel T and Parekh R (1993) Effect of culture conditions on IgM antibody structure, pharmacokinetics and activity. Bio/Technol. 11: 387–392.

Moreira JL, Alves PM, Rodrigues JM, Cruz PE, Aunins JG and Carrondo MJT (1994a) Studies of Baby Hamster Kidney (BHK) natural cell aggregation in suspended batch cultures. Annals N.Y. Acad. Sci. In Press.

Moreira JM, Alves PM, Feliciano AS and Carrondo MJT (1994b) Serum free and containing medium for growth of BHK suspended natural aggregates in stirred vessels. Enz. Microb. Biotechnol. In press.

Moreira JL, Feliciano AS, Aunins JG and Carrondo MJT (1994c) Long term cultures of BHK suspended aggregates. In: Spier RE, Griffiths JB and Berthold W (eds.) Animal Cell Technology: Products of Today, Prospects for Tomorrow. (pp. 507–510) Butterworth-Heinemann, Oxford.

Moreira JL, Alves PM, Aunins JG and Carrondo MJT (1994d) Changes in animal cell natural aggregates in suspended batch cultures. Appl. Microb. Biotechnol. 41: 203–209.

Moreira JL, Alves PM, Aunins JG and Carrondo MJT (1994e) Hydrodynamic effects on Baby Hamster Kidney cells grown as suspended natural aggregates. Biotechnol. Bioeng. Submitted for publication.

Mueller-Klieser W, Freyer JP and Sutherland RM (1986) Influence of glucose and oxygen supply conditions on the oxygenation of multicellular spheroids J. Cancer 53: 345–353.

Nyberg SL, Shafford RA, Peshwa MV, White JG, Cerra FB and Hu W-S (1993) Evaluation of a hepatocyte-entrapment hollow fibber bioreactor: a potential bioartificial liver. Biotechnol. Bioeng. 41: 194–203.

Ogata M, Wakita K Kimura K, Marutomo Y, Oh-i K and Shimizu S (1993) High-level expression of recombinant human soluble thrombomodulin in serum-free medium by CHO-K1 cells. Appl. Microbiol. Biotechnol. 38: 520–525.

Pay TWF, Boge A, Menard FJRR and Radlett PJ (1985) Production of rabies vaccine by an industrial scale BHK 21 suspension cell culture process. Develop. Bio. Standard. 60: 171–174.

Perusich CM, Goetghebeur S and Hu W-S (1991) Virus production in microsphere-induced aggregate culture of animal cells. Biotechnol. Techn. 5 (2): 145–148.

Peshwa MV, Kyung Y-S, McClure DB and Hu W-S (1993) Cultivation of mammalian cells as aggregates in bioreactors: effect of calcium concentration on spatial distribution of viability. Biotechnol. Bioeng. 41: 179–187.

Racher AJ, Moreira JL, Alves PM, Wirth M, Weidle UH, Hauser H, Carrondo MJT and Griffiths JB (1994). Expression of recombinant antibody and secreted alkaline phosphatase in mammalian cells. Influence of cell line and culture system upon production kinetics. Appl. Microb. Biotechnol. (in press).

Radlet PJ (1987) The use of BHK suspension cells for the production of foot and mouth disease vaccines. In: Fiechter (ed.) Advances in Biochemical Engineering/Biotechnology. Vol. 34 (pp. 130–146) Springer-Verlag, Berlin.

Renner WA, Jordan M, Eppenberger HM and Leist C (1993) Cell-cell adhesion and aggregation: influence on the growth behaviour of CHO cells. Biotechnol. Bioeng. 41: 188–193.

Sutherland RM (1988) Cell and environment interactions in tumour microregions: the multicell spheroid model. Science. 240: 177–184.

Tolbert WR, Hitt MM and Feder J (1980) Cell aggregate suspension culture for large-scale production of biomolecules. In Vitro 16 (6): 486–490.

Address for offprints: M.J.T. Carrondo, Instituto de Biologia Experimental e Tecnológica / Instituto de Tecnologia Quimíca e Biológica (IBET/ITQB), Apartado 12, 2780 Oeiras, Portugal.

Cytotechnology **15:** 351–363, 1994.

Towards the development of a bioartificial pancreas: immunoisolation and NMR monitoring of mouse insulinomas

A. Sambanis[1], K.K. Papas[1], P.C. Flanders[2], R.C. Long Jr.[2], H. Kang[2] & I. Constantinidis[2]
[1] *School of Chemical Engineering, Georgia Institute of Technology, Atlanta, GA 30332-0100, USA;* [2] *Frederik Philips Magnetic Resonance Research Center, Department of Radiology, Emory University, School of Medicine, Atlanta, GA 30322, USA*

Key words: bioartificial pancreas, βTC3 cells, immunoisolation, NMR

Abstract

A promising method for diabetes treatment is the implantation of immunoisolated cells secreting insulin in response to glucose. Cell availability limits the application of this approach at a medically-relevant scale. We explore the use of transformed cells that can be grown to large homogeneous populations in developing artificial pancreatic tissues. We also investigate the use of NMR in evaluating, non-invasively, cellular bioenergetics in the tissue environment. The system employed in this study consisted of mouse insulinoma βTC3 cells entrapped in calcium alginate/poly-L-lysine (PPL)/alginate beads. The PPL layer imposed a molecular weight cutoff of approximately 60 kDa, allowing nutrients and insulin to diffuse through but excluding high molecular weight antibodies and cytotoxic cells of the host. We fabricated a radiofrequency coil that can be double-tuned to ^{1}H and ^{31}P, and an NMR-compatible perfusion bioreactor and support circuit that can maintain cells viable during prolonged studies. The bioreactor operated differentially, was macroscopically homogeneous and allowed the acquisition of ^{1}H images and ^{31}P NMR spectra in reasonable time intervals. Results indicated that entrapment had little effect on cell viability; that insulin secretion from beads was responsive to glucose; and that the bioenergetics of perfused, entrapped cells were not grossly different from those of cells never subjected to the immobilization procedure. These findings offer promise for developing an artificial pancreatic tissue for diabetes treatment based on continuous cell lines.

Introduction

Diabetes mellitus, or diabetes type I, is a serious disease resulting from the autoimmune destruction of the insulin-producing pancreatic islets. The disease affects an estimated 14 million people in the U.S. alone. Diabetes is commonly treated by daily insulin injections which provide some, but not physiologic, control of blood glucose levels. Normal β cells exhibit an elaborate secretory response to changing glucose levels, which cannot be mimicked by conventional drug administration procedures. As a result, diabetics who do receive insulin still develop serious complications, including eye, kidney and cardiovascular disease. In the U.S., the annual cost for direct treatment of diabetes, and for attending its long-term side effects and social consequences, such as work loss, exceeds 14 billion dollars. A treatment providing better control of blood glucose levels, thus reducing complications, would result in substantial economic benefit (Peura, 1988; Sefton, 1989; Stinson, 1991).

An alternative method for diabetes treatment is based on implanting a pancreatic tissue analog containing insulin-secreting cells. The cells are surrounded by a semipermeable membrane imposing a molecular weight cutoff of approximately 60 kDa. The membrane allows cellular nutrients and metabolites, including insulin, to diffuse in and out, respectively, but it excludes larger molecular weight antibodies and cytotoxic cells of the host, thus immunoprotecting the implant (Darquy and Reach, 1985; Soon-Shiong *et al.*, 1990). Implantation of such artificial tissues fabricated with cells isolated from animal glands has been successful in restoring normoglycemia for various time periods in both small (Fan *et al.*, 1990; Lacy *et al.*, 1991; Lanza *et al.*, 1991; Lim and Sun,, 1980) and large (Soon-Shiong *et al.*, 1992; Sullivan *et al.*, 1992) diabetic animal models. Recently, limited human trials have also begun (Calafiore, 1992; Soon-Shiong *et al.*, 1994). Use of cells secreting insulin in response to glucose may achieve a much more physiologic regulation of blood glucose levels than simple insulin injections. If artificial tissues can be fabricated inexpensively, their use would reduce the cost of both the direct treatment of diabetes and its long-term complications.

Although the feasibility of restoring normoglycemia by immunoisolated cell implants has been demonstrated, it is clear that obstacles related to cells, biomaterials, the three-dimensional tissue architecture, and the monitoring of cell function need to be overcome before any routine clinical application. The semipermeable membrane should provide adequate immunoprotection while not causing inflammatory responses. It may be necessary for the cells to be imbedded in a biopolymer, instead of simply floating in a capsule defined by the membrane; in such a case, the biopolymer should ensure proper cell function and stability. A method, particularly a non-invasive one, for monitoring cell biochemistry would be invaluable in characterizing and optimizing the artificial devices, and in understanding cell function in the sequestered environment.

The most pressing challenge, however, is cell availability. Although cells from human and animal pancreata are adequate for feasibility studies, they are in short supply for any medical-scale applications. Human and animal β cells and islets cannot be grown in culture while retaining their differentiated properties, primarily insulin secretion, so a collection of such cells cannot be effectively amplified. Thus, with human cells, one donor is theoretically needed for a single treatment of one recipient. Although there is a virtually inexhaustible supply of porcine and bovine β cells and islets, it is particularly difficult to reproducibility isolate such cells on a large scale. Furthermore, there exist potential incompatibility problems of animal insulin in humans. An alternative approach involves the use of transformed, continuous cell lines that retain in culture key differentiated properties of normal β cells and can be grown to large, homogeneous populations. Promising lines, such as the family of βTC insulinomas, have been developed and could be used in engineering artificial tissues (Efrat *et al.*, 1988; Efrat *et al.*, 1993).

Certain aspects of the function of a bioartificial device can be obtained by following metabolic and secretory indicators, such as glucose uptake and insulin secretion. Equally important, however, is to develop a fundamental understanding of cell behavior in the immunoisolated environment. Nuclear magnetic resonance (NMR) spectroscopy is a powerful non-invasive technique that can detect changes in cellular metabolism under various culture conditions. Previous NMR studies have focused on characterizing the biochemistry of tumor cell lines and their response to cytotoxic agents, as well as on studying industrially important cell lines, primarily hybridomas. ^{31}P NMR spectroscopy can be used to monitor intracellular pH and the levels of high- and low-energy phosphorous-containing compounds, such as nucleotide triphosphates and membrane-related phosphomonoesters and phosphodiesters (Fernandez *et al.*, 1990; Gillies *et al.*, 1993; Minichiello *et al.*, 1989; Narayan *et al.*, 1990; Ronen and Degani, 1989). With ^{13}C NMR spectroscopy, one can follow non-invasively biochemical pathways in tumor cells (Constantinidis *et al.*, 1991), hybridomas (Fernandez *et al.*, 1990) and tissues (Jans and Willem, 1988; Sherry *et al.*, 1988). NMR methods may also be used to collect information on the bioenergetics of cells in artificial tissue constructs (Constantinidis *et al.*, 1994; Sambanis *et al.*, 1993, 1994).

In this paper, we address issues relating to developing a bioartificial endocrine pancreas based on transformed, insulin-secreting cells, and to monitoring cell bioenergetics, non-invasively, by NMR spectroscopy. We present results on the effect of entrapment on the viability and secretory capacity of mouse insulinomas; we describe a fixed-bed, NMR-compatible bioreactor, radiofrequency coil, and supporting medium perfusion circuit that can maintain immunoisolated cells under various conditions; and we present ^{1}H NMR images of packed microbeads and ^{31}P NMR spectra of entrapped cells and cell extracts that provide information on the function of the bioreactor and on cellular bioenergetics.

Materials and methods

Cells and cell culture

Mouse pituitary tumor AtT-20 cells expressing recombinant human proinsulin were obtained from Dr. R.B. Kelly, Department of Biochemistry and Biophysics,

University of California, San Francisco, CA. Proinsulin is cleaved intracellularly to an insulin-like peptide, which is stored in secretory granules and secreted at a high rate only when cells are triggered with secretagogues. The cells are responsive to a variety of non-metabolic secretagogues, but not to glucose. Proinsulin is secreted both in the absence and presence of secretagogues (Moore *et al.*, 1983). βCT3 cells were obtained from the laboratory of Dr. Shimon Efrat, Department of Molecular Pharmacology, Albert Einstein College of Medicine, Bronx, NY. These cells are glucose-responsive, but they are also glucose hypersensitive, exhibiting half-saturation of the glucose stimulus-secretion response at 0.1–0.2 mM glucose, as opposed to 8 mM for normal islets (Efrat *et al*, 1993; Tal *et al.*, 1992).

Cells were cultivated as described by Papas *et al.* (1993) and Dyken and Sambanis (1994). In short, AtT-20 cells were propagated in complete growth medium (C-DMEM) consisting of Dulbecco's Modified Eagle's Medium (DMEM) with 25 mM glucose and 4.0 mM L-glutamine and supplemented with 10% fetal bovine serum (FBS) and glutamine to a final concentration of 6.0 mM (all from Sigma Chemical Company, St. Louis, MO). βTC3 cells were cultivated in β-DMEM medium consisting of DMEM supplemented with glutamine to 6.0 mM, 15% heat-inactivated horse serum (HIHS, Sigma) and 2.5% FBS. All cultures were maintained at 37 °C in a humidified 5% CO_2/95% air atmosphere. Cultures consistently tested negative for mycoplasma contamination using a nucleic acid hybridization test kit (GenProbe Inc., San Diego, CA).

Entrapment procedure

βTC3 cells were entrapped in calcium alginate/poly-L-lysine (PLL)/alginate beads following the procedure developed by Lim and Sun (1980) and Sun (1988) and somewhat modified by Papas *et al.* (1993). Cells were detached from T flasks by trypsinization, the trypsin neutralized by β-DMEM, the cells spun down by mild centrifugation and resuspended at densities of 3–7 × 10^7 cells ml^{-1} in 2% sodium alginate (Keltone LV, Kelco, Chicago, IL) sterilized by filtration. The suspension was transferred to a syringe positioned on a syringe pump (Razel Scientific, Stanford, CT) and connected to an 18 gauge needle. Air was blown parallel to the needle at a rate necessary to obtain beads of 1 mm average diameter, as originally described by Vorlop and Klein (1983). Droplets falling in a 1.1 % $CaCl_2$ solution produced calcium alginate beads containing entrapped cells. Beads were left overnight in β-DMEM, then washed with $CaCl_2$, treated with 2-(N-cyclohexylamino) ethanesulfonic acid (CHES), suspended for six minutes under continuous agitation in 0.05% PLL hydrobromide (molecular weight 18,000–20,900, Product # P7890, Sigma), washed with CHES, $CaCl_2$, and NaCl, and incubated for 4 minutes in 0.15% sodium alginate to neutralize any residual charges from PLL. Beads were finally washed twice with NaCl, washed once with β-DMEM, and suspended in β-DMEM for incubation at 37 °C in a humidified, 5% CO_2 atmosphere. Beads prepared according to the above protocol had a MW cutoff between 45 and 67 kDa (Tziampazis, 1993).

Beads were sized by being transferred into 25 cm^2T flasks in β-DMEM and placed under an inverted Diaphot-TMT Nikon microscope (Nikon Company, Tokyo, Japan). The microscope was connected to a video camera (SIT-66X, Dage-MTI, Michigan City, IN); the camera outputs were transmitted to an acquisition-and-processing workstation (Perceptics Company, Knoxville, TN) from which micrographs could be printed. The micrograph of a hemocytometer was also taken under the same conditions for size conditions calibration. For each bead, two orthogonal diameters D_1 and D_2 were measured on pictures with a ruler, and size was represented by the geometric mean $\sqrt{D_1 D_2}$. The absolute difference of the two diameters was also recorded and compared to the geometric mean to estimate the degree to which each bead was actually a sphere. Differences were on the average about 5% of geometric means.

NMR spectroscopy

^{31}P NMR spectra and ^1H images were obtained with a SISCO 200/33 spectrometer operating at 200.057 MHz for ^1H and 80.984 MHz for ^{31}P. The usable diameter of the horizontal bore was 12.5 cm and the maximum gradient strength 10 Gauss/cm. Data were acquired using a double-tuned $^{31}P/^1H$ home-built probe based on the modifications proposed by Chang *et al.* (1987) on the original design of Schnall *et al.* (1985). The radiofrequency antenna consisted of a loop-gap resonator of 3 cm diameter by 4 cm length that surrounded the cylindrical shape of the bioreactor. Typical acquisition parameters for both *in vitro* and perchloric acid extract ^{31}P experiments were: 3000 Hz spectral width, 4096 complex points per free induction decay (FID), 90° pulses, 3 seconds relaxation delay and 2048 transients. ^1H NMR images were acquired with a spin echo

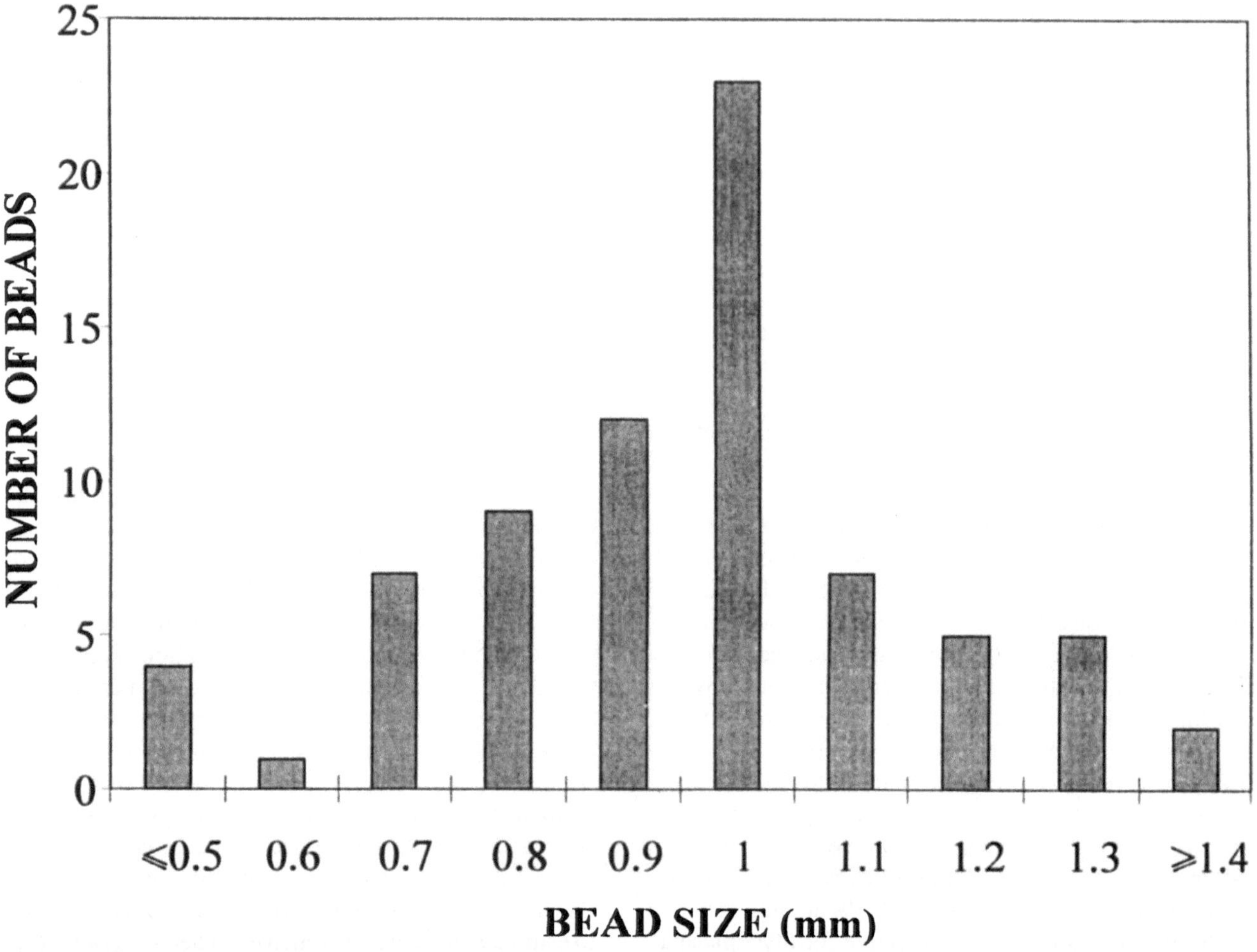

Fig. 1. Typical size distribution of calcium alginate/PLL/alginate beads containing βTC3 cells at 6 × 10^7 cells/ml alginate. The histogram was constructed by sizing 75 beads. The average bead diameter was 0.95 mm and the standard deviation of the distribution 0.20 mm.

sequence using an echo time of 50 ms, a repetition time of 2 s, a field of view of 6 cm, and a matrix of 128 × 128 phase-encoding steps.

Perchloric acid (PCA) extraction

Cell extraction was carried out using a protocol similar to that of Evanochko *et al.* (1984). Cells detached from monolayers by trypsinization were pelleted by centrifugation and placed directly in cold (4 °C) 0.5 N perchloric acid of five times the volume of the cell pellet. The mixture was blended with a homogenizer, and the suspension was allowed to stand for approximately 5 min at 4 °C. Debris was removed by centrifugation at 4,000 g for 5 min. The supernatant was removed and stored, while the pellet was washed with 1:5 solution of ethanol:water. The original supernatant and that from the wash were mixed and neutralized with KOH. The mixture was centrifuged to remove precipitated potassium perchlorate and the supernatant was passed through a 5 ml Chelex 100 column (Bio-Rad, Richmond, CA) to remove any metal ions present in solution. The final solution was lyophilized to dryness and stored frozen in a desiccator. At a later time, the sample was thawed and evaluated by NMR.

Assays

Total cell numbers were measured by treating pelleted cells with a crystal violet/citric acid solution. Citric acid lysed the plasma membranes and crystal violet stained nuclei which were then counted on a hemocytometer. Cell viability was estimated by exposing suspended cells to a solution of trypan blue; cells that

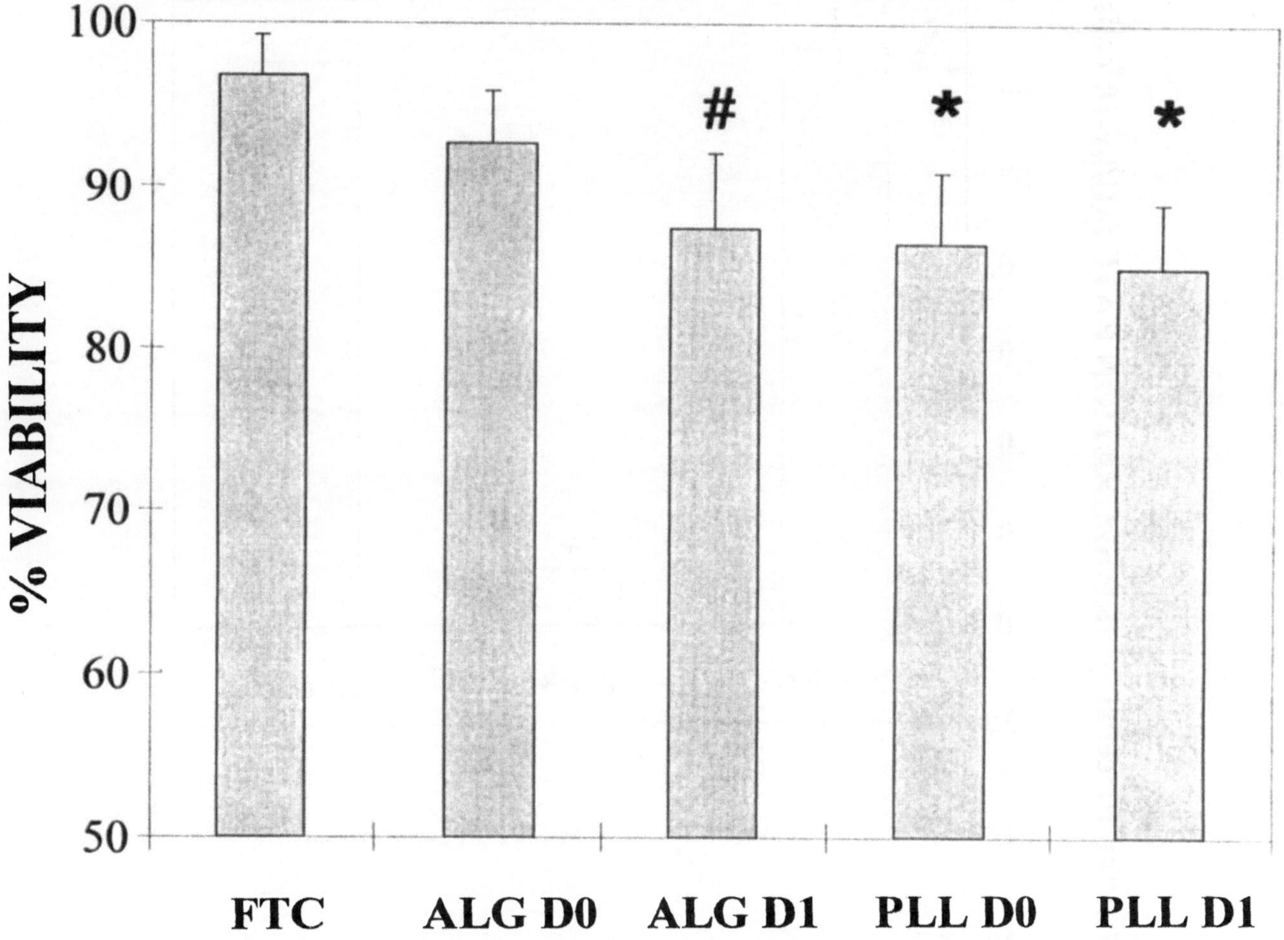

Fig. 2. Viability of βTC3 cells after each step in the bead preparation procedure. Cells entrapped at a density of 6×10^7 cells ml^{-1} alginate. FTC: cells from a freshly trypsinized monolayer; ALGD0 and ALGD1: cells in calcium alginate beads within one hour and after one day, respectively, from alginate entrapment; PLLD0 and PLLD1: cells in calcium alginate beads coated with polylysine (PLL)/alginate within 1.5 hours and after one day from addition of the PLL/alginate layers. #, $*$: Statistically different from the viability of FTC at the 90% and 95% confidence level, respectively. There is no statistically significant difference between viabilities of ALGD1 and PLLD0 or PLLD1.

were stained were counted as dead and those that were not as viable. For cells entrapped in calcium alginate, the gel was first solubilized through treatment with a 2.2% sodium citrate solution at pH = 7.4. PLL-coated calcium alginate beads were solubilized with sodium citrate and mechanical shearing. Unsolubilized PLL fragments interfered with nuclei counts, so in this case only viability measurements were performed. Control experiments showed that viability loss during calcium alginate/PLL solubilization was minimal.

Insulin concentrations were measured with a double-antibody radioimmunoassay kit (Binax, South Portland, ME) involving competitite binding of sample and ^{125}I-insulin to antibody sites. The manufacturer reports a 33% cross-reactivity of proinsulin against the antibody. The term insulin-related peptides (IRP) is used to describe the total amount of insulin and proinsulin reacting with the antibody according to their respective activities.

Results

A typical size distribution of cell-loaded beads is shown in Fig. 1. The size distribution of beads was quite reproducible: histograms constructed for three different preparations yielded average bead diameters ranging from 0.95 to 1.05 mm and standard deviations of 0.18–0.24 mm.

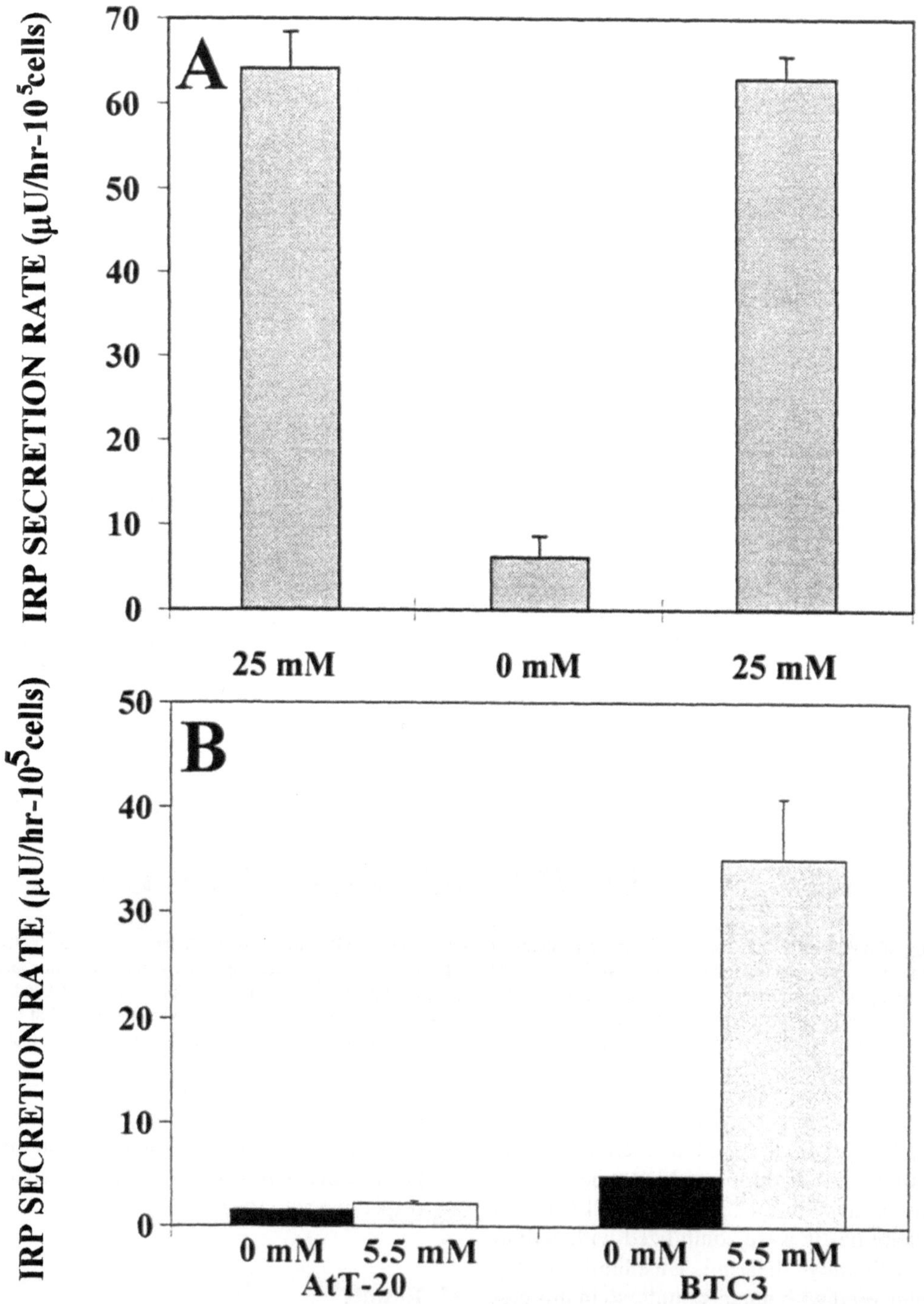

Fig. 3. Rates of IRP secretion from entrapped βTC3 cells and βCT3 and AtT-20 monolayers. A. βTC3 cells entrapped in 1 mm calcium alginate/PLL/alginate beads at a density of 6×10^7 cells ml^{-1} alginate were cultured in complete growth medium in spinner flasks. The beads were subsequently exposed for 1.2-hour periods to serum-free DMEM with 25, 0, and 25 mM glucose, in this sequence. All media contained 6 mM glutamine. The reported rates of secretion were averaged over each 1.2-hour period and were normalized to the total number of viable cells measured after entrapment in alginate and before addition of PLL. Bars represent standard deviations of multiple IRP assays and cell counts on collected samples. B. AtT-20 and βTC3 monolayers were exposed for two-hour periods to serum-free DMEM with 0 and 5.5 mM glucose. All media contained 6 mM glutamine. Rates of IRP secretion were averaged over the period of each secretion episode and normalized to the total number of cells in culture. Bars represent standard deviations of multiple IRP assays on samples from two different AtT-20 and βTC3 cultures.

To characterize the possible damage inflicted to cells by the entrapment procedure, cell viability was measured after each step in the preparation by using the trypan blue exclusion method. Results are shown in Fig. 2. Viability decreased from ~97% to ~85% in PLL-coated beads one day after addition of the PLL layer. This decrease is statistically significant but not detrimental to the usefulness of the preparation. A preliminary assessment of long-term viability was carried out in a fed-batch spinner flask experiment. Entrapped βTC3 cells propagated in β-DMEM and fed roughly every two days to maintain the glucose concentration above 1 mg mL^{-1} had a viability of ~73% after 30 days in culture.

To restore normoglycemia after implantation, entrapped cells should secrete insulin in response to glucose. It is unknown how closely the implant should mimic the biphasic glucose stimulus – insulin secretion response of normal islets. It is clear, however, that beads should secrete insulin at high glucose concentrations (above 5 mM) and should stop secreting at low glucose levels (below 1–2 mM). Due to their hypersensitivity to glucose, the low glucose level used in testing the secretory response of entrapped βTC3 cells was 0 mM; the high glucose level employed was 25 mM. The specific rates of IRP secretion from entrapped βTC3 cells in media with 0 and 25 mM glucose are shown in Fig. 3A. Similar trends were observed in a different experiment with 1 mm beads containing 3×10^7 cells ml^{-1} alginate exposed over 2-hour periods in media with 0 and 5.5 mM glucose (data not shown).

Clearly, secretion in glucose-free medium occurs at a much lower rate than secretion in medium with 25 mM glucose. To evaluate whether a similar response is exhibited by a glucose-unresponsive endocrine cell line, in particular AtT-20 cells, βCT3 and AtT-20 monolayers were exposed to media with 0 and 5.5 mM glucose and secretion was measured over 2-hour periods. Results are shown in Fig. 3B. Secretion from βTC3 cells in the presence of glucose was 8.6-fold higher than secretion in its absence, whereas this factor was only 1.4 in the case of AtT-20 cells. Thus, secretion from βTC3 cells is much more sensitive to glucose than secretion from AtT-20 cells under the same culture conditions.

NMR studies

Perfusion system

The system used to maintain entrapped cells under medium perfusion is shown schematically in Fig. 4. Beads were packed in a cylindrical fixed-bed reactor of 2.6 cm internal diameter and 4.1 cm length (22 ml in volume). The bed was loaded in a laminar flow hood by adding beads suspended in DMEM through the top. When the beads had settled by gravity in the reactor, some medium was removed and more beads were added. During this procedure, the beads were always covered by medium. The bioreactor was supported by a perfusion circuit containing two medium circulation loops: the first (perfusion loop) recycled medium from a reservoir to the reactor and back to the reservoir; the second (replenishment loop) replenished spent medium with fresh. The flow rate in the perfusion loop was maintained at 30 ml min^{-1}; under these conditions, the reactor operated differentially, i.e. concentrations of nutrients and metabolites were virtually the same at the entrance and exit of the bed (Sambanis *et al.*, 1993). Temperature sensors in flow cells positioned approximately 30 cm upstream and downstream of the reactor were used to monitor the temperature, which was maintained at 37.5 °C and 36.5 °C, respectively. The dissolved oxygen in the medium was maintained constant at a level between 40 and 50% air saturation by aerating the head space of the medium reservoir.

Bioreactor characteristics

The large size of the bioreactor and, consequently, of the receiving coil warranted an investigation of the radiofrequency homogeneity of this probe. This issue was addressed by acquiring ^1H NMR images transverse and parallel to the bioreactor axis without medium flow. All images demonstrated constant signal intensity within the sensitive volume of the coil (Constantinidis *et al.*, 1994).

To determine the packing uniformity of the bioreactor, multiple ^1H NMR images were obtained both transverse and parallel to the bioreactor axis while cells were being perfused. Typical images are shown in Figs. 5A and 5B. Perfusion resulted in diminution of signal from longitudinal flow, which in turn better outlined the alginate beads.

Due to intrabead diffusion and reaction, the cells in a bead are not all exposed to the same environment: cells at the bead center experience lower nutrient and higher metabolite concentrations than cells at

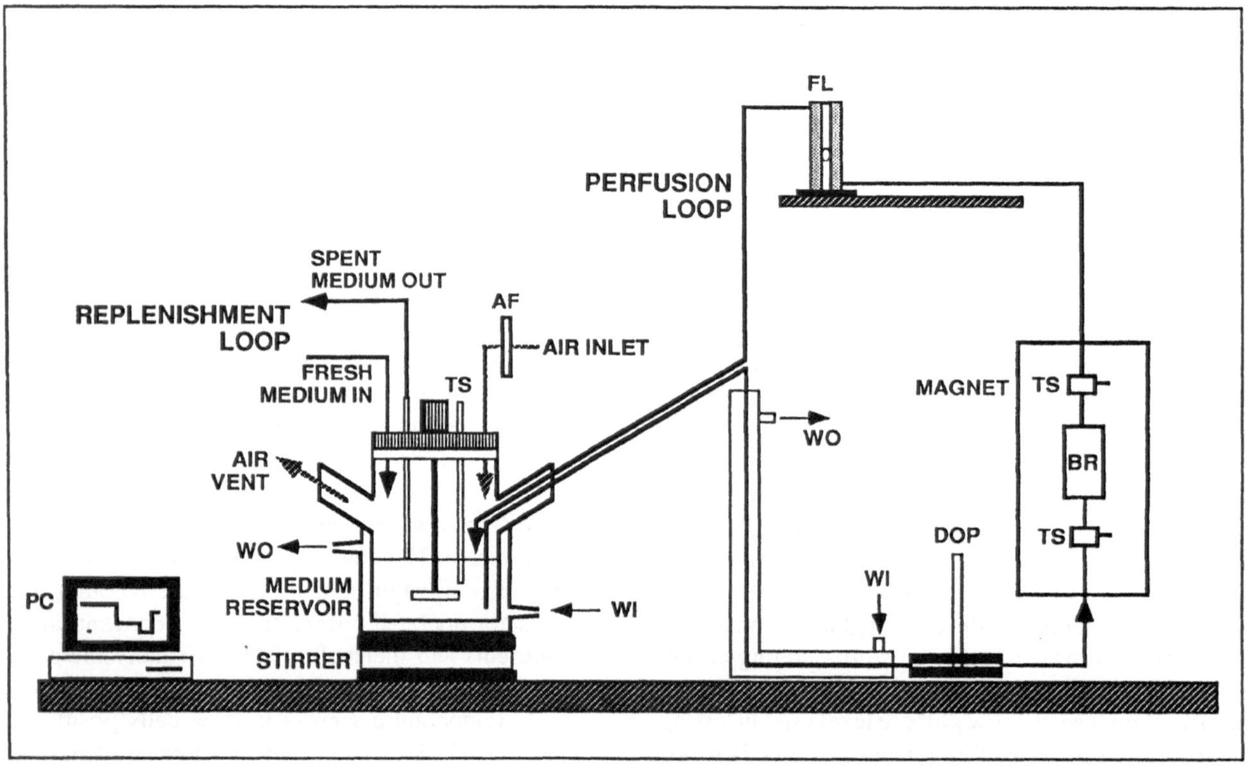

Fig. 4. Perfusion system used to maintain entrapped cells. AF: sterile air filters; BR: bioreactor; DOP: dissolved oxygen probe; FL: flowmeter; PC: personal computer for data acquisition from DOP and TS; TS: temperature sensor; WI: water in from constant temperature recirculating water bath; WO: water out, back to the bath. The pumps and medium reservoir were positioned approximately 12 ft. away from the magnet. In the perfusion loop, medium was circulated in the inner tube of coaxial tubing, with water from the bath flowing in the outer annulus. For the sake of clarity, the various pieces of equipment are not drawn to scale.

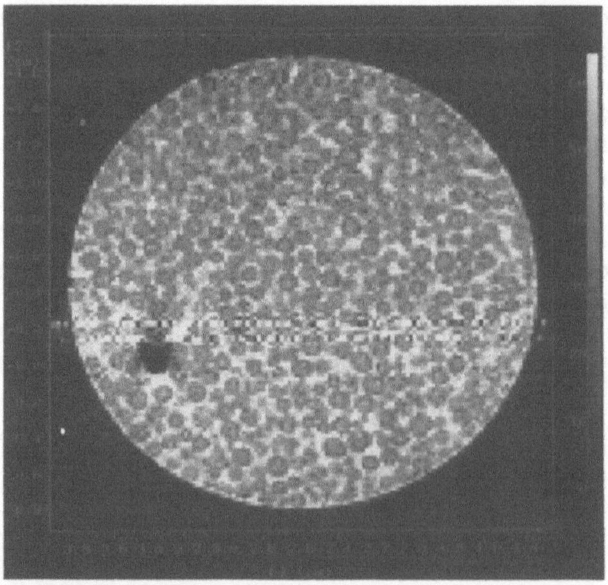

Fig. 5a.

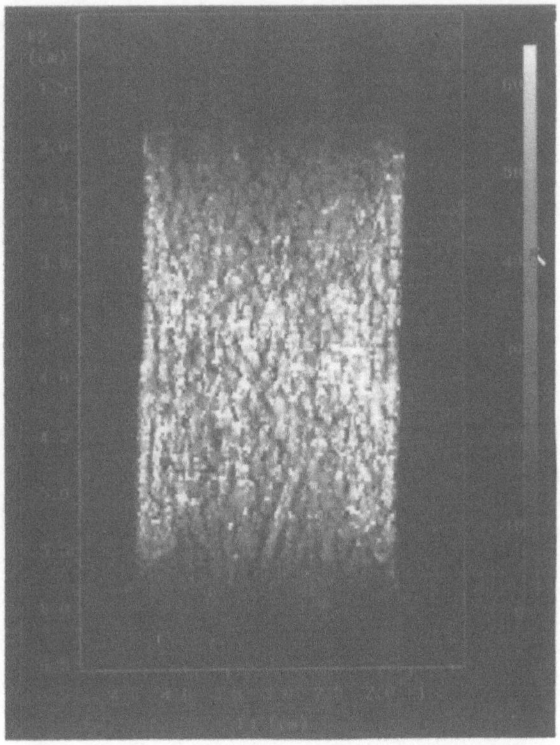

Fig. 5b.

Figs. 5(a)–(b). ^{1}H spin echo images of the bioreactor loaded with 1 mm beads containing βTC3 cells. A. Image transverse to the bioreactor axis. The dark circle in the lower left-hand corner of the image is due to a capillary tube present in the bed. B. Image parallel to the bioreactor axis.

the periphery. Intrabead heterogeneities are inherent to this system and can be reduced but not eliminated (Tziampazis, 1993; Tziampazis and Sambanis, 1994). Interbead heterogeneities can be minimized, however, if all beads are exposed to the same environment in the reactor. Under such conditions, the average behavior of cells in a bead should be the same at all positions in the bed. This question was addressed in experiments with AtT-20 spheroids in 3.2 mm calcium alginate/PLL/alginate beads. ^{31}P 1-Dimensional Chemical Shift Imaging (1D-CSI) spectra were acquired from approximately 7.5 mm thick slabs transverse and parallel to the bioreactor axis. All spectra were processed and metabolite ratios determined; results indicated that entrapped spheroids throughout the bioreactor possessed the same average bioenergetic charge (Constantinidis *et al.*, 1994; Sambanis *et al.*, 1993). Preliminary results with βTC3 cells entrapped in 1 mm beads indicated that this is the case with this system as well. Thus, the bioreactor was macroscopically homogeneous, in the sense that there existed no significant heterogeneities over a length of a few bead diameters.

βTC3 cell studies

^{31}P NMR spectra obtained from perfused immunoisolated βCT3 cells and from a PCA extract of freshly trypsinized βTC3 cells are shown in Figs. 6A and 6B, respectively. The spectral characteristics of the βTC3 line are similar to those of AtT-20 cells (Constantinidis *et al.*, 1994; Sambanis *et al.*, 1993) with the notable exception of phosphocreatine (PCr). Lack of PCr is not an unusual observation in abdominal organs like liver and kidneys or in some tumor cell lines.

Spectra were analyzed by integrating the resonance peaks. A direct quantification of metabolite concentrations was not performed due to the absence of a standard compound on the spectra. However, ratios of metabolites considered not to be affected by the extraction procedure (NTP and PME) were similar in the two spectra.

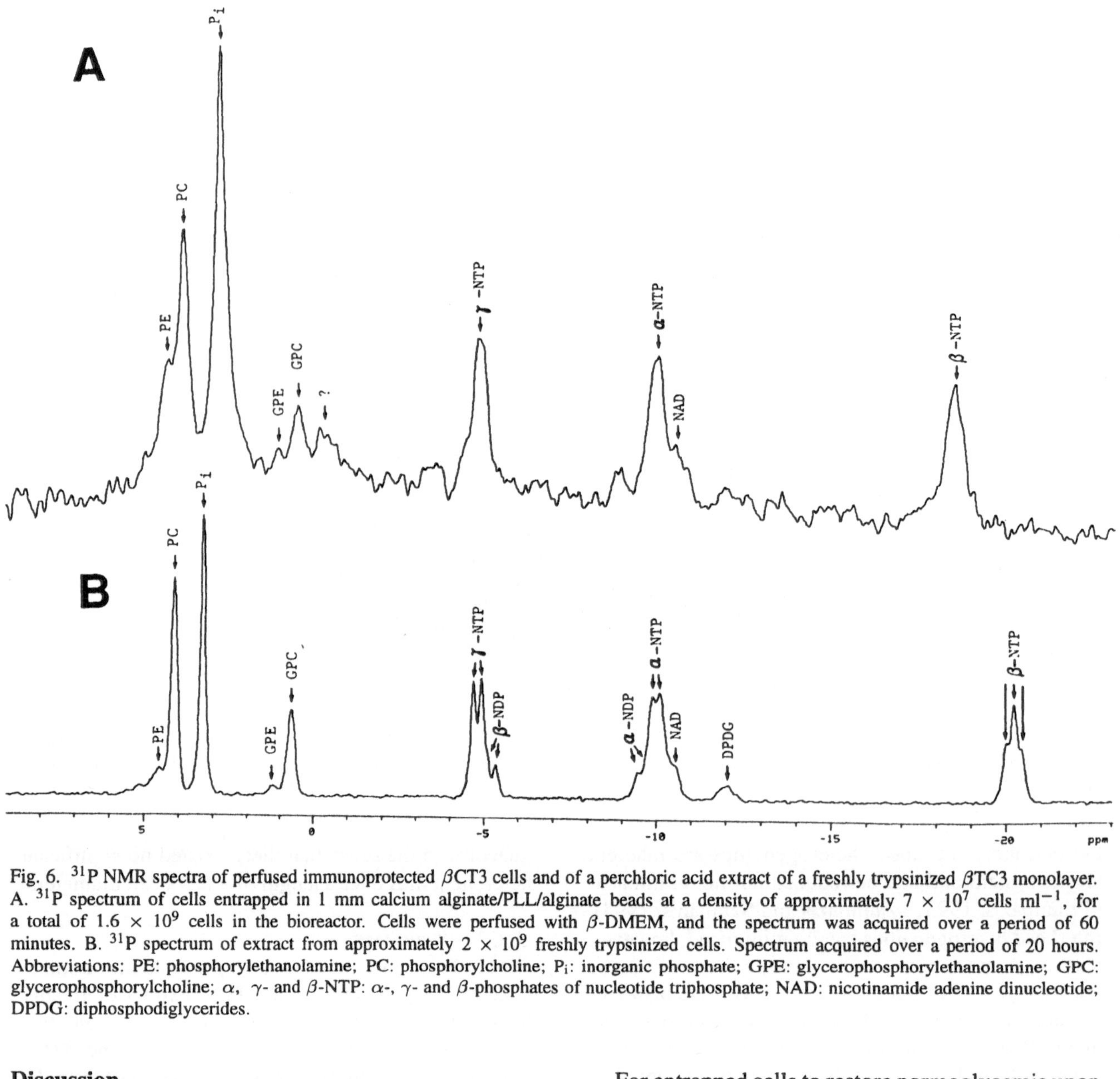

Fig. 6. ^{31}P NMR spectra of perfused immunoprotected βCT3 cells and of a perchloric acid extract of a freshly trypsinized βTC3 monolayer. A. ^{31}P spectrum of cells entrapped in 1 mm calcium alginate/PLL/alginate beads at a density of approximately 7×10^7 cells ml^{-1}, for a total of 1.6×10^9 cells in the bioreactor. Cells were perfused with β-DMEM, and the spectrum was acquired over a period of 60 minutes. B. ^{31}P spectrum of extract from approximately 2×10^9 freshly trypsinized cells. Spectrum acquired over a period of 20 hours. Abbreviations: PE: phosphorylethanolamine; PC: phosphorylcholine; P_i: inorganic phosphate; GPE: glycerophosphorylethanolamine; GPC: glycerophosphorylcholine; α, γ- and β-NTP: α-, γ- and β-phosphates of nucleotide triphosphate; NAD: nicotinamide adenine dinucleotide; DPDG: diphosphodiglycerides.

Discussion

Mouse insulinoma βTC3 cells can be entrapped with little loss of viability in calcium alginate/PLL/alginate beads of a narrow size distribution. With secretion rates averaged over 1.2 hour periods, the preparations were found to be glucose-responsive. Entrapped cells can be maintained in an NMR-compatible perfusion bioreactor that operates differentially and is macroscopically homogeneous. ^{31}P NMR spectra of perfused immunoisolated cells provide a fingerprint of the cell bioenergetics in the sequestered environment.

For entrapped cells to restore normoglycemia upon implantation, beads should be stable in the long-term and responsive to glucose. Using spheroids of AtT-20 cells, we have shown that it is possible to fabricate beads that exhibit stable metabolic and secretory characteristics for at least a month (Constantinidis *et al.*, 1994; Papas, 1993; Sambanis *et al.*, 1994). The results reported in this article with βTC3 cells, show that it is also possible to develop a glucose-responsive system based on transformed cells. Glucose responsiveness was measured over 1.2 hour periods, so it is unknown exactly how fast entrapped cells respond to changes in glucose levels. The usefulness of βTC3 cells in developing artificial tissues may be compromised by their hypersensitivity to glucose. If implants do not stop secreting at low glucose levels, they may revert diabetes to hyperinsulinemia and hypoglycemia, a serious pathological condition. Modeling results indicate that hypersensitivity is not alleviated by entrapment alone; engineering design beyond simple entrapment is needed to produce normally-sensitive beads with hypersensitive cells (Tziampazis, 1993; Tziampazis and Sambanis, 1994). The development of normally-responsive cell lines would also solve the hypersensitivity problem. Indeed, a newer member of the βTC family of insulinomas does not initially exhibit glucose hypersensitivity, although it becomes hypersensitive upon continuous propagation in culture (Efrat *et al.*, 1993).

The NMR images demonstrate that the bioreactor is maintained uniformly packed with beads during perfusion. The instrument used does not have the capability to image spheroids or cells in beads. Heterogeneities within beads are inherently present in the system due to the diffusional resistance imposed by the alginate matrix. Human pancreatic islets, each consisting of several thousand cells, have diameters ranging from 160 μm to more than 250 μm. Islets are well vascularized, receiving a disproportionately high blood flow relative to the rest of the gland (Carroll, 1992). Thus, the maximum diffusional length in pancreatic islets is considerably less than the 0.5 mm in alginate beads, and cellular heterogeneities in normal pancreatic tissue are expected to be minimal.

Beads are exposed to a homogeneous environment throughout the bioreactor if the reactor operates differentially without domains of stagnant medium. Under such conditions, each bead should exhibit the same type of behavior. The 1D-CSI experiments with entrapped AtT-20 spheroids (Constantinidis *et al.*, 1994; Sambanis *et al.*, 1993) demonstrated the same average bioenergetic status of cells at all positions within the reactor. Thus, NMR experiments can be performed without spatial localization, which, although feasible, prolongs the necessary acquisition time for a similar signal-to-noise ratio.

Previous results with AtT-20 spheroids indicate that our immunoisolation protocols do not significantly compromise cellular metabolism and secretion relative to spheroids in suspension (Papas, 1993; Papas *et al.*, 1993; Sambanis *et al.*, 1994). With islets, however, it has been suggested that encapsulation may irreversibly damage cell function (Levesque *et al.*, 1992), and that a larger quantity of immunoprotected islets is needed to restore normoglycemia in rats, than a similar implantation of free islets (Clayton *et al.*, 1992). The IRP secretion rate of 60–70 μU/10^5 cells-hr reported in Fig. 3A for entrapped βTC3 cells lies in the general range of values reported for βTC3 monolayers (Grampp *et al.*, 1992). Experiments thoroughly comparing the secretory behavior of βTC3 cells entrapped in microbeads, free in suspension, and in monolayer cultures remain to be completed in our laboratory.

Major differences between the spectrum of perfused entrapped cells and that of the PCA extract included the chemical shifts of β-NTP, P_i, PC and PE. The chemical shift difference of β-NTP is attributed to lack of Mg^{+2} in the extract (Mosher *et al.*, 1992). Indeed, in preparing the extract, all metal ions were removed by passing the solution through a chelating column in order to make the resonances sharper. The differences in P_i, PC and PE chemical shifts are attributed to the difference in the pH of the extract solution (pH = 7.55) from that of the perfusion medium (pH = 7.20) (Moon and Richards, 1973). Ratios involving P_i and/or PDE were different for the two spectra. Both P_i and PDE were higher in the perfused system relative to the extract due to the presence of P_i in the culture medium and the low water solubility of PDE which is thus poorly extracted by the PCA method. The ratios of PME/β-NTP, γ-NTP/β-NTP, γ-NTP/α-NTP, and α-NTP/β-NTP were determined to be similar in the spectra from perfused entrapped cells and from PCA-extracted trypsinized cells. Thus, although the comparison of the two spectra is only partially quantitative, the bioenergetics of entrapped, perfused cells do not appear grossly different from those of cells from freshly-trypsinized monolayers.

In conclusion, the reported results are promising in terms of developing artificial tissues based on transformed cells. Entrapment in calcium alginate/PLL/alginate beads causes little damage to βTC3 cells, as evidenced by viability, secretion, and ^{31}P NMR spectroscopic studies. NMR spectroscopy is expected to be particularly useful in evaluating cell bioenergetics and function in artificial tissues, especially in the long-term.

Acknowledgements

This work was partially supported by grants from the Whitaker Foundation and the Emory/Georgia Tech Biomedical Technology Research Center. This support is gratefully acknowledged. The authors also wish to thank Tom Maier and Pete Noel for technical assistance in developing the data acquisition interface and components of the perfusion apparatus.

References

Calafiore R (1992) Transplantation of microencapsulated pancreatic human islets for therapy of diabetes mellitus. ASAIO J. 38: 34–37.

Carroll P (1992) Anatomy and physiology of islets of Langerhans. In: Ricordi C (ed.) Pancreatic Islet Cell Transplantation, pp. 7–18. Austin: R.G. Landes Co.

Chang L-H, Chew WM, Weinstein PR and James TL (1987) A balanced-matched double tuned probe for in vivo ^{1}H and ^{31}P NMR. J. Magn. Reson. 72:168–172.

Clayton HA, London NJM, Bell PRF and James RFL (1992) The transplantation of encapsulated islets of Langerhans into the peritoneal cavity of the biobreeding rat. Transplantation 54: 558–560.

Constantinidis I, Chatham JC, Wehrle JP and Glickson JD (1991) *In vivo* ^{13}C-NMR studies of the RIF-1 tumor after glucose administration. Magn. Res. Med. 20: 17–26.

Constantinidis I, Long RC Jr, Erickson M and Sambanis A (1994) Towards the development of artificial endocrine tissues: II. ^{31}P NMR spectroscopic studies of entrapped insulin-secreting cells. Submitted.

Darquy S and Reach G (1985) Immunoisolation of pancreatic B cells by microencapsulation. An *in vitro* study. Diabetologia 28: 776–780.

Dyken JJ and Sambanis A (1994) Ammonium selectively inhibits the regulated pathway of protein secretion in two endocrine cell lines. Enz. Microb. Technol. 16: 90–98.

Efrat S, Leiser M, Surana M, Tal M, Fusco-Demane D and Fleischer N (1993) Murine insulinoma cell line with normal glucose-regulated insulin secretion. Diabetes 42: 901–907.

Efrat S, Linde S, Kofod H, Spector D, Delannoy M, Grant S, Hanahan D and Baekkeskov S (1988) Beta-cell lines derived from transgenic mice expressing a hybrid insulin gene-oncogene. Proc. Natl. Acad. Sci. USA 85: 9037–9041.

Evanochko WT, Sakai TT, Ng TC, Krishna NR, Kim HD, Zeidler RB, Ghanta VK, Brockman RW, Schiffer LM, Braunschweiger PG and Glickson JD (1984) NMR studies of *in vivo* RIF-1 tumors. Analysis of perchloric acid extracts and identification of ^{1}H, ^{31}P and ^{13}C resonances. Biochim. Biophys. Acta 805: 104–116.

Fan M-Y, Lum Z-P, Fu X-W, Levesque L, Tai IT and Sun AM (1990) Reversal of diabetes in BB rats by transplantation of encapsulated pancreatic islets. Diabetes 39: 519–522.

Fernandez EJ, Mancuso A, Murphy MK, Blanch HW and Clark DS (1990) Nuclear magnetic resonance methods for observing the intracellular environment of mammalian cells. Ann. NY Acad. Sci. 589: 458–475.

Gillies RF, Galons J-P, McGovern KA, Scherer PG, Lien Y-H, Job C, Ratcliff R, Chapa F, Cerdan S and Dale BE (1993) Design and application of NMR-compatible bioreactor circuits for extended perfusion of high-density mammalian cell cultures. NMR Biomedicine 6: 95–104.

Grampp GE, Sambanis A and Stephanopoulos GN (1992) Use of regulated secretion in protein production from animal cells: an overview. Adv. Biochem. Eng. Biotechnol. 46: 35–62.

Jans AWH and Willem R (1988) ^{13}C-NMR study of glycerol metabolism in rabbit renal cells of proximal convoluted tubulus. Eur. J. Biochem. 174: 64–73.

Lacy PE, HEgre OD, Gerasimidi-Vazeou A, Gentile FT and Dionne KE (1991) Maintenance of normoglycemia in diabetic mice by subcateneous xenografts of encapsulated islets. Science 254: 1782–1784.

Lanza RP, Butler DH, Borland KM, Staruk JE, Faustman DL, Solomon BA, Muller TE, Rupp RG, Maki T, Monaco AP and Chick WL (1991) Xenotransplantation of canine, bovine, and procine islets in diabetic rats without immunosuppression. Proc. Natl. Acad. Sci USA 88: 11100–11104.

Levesque L, Brubaker PL and Sun AM (1992) Maintenance of long-term secretory function by microencapsulated islets of Langerhands. Endocrinology 130: 644–650.

Lim F and Sun AM (1980) Microencapsulated islets as bioartificial endocrine pancreas. Science 210: 908–910.

Minichiello MM, Albert DM, Kolodny NH, Lee M-N and Craft JL (1989) A perfusion system developed for ^{31}P NMR study of melanoma cells at tissue-like density. Magn. Res. Med. 10: 96–107.

Moon RB and Richards JEL (1973) Determination of intracellular pH by ^{31}P magnetic resonance. J. Biol. Chem. 248: 7276–7278.

Moore H-PH, Walker MD, Lee F and Kelly RB (1983) Expressing a human proinsulin cDNA in a mouse ACTH-secreting cell. Intracellular storage, proteolytic processing, and secretion on stimulation. Cell 35: 531–538.

Mosher TJ, Williams GD, Doumen C, LaNoue KF and Smith MB (1992) Error in the calibration of the MgATP chemical-shift limit: effects on the determination of free magnesium by ^{31}P NMR spectroscopy. Magn. Res. Med. 24: 163–169.

Narayan KS, Mores EA, Chatham JC and Barker PB (1990) ^{31}P NMR of mammalian cells encapsulated in alginate gels utilizing a new phosphate-free perfusion medium. NMR Biomed. 3: 23–26.

Neeman M and Degani H (1989) Early estrogen-induced metabolic changes and their inhibition by actinomycin D and cycloheximide in human breast cancer cells: ^{31}P and ^{13}C NMR studies. Proc. Natl. Acad. Sci. USA 86: 5585–5589.

Papas KK (1992) Characterization of the metabolic and secretory behavior of suspended free and entrapped AtT-20 spheroids in fed-batch and perfusion cultures. MS thesis, Georgia Institute of Technology, Atlanta GA.

Papas KK, Constantinidis I and Sambanis A (1993) Cultivation of recombinant, insulin-secreting AtT-20 cells as free and entrapped spheroids. Cytotechnology 13: 1–12.

Peura RA (1988) Comparison of artificial pancreas devices. In: Skalak R and Fox CF (eds.). Tissue Engineering: Proceedings of a Workshop Held at Granlibakken, Lake Tahoe, California, February 26–29, pp. 223–230.

Ronen S and Degani H (1989) Studies of the metabolism of human breast cancer spheroids by NMR. Magn. Res. Med. 12: 274–281.

Sambanis A, Papas KK, Constantinidis I and Long RC Jr (1993) Towards the development of a bioartificial pancreas: Fabrication and non-invasive monitoring of microbeads containing insulin-secreting, transformed cells. Presented at Annual Meeting of the American Institute of Chemical Engineers, St. Louis, Missouri, November 1993.

Sambanis A, Papas KK and Constantinidis I (1994) Towards the development of artificial endocrine tissues: I. Stability and function of immunoisolated, transformed insulin-secreting cells. Submitted.

Schnall MD, Subramanian VH, Leigh JS and Chance B (1985) A new double-tuned probe for concurrent ^{1}H and ^{31}P NMR. J. Magn. Reson. 65: 122–129.

Sefton MV (1989) Blood, guts and chemical engineering. Can. J. Chem. Eng. 67: 705–712.

Sherry AD, Malloy CR, Roby RE, Rajagopal A and Jeffrey FMH (1988) Propionate metabolism in the rat heart by ^{13}C NMR spectroscopy. Biochem. J. 254: 593–598.

Soon-Shiong P, Feldman E, Nelson R, Komtebedde J, Smidsrod O, Skjak-Braek G, Espevik T, Heintz R and Lee M (1992) Successful reversal of spontaneous diabetes in dogs by intraperitoneal microencapsulated islets. Transplantation 54: 769–774.

Soon-Shiong P, Lu ZN, Grewal I, Lanza RP and Clark W (1990) An in vitro method of assessing the immunoprotective properties of microcapsule membranes using pancreatic and tumor cell targets. Transplantation Proceedings 22: 754–755.

Soon-Shiong P, Heintz RE, Merideth N, Yao QX, Yao Z, Zheng T, Murphy M, Moloney MK, Schmehl M, Harris M, Mendez R, Mendez R and Sandford PA (1994) Insulin independence in a type I diabetic patient after encapsulated islet transplantation. Lancet 343: 950–951.

Stinson SC (1991) New drugs under development for diabetes. Chem. Eng. News, September 30: 35–59.

Sullivan SJ, Maki T, Carretta M, Ozato H, Borland K, Mahoney MD, Muller TE, Solomon BA, Monaco AP and Chick WL (1992) Evaluation of the hybrid artificial pancreas in diabetic dogs. ASAIO Journal 38: 29–33.

Sun AM (1988) Microencapsulation of pancreatic islet cells: a bioartificial endocrine pancreas. Methods Enzymol. 137: 575–580.

Tal M, Thorens B, Surana M, Fleischer N, Lodish HF, Hanahan D and Efrat S (1992) Glucose transporter isotypes switch in T-antigen-transformed pancreatic β cells growing in culture and in mice. Molec. Cell. Biol. 12: 422–432.

Tziampazis E (1993) Engineering functional, insulin-secreting cell systems: Effect of entrapment on cellular environment and secretory response. MS thesis, Georgia Institute of Technology, Atlanta GA.

Tziampazis E and Sambanis A (1994) Tissue engineering a bioartificial pancreas: Modeling the cell environment and device function. Biotechnol. Progress, in press.

Vorlop K-D and Klein J (1983) New developments in the field of cell immobilization-formation of biocatalysis by ionotropic gelation. In: Lafferty RM (ed.) Enzyme Technology. III. Rotenburg Fermentation Symposium 1982, 22–24 September 1982, pp. 219–235. New York: Springer-Verlag.

Address for correspondence: A. Sambanis, School of Chemical Engineering, Georgia Institute of Technology, Atlanta, GA 30332-0100, USA.

Cytotechnology **15:** 365–372, 1994.

Overview of a quality assurance/quality control compliance program consistent with FDA regulations and policies for somatic cell and gene therapies: a four year experience

Gary C. du Moulin, Zorina Pitkin, Yuan-Jin Shen, Evelyn Conti, Jean Ko Stewart, Carla Charles & Dylan Hamilton
Department of Quality Control, Cellcor, Inc. 200 Wells Avenue Newton, MA 02159

Key words: Autolymphocyte therapy, somatic cellular therapy, gene therapy, compliance, FDA, quality assurance, regulations

Abstract

Somatic cell and gene therapy involve the application of biological technologies to an individual patient through the use of living cells which provide a therapeutic benefit (Aliski, 1991). Various forms of cellular and gene therapies are being developed and evaluated in an increasing number of clinical trials for congenital and acquired disorders. The potential and progress of these therapeutic applications have resulted in an increasing effort by the Food and Drug Administration (FDA) to develop the regulatory framework under which these therapeutic approaches would insure safety and efficacy, the primary mandate of the FDA.

Over five years ago Cellcor began to define the parameters, specifications, and conditions relevant to a Quality Assurance / Quality Control (QA/QC) program that has evolved to insure safety and maximize the efficacy of applications of the company's *ex vivo* technology, autolymphocyte therapy. Autolymphocyte therapy is an outpatient form of somatic cell immunotherapy based upon the infusion of T cells that have been activated *ex vivo* using a combination of previously generated autologous cytokines and an anti-CD3 monoclonal antibody.

We have been able to demonstrate the feasibility for the safe, controlled, and consistent preparation and delivery of a cellular therapy by application of relevant GMP regulations. This presentation reviews aspects of this program and chronicles our experience which at present amounts to over 4400 infusions for over 700 patients. This program provides a high degree of assurance that a cellular therapy program can be carried out in a multisite mode involving hundreds of patients through the strict adherence to cGMP as set forth in existing regulations. It would be prudent that developers of cellular and *ex vivo* gene therapies establish a similar cell processing and QA/QC infrastructure at an early developmental stage to optimize safety and reproducibility and facilitate regulatory review.

Introduction

The Food and Drug Administration has through the Center for Biologics Research and Evaluation (CBER) Division of Cellular and Gene Therapies articulated a regulatory position regarding the therapeutic use of living tissues or cells (Epstein, 1991; FDA 1993). If cells or tissues are intended for use in the diagnosis, cure, mitigation, treatment or prevention of disease or illness in man or animal, then the product meets the definition of a drug, as defined in section 201 of the Federal Food Drug and Cosmetic Act (FD&C Act) [21 USC, section 321]. Regulations implementing the FD&C Act, as applicable to the product are found in the U.S. Code of Federal Regulations, title 21-Food and Drugs (21 CFR) Chapter 5 in Subchapters C, D, and F, and particularly in 21 CFR, Parts 200 and 600. The establishment of the Division of Cellular and Gene Therapies under the Center for Biologics Evaluation and Research requires that regulatory submissions in the form of Product License Applications (PLA) and Establishment License Applications (ELA) be submitted after appropriate pre-clinical and clinical testing. This approach would require that the operations of a facility where cells are isolated, cultured, modified or processed in any way comply with the current good manufacturing practice regulations (cGMPs), as applicable, under 21 CFR Parts 210 through 226, and 600 through 680.

Five years ago Cellcor began to define the parameters, specifications, and conditions relevant to a QA/QC program that would insure safety and maximize the efficacy of applications of the company's technology, autolymphocyte therapy (du Moulin, 1990). Autolymphocyte therapy is an outpatient form of adoptive immunotherapy based upon the infusion of polyclonally activated T cells that have been activated *ex vivo* using a combination of previously generated autologous cytokines and an anti-CD3 monoclonal antibody. Autolymphocyte therapy in the treatment of metastatic renal cell carcinoma has been previously shown to significantly prolong survival 2.3 fold and induce durable tumor responses in treated patients, and is accompanied by only minimal toxicity (Osband, 1990). When we looked for an existing QA/QC program we found no academic or industrial model. At that time there was no specific written guidance from the FDA. Cellular therapy programs were based upon a 'Practice of Medicine' regulatory model, i.e., bone marrow transplantation, *in vitro* fertilization. Under this construct, some academic centers conducting these programs reported up to a 25% post infusion infection rate (Arnow, 1991; FDA, 1989; Taylor, 1994). As a regulated therapeutic modality significant patient risk such as post infusion infection would be unacceptable. In contrast, we have been able to demonstrate the feasibility for the safe, controlled, and consistent preparation and delivery of a cellular therapy by application of relevant GMP regulations and 'Points to Consider' memoranda that have recently been made available (FDA, 1989; FDA, 1991). We have previously presented aspects of this program and review of our experience (du Moulin, 1992, 1993, 1994).

A number of issues and challenges face the development of cellular therapies. Key among them is the need to reconcile the inherent variability of the critical raw materials, e.g. cells derived by lymphocytopheresis from a heterogeneous patient population. Secondly, because of the limited shelf life of living cells, there are unique aspects of quality control testing that must insure safe release. Finally, the product must be safely delivered and preserved during transport with no significant changes in potency, viability, and sterility over the limited shelf-life. This is particularly a concern since the infusion takes place under the aegis of the 'practice of medicine'. The ultimate challenge is to develop a program that convincingly demonstrates compliance under a biologic regulatory framework.

With an understanding of these challenges and having anticipated the increase in IND's related to Cellular and Gene Therapies, the FDA has published two 'Points to Consider' documents that have provided guidance to the industry (FDA, 1989; FDA, 1991). These guidelines address product development, cell processing and quality control issues, as well as safety testing for lots of cellular materials processed for clinical use. The 'Points to Consider' are not regulations but rather offer the 'current thinking' of the CBER staff regarding important issues in an emerging area of cell based therapies, product development, and testing.

Materials and methods

Quality Assurance Program for a Cellular Therapy

Key elements of this multi-dimensional program have been previously presented (du Moulin, 1993). The master record format was adapted from the classic pharmaceutical model and modified to encompass the additional requirements of a cellular therapy. All areas in the FDA's 'Points to Consider' were addressed in the development of this document. Among these areas were cell collection and separation, the materials management system, the cell culture procedures the testing and quality control of the cells, as well as the monitoring of the patients who receive the cells.

Standard Operating Procedures (SOP), each covering a specific component of cell activity were written by a multidisciplinary task force. This task force was comprised of personnel with a detailed knowledge of the process and included research and process development scientists, cell processing technologists, and quality control staff members. Cell processing batch records, which were based on the pharmaceutical model, were developed to accompany the SOPs and to provide personnel direction and accountability during each cell processing step. Each cell processing activity was divided into separate procedural protocols, which included preparation of the working area, gathering the appropriate raw materials, reagents, components, and devices, the actual manipulation of the cells, and finally the completion of the procedure and the subsequent clean-up. Lot numbers and expiration dates of critical components were recorded on the batch record. Critical steps and calculations were checked by another cell processor. Completed batch records from these protocols were separately reviewed and verified and then assembled into the patient's cell processing history record.

Patient and cell lot identification

Procedures were developed to insure the rigorous monitoring of identification of the patient's cell lot. Patient specific labels and tags, generated on the patient's first visit were attached to cell bags, sample tubes, and final products. In addition, the name, ID number, and date were written on the top of each batch record page.

Specifications in ALT

Specifications were developed to monitor the quality and consistency of cellular products. Specifications on cell yield, viability, sterility, purity, and potency were based on the FDA's 'Points to Consider' document. Release specificati- ons were applied to all cell lots. 'Cell lots' were defined by the FDA as the cells from a single collection from a single donor. Before a lot of cells was infused into a patient, a formal and methodical inspection of the product and testing was performed by Quality Control technologists; specifications had to be met before an infusion product could be released. A certificate of quality control and release, signed by the QC technologist, ensuring that the release specifications had been met, was issued along with the infusion product.

Nonconformance analysis

Nonconformances constituted any event that deviated from approved specifications and/or standard procedures. The quality control program to measure and categorize nonconformances was developed to insure integrity of the product while allowing for flexibility based on the inherent variability of apheresis cell products. Based upon an analysis of all nonconformances recorded during 1200 cell processing procedures during Phase II clinical trials, a classification system was developed with 15 categories that assessed the relative risk of product rejection. These nonconformance categories include such issues as cell yield, cell viability, equipment failures, and inconsistencies in documentation. For example, a nonconformance with a Category I risk indicates a deviation from procedure that would be unlikely to lead to rejection pending additional quality control testing. An example of Category I risk might be the temporary malfunction of a device required for cell processing. A nonconformance with a Category II risk was a deviation from procedure or specification that was more likely to lead to rejection, such as a leak in a cell product bag.

A nonconformance with a Category III risk was one in which a cell product must be rejected and destroyed, for example, overt bacterial contamination of the cell culture. A set of criteria describing nonconformance categories, risk categories, and proper responses for each nonconformance was provided in an approved SOP with a listing of additional quality control tests required to requalify the product if appropriate. Quality control technologists coordinated each nonconformance review, insured the completion of the proper documentation, and certified the expeditious resolution of the event.

To comply with Category III failures, complete investigation as to cause is required and must be fully documented and reviewed by a multidisciplinary committee. For example, we recently reported the observation of bacterial contamination during the preparation of *ex-vivo* activated (EVA cells) for patient administration (Pitkin, 1994). On the day of scheduled cell infusion, bacterial growth was observed in a biphasic blood culture system inoculated with apheresis cell product. All other sterility tests collected during cell processing showed no growth including samples of the final cell infusion product. Further investigation was conducted to: 1) confirm the sterility of the infusion product, 2) identify the contaminant, and 3) evaluate the ability of the contaminant to grow in cell culture medium. After continual analysis and subculture of all sterility tests it became evident that the infusion product was sterile. The patient was subsequently infused with no adverse reaction within the specified time period. Follow up investigation revealed that the ACP was contaminated. The contaminant was identified as *Propionibacterium acnes* through biochemical analysis. Additional experiments were conducted to investigate the potential of *P. acnes* to grow in medium used for EVA of lymphocytes. A suspension of *P. acnes* (1×10^8 CFU/ml) was inoculated into simulated cell culture. Samples were periodically collected and examined for bacterial growth. After three days of incubation recovery of *P. acnes* declined to 100 CFU/ml. There was no evidence of *P. acnes* in the cell culture media by the end of the incubation period. This investigation pointed out the need for sensitive methods of microbial detection due to the possibility of contamination during the procurement of blood cells for cellular therapy.

Quality control testing program for a cellular therapy

The quality control testing program has been well described and can be divided into the following six categories: (1) cell yield, (2) cell viability, (3) sterility testing, (4) pyrogen testing, (5) cell function, and (6) phenotypic characterization of lymphocytes (du Moulin, 1992, 1993; Hamilton, 1994).

On a monthly basis statistical analysis is performed on all cell products prepared during the month. Key cell quality parameters are compared to data collected during previous months or years. Fluctuations and trends allow an examination of the consistency of overall therapy. Software has been designed to be able to filter data specific to patient, clinical site, cell separator source and other variables of interest. This ability allows a precise evaluation of the therapy and expedites the identification of unexplained shifts in cell product quality within acceptable lot to lot variability.

Preparation of activated T-cells

Ex vivo activated (EVA) cells used in ALT were prepared using a two-stage process. In the first stage, patients underwent an initial lymphopheresis during which approximately 2×10^9 lymphocytes were collected and transported to one of three regional cell processing centers. Peripheral blood mononuclear cells (PBM) were separated using density-gradient centrifugation in a closed system and then incubated for 3 days in the presence of OKT3, a monoclonal antibody directed against the CD3 portion of the T-cell receptor (Orthoclone OKT3, Ortho Biotech, Raritan, NJ). After three days, the supernatant (T3CS - OKT3 derived culture supernatant) was collected, aliquoted × 6, and frozen at –70 °C for future use in cell processing. The T3CS contains a mixture of autologous cytokines, including interleukins (IL-I alpha, IL-1 beta, IL-6, IL-8), granulocyte-macrophage colony-stimulating factor (GM-CSF), interferon (IFN-gamma), and tumor necrosis factor (TFN-alpha and beta).

In the second stage, approximately 2×10^9 lymphocytes were collected 7–10 days later by lymphopheresis. Peripheral blood mononuclear cells were prepared as previously decribed and suspended at 2×10^6/ml in media (AIMV, Life Technologies, Inc, Grand Island NY) containing 25% (vol/vol) T3CS, as well as cimetidine and indomethacin; the latter two drugs were used to reduce suppressor cell activity (Osband, 1990). Cells were incubated for 5 days at 37 °C in a moist air incubator containing 5% CO_2. After 5 days in culture, EVA cells were washed extensively and resuspended in infusion media containing lactated ringers, dextrose, and human serum albumin at a concentration of 10^7 cells/ml and stored or shipped overnight under controlled conditions for infusion into the patient the following day. Validation of shelflife has previously indicated stability in cell yield, cell viability, and cell function over 48 hours after cell harvesting. Each EVA infusion product was infused intravenously over 30 minutes through a standard blood administration filter. All patients are pheresed at monthly intervals for 6 months, and infused with approximately 10^9 cells/infusion. Patients were also instructed to take oral cimetidine, 2400 mg/d, to reduce *in vivo* suppressor T-cell activity (Osband, 1994).

QA/QC program assessment

The experience and effectiveness of this QA/QC program was assessed over a four year period. Cell processing activity was evaluated using the following time intervals:

Year One (1990): Standard operating procedures, batch records, and training programs were developed, written, and implemented during this time. The pilot facility was closed and all operations were moved to a location where laboratory space was better designed to fully comply with cGMP regulations. 1160 procedures were performed and 432 patients were treated during Year One.

Year Two (1991): Two additional cell processing centers were opened in Atlanta, GA and Orange, CA. Additional personnel were centrally trained and certified to work at these sites. Equipment and calibration standard operating procedures were developed, written, and implemented. 3346 procedures were performed and 1157 treatments were administered during Year Two.

Years Three and Four (1992–1993): Three cell processing centers were completely operational under full cGMP compliance. A computer program was instituted in late 1992 that allowed entry and statistical analysis of all cell processing and QC data from all centers in a timely manner. 7556 procedures were performed and 2844 patient treatments administered during 1992–1993.

Table 1. Nonconformance reporting during autolymphocyte therapy (1990–1993)

	Year				
	1990	1991	1992	1993	TOTAL
Total procedures	1160	3346	5419	2137	12062
Patient Treatments	432	1157	2067	777	4433
Nonconformances Reported	97	372	400	156	1025
Nonconformance rate (%)	8.3	11.1	7.4	7.3	8.5

Results

Between May, 1990 and December, 1993 a total of 12,062 cell procedures were conducted in three cell processing facilities (Boston, MA; Atlanta, GA; and Orange, CA). This activity resulted in the delivery of over 4,400 cell infusions to seven clinical sites nationwide. Approximately 700 patients with metastatic renal cell carcinoma received autolymphocyte therapy. No patient experienced unanticipated adverse reactions from the preparation of activated cells. 91 of 4400 (2.1%) cell cultures prepared necessitated a repeat apheresis procedure because cell products failed either in process or release specification. 1025 nonconformances were reported during the period of analysis for an overall nonconformance rate of 8.5% (Table 1). 91 percent of all nonconformances reported resulted in infusion products that met lot release specifications. No infusion product in which a nonconformance was observed and evaluated during cell processing resulted in any post infusion complication.

Nonconformance activity was consistently evaluated to monitor process control and patient safety (Table 1). In 1990 the nonconformance rate was 8.3% and reflected the activity of a single cell processing facility with an experienced staff using approved standard operating procedures and batch records. A period of active expansion followed with the opening of two additional cell processing facilities with concurrent training of new personnel. During this phase of development the nonconformance rate rose to 11.1%. During 1992 three facilities were actively processing cells each with an experienced staff. Nonconformance rates decreased to 7.4% and remained at this level throughout 1993 indicating a marked improvement in the quality of cell processing activity throughout the network of laboratory facilities. Rejection of cell products necessitating a repeat apheresis procedure consistently decreased over time (Table 2).

Table 2. Rejection of cell products during autolymphocyte therapy (1990–1993)

	Year			
	1990	1991	1992	1993
Cultures processed	423	1157	2067	777
No. of Rejections	28	27	27	9
Rejection rate (%)	6.50	2.30	1.30	1.16

Table 3. Cell products rejected: a two year experience

Reason for rejection	No.	Total rejected (%)
• Low cell count	16	44
• Low T3CS activation	6	17
• Cell culture contaminated	5	14
• Low cell viability	4	11
• Other	5	14

Total cell cultures processed = 2,844
Total products rejected = 36

Causes of infusion product rejection

Over the 4 years of review a total of 91 cell products were rejected by three cell processing facilities for failing to meet established specifications. Causes of cell product rejections for 1992–1993 are shown in Table 3. Low cell yields accounted for the greatest number of rejections (44.0%), However, activities of the Process Development group resulted in a significantly reducing of low cell yield as a cause for rejection (86%).

Maintaining sterility in multiple patient specific cell cultures was a crucial measure of process quality in large scale cell therapy operations. In 1992–1993, 5 of

Table 4. Analysis of EVA lymphocyte cell surface markers (N=300)

CD3 %(+)	HLA DR %(+)	CD25 %(+)
Mean ± S.D.	Mean ± S.D.	Mean ± S.D.
75.7 ± 20.9	22. 1 ± 11.8	46.9 ± 18.7

2844 cell cultures prepared were rejected due to cell product contamination. Investigation of these sterility failures revealed that two apheresis cell products were contaminated at the time of apheresis cell product collection. Contaminating microorganisms were due to *Staphylococcus epidermidis, Staphylococus aureus* and *Propionibacterium acnes*. Two cell products were apparently contaminated during cell processing. The organisms responsible were *Bacillus subtilis* and *Streptococcus sp.*. This low rate of processing related contamination was remarkable in light of the fact that cell culture media is formulated *without* antimicrobial agents. Maintenance of sterile conditions on a routine basis is essential to uninterrupted cell therapy and reflects directly upon the level of training of personnel and the appropriateness of the laboratory environment designed for cell therapy activities.

Expression of pertinent cell phenotypes after polyclonal cell activation

Polyclonally activated T cells are generated in a conditioned medium enriched in autologous cytokines and containing ng/ml quantities of OKT3 over a 5 day incubation period. While there is no significant cell expansion there is demonstrable production of cells populations enriched in T cells expressing enhanced levels of CD25 (IL-2 receptor) and MHC Class II molecules (HLADR). Table 4 shows these data among 300 cell infusion products prepared during 1993.

Effect of process development and validation on cell product quality

Prior to initiation of a current pivotal phase III clinical trial, validation of cell process changes to enhance infusion product quality was conducted. After formal approval of process changes and implementation of modified cell processing SOPs key cell quality parameters were monitored. Maintenance of cGMP within cell processing allowed the observation of overall improvements in quality and quantity. For example, infusion product concentration rose from a mean of 6.7×10^6/ml to 9.5×10^6/ml (Figure 1) significantly improving the dose of cells patients were receiving.

Discussion

Cellular therapy, a novel approach to the treatment of certain diseases, involves 1) the selection of targeted cells, 2) the *ex-vivo* activation, modification, or enhancement of these cells, and 3) the return of these modulated cells to the patient. Cellular therapies can be an effective treatment modality. Because these enhanced cells are alive, they can specifically interact with other cell types through chemical mediators, respond to subtle changes in medical condition and generate a series of biologically active molecules in response to the disease.

These molecules are more likely to be secreted in optimal concentrations, in the proper sequence, and at the proper site in the body. Because cellular therapies enhances the body's own response to the disease, it is a dynamic, self-regulating treatment.

Various forms of cellular therapies are being developed and evaluated through clinical trials. While not presently regulated, bone marrow transplantation and *in-vitro* fertilization are well established and are currently being reimbursed. The use of laboratory grown skin, and adoptive immunotherapy for cancer utilizing *ex-vivo* activated lymphocytes are being evaluated in Phase III-trials. Newer technologies including gene therapy, transplantation of living cells including pancreatic islet cells for the treatment of diabetes, expansion and differentiation of peripheral blood stem cells, and myoblasts for the treatment of muscular dystrophy are in early development.

Demonstration of consistency in sterility, viability, cell yield, cell function characteristics in EVA cells produced at multiple cell processing facilities validates the requirement for compliance to a rigorous set of procedures and practices required under the Good Manufacturing Practice Regulations. The combination of sound process development coupled with controls of personnel, facility, environment, raw materials, and equipment are essential for safe and reproducible cell therapy especially when there is a dependence on a heterogeneous source of patient derived raw materials. Anticipation of multiple cell processing facilities for other developing cell therapy indications requires

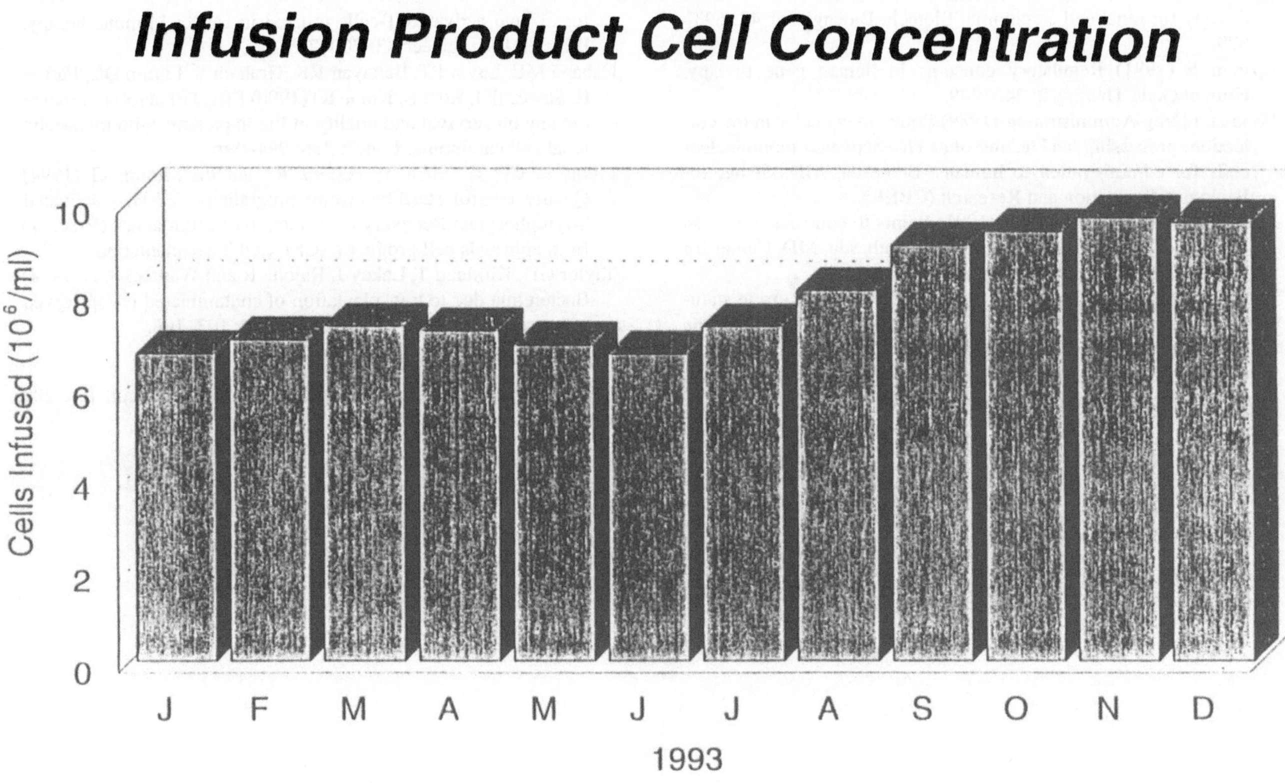

Fig. 1. Effect of Process Development activities on the improvement of cell concentration of *ex vivo* activated (EVA) T-lymphocytes infusion product. Improved methods of culture and harvesting were validated and implemented in August. Average concentration of the infusion product improved from 6.7×10^6/ml to 9.5×10^6/ml.

a sound QA/QC infrastructure predicated on the principles of cGMP.

In light of the current stringent regulatory climate QA/QC mechanisms need to be established coincident with the development of any cellular therapy. These systems should be employed during the conduct of clinical trials. Development and implementation of quality systems should not be considered *after* the technology has been developed. Feasibility of a cell therapy cannot be easily validated in a multisite mode unless all the elements of cGMP are included in early planning stages. The FDA's 'Points to Consider' on *ex vivo* activation and somatic cell and gene therapy are useful documents and should be consulted in context with each developer's goals. Early attention to the development of a QA/QC infrastructure will help insure consistency and reproducibility of cellular therapy programs, optimize the clinical result of these novel modalities and ultimately contribute to a successful regulatory submission.

References

Aliski WE, Cashon GW, and Osband ME (1991) The regulation of living cell therapies (Biocare). Regulatory Affairs 3: 639–650.

Arnow PM, Houchins SG, Richards JM, and Chudy R (1991) *Aspirgillus fumigatus* contamination of lymphokine – activated killer cells infused into cancer patients. J. Clin. Microbiol. 29: 1038–1041.

du Moulin GC, Hamilton DL, Price JE, Kosik ZJ, Liu VYS, Osband ME (1990) Quality assurance and quality control (QA/QC) in biocare. Autolymphocyte therapy as a model. Biopharm. 3: 30–35.

du Moulin GC, Price JE, Shen YJ, and Osband ME (1992) Use of the *Limulus* amebocyte lysate (LAL) assay in living cell therapies. Biopharm. 5: 32–54.

du Moulin GC, Liu V, and Osband ME (1992) Implementation of an effective program for quality assurance and quality control in living cell therapy: A two year experience with autolymphocyte therapy. Transplantation Proceedings 24: 2803–2808.

du Moulin GC, Stack J, Kruger R, and Liu V (1993) Quality assurance/quality control compliance program: A cellular therapy developer's perspective. Regulatory Affairs 5: 461–472.

du Moulin GC, Stack J, Pitkin Z, Chew-Darke J, Cyr C, White A, Ho L. Shen Y, Hamilton D, Davies B, Charles C, Conti E, and Liu V (1994) A 3 year experience of Quality control and quality assurance in the multisite delivery of a lymphocyte based cellular therapy for renal cell carcinoma. Biotech. Bioengineer 43: 693–699.

Epstein S (1991) Regulatory concerns in human gene therapy. Human Gene Therapy 2: 243–249.

Food and Drug Administration (1989) Points to consider in the collection, processing, and testing of *ex vivo* activated mononuclear cells for administration to humans. Bethesda, MD. Center for Biologics Evaluation and Research (CBER).

Food and Drug Administration (1991) Points to consider in human somatic cell therapy and gene therapy. Bethesda, MD. Center for Biologics Evaluation and Research (CBER).

Food and Drug Administration (1993) Application of current statutory authorities to human somatic cell therapy products and gene therapy products. Federal Register 58: 53248–53251.

Hamilton D, Goodwin J, Clarke MB, du Moulin GC, Liu V, Caplan B, and Babbitt B (1994) Preliminary validation of an activation assay for *ex-vivo* activated T-cells utilized in cancer immunotherapy. Biotech. Bioengineer 43: 700–705.

Osband ME, Lavin PT, Babayan RK, Graham S, Lamm DL, Parker B, Sawczuk I, Ross S, Krane RT(1990) Effect of autolymphocyte therapy on survival and quality of life in patients with metastatic renal cell carcinoma. Lancet 335: 994–998.

Pitkin Z, Cyr C, Shen Y, Perkins R, and du Moulin G (1994) Quality control sterility testing program for *ex vivo* activated T-lymphocytes: Recovery of Propionibacterium acnes (P. acnes) from apheresis cell product (ACP). Cell Transplantation 3: 222.

Taylor GD, Kirkland T, Lakey J, Rajotte R and Warnock GL (1994) Bacteremia due to transplantation of contaminated cryoprserved pancreatic islets. Cell Transplantation 3: 103–106.

Address for offprints: Gary C. du Moulin, Ph.D. Cellcor, Inc. 200 Wells Avenue, Newton, MA 02159, U.S.A.

Cytotechnology **15**: 373–374, 1994.

Author index

Cytotechnology **15**: 375–376, 1994.

Key word index